Introduction to Mathematical Statistics

Robert V. Hogg
Allen T. Craig

THE UNIVERSITY OF IOWA

Introduction to Mathematical Statistics

Fourth Edition

Macmillan Publishing Co., Inc.
NEW YORK
Collier Macmillan Publishers
LONDON

Copyright © 1978, Macmillan Publishing Co., Inc.

Printed in the United States of America

Earlier editions © 1958 and 1959 and copyright © 1965 and 1970 by Macmillan Publishing Co., Inc.

Macmillan Publishing Co., Inc.
866 Third Avenue, New York, New York 10022

Collier Macmillan Canada, Ltd.

Library of Congress Cataloging in Publication Data

Hogg, Robert V
  Introduction to mathematical statistics.

  Bibliography: p.
  Includes index.
  1. Mathematical statistics. I. Craig, Allen
Thornton, (date)   joint author. II. Title.
QA276.H59   1978   519   77-2884
ISBN 0-02-355710-9 (Hardbound)
ISBN 0-02-978990-7 (International Edition)

Printing: 9 10   Year: 3 4

ISBN   0-02-355710-9

Preface

We are much indebted to our colleagues throughout the country who have so generously provided us with suggestions on both the order of presentation and the kind of material to be included in this edition of *Introduction to Mathematical Statistics*. We believe that you will find the book much more adaptable for classroom use than the previous edition. Again, essentially all the distribution theory that is needed is found in the first five chapters. Estimation and tests of statistical hypotheses, including nonparameteric methods, follow in Chapters 6, 7, 8, and 9, respectively. However, sufficient statistics can be introduced earlier by considering Chapter 10 immediately after Chapter 6 on estimation. Many of the topics of Chapter 11 are such that they may also be introduced sooner: the Rao–Cramér inequality (11.1) and robust estimation (11.7) after measures of the quality of estimators (6.2), sequential analysis (11.2) after best tests (7.2), multiple comparisons (11.3) after the analysis of variance (8.5), and classification (11.4) after material on the sample correlation coefficient (8.7). With this flexibility the first eight chapters can easily be covered in courses of either six semester hours or eight quarter hours, supplementing with the various topics from Chapters 9 through 11 as the teacher chooses and as the time permits. In a longer course, we hope many teachers and students will be interested in the topics of stochastic independence (11.5), robustness (11.6 and 11.7), multivariate normal distributions (12.1), and quadratic forms (12.2 and 12.3).

We are obligated to Catherine M. Thompson and Maxine Merrington and to Professor E. S. Pearson for permission to include Tables II and V, which are abridgments and adaptations of tables published in *Biometrika*. We wish to thank Oliver & Boyd Ltd., Edinburgh, for permission to include Table IV, which is an abridgment and adaptation of Table III from the book *Statistical Tables for Biological, Agricultural, and Medical Research* by the late Professor Sir Ronald A. Fisher, Cambridge, and Dr. Frank Yates, Rothamsted. Finally, we wish to thank Mrs. Karen Horner for her first-class help in the preparation of the manuscript.

R. V. H.
A. T. C.

Contents

Chapter 4

Distributions of Functions of Random Variables   **122**

Chapter 5

Limiting Distributions   **181**

Chapter 6

Estimation   **200**

Chapter 7

Statistical Hypotheses   **235**

Chapter I

Distributions of Random Variables

1.1 Introduction

Many kinds of investigations may be characterized in part by the fact that repeated experimentation, under essentially the same conditions, is more or less standard procedure. For instance, in medical research, interest may center on the effect of a drug that is to be administered; or an economist may be concerned with the prices of three specified commodities at various time intervals; or the agronomist may wish to study the effect that a chemical fertilizer has on the yield of a cereal grain. The only way in which an investigator can elicit information about any such phenomenon is to perform his experiment. Each experiment terminates with an *outcome*. But it is characteristic of these experiments that the outcome cannot be predicted with certainty prior to the performance of the experiment.

Suppose that we have such an experiment, the outcome of which cannot be predicted with certainty, but the experiment is of such a nature that the collection of every possible outcome can be described prior to its performance. If this kind of experiment can be repeated under the same conditions, it is called a *random experiment*, and the collection of every possible outcome is called the experimental space or the *sample space*.

Example 1. In the toss of a coin, let the outcome tails be denoted by T and let the outcome heads be denoted by H. If we assume that the coin may be repeatedly tossed under the same conditions, then the toss of this coin is an example of a random experiment in which the outcome is one of

1

the two symbols T and H; that is, the sample space is the collection of these two symbols.

Example 2. In the cast of one red die and one white die, let the outcome be the ordered pair (number of spots up on the red die, number of spots up on the white die). If we assume that these two dice may be repeatedly cast under the same conditions, then the cast of this pair of dice is a random experiment and the sample space consists of the 36 order pairs $(1, 1), \ldots,$ $(1, 6), (2, 1), \ldots, (2, 6), \ldots, (6, 6)$.

Let $\mathscr{C}$ denote a sample space, and let C represent a part of $\mathscr{C}$. If, upon the performance of the experiment, the outcome is in C, we shall say that the *event C* has occurred. Now conceive of our having made N repeated performances of the random experiment. Then we can count the number f of times (the frequency) that the event C actually occurred throughout the N performances. The ratio f/N is called the *relative frequency* of the event C in these N experiments. A relative frequency is usually quite erratic for small values of N, as you can discover by tossing a coin. But as N increases, experience indicates that relative frequencies tend to stabilize. This suggests that we associate with the event C a number, say p, that is equal or approximately equal to that number about which the relative frequency seems to stabilize. If we do this, then the number p can be interpreted as that number which, in future performances of the experiment, the relative frequency of the event C will either equal or approximate. Thus, although we *cannot* predict the outcome of a random experiment, we *can*, for a large value of N, predict approximately the relative frequency with which the outcome will be in C. The number p associated with the event C is given various names. Sometimes it is called the *probability* that the outcome of the random experiment is in C; sometimes it is called the *probability* of the event C; and sometimes it is called the *probability measure* of C. The context usually suggests an appropriate choice of terminology.

Example 3. Let $\mathscr{C}$ denote the sample space of Example 2 and let C be the collection of every ordered pair of $\mathscr{C}$ for which the sum of the pair is equal to seven. Thus C is the collection $(1, 6), (2, 5), (3, 4), (4, 3), (5, 2),$ and $(6, 1)$. Suppose that the dice are cast $N = 400$ times and let f, the frequency of a sum of seven, be $f = 60$. Then the relative frequency with which the outcome was in C is $f/N = \frac{60}{400} = 0.15$. Thus we might associate with C a number p that is close to 0.15, and p would be called the probability of the event C.

Remark. The preceding interpretation of probability is sometimes referred to as the *relative frequency approach*, and it obviously depends upon the fact that an experiment can be repeated under essentially identical conditions. However, many persons extend probability to other situations by treating it as rational measure of belief. For example, the statement $p = \frac{2}{5}$ would mean to them that their *personal* or *subjective* probability of the event C is equal to $\frac{2}{5}$. Hence, if they are not opposed to gambling, this could be interpreted as a willingness on their part to bet on the outcome of C so that the two possible payoffs are in the ratio $p/(1 - p) = \frac{2}{5}/\frac{3}{5} = \frac{2}{3}$. Moreover, if they truly believe that $p = \frac{2}{5}$ is correct, they would be willing to accept either side of the bet: (a) win 3 units if C occurs and lose 2 if it does not occur, or (b) win 2 units if C does not occur and lose 3 if it does. However, since the mathematical properties of probability given in Section 1.4 are consistent with either of these interpretations, the subsequent mathematical development does not depend upon which approach is used.

The primary purpose of having a mathematical theory of statistics is to provide mathematical models for random experiments. Once a model for such an experiment has been provided and the theory worked out in detail, the statistician may, within this framework, make inferences (that is, draw conclusions) about the random experiment. The construction of such a model requires a theory of probability. One of the more logically satisfying theories of probability is that based on the concepts of sets and functions of sets. These concepts are introduced in Sections 1.2 and 1.3.

EXERCISES

1.1. In each of the following random experiments, describe the sample space $\mathscr{C}$. Use any experience that you may have had (or use your intuition) to assign a value to the probability p of the event C in each of the following instances:

(a) The toss of an unbiased coin where the event C is tails.

(b) The cast of an honest die where the event C is a five or a six.

(c) The draw of a card from an ordinary deck of playing cards where the event C occurs if the card is a spade.

(d) The choice of a number on the interval zero to 1 where the event C occurs if the number is less than $\frac{1}{3}$.

(e) The choice of a point from the interior of a square with opposite vertices $(-1, -1)$ and $(1, 1)$ where the event C occurs if the sum of the coordinates of the point is less than $\frac{1}{2}$.

1.2. A point is to be chosen in a haphazard fashion from the interior of a fixed circle. Assign a probability p that the point will be inside another circle, which has a radius of one-half the first circle and which lies entirely within the first circle.

1.3. An unbiased coin is to be tossed twice. Assign a probability p_1 to the event that the first toss will be a head and that the second toss will be a tail. Assign a probability p_2 to the event that there will be one head and one tail in the two tosses.

1.2 Algebra of Sets

The concept of a *set* or a *collection* of objects is usually left undefined. However, a particular set can be described so that there is no misunderstanding as to what collection of objects is under consideration. For example, the set of the first 10 positive integers is sufficiently well described to make clear that the numbers $\frac{3}{4}$ and 14 are not in the set, while the number 3 is in the set. If an object belongs to a set, it is said to be an *element* of the set. For example, if A denotes the set of real numbers x for which $0 \leq x \leq 1$, then $\frac{3}{4}$ is an element of the set A. The fact that $\frac{3}{4}$ is an element of the set A is indicated by writing $\frac{3}{4} \in A$. More generally, $a \in A$ means that a is an element of the set A.

The sets that concern us will frequently be *sets of numbers*. However, the language of sets of *points* proves somewhat more convenient than that of sets of numbers. Accordingly, we briefly indicate how we use this terminology. In analytic geometry considerable emphasis is placed on the fact that to each point on a line (on which an origin and a unit point have been selected) there corresponds one and only one number, say x; and that to each number x there corresponds one and only one point on the line. This one-to-one correspondence between the numbers and points on a line enables us to speak, without misunderstanding, of the "point x" instead of the "number x." Furthermore, with a plane rectangular coordinate system and with x and y numbers, to each symbol (x, y) there corresponds one and only one point in the plane; and to each point in the plane there corresponds but one such symbol. Here again, we may speak of the "point (x, y)," meaning the "ordered number pair x and y." This convenient language can be used when we have a rectangular coordinate system in a space of three or more dimensions. Thus the "point $(x_1, x_2, \ldots, x_n)$" means the numbers $x_1, x_2, \ldots, x_n$ in the order stated. Accordingly, in describing our sets, we frequently speak of a set of points (a set whose elements are points), being careful, of course, to describe the set so as to avoid any ambiguity. The notation $A = \{x; 0 \leq x \leq 1\}$ is read "A is the one-dimensional set of points x for which $0 \leq x \leq 1$." Similarly, $A = \{(x, y); 0 \leq x \leq 1, 0 \leq y \leq 1\}$ can be read "A is the two-dimensional set of points (x, y) that are interior to, or on the boundary of, a square with opposite vertices at $(0, 0)$ and $(1, 1)$." We now give some definitions (together with illustrative examples) that lead to an elementary algebra of sets adequate for our purposes.

Definition 1. If each element of a set A_1 is also an element of set A_2, the set A_1 is called a *subset* of the set A_2. This is indicated by writing $A_1 \subset A_2$. If $A_1 \subset A_2$ and also $A_2 \subset A_1$, the two sets have the same elements, and this is indicated by writing $A_1 = A_2$.

Example 1. Let $A_1 = \{x; 0 \le x \le 1\}$ and $A_2 = \{x; -1 \le x \le 2\}$. Here the one-dimensional set A_1 is seen to be a subset of the one-dimensional set A_2; that is, $A_1 \subset A_2$. Subsequently, when the dimensionality of the set is clear, we shall not make specific reference to it.

Example 2. Let $A_1 = \{(x,y); 0 \le x = y \le 1\}$ and $A_2 = \{(x,y); 0 \le x \le 1, 0 \le y \le 1\}$. Since the elements of A_1 are the points on one diagonal of the square, then $A_1 \subset A_2$.

Definition 2. If a set A has no elements, A is called the *null set.* This is indicated by writing $A = \varnothing$.

Definition 3. The set of all elements that belong to at least one of the sets A_1 and A_2 is called the *union* of A_1 and A_2. The union of A_1 and A_2 is indicated by writing $A_1 \cup A_2$. The union of several sets $A_1, A_2, A_3, \ldots$ is the set of all elements that belong to at least one of the several sets. This union is denoted by $A_1 \cup A_2 \cup A_3 \cup \cdots$ or by $A_1 \cup A_2 \cup \cdots \cup A_k$ if a finite number k of sets is involved.

Example 3. Let $A_1 = \{x; x = 0, 1, \ldots, 10\}$ and $A_2 = \{x; x = 8, 9, 10, 11,$ or $11 < x \le 12\}$. Then $A_1 \cup A_2 = \{x; x = 0, 1, \ldots, 8, 9, 10, 11,$ or $11 < x \le 12\} = \{x; x = 0, 1, \ldots, 8, 9, 10,$ or $11 \le x \le 12\}$.

Example 4. Let A_1 and A_2 be defined as in Example 1. Then $A_1 \cup A_2 = A_2$.

Example 5. Let $A_2 = \varnothing$. Then $A_1 \cup A_2 = A_1$ for every set A_1.

Example 6. For every set A, $A \cup A = A$.

Example 7. Let $A_k = \{x; 1/(k + 1) \le x \le 1\}$, $k = 1, 2, 3, \ldots$. Then $A_1 \cup A_2 \cup A_3 \cup \cdots = \{x; 0 < x \le 1\}$. Note that the number zero is not in this set, since it is not in one of the sets $A_1, A_2, A_3, \ldots$.

Definition 4. The set of all elements that belong to each of the sets A_1 and A_2 is called the *intersection* of A_1 and A_2. The intersection of A_1 and A_2 is indicated by writing $A_1 \cap A_2$. The intersection of several sets $A_1, A_2, A_3, \ldots$ is the set of all elements that belong to each of the sets $A_1, A_2, A_3, \ldots$. This intersection is denoted by $A_1 \cap A_2 \cap A_3 \cap \cdots$ or by $A_1 \cap A_2 \cap \cdots \cap A_k$ if a finite number k of sets is involved.

Example 8. Let $A_1 = \{(x, y); (x, y) = (0, 0), (0, 1), (1, 1)\}$ and $A_2 = \{(x, y); (x, y) = (1, 1), (1, 2), (2, 1)\}$. Then $A_1 \cap A_2 = \{(x, y); (x, y) = (1, 1)\}$.

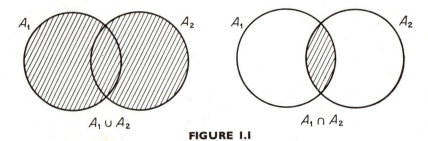

$$A_1 \cup A_2 \qquad\qquad A_1 \cap A_2$$

FIGURE 1.1

Example 9. Let $A_1 = \{(x, y); 0 \le x + y \le 1\}$ and $A_2 = \{(x, y); 1 < x + y\}$. Then A_1 and A_2 have no points in common and $A_1 \cap A_2 = \varnothing$.

Example 10. For every set A, $A \cap A = A$ and $A \cap \varnothing = \varnothing$.

Example 11. Let $A_k = \{x; 0 < x < 1/k\}$, $k = 1, 2, 3, \ldots$. Then $A_1 \cap A_2 \cap A_3 \cdots$ is the null set, since there is no point that belongs to each of the sets $A_1, A_2, A_3, \ldots$.

Example 12. Let A_1 and A_2 represent the sets of points enclosed, respectively, by two intersecting circles. Then the sets $A_1 \cup A_2$ and $A_1 \cap A_2$ are represented, respectively, by the shaded regions in the *Venn diagrams* in Figure 1.1.

Example 13. Let A_1, A_2, and A_3 represent the sets of points enclosed, respectively, by three intersecting circles. Then the sets $(A_1 \cup A_2) \cap A_3$ and $(A_1 \cap A_2) \cup A_3$ are depicted in Figure 1.2.

Definition 5. In certain discussions or considerations the totality of all elements that pertain to the discussion can be described. This set of all elements under consideration is given a special name. It is called the *space*. We shall often denote spaces by capital script letters such as $\mathscr{A}$, $\mathscr{B}$, and $\mathscr{C}$.

Example 14. Let the number of heads, in tossing a coin four times, be

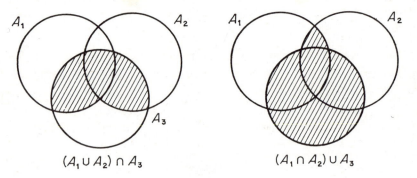

$$(A_1 \cup A_2) \cap A_3 \qquad\qquad (A_1 \cap A_2) \cup A_3$$

FIGURE 1.2

denoted by x. Of necessity, the number of heads will be one of the numbers 0, 1, 2, 3, 4. Here, then, the space is the set $\mathscr{A} = \{x; x = 0, 1, 2, 3, 4\}$.

Example 15. Consider all nondegenerate rectangles of base x and height y. To be meaningful, both x and y must be positive. Thus the space is the set $\mathscr{A} = \{(x, y); x > 0, y > 0\}$.

Definition 6. Let $\mathscr{A}$ denote a space and let A be a subset of the set $\mathscr{A}$. The set that consists of all elements of $\mathscr{A}$ that are not elements of A is called the *complement* of A (actually, with respect to $\mathscr{A}$). The complement of A is denoted by A^*. In particular, $\mathscr{A}^* = \varnothing$.

Example 16. Let $\mathscr{A}$ be defined as in Example 14, and let the set $A = \{x; x = 0, 1\}$. The complement of A (with respect to $\mathscr{A}$) is $A^* = \{x; x = 2, 3, 4\}$.

Example 17. Given $A \subset \mathscr{A}$. Then $A \cup A^* = \mathscr{A}, A \cap A^* = \varnothing, A \cup \mathscr{A} = \mathscr{A}$, $A \cap \mathscr{A} = A$, and $(A^*)^* = A$.

EXERCISES

1.4. Find the union $A_1 \cup A_2$ and the intersection $A_1 \cap A_2$ of the two sets A_1 and A_2, where:
(a) $A_1 = \{x; x = 0, 1, 2\}, A_2 = \{x; x = 2, 3, 4\}$.
(b) $A_1 = \{x; 0 < x < 2\}, A_2 = \{x; 1 \le x < 3\}$.
(c) $A_1 = \{(x, y); 0 < x < 2, 0 < y < 2\}, A_2 = \{(x, y); 1 < x < 3, 1 < y < 3\}$.

1.5. Find the complement A^* of the set A with respect to the space $\mathscr{A}$ if:
(a) $\mathscr{A} = \{x; 0 < x < 1\}, A = \{x; \frac{5}{8} \le x < 1\}$.
(b) $\mathscr{A} = \{(x, y, z); x^2 + y^2 + z^2 \le 1\}, A = \{(x, y, z); x^2 + y^2 + z^2 = 1\}$.
(c) $\mathscr{A} = \{(x, y); |x| + |y| \le 2\}, A = \{(x, y); x^2 + y^2 < 2\}$.

1.6. List all possible arrangements of the four letters m, a, r, and y. Let A_1 be the collection of the arrangements in which y is in the last position. Let A_2 be the collection of the arrangements in which m is in the first position. Find the union and the intersection of A_1 and A_2.

1.7. By use of Venn diagrams, in which the space $\mathscr{A}$ is the set of points enclosed by a rectangle containing the circles, compare the following sets:
(a) $A_1 \cap (A_2 \cup A_3)$ and $(A_1 \cap A_2) \cup (A_1 \cap A_3)$.
(b) $A_1 \cup (A_2 \cap A_3)$ and $(A_1 \cup A_2) \cap (A_1 \cup A_3)$.
(c) $(A_1 \cup A_2)^*$ and $A_1^* \cap A_2^*$.
(d) $(A_1 \cap A_2)^*$ and $A_1^* \cup A_2^*$.

1.8. If a sequence of sets $A_1, A_2, A_3, \ldots$ is such that $A_k \subset A_{k+1}$, $k = 1, 2, 3, \ldots$, the sequence is said to be a *nondecreasing sequence*. Give an example of this kind of sequence of sets.

1.9. If a sequence of sets $A_1, A_2, A_3, \ldots$ is such that $A_k \supset A_{k+1}$, $k = 1, 2, 3, \ldots$, the sequence is said to be a *nonincreasing sequence*. Give an example of this kind of sequence of sets.

1.10. If $A_1, A_2, A_3, \ldots$ are sets such that $A_k \subset A_{k+1}$, $k = 1, 2, 3, \ldots$, $\lim_{k \to \infty} A_k$ is defined as the union $A_1 \cup A_2 \cup A_3 \cup \cdots$. Find $\lim_{k \to \infty} A_k$ if:

(a) $A_k = \{x; 1/k \leq x \leq 3 - 1/k\}$, $k = 1, 2, 3, \ldots$;

(b) $A_k = \{(x, y); 1/k \leq x^2 + y^2 \leq 4 - 1/k\}$, $k = 1, 2, 3, \ldots$.

1.11. If $A_1, A_2, A_3, \ldots$ are sets such that $A_k \supset A_{k+1}$, $k = 1, 2, 3, \ldots$, $\lim_{k \to \infty} A_k$ is defined as the intersection $A_1 \cap A_2 \cap A_3 \cap \cdots$. Find $\lim_{k \to \infty} A_k$ if:

(a) $A_k = \{x; 2 - 1/k < x \leq 2\}$, $k = 1, 2, 3, \ldots$.

(b) $A_k = \{x; 2 < x \leq 2 + 1/k\}$, $k = 1, 2, 3, \ldots$.

(c) $A_k = \{(x, y); 0 \leq x^2 + y^2 \leq 1/k\}$, $k = 1, 2, 3, \ldots$.

1.3 Set Functions

In the calculus, functions such as

$$f(x) = 2x, \qquad -\infty < x < \infty,$$

or

$$g(x, y) = e^{-x-y}, \qquad 0 < x < \infty, \quad 0 < y < \infty,$$

$$= 0 \text{ elsewhere,}$$

or possibly

$$h(x_1, x_2, \ldots, x_n) = 3x_1 x_2 \cdots x_n, \qquad 0 \leq x_i \leq 1, \quad i = 1, 2, \ldots, n,$$

$$= 0 \text{ elsewhere,}$$

were of common occurrence. The value of $f(x)$ at the "point $x = 1$" is $f(1) = 2$; the value of $g(x, y)$ at the "point $(-1, 3)$" is $g(-1, 3) = 0$; the value of $h(x_1, x_2, \ldots, x_n)$ at the "point $(1, 1, \ldots, 1)$" is 3. Functions such as these are called functions of a point or, more simply, *point functions* because they are evaluated (if they have a value) at a point in a space of indicated dimension.

There is no reason why, if they prove useful, we should not have functions that can be evaluated, not necessarily at a point, but for an entire set of points. Such functions are naturally called functions of a set or, more simply, *set functions*. We shall give some examples of set functions and evaluate them for certain simple sets.

Example 1. Let A be a set in one-dimensional space and let $Q(A)$ be equal to the number of points in A which correspond to positive integers. Then $Q(A)$ is a function of the set A. Thus, if $A = \{x; 0 < x < 5\}$, then $Q(A) = 4$; if $A = \{x; x = -2, -1\}$, then $Q(A) = 0$; if $A = \{x; -\infty < x < 6\}$, then $Q(A) = 5$.

Example 2. Let A be a set in two-dimensional space and let $Q(A)$ be the area of A, if A has a finite area; otherwise, let $Q(A)$ be undefined. Thus, if $A = \{(x, y); x^2 + y^2 \le 1\}$, then $Q(A) = \pi$; if $A = \{(x, y); (x, y) = (0, 0), (1, 1), (0, 1)\}$, then $Q(A) = 0$; if $A = \{(x, y); 0 \le x, 0 \le y, x + y \le 1\}$, then $Q(A) = \frac{1}{2}$.

Example 3. Let A be a set in three-dimensional space and let $Q(A)$ be the volume of A, if A has a finite volume; otherwise, let $Q(A)$ be undefined. Thus, if $A = \{(x, y, z); 0 \le x \le 2, 0 \le y \le 1, 0 \le z \le 3\}$, then $Q(A) = 6$; if $A = \{(x, y, z); x^2 + y^2 + z^2 \ge 1\}$, then $Q(A)$ is undefined.

At this point we introduce the following notations. The symbol

$$\int_A f(x)\, dx$$

will mean the ordinary (Riemann) integral of $f(x)$ over a prescribed one-dimensional set A; the symbol

$$\int_A \int g(x, y)\, dx\, dy$$

will mean the Riemann integral of $g(x, y)$ over a prescribed two-dimensional set A; and so on. To be sure, unless these sets A and these functions $f(x)$ and $g(x, y)$ are chosen with care, the integrals will frequently fail to exist. Similarly, the symbol

$$\sum_A f(x)$$

will mean the sum extended over all $x \in A$; the symbol

$$\sum_A \sum g(x, y)$$

will mean the sum extended over all $(x, y) \in A$; and so on.

Example 4. Let A be a set in one-dimensional space and let $Q(A) = \sum_A f(x)$, where

$$f(x) = (\tfrac{1}{2})^x, \qquad x = 1, 2, 3, \ldots,$$
$$= 0 \text{ elsewhere.}$$

If $A = \{x; 0 \le x \le 3\}$, then

$$Q(A) = \tfrac{1}{2} + (\tfrac{1}{2})^2 + (\tfrac{1}{2})^3 = \tfrac{7}{8}.$$

Example 5. Let $Q(A) = \sum_A f(x)$, where

$$f(x) = p^x(1 - p)^{1-x}, \qquad x = 0, 1,$$
$$= 0 \text{ elsewhere.}$$

If $A = \{x; x = 0\}$, then

$$Q(A) = \sum_{x=0}^{x=0} p^x (1 - p)^{1-x} = 1 - p;$$

if $A = \{x; 1 \le x \le 2\}$, then $Q(A) = f(1) = p$.

Example 6. Let A be a one-dimensional set and let

$$Q(A) = \int_A e^{-x} \, dx.$$

Thus, if $A = \{x; 0 \le x < \infty\}$, then

$$Q(A) = \int_0^\infty e^{-x} \, dx = 1;$$

if $A = \{x; 1 \le x \le 2\}$, then

$$Q(A) = \int_1^2 e^{-x} \, dx = e^{-1} - e^{-2};$$

if $A_1 = \{x; 0 \le x \le 1\}$ and $A_2 = \{x; 1 < x \le 3\}$, then

$$Q(A_1 \cup A_2) = \int_0^3 e^{-x} \, dx$$

$$= \int_0^1 e^{-x} \, dx + \int_1^3 e^{-x} \, dx$$

$$= Q(A_1) + Q(A_2);$$

if $A = A_1 \cup A_2$, where $A_1 = \{x; 0 \le x \le 2\}$ and $A_2 = \{x; 1 \le x \le 3\}$, then

$$Q(A) = Q(A_1 \cup A_2) = \int_0^3 e^{-x} \, dx$$

$$= \int_0^2 e^{-x} \, dx + \int_1^3 e^{-x} \, dx - \int_1^2 e^{-x} \, dx$$

$$= Q(A_1) + Q(A_2) - Q(A_1 \cap A_2).$$

Example 7. Let A be a set in n-dimensional space and let

$$Q(A) = \int \cdots \int_A dx_1 \, dx_2 \cdots dx_n.$$

If $A = \{(x_1, x_2, \ldots, x_n); 0 \le x_1 \le x_2 \le \cdots \le x_n \le 1\}$, then

$$Q(A) = \int_0^1 \int_0^{x_n} \cdots \int_0^{x_3} \int_0^{x_2} dx_1 \, dx_2 \cdots dx_{n-1} \, dx_n$$

$$= \frac{1}{n!}, \qquad \text{where } n! = n(n-1) \cdots 3 \cdot 2 \cdot 1.$$

EXERCISES

1.12. For every one-dimensional set A, let $Q(A) = \sum_A f(x)$, where $f(x) = (\frac{2}{3})(\frac{1}{3})^x$, $x = 0, 1, 2, \ldots$, zero elsewhere. If $A_1 = \{x; x = 0, 1, 2, 3\}$ and $A_2 = \{x; x = 0, 1, 2, \ldots\}$, find $Q(A_1)$ and $Q(A_2)$. *Hint.* Recall that $S_n = $

$a + ar + \cdots + ar^{n-1} = a(1 - r^n)/(1 - r)$ and $\lim_{n \to \infty} S_n = a/(1 - r)$ provided that $|r| < 1$.

1.13. For every one-dimensional set A for which the integral exists, let $Q(A) = \int_A f(x) \, dx$, where $f(x) = 6x(1 - x)$, $0 < x < 1$, zero elsewhere; otherwise, let $Q(A)$ be undefined. If $A_1 = \{x; \frac{1}{4} < x < \frac{3}{4}\}$, $A_2 = \{x; x = \frac{1}{2}\}$, and $A_3 = \{x; 0 < x < 10\}$, find $Q(A_1)$, $Q(A_2)$, and $Q(A_3)$.

1.14. For every one-dimensional set A, let $Q(A)$ be equal to the number of points in A that correspond to positive integers. If $A_1 = \{x; x$ a multiple of 3, less than or equal to 50\} and $A_2 = \{x; x$ a multiple of 7, less than or equal to 50\}, find $Q(A_1), Q(A_2), Q(A_1 \cup A_2)$, and $Q(A_1 \cap A_2)$. Show that $Q(A_1 \cup A_2) = Q(A_1) + Q(A_2) - Q(A_1 \cap A_2)$.

1.15. For every two-dimensional set A, let $Q(A)$ be equal to the number of points (x, y) in A for which both x and y are positive integers. Find $Q(A_1)$ and $Q(A_2)$, where $A_1 = \{(x, y); x^2 + y^2 \leq 4\}$ and $A_2 = \{(x, y); x^2 + y^2 \leq 9\}$. Note that $A_1 \subset A_2$ and that $Q(A_1) \leq Q(A_2)$.

1.16. Let $Q(A) = \int_A \int (x^2 + y^2) \, dx \, dy$ for every two-dimensional set A for which the integral exists; otherwise, let $Q(A)$ be undefined. If $A_1 = \{(x, y); -1 \leq x \leq 1, -1 \leq y \leq 1\}$, $A_2 = \{(x, y); -1 \leq x = y \leq 1\}$, and $A_3 = \{(x, y); x^2 + y^2 \leq 1\}$, find $Q(A_1), Q(A_2)$, and $Q(A_3)$. *Hint.* In evaluating $Q(A_2)$, recall the definition of the double integral (or consider the volume under the surface $z = x^2 + y^2$ above the line segment $-1 \leq x = y \leq 1$ in the xy-plane). Use polar coordinates in the calculation of $Q(A_3)$.

1.17. Let $\mathscr{A}$ denote the set of points that are interior to or on the boundary of a square with opposite vertices at the point $(0, 0)$ and at the point $(1, 1)$. Let $Q(A) = \int_A \int dy \, dx$. (a) If $A \subset \mathscr{A}$ is the set $\{(x, y); 0 < x < y < 1\}$, compute $Q(A)$. (b) If $A \subset \mathscr{A}$ is the set $\{(x, y); 0 < x = y < 1\}$, compute $Q(A)$. (c) If $A \subset \mathscr{A}$ is the set $\{(x, y); 0 < x/2 \leq y \leq 3x/2 < 1\}$, compute $Q(A)$.

1.18. Let $\mathscr{A}$ be the set of points interior to or on the boundary of a cube with edge 1. Moreover, say that the cube is in the first octant with one vertex at the point $(0, 0, 0)$ and an opposite vertex is at the point $(1, 1, 1)$. Let $Q(A) = \iiint_A dx \, dy \, dz$. (a) If $A \subset \mathscr{A}$ is the set $\{(x, y, z); 0 < x < y < z < 1\}$, compute $Q(A)$. (b) If A is the subset $\{(x, y, z); 0 < x = y = z < 1\}$, compute $Q(A)$.

1.19. Let A denote the set $\{(x, y, z); x^2 + y^2 + z^2 \leq 1\}$. Evaluate $Q(A) = \iiint_A \sqrt{x^2 + y^2 + z^2} \, dx \, dy \, dz$. *Hint.* Change variables to spherical coordinates.

1.20. To join a certain club, a person must be either a statistician or a mathematician or both. Of the 25 members in this club, 19 are statisticians and 16 are mathematicians. How many persons in the club are both a statistician and a mathematician?

1.21. After a hard-fought football game, it was reported that, of the 11 starting players, 8 hurt a hip, 6 hurt an arm, 5 hurt a knee, 3 hurt both a hip and an arm, 2 hurt both a hip and a knee, 1 hurt both an arm and a knee, and no one hurt all three. Comment on the accuracy of the report.

1.4 The Probability Set Function

Let $\mathscr{C}$ denote the set of every possible outcome of a random experiment; that is, $\mathscr{C}$ is the sample space. It is our purpose to define a set function $P(C)$ such that if C is a subset of $\mathscr{C}$, then $P(C)$ is the probability that the outcome of the random experiment is an element of C. Henceforth it will be tacitly assumed that the structure of each set C is sufficiently simple to allow the computation. We have already seen that advantages accrue if we take $P(C)$ to be that number about which the relative frequency f/N of the event C tends to stabilize after a long series of experiments. This important fact suggests some of the properties that we would surely want the set function $P(C)$ to possess. For example, no relative frequency is ever negative; accordingly, we would want $P(C)$ to be a nonnegative set function. Again, the relative frequency of the whole sample space $\mathscr{C}$ is always 1. Thus we would want $P(\mathscr{C}) = 1$. Finally, if $C_1, C_2, C_3, \ldots$ are subsets of $\mathscr{C}$ such that no two of these subsets have a point in common, the relative frequency of the union of these sets is the sum of the relative frequencies of the sets, and we would want the set function $P(C)$ to reflect this additive property. We now formally define a probability set function.

Definition 7. If $P(C)$ is defined for a type of subset of the space $\mathscr{C}$, and if
(a) $P(C) \geq 0$,
(b) $P(C_1 \cup C_2 \cup C_3 \cup \cdots) = P(C_1) + P(C_2) + P(C_3) + \cdots$, where the sets C_i, $i = 1, 2, 3, \ldots$, are such that no two have a point in common, (that is, where $C_i \cap C_j = \varnothing$, $i \neq j$),
(c) $P(\mathscr{C}) = 1$,
then $P(C)$ is called the *probability set function* of the outcome of the random experiment. For each subset C of $\mathscr{C}$, the number $P(C)$ is called the probability that the outcome of the random experiment is an element of the set C, or the probability of the event C, or the probability measure of the set C.

A probability set function tells us how the probability is distributed over various subsets C of a sample space $\mathscr{C}$. In this sense we speak of a distribution of probability.

Remark. In the definition, the phrase "a type of subset of the space $\mathscr{C}$" would be explained more fully in a more advanced course. Nevertheless, a few observations can be made about the collection of subsets that are of the type. From condition (c) of the definition, we see that the space $\mathscr{C}$ must be in the collection. Condition (b) implies that if the sets $C_1, C_2, C_3, \ldots$ are in the collection, their union is also one of that type. Finally, we observe from the following theorems and their proofs that if the set C is in the collection, its complement must be one of those subsets. In particular, the null set, which is the complement of $\mathscr{C}$, must be in the collection.

The following theorems give us some other properties of a probability set function. In the statement of each of these theorems, $P(C)$ is taken, tacitly, to be a probability set function defined for a certain type of subset of the sample space $\mathscr{C}$.

Theorem 1. *For each* $C \subset \mathscr{C}$, $P(C) = 1 - P(C^*)$.

Proof. We have $\mathscr{C} = C \cup C^*$ and $C \cap C^* = \varnothing$. Thus, from (c) and (b) of Definition 7, it follows that

$$1 = P(C) + P(C^*),$$

which is the desired result.

Theorem 2. *The probability of the null set is zero; that is,* $P(\varnothing) = 0$.

Proof. In Theorem 1, take $C = \varnothing$ so that $C^* = \mathscr{C}$. Accordingly, we have

$$P(\varnothing) = 1 - P(\mathscr{C}) = 1 - 1 = 0,$$

and the theorem is proved.

Theorem 3. *If* C_1 *and* C_2 *are subsets of* $\mathscr{C}$ *such that* $C_1 \subset C_2$, *then* $P(C_1) \leq P(C_2)$.

Proof. Now $C_2 = C_1 \cup (C_1^* \cap C_2)$ and $C_1 \cap (C_1^* \cap C_2) = \varnothing$. Hence, from (b) of Definition 7,

$$P(C_2) = P(C_1) + P(C_1^* \cap C_2).$$

However, from (a) of Definition 7, $P(C_1^* \cap C_2) \geq 0$; accordingly, $P(C_2) \geq P(C_1)$.

Theorem 4. *For each* $C \subset \mathscr{C}$, $0 \leq P(C) \leq 1$.

Proof. Since $\varnothing \subset C \subset \mathscr{C}$, we have by Theorem 3 that

$$P(\varnothing) \leq P(C) \leq P(\mathscr{C}) \qquad \text{or} \qquad 0 \leq P(C) \leq 1,$$

the desired result.

Theorem 5. *If C_1 and C_2 are subsets of $\mathscr{C}$, then*

$$P(C_1 \cup C_2) = P(C_1) + P(C_2) - P(C_1 \cap C_2).$$

Proof. Each of the sets $C_1 \cup C_2$ and C_2 can be represented, respectively, as a union of nonintersecting sets as follows:

$$C_1 \cup C_2 = C_1 \cup (C_1^* \cap C_2) \qquad \text{and} \qquad C_2 = (C_1 \cap C_2) \cup (C_1^* \cap C_2).$$

Thus, from (b) of Definition 7,

$$P(C_1 \cup C_2) = P(C_1) + P(C_1^* \cap C_2)$$

and

$$P(C_2) = P(C_1 \cap C_2) + P(C_1^* \cap C_2).$$

If the second of these equations is solved for $P(C_1^* \cap C_2)$ and this result substituted in the first equation, we obtain

$$P(C_1 \cup C_2) = P(C_1) + P(C_2) - P(C_1 \cap C_2).$$

This completes the proof.

Example 1. Let $\mathscr{C}$ denote the sample space of Example 2 of Section 1.1. Let the probability set function assign a probability of $\frac{1}{36}$ to each of the 36 points in $\mathscr{C}$. If $C_1 = \{c; c = (1, 1), (2, 1), (3, 1), (4, 1), (5, 1)\}$ and $C_2 = \{c; c = (1, 2), (2, 2), (3, 2)\}$, then $P(C_1) = \frac{5}{36}$, $P(C_2) = \frac{3}{36}$, $P(C_1 \cup C_2) = \frac{8}{36}$, and $P(C_1 \cap C_2) = 0$.

Example 2. Two coins are to be tossed and the outcome is the ordered pair (face on the first coin, face on the second coin). Thus the sample space may be represented as $\mathscr{C} = \{c; c = (H, H), (H, T), (T, H), (T, T)\}$. Let the probability set function assign a probability of $\frac{1}{4}$ to each element of $\mathscr{C}$. Let $C_1 = \{c; c = (H, H), (H, T)\}$ and $C_2 = \{c; c = (H, H), (T, H)\}$. Then $P(C_1) = P(C_2) = \frac{1}{2}$, $P(C_1 \cap C_2) = \frac{1}{4}$, and, in accordance with Theorem 5, $P(C_1 \cup C_2) = \frac{1}{2} + \frac{1}{2} - \frac{1}{4} = \frac{3}{4}$.

Let $\mathscr{C}$ denote a sample space and let $C_1, C_2, C_3, \ldots$ denote subsets of $\mathscr{C}$. If these subsets are such that no two have an element in common, they are called mutually disjoint sets and the corresponding events $C_1, C_2, C_3, \ldots$ are said to be *mutually exclusive events*. Then, for example, $P(C_1 \cup C_2 \cup C_3 \cup \cdots) = P(C_1) + P(C_2) + P(C_3) + \cdots$, in accordance with (b) of Definition 7. Moreover, if $\mathscr{C} = C_1 \cup C_2 \cup C_3 \cup \cdots$, the mutually exclusive events are further characterized as being *exhaustive* and the probability of their union is obviously equal to 1.

EXERCISES

1.22. A positive integer from one to six is to be chosen by casting a die. Thus the elements c of the sample space $\mathscr{C}$ are 1, 2, 3, 4, 5, 6. Let $C_1 = \{c; c = 1, 2, 3, 4\}$, $C_2 = \{c; c = 3, 4, 5, 6\}$. If the probability set function P assigns a probability of $\frac{1}{6}$ to each of the elements of $\mathscr{C}$, compute $P(C_1)$, $P(C_2)$, $P(C_1 \cap C_2)$, and $P(C_1 \cup C_2)$.

1.23. A random experiment consists in drawing a card from an ordinary deck of 52 playing cards. Let the probability set function P assign a probability of $\frac{1}{52}$ to each of the 52 possible outcomes. Let C_1 denote the collection of the 13 hearts and let C_2 denote the collection of the 4 kings. Compute $P(C_1)$, $P(C_2)$, $P(C_1 \cap C_2)$, and $P(C_1 \cup C_2)$.

1.24. A coin is to be tossed as many times as is necessary to turn up one head. Thus the elements c of the sample space $\mathscr{C}$ are H, TH, TTH, $TTTH$, and so forth. Let the probability set function P assign to these elements the respective probabilities $\frac{1}{2}$, $\frac{1}{4}$, $\frac{1}{8}$, $\frac{1}{16}$, and so forth. Show that $P(\mathscr{C}) = 1$. Let $C_1 = \{c; c$ is H, TH, TTH, $TTTH$, or $TTTTH\}$. Compute $P(C_1)$. Let $C_2 = \{c; c$ is $TTTTH$ or $TTTTTH\}$. Compute $P(C_2)$, $P(C_1 \cap C_2)$, and $P(C_1 \cup C_2)$.

1.25. If the sample space is $\mathscr{C} = C_1 \cup C_2$ and if $P(C_1) = 0.8$ and $P(C_2) = 0.5$, find $P(C_1 \cap C_2)$.

1.26. Let the sample space be $\mathscr{C} = \{c; 0 < c < \infty\}$. Let $C \subset \mathscr{C}$ be defined by $C = \{c; 4 < c < \infty\}$ and take $P(C) = \int_C e^{-x}\, dx$. Evaluate $P(C)$, $P(C^*)$, and $P(C \cup C^*)$.

1.27. If the sample space is $\mathscr{C} = \{c; -\infty < c < \infty\}$ and if $C \subset \mathscr{C}$ is a set for which the integral $\int_C e^{-|x|}\, dx$ exists, show that this set function is not a probability set function. What constant could we multiply the integral by to make it a probability set function?

1.28. If C_1 and C_2 are subsets of the sample space $\mathscr{C}$, show that

$$P(C_1 \cap C_2) \le P(C_1) \le P(C_1 \cup C_2) \le P(C_1) + P(C_2).$$

1.29. Let C_1, C_2, and C_3 be three mutually disjoint subsets of the sample space $\mathscr{C}$. Find $P[(C_1 \cup C_2) \cap C_3]$ and $P(C_1^* \cup C_2^*)$.

1.30. If C_1, C_2, and C_3 are subsets of $\mathscr{C}$, show that

$$
\begin{aligned}
P(C_1 \cup C_2 \cup C_3) = {} & P(C_1) + P(C_2) + P(C_3) - P(C_1 \cap C_2) \\
& - P(C_1 \cap C_3) - P(C_2 \cap C_3) + P(C_1 \cap C_2 \cap C_3).
\end{aligned}
$$

What is the generalization of this result to four or more subsets of $\mathscr{C}$?
Hint. Write $P(C_1 \cup C_2 \cup C_3) = P[C_1 \cup (C_2 \cup C_3)]$ and use Theorem 5.

1.5 Random Variables

The reader will perceive that a sample space $\mathscr{C}$ may be tedious to
describe if the elements of $\mathscr{C}$ are not numbers. We shall now discuss
how we may formulate a rule, or a set of rules, by which the elements c
of $\mathscr{C}$ may be represented by numbers x or ordered pairs of numbers
(x_1, x_2) or, more generally, ordered n-tuplets of numbers $(x_1, \ldots, x_n)$.
We begin the discussion with a very simple example. Let the random
experiment be the toss of a coin and let the sample space associated with
the experiment be $\mathscr{C} = \{c;$ where c is T or c is $H\}$ and T and H represent, respectively, tails and heads. Let X be a function such that
$X(c) = 0$ if c is T and let $X(c) = 1$ if c is H. Thus X is a real-valued
function defined on the sample space $\mathscr{C}$ which takes us from the sample
space $\mathscr{C}$ to a space of real numbers $\mathscr{A} = \{x; x = 0, 1\}$. We call X a
random variable and, in this example, the space associated with X is
$\mathscr{A} = \{x; x = 0, 1\}$. We now formulate the definition of a random
variable and its space.

Definition 8. Given a random experiment with a sample space $\mathscr{C}$.
A function X, which assigns to each element $c \in \mathscr{C}$ one and only one
real number $X(c) = x$, is called a *random variable*. The *space* of X is the
set of real numbers $\mathscr{A} = \{x; x = X(c), c \in \mathscr{C}\}$.

It may be that the set $\mathscr{C}$ has elements which are themselves real
numbers. In such an instance we could write $X(c) = c$ so that $\mathscr{A} = \mathscr{C}$.

Let X be a random variable that is defined on a sample space $\mathscr{C}$,
and let $\mathscr{A}$ be the space of X. Further, let A be a subset of $\mathscr{A}$. Just as we
used the terminology "the event C," with $C \subset \mathscr{C}$, we shall now speak
of "the event A." The probability $P(C)$ of the event C has been defined.
We wish now to define the probability of the event A. This probability
will be denoted by $\Pr (X \in A)$, where $\Pr$ is an abbreviation of "the
probability that." With A a subset of $\mathscr{A}$, let C be that subset of $\mathscr{C}$ such
that $C = \{c; c \in \mathscr{C}$ and $X(c) \in A\}$. Thus C has as its elements all outcomes in $\mathscr{C}$ for which the random variable X has a value that is in A.
This prompts us to define, as we now do, $\Pr (X \in A)$ to be equal to
$P(C)$, where $C = \{c; c \in \mathscr{C}$ and $X(c) \in A\}$. Thus $\Pr (X \in A)$ is an assignment of probability to a set A, which is a subset of the space $\mathscr{A}$ associated
with the random variable X. This assignment is determined by the

probability set function P and the random variable X and is sometimes denoted by $P_X(A)$. That is,

$$\Pr\,(X \in A) = P_X(A) = P(C),$$

where $C = \{c; c \in \mathscr{C} \text{ and } X(c) \in A\}$. Thus a random variable X is a function that carries the probability from a sample space $\mathscr{C}$ to a space $\mathscr{A}$ of real numbers. In this sense, with $A \subset \mathscr{A}$, the probability $P_X(A)$ is often called an *induced probability*.

The function $P_X(A)$ satisfies the conditions (a), (b), and (c) of the definition of a probability set function (Section 1.4). That is, $P_X(A)$ is also a probability set function. Conditions (a) and (c) are easily verified by observing, for an appropriate C, that

$$P_X(A) = P(C) \geq 0,$$

and that $\mathscr{C} = \{c; c \in \mathscr{C} \text{ and } X(c) \in \mathscr{A}\}$ requires

$$P_X(\mathscr{A}) = P(\mathscr{C}) = 1.$$

In discussing condition (b), let us restrict our attention to two mutually exclusive events A_1 and A_2. Here $P_X(A_1 \cup A_2) = P(C)$, where $C = \{c; c \in \mathscr{C} \text{ and } X(c) \in A_1 \cup A_2\}$. However,

$$C = \{c; c \in \mathscr{C} \text{ and } X(c) \in A_1\} \cup \{c; c \in \mathscr{C} \text{ and } X(c) \in A_2\},$$

or, for brevity, $C = C_1 \cup C_2$. But C_1 and C_2 are disjoint sets. This must be so, for if some c were common, say c_i, then $X(c_i) \in A_1$ and $X(c_i) \in A_2$. That is, the same number $X(c_i)$ belongs to both A_1 and A_2. This is a contradiction because A_1 and A_2 are disjoint sets. Accordingly,

$$P(C) = P(C_1) + P(C_2).$$

However, by definition, $P(C_1)$ is $P_X(A_1)$ and $P(C_2)$ is $P_X(A_2)$ and thus

$$P_X(A_1 \cup A_2) = P_X(A_1) + P_X(A_2).$$

This is condition (b) for two disjoint sets.

Thus each of $P_X(A)$ and $P(C)$ is a probability set function. But the reader should fully recognize that the probability set function P is defined for subsets C of $\mathscr{C}$, whereas P_X is defined for subsets A of $\mathscr{A}$, and, in general, they are not the same set function. Nevertheless, they are closely related and some authors even drop the index X and write $P(A)$ for $P_X(A)$. They think it is quite clear that $P(A)$ means the probability of A, a subset of $\mathscr{A}$, and $P(C)$ means the probability of C, a subset of $\mathscr{C}$. From this point on, we shall adopt this convention and simply write $P(A)$.

Perhaps an additional example will be helpful. Let a coin be tossed twice and let our interest be in the *number* of heads to be observed. Thus the sample space is $\mathscr{C} = \{c;$ where c is TT or TH or HT or $HH\}$. Let $X(c) = 0$ if c is TT; let $X(c) = 1$ if c is either TH or HT; and let $X(c) = 2$ if c is HH. Thus the space of the random variable X is $\mathscr{A} = \{x; x = 0, 1, 2\}$. Consider the subset A of the space $\mathscr{A}$, where $A = \{x; x = 1\}$. How is the probability of the event A defined? We take the subset C of $\mathscr{C}$ to have as its elements all outcomes in $\mathscr{C}$ for which the random variable X has a value that is an element of A. Because $X(c) = 1$ if c is either TH or HT, then $C = \{c;$ where c is TH or $HT\}$. Thus $P(A) = \Pr(X \in A) = P(C)$. Since $A = \{x; x = 1\}$, then $P(A) = \Pr(X \in A)$ can be written more simply as $\Pr(X = 1)$. Let $C_1 = \{c; c$ is $TT\}$, $C_2 = \{c; c$ is $TH\}$, $C_3 = \{c; c$ is $HT\}$, and $C_4 = \{c; c$ is $HH\}$ denote subsets of $\mathscr{C}$. Suppose that our probability set function $P(C)$ assigns a probability of $\frac{1}{4}$ to each of the sets C_i, $i = 1, 2, 3, 4$. Then $P(C_1) = \frac{1}{4}$, $P(C_2 \cup C_3) = \frac{1}{4} + \frac{1}{4} = \frac{1}{2}$, and $P(C_4) = \frac{1}{4}$. Let us now point out how much simpler it is to couch these statements in a language that involves the random variable X. Because X is the number of heads to be observed in tossing a coin two times, we have

$$\Pr(X = 0) = \tfrac{1}{4}, \text{ since } P(C_1) = \tfrac{1}{4};$$

$$\Pr(X = 1) = \tfrac{1}{2}, \text{ since } P(C_2 \cup C_3) = \tfrac{1}{2};$$

and

$$\Pr(X = 2) = \tfrac{1}{4}, \text{ since } P(C_4) = \tfrac{1}{4}.$$

This may be further condensed in the following table:

x	0	1	2
$\Pr(X = x)$	$\frac{1}{4}$	$\frac{1}{2}$	$\frac{1}{4}$

This table depicts the distribution of probability over the elements of $\mathscr{A}$, the space of the random variable X.

We shall now discuss two random variables. Again, we start with an example. A coin is to be tossed three times and our interest is in the ordered number pair (number of H's on first two tosses, number H's on all three tosses). Thus the sample space is $\mathscr{C} = \{c; c = c_i, i = 1, 2, \ldots, 8\}$, where c_1 is TTT, c_2 is TTH, c_3 is THT, c_4 is HTT, c_5 is THH, c_6 is HTH, c_7 is HHT, and c_8 is HHH. Let X_1 and X_2 be two functions such that $X_1(c_1) = X_1(c_2) = 0$, $X_1(c_3) = X_1(c_4) = X_1(c_5) = X_1(c_6) = 1$, $X_1(c_7) = X_1(c_8) = 2$; and $X_2(c_1) = 0$, $X_2(c_2) = X_2(c_3) = X_2(c_4) = 1$, $X_2(c_5) = X_2(c_6) = X_2(c_7) = 2$, $X_2(c_8) = 3$. Thus X_1 and X_2 are real-valued functions defined on the sample space $\mathscr{C}$, which takes us from that sample space to the space of ordered number pairs

$$\mathscr{A} = \{(x_1, x_2); (x_1, x_2) = (0, 0), (0, 1), (1, 1), (1, 2), (2, 2), (2, 3)\}.$$

Thus X_1 and X_2 are two random variables defined on the space $\mathscr{C}$, and, in this example, the space of these random variables is the two-dimensional set $\mathscr{A}$ given immediately above. We now formulate the definition of the space of two random variables.

Definition 9. Given a random experiment with a sample space $\mathscr{C}$. Consider two random variables X_1 and X_2, which assign to each element c of $\mathscr{C}$ one and only one ordered pair of numbers $X_1(c) = x_1$, $X_2(c) = x_2$. The *space* of X_1 and X_2 is the set of ordered pairs $\mathscr{A} = \{(x_1, x_2); x_1 = X_1(c), x_2 = X_2(c), c \in \mathscr{C}\}$.

Let $\mathscr{A}$ be the space associated with the two random variables X_1 and X_2 and let A be a subset of $\mathscr{A}$. As in the case of one random variable, we shall speak of the event A. We wish to define the probability of the event A, which we denote by $\Pr[(X_1, X_2) \in A]$. Take $C = \{c; c \in \mathscr{C}$ and $[X_1(c), X_2(c)] \in A\}$, where $\mathscr{C}$ is the sample space. We then define $\Pr[(X_1, X_2) \in A] = P(C)$, where P is the probability set function defined for subsets C of $\mathscr{C}$. Here again we could denote $\Pr[(X_1, X_2) \in A]$ by the probability set function $P_{X_1, X_2}(A)$; but, with our previous convention, we simply write

$$P(A) = \Pr[(X_1, X_2) \in A].$$

Again it is important to observe that this function is a probability set function defined for subsets A of the space $\mathscr{A}$.

Let us return to the example in our discussion of two random variables. Consider the subset A of $\mathscr{A}$, where $A = \{(x_1, x_2); (x_1, x_2) = (1, 1), (1, 2)\}$. To compute $\Pr[(X_1, X_2) \in A] = P(A)$, we must include as elements of C all outcomes in $\mathscr{C}$ for which the random variables X_1 and X_2 take values (x_1, x_2) which are elements of A. Now $X_1(c_3) = 1$, $X_2(c_3) = 1$, $X_1(c_4) = 1$, and $X_2(c_4) = 1$. Also, $X_1(c_5) = 1$, $X_2(c_5) = 2$, $X_1(c_6) = 1$, and $X_2(c_6) = 2$. Thus $P(A) = \Pr[(X_1, X_2) \in A] = P(C)$, where $C = \{c; c = c_3, c_4, c_5,$ or $c_6\}$. Suppose that our probability set function $P(C)$ assigns a probability of $\frac{1}{8}$ to each of the eight elements of $\mathscr{C}$. Then $P(A)$, which can be written as $\Pr(X_1 = 1, X_2 = 1$ or $2)$, is equal to $\frac{4}{8} = \frac{1}{2}$. It is left for the reader to show that we can tabulate the probability, which is then assigned to each of the elements of $\mathscr{A}$, with the following result:

(x_1, x_2)	$(0, 0)$	$(0, 1)$	$(1, 1)$	$(1, 2)$	$(2, 2)$	$(2, 3)$
$\Pr[(X_1, X_2) = (x_1, x_2)]$	$\frac{1}{8}$	$\frac{1}{8}$	$\frac{2}{8}$	$\frac{2}{8}$	$\frac{1}{8}$	$\frac{1}{8}$

This table depicts the distribution of probability over the elements of $\mathscr{A}$, the space of the random variables X_1 and X_2.

The preceding notions about one and two random variables can be immediately extended to n random variables. We make the following definition of the space of n random variables.

Definition 10. Given a random experiment with the sample space $\mathscr{C}$. Let the random variable X_i assign to each element $c \in \mathscr{C}$ one and only one real number $X_i(c) = x_i$, $i = 1, 2, \ldots, n$. The *space* of these random variables is the set of ordered n-tuplets $\mathscr{A} = \{(x_1, x_2, \ldots, x_n);$ $x_1 = X_1(c), \ldots, x_n = X_n(c), c \in \mathscr{C}\}$. Further, let A be a subset of $\mathscr{A}$. Then $\Pr[(X_1, \ldots, X_n) \in A] = P(C)$, where $C = \{c; c \in \mathscr{C}$ and $[X_1(c), X_2(c), \ldots, X_n(c)] \in A\}$.

Again we should make the comment that $\Pr[(X_1, \ldots, X_n) \in A]$ could be denoted by the probability set function $P_{X_1, \ldots, X_n}(A)$. But, if there is no chance of misunderstanding, it will be written simply as $P(A)$.

Up to this point, our illustrative examples have dealt with a sample space $\mathscr{C}$ that contains a finite number of elements. We now give an example of a sample space $\mathscr{C}$ that is an interval.

Example 1. Let the outcome of a random experiment be a point on the interval $(0, 1)$. Thus, $\mathscr{C} = \{c; 0 < c < 1\}$. Let the probability set function be given by

$$P(C) = \int_C dz.$$

For instance, if $C = \{c; \frac{1}{4} < c < \frac{1}{2}\}$, then

$$P(C) = \int_{1/4}^{1/2} dz = \frac{1}{4}.$$

Define the random variable X to be $X = X(c) = 3c + 2$. Accordingly, the space of X is $\mathscr{A} = \{x; 2 < x < 5\}$. We wish to determine the probability set function of X, namely $P(A)$, $A \subset \mathscr{A}$. At this time, let A be the set $\{x; 2 < x < b\}$, where $2 < b < 5$. Now $X(c)$ is between 2 and b when and only when $c \in C = \{c; 0 < c < (b - 2)/3\}$. Hence

$$P_X(A) = P(A) = P(C) = \int_0^{(b-2)/3} dz.$$

In the integral, make the change of variable $x = 3z + 2$ and obtain

$$P_X(A) = P(A) = \int_2^b \tfrac{1}{3}\, dx = \int_A \tfrac{1}{3}\, dx,$$

since $A = \{x; 2 < x < b\}$. This kind of argument holds for every set $A \subset \mathscr{A}$ for which the integral

$$P(A) = \int_A \tfrac{1}{3} \, dx$$

exists. Thus the probability set function of X is this integral.

In statistics we are usually more interested in the probability set function of the random variable X than we are in the sample space $\mathscr{C}$ and the probability set function $P(C)$. Therefore, in most instances, we begin with an assumed distribution of probability for the random variable X. Moreover, we do this same kind of thing with two or more random variables. Two illustrative examples follow.

Example 2. Let the probability set function $P(A)$ of a random variable X be

$$P(A) = \int_A f(x) \, dx, \qquad \text{where } f(x) = \frac{3x^2}{8}, \quad x \in \mathscr{A} = \{x; 0 < x < 2\}.$$

Let $A_1 = \{x; 0 < x < \tfrac{1}{2}\}$ and $A_2 = \{x; 1 < x < 2\}$ be two subsets of $\mathscr{A}$. Then

$$P(A_1) = \Pr(X \in A_1) = \int_{A_1} f(x) \, dx = \int_0^{1/2} \frac{3x^2}{8} \, dx = \frac{1}{64}$$

and

$$P(A_2) = \Pr(X \in A_2) = \int_{A_2} f(x) \, dx = \int_1^2 \frac{3x^2}{8} \, dx = \frac{7}{8}.$$

To compute $P(A_1 \cup A_2)$, we note that $A_1 \cap A_2 = \varnothing$; then we have $P(A_1 \cup A_2) = P(A_1) + P(A_2) = \frac{57}{64}$.

Example 3. Let $\mathscr{A} = \{(x, y); 0 < x < y < 1\}$ be the space of two random variables X and Y. Let the probability set function be

$$P(A) = \int_A \int 2 \, dx \, dy.$$

If A is taken to be $A_1 = \{(x, y); \tfrac{1}{2} < x < y < 1\}$, then

$$P(A_1) = \Pr[(X, Y) \in A_1] = \int_{1/2}^1 \int_{1/2}^y 2 \, dx \, dy = \tfrac{1}{4}.$$

If A is taken to be $A_2 = \{(x, y); x < y < 1, 0 < x \leq \tfrac{1}{2}\}$, then $A_2 = A_1^*$, and

$$P(A_2) = \Pr[(X, Y) \in A_2] = P(A_1^*) = 1 - P(A_1) = \tfrac{3}{4}.$$

EXERCISES

1.31. Let a card be selected from an ordinary deck of playing cards. The outcome c is one of these 52 cards. Let $X(c) = 4$ if c is an ace, let $X(c) = 3$ if

c is a king, let $X(c) = 2$ if c is a queen, let $X(c) = 1$ if c is a jack, and let $X(c) = 0$ otherwise. Suppose that $P(C)$ assigns a probability of $\frac{1}{52}$ to each outcome c. Describe the induced probability $P_X(A)$ on the space $\mathscr{A} = \{x; x = 0, 1, 2, 3, 4\}$ of the random variable X.

1.32. Let a point be selected from the sample space $\mathscr{C} = \{c; 0 < c < 10\}$. Let $C \subset \mathscr{C}$ and let the probability set function be $P(C) = \int_C \frac{1}{10} dz$. Define the random variable X to be $X = X(c) = 2c - 10$. Find the probability set function of X. *Hint.* If $-10 < a < b < 10$, note that $a < X(c) < b$ when and only when $(a + 10)/2 < c < (b + 10)/2$.

1.33. Let the probability set function $P(A)$ of two random variables X and Y be $P(A) = \sum_A \sum f(x, y)$, where $f(x, y) = \frac{1}{52}$, $(x, y) \in \mathscr{A} = \{(x, y); (x, y) = (0, 1), (0, 2), \ldots, (0, 13), (1, 1), \ldots, (1, 13), \ldots, (3, 13)\}$. Compute $P(A) = \Pr[(X, Y) \in A]$: (a) when $A = \{(x, y); (x, y) = (0, 4), (1, 3), (2, 2)\}$; (b) when $A = \{(x, y); x + y = 4, (x, y) \in \mathscr{A}\}$.

1.34. Let the probability set function $P(A)$ of the random variable X be $P(A) = \int_A f(x) dx$, where $f(x) = 2x/9$, $x \in \mathscr{A} = \{x; 0 < x < 3\}$. Let $A_1 = \{x; 0 < x < 1\}$, $A_2 = \{x; 2 < x < 3\}$. Compute $P(A_1) = \Pr[X \in A_1]$, $P(A_2) = \Pr(X \in A_2)$, and $P(A_1 \cup A_2) = \Pr(X \in A_1 \cup A_2)$.

1.35. Let the space of the random variable X be $\mathscr{A} = \{x; 0 < x < 1\}$. If $A_1 = \{x; 0 < x < \frac{1}{2}\}$ and $A_2 = \{x; \frac{1}{2} \leq x < 1\}$, find $P(A_2)$ if $P(A_1) = \frac{1}{4}$.

1.36. Let the space of the random variable X be $\mathscr{A} = \{x; 0 < x < 10\}$ and let $P(A_1) = \frac{3}{8}$, where $A_1 = \{x; 1 < x < 5\}$. Show that $P(A_2) \leq \frac{5}{8}$, where $A_2 = \{x; 5 \leq x < 10\}$.

1.37. Let the subsets $A_1 = \{x; \frac{1}{4} < x < \frac{1}{2}\}$ and $A_2 = \{x; \frac{1}{2} \leq x < 1\}$ of the space $\mathscr{A} = \{x; 0 < x < 1\}$ of the random variable X be such that $P(A_1) = \frac{1}{8}$ and $P(A_2) = \frac{1}{2}$. Find $P(A_1 \cup A_2)$, $P(A_1^*)$, and $P(A_1^* \cap A_2^*)$.

1.38. Let $A_1 = \{(x, y); x \leq 2, y \leq 4\}$, $A_2 = \{(x, y); x \leq 2, y \leq 1\}$, $A_3 = \{(x, y); x \leq 0, y \leq 4\}$, and $A_4 = \{(x, y); x \leq 0, y \leq 1\}$ be subsets of the space $\mathscr{A}$ of two random variables X and Y, which is the entire two-dimensional plane. If $P(A_1) = \frac{7}{8}$, $P(A_2) = \frac{4}{8}$, $P(A_3) = \frac{3}{8}$, and $P(A_4) = \frac{2}{8}$, find $P(A_5)$, where $A_5 = \{(x, y); 0 < x \leq 2, 1 < y \leq 4\}$.

1.39. Given $\int_A [1/\pi(1 + x^2)] dx$, where $A \subset \mathscr{A} = \{x; -\infty < x < \infty\}$. Show that the integral could serve as a probability set function of a random variable X whose space is $\mathscr{A}$.

1.40. Let the probability set function of the random variable X be

$$P(A) = \int_A e^{-x} dx, \qquad \text{where } \mathscr{A} = \{x; 0 < x < \infty\}.$$

Let $A_k = \{x; 2 - 1/k < x \le 3\}, \quad k = 1, 2, 3, \ldots.$ Find $\lim_{k \to \infty} A_k$ and $P\left(\lim_{k \to \infty} A_k\right)$. Find $P(A_k)$ and $\lim_{k \to \infty} P(A_k)$. Note that $\lim_{k \to \infty} P(A_k) = P\left(\lim_{k \to \infty} A_k\right)$.

1.6 The Probability Density Function

Let X denote a random variable with space $\mathscr{A}$ and let A be a subset of $\mathscr{A}$. If we know how to compute $P(C), C \subset \mathscr{C}$, then for each A under consideration we can compute $P(A) = \mathrm{Pr}\,(X \in A)$; that is, we know how the probability is distributed over the various subsets of $\mathscr{A}$. In this sense, we speak of the distribution of the random variable X, meaning, of course, the distribution of probability. Moreover, we can use this convenient terminology when more than one random variable is involved and, in the sequel, we shall do this.

In this section, we shall investigate some random variables whose distributions can be described very simply by what will be called the *probability density function*. The two types of distributions that we shall consider are called, respectively, the *discrete type* and the *continuous type*. For simplicity of presentation, we first consider a distribution of one random variable.

(a) *The discrete type of random variable.* Let X denote a random variable with one-dimensional space $\mathscr{A}$. Suppose that the space $\mathscr{A}$ is a set of points such that there is at most a finite number of points of $\mathscr{A}$ in every finite interval. Such a set $\mathscr{A}$ will be called a set of discrete points. Let a function $f(x)$ be such that $f(x) > 0, x \in \mathscr{A}$, and that

$$\sum_{\mathscr{A}} f(x) = 1.$$

Whenever a probability set function $P(A), A \subset \mathscr{A}$, can be expressed in terms of such an $f(x)$ by

$$P(A) = \mathrm{Pr}\,(X \in A) = \sum_{A} f(x),$$

then X is called a random variable of the *discrete type*, and X is said to have a distribution of the discrete type.

Example 1. Let X be a random variable of the discrete type with space $\mathscr{A} = \{x; x = 0, 1, 2, 3, 4\}$. Let

$$P(A) = \sum_{A} f(x),$$

where

$$f(x) = \frac{4!}{x!\,(4 - x)!} \left(\frac{1}{2}\right)^4, \qquad x \in \mathscr{A},$$

and, as usual, $0! = 1$. Then if $A = \{x; x = 0, 1\}$, we have

$$\Pr(X \in A) = \frac{4!}{0!\,4!}\left(\frac{1}{2}\right)^4 + \frac{4!}{1!\,3!}\left(\frac{1}{2}\right)^4 = \frac{5}{16}.$$

Example 2. Let X be a random variable of the discrete type with space $\mathscr{A} = \{x; x = 1, 2, 3, \ldots\}$, and let

$$f(x) = (\tfrac{1}{2})^x, \qquad x \in \mathscr{A}.$$

Then

$$\Pr(X \in A) = \sum_A f(x).$$

If $A = \{x; x = 1, 3, 5, 7, \ldots\}$, we have

$$\Pr(X \in A) = (\tfrac{1}{2}) + (\tfrac{1}{2})^3 + (\tfrac{1}{2})^5 + \cdots = \tfrac{2}{3}.$$

(b) *The continuous type of random variable.* Let the one-dimensional set $\mathscr{A}$ be such that the Riemann integral

$$\int_{\mathscr{A}} f(x)\,dx = 1,$$

where (1) $f(x) > 0$, $x \in \mathscr{A}$, and (2) $f(x)$ has at most a finite number of discontinuities in every finite interval that is a subset of $\mathscr{A}$. If $\mathscr{A}$ is the space of the random variable X and if the probability set function $P(A)$, $A \subset \mathscr{A}$, can be expressed in terms of such an $f(x)$ by

$$P(A) = \Pr(X \in A) = \int_A f(x)\,dx,$$

then X is said to be a random variable of the *continuous type* and to have a distribution of that type.

Example 3. Let the space $\mathscr{A} = \{x; 0 < x < \infty\}$, and let

$$f(x) = e^{-x}, \qquad x \in \mathscr{A}.$$

If X is a random variable of the continuous type so that

$$\Pr(X \in A) = \int_A e^{-x}\,dx,$$

we have, with $A = \{x; 0 < x < 1\}$,

$$\Pr(X \in A) = \int_0^1 e^{-x}\,dx = 1 - e^{-1}.$$

Note that $\Pr(X \in A)$ is the area under the graph of $f(x) = e^{-x}$, which lies above the x-axis and between the vertical lines $x = 0$ and $x = 1$.

Example 4. Let X be a random variable of the continuous type with space $\mathscr{A} = \{x; 0 < x < 1\}$. Let the probability set function be

$$P(A) = \int_A f(x)\,dx,$$

where

$$f(x) = cx^2, \qquad x \in \mathscr{A}.$$

Since $P(A)$ is a probability set function, $P(\mathscr{A}) = 1$. Hence the constant c is determined by

$$\int_0^1 cx^2 \, dx = 1,$$

or $c = 3$.

It is seen that whether the random variable X is of the discrete type or of the continuous type, the probability $\Pr (X \in A)$ is completely determined by a function $f(x)$. In either case $f(x)$ is called the *probability density function* (hereafter abbreviated p.d.f.) of the random variable X. If we restrict ourselves to random variables of either the discrete type or the continuous type, we may work exclusively with the p.d.f. $f(x)$. This affords an enormous simplification; but it should be recognized that this simplification is obtained at considerable cost from a mathematical point of view. Not only shall we exclude from consideration many random variables that do not have these types of distributions, but we shall also exclude many interesting subsets of the space. In this book, however, we shall in general restrict ourselves to these simple types of random variables.

Remarks. Let X denote the number of spots that show when a die is cast. We can assume that X is a random variable with $\mathscr{A} = \{x; x = 1, 2, \ldots, 6\}$ and with a p.d.f. $f(x) = \frac{1}{6}, x \in \mathscr{A}$. Other assumptions can be made to provide different mathematical models for this experiment. Experimental evidence can be used to help one decide which model is the more realistic. Next, let X denote the point at which a balanced pointer comes to rest. If the circumference is graduated $0 \leq x < 1$, a reasonable mathematical model for this experiment is to take X to be a random variable with $\mathscr{A} = \{x; 0 \leq x < 1\}$ and with a p.d.f. $f(x) = 1, x \in \mathscr{A}$.

Both types of probability density functions can be used as distributional models for many random variables found in real situations. For illustrations consider the following. If X is the number of automobile accidents during a given day, then $f(0), f(1), f(2), \ldots$ represent the probabilities of $0, 1, 2, \ldots$ accidents. On the other hand, if X is length of life of a female born in a certain community, the integral [area under the graph of $f(x)$ that lies above the x-axis and between the vertical lines $x = 40$ and $x = 50$]

$$\int_{40}^{50} f(x) \, dx$$

represents the probability that she dies between 40 and 50 (or the percentage of these females dying between 40 and 50). A particular $f(x)$ will be suggested later for each of these situations, but again experimental evidence must be used to decide whether we have realistic models.

The notion of the p.d.f. of one random variable X can be extended to the notion of the p.d.f. of two or more random variables. Under certain restrictions on the space $\mathscr{A}$ and the function $f > 0$ on $\mathscr{A}$ (restrictions that will not be enumerated here), we say that the two random variables X and Y are of the discrete type or of the continuous type, and have a distribution of that type, according as the probability set function $P(A)$, $A \subset \mathscr{A}$, can be expressed as

$$P(A) = \Pr[(X, Y) \in A] = \sum_A \sum f(x, y),$$

or as

$$P(A) = \Pr[(X, Y) \in A] = \int_A \int f(x, y) \, dx \, dy.$$

In either case f is called the p.d.f. of the two random variables X and Y. Of necessity, $P(\mathscr{A}) = 1$ in each case. More generally, we say that the n random variables $X_1, X_2, \ldots, X_n$ are of the discrete type or of the continuous type, and have a distribution of that type, according as the probability set function $P(A)$, $A \subset \mathscr{A}$, can be expressed as

$$P(A) = \Pr[(X_1, \ldots, X_n) \in A] = \sum_A \cdots \sum f(x_1, \ldots, x_n),$$

or as

$$P(A) = \Pr[(X_1, \ldots, X_n) \in A] = \int_A \cdots \int f(x_1, \ldots, x_n) \, dx_1 \cdots dx_n.$$

The idea to be emphasized is that a function f, whether in one or more variables, essentially satisfies the conditions of being a p.d.f. if $f > 0$ on a space $\mathscr{A}$ and if its integral [for the continuous type of random variable(s)] or its sum [for the discrete type of random variable(s)] over $\mathscr{A}$ is one.

Our notation can be considerably simplified when we restrict ourselves to random variables of the continuous or discrete types. Suppose that the space of a continuous type of random variable X is $\mathscr{A} = \{x; 0 < x < \infty\}$ and that the p.d.f. of X is e^{-x}, $x \in \mathscr{A}$. We shall in no manner alter the distribution of X [that is, alter any $P(A)$, $A \subset \mathscr{A}$] if we extend the definition of the p.d.f. of X by writing

$$f(x) = e^{-x}, \qquad 0 < x < \infty,$$

$$= 0 \text{ elsewhere,}$$

and then refer to $f(x)$ as the p.d.f. of X. We have

$$\int_{-\infty}^{\infty} f(x)\ dx = \int_{-\infty}^{0} 0\ dx + \int_{0}^{\infty} e^{-x}\ dx = 1.$$

Thus we may treat the entire axis of reals as though it were the space of X. Accordingly, we now replace

$$\int_{\mathscr{A}} f(x)\ dx \qquad \text{by} \qquad \int_{-\infty}^{\infty} f(x)\ dx.$$

Similarly, we may extend the definition of a p.d.f. $f(x, y)$ over the entire xy-plane, or a p.d.f. $f(x, y, z)$ throughout three-dimensional space, and so on. We shall do this consistently so that tedious, repetitious references to the space $\mathscr{A}$ can be avoided. Once this is done, we replace

$$\int_{\mathscr{A}} \int f(x, y)\ dx\ dy \qquad \text{by} \qquad \int_{-\infty}^{\infty} \int_{-\infty}^{\infty} f(x, y)\ dx\ dy,$$

and so on. Similarly, after extending the definition of a p.d.f. of the discrete type, we replace, for one random variable,

$$\sum_{\mathscr{A}} f(x) \qquad \text{by} \qquad \sum_{x} f(x),$$

and, for two random variables,

$$\sum_{\mathscr{A}} \sum f(x, y) \qquad \text{by} \qquad \sum_{y} \sum_{x} f(x, y),$$

and so on.

In accordance with this convention (of extending the definition of a p.d.f.), it is seen that a point function f, whether in one or more variables, essentially satisfies the conditions of being a p.d.f. if (a) f is defined and is not negative for all real values of its argument(s) and if (b) its integral [for the continuous type of random variable(s)], or its sum [for the discrete type of random variable(s)] over all real values of its argument(s) is 1.

If $f(x)$ is the p.d.f. of a continuous type of random variable X and if A is the set $\{x;\ a < x < b\}$, then $P(A) = \Pr\,(X \in A)$ can be written as

$$\Pr\,(a < X < b) = \int_{a}^{b} f(x)\ dx.$$

Moreover, if $A = \{x;\ x = a\}$, then

$$P(A) = \Pr\,(X \in A) = \Pr\,(X = a) = \int_{a}^{a} f(x)\ dx = 0,$$

since the integral $\int_{a}^{a} f(x)\ dx$ is defined in calculus to be zero. That is, if X is a random variable of the continuous type, the probability of every set consisting of a single point is zero. This fact enables us to write, say,

$$\Pr\,(a < X < b) = \Pr\,(a \le X \le b).$$

More important, this fact allows us to change the value of the p.d.f. of a continuous type of random variable X at a single point without altering the distribution of X. For instance, the p.d.f.

$$f(x) = e^{-x}, \qquad 0 < x < \infty,$$
$$= 0 \text{ elsewhere,}$$

can be written as

$$f(x) = e^{-x}, \qquad 0 \le x < \infty,$$
$$= 0 \text{ elsewhere,}$$

without changing any $P(A)$. We observe that these two functions differ only at $x = 0$ and $\Pr\,(X = 0) = 0$. More generally, if two probability density functions of random variables of the continuous type differ only on a set having probability zero, the two corresponding probability set functions are exactly the same. Unlike the continuous type, the p.d.f. of a discrete type of random variable may not be changed at any point, since a change in such a p.d.f. alters the distribution of probability.

Finally, if a p.d.f. in one or more variables is explicitly defined, we can see by inspection whether the random variables are of the continuous or discrete type. For example, it seems obvious that the p.d.f.

$$f(x, y) = \frac{9}{4^{x+y}}, \qquad x = 1, 2, 3, \ldots, y = 1, 2, 3, \ldots,$$
$$= 0 \text{ elsewhere,}$$

is a p.d.f. of two discrete-type random variables X and Y, whereas the p.d.f.

$$f(x, y) = 4xye^{-x^2 - y^2}, \qquad 0 < x < \infty, 0 < y < \infty,$$
$$= 0 \text{ elsewhere,}$$

is clearly a p.d.f. of two continuous-type random variables X and Y. In such cases it seems unnecessary to specify which of the two simpler types of random variables is under consideration.

Example 5. Let the random variable X have the p.d.f.

$$f(x) = 2x, \qquad 0 < x < 1,$$
$$= 0 \text{ elsewhere.}$$

Find Pr $(\frac{1}{2} < X < \frac{3}{4})$ and Pr $(-\frac{1}{2} < X < \frac{1}{2})$. First,

$$\Pr\left(\tfrac{1}{2} < X < \tfrac{3}{4}\right) = \int_{1/2}^{3/4} f(x)\,dx = \int_{1/2}^{3/4} 2x\,dx = \tfrac{5}{16}.$$

Next,

$$\Pr\left(-\tfrac{1}{2} < X < \tfrac{1}{2}\right) = \int_{-1/2}^{1/2} f(x)\,dx$$

$$= \int_{-1/2}^{0} 0\,dx + \int_{0}^{1/2} 2x\,dx$$

$$= 0 + \tfrac{1}{4} = \tfrac{1}{4}.$$

Example 6. Let

$$f(x, y) = 6x^2 y, \qquad 0 < x < 1,\ 0 < y < 1,$$
$$= 0 \text{ elsewhere,}$$

be the p.d.f. of two random variables X and Y. We have, for instance,

$$\Pr\left(0 < X < \tfrac{3}{4}, \tfrac{1}{3} < Y < 2\right) = \int_{1/3}^{2} \int_{0}^{3/4} f(x, y)\,dx\,dy$$

$$= \int_{1/3}^{1} \int_{0}^{3/4} 6x^2 y\,dx\,dy + \int_{1}^{2} \int_{0}^{3/4} 0\,dx\,dy$$

$$= \tfrac{3}{8} + 0 = \tfrac{3}{8}.$$

Note that this probability is the volume under the surface $f(x, y) = 6x^2 y$ and above the rectangular set $\{(x, y);\ 0 < x < \frac{3}{4}, \frac{1}{3} < y < 1\}$ in the xy-plane.

EXERCISES

1.41. For each of the following, find the constant c so that $f(x)$ satisfies the conditions of being a p.d.f. of one random variable X.
(a) $f(x) = c(\frac{2}{3})^x$, $x = 1, 2, 3, \ldots$, zero elsewhere.
(b) $f(x) = cxe^{-x}$, $0 < x < \infty$, zero elsewhere.

1.42. Let $f(x) = x/15$, $x = 1, 2, 3, 4, 5$, zero elsewhere, be the p.d.f. of X. Find Pr $(X = 1$ or $2)$, Pr $(\frac{1}{2} < X < \frac{5}{2})$, and Pr $(1 \le X \le 2)$.

1.43. For each of the following probability density functions of X, compute Pr $(|X| < 1)$ and Pr $(X^2 < 9)$.
(a) $f(x) = x^2/18$, $-3 < x < 3$, zero elsewhere.
(b) $f(x) = (x + 2)/18$, $-2 < x < 4$, zero elsewhere.

1.44. Let $f(x) = 1/x^2$, $1 < x < \infty$, zero elsewhere, be the p.d.f. of X. If $A_1 = \{x; 1 < x < 2\}$ and $A_2 = \{x; 4 < x < 5\}$, find $P(A_1 \cup A_2)$ and $P(A_1 \cap A_2)$.

1.45. Let $f(x_1, x_2) = 4x_1 x_2$, $0 < x_1 < 1$, $0 < x_2 < 1$, zero elsewhere, be the p.d.f. of X_1 and X_2. Find Pr $(0 < X_1 < \frac{1}{2}, \frac{1}{4} < X_2 < 1)$, Pr $(X_1 = X_2)$,

Pr $(X_1 < X_2)$, and Pr $(X_1 \leq X_2)$. *Hint.* Recall that Pr $(X_1 = X_2)$ would be the volume under the surface $f(x_1, x_2) = 4x_1x_2$ and above the line segment $0 < x_1 = x_2 < 1$ in the x_1x_2-plane.

1.46. Let $f(x_1, x_2, x_3) = \exp[-(x_1 + x_2 + x_3)]$, $0 < x_1 < \infty$, $0 < x_2 < \infty$, $0 < x_3 < \infty$, zero elsewhere, be the p.d.f. of X_1, X_2, X_3. Compute Pr $(X_1 < X_2 < X_3)$ and Pr $(X_1 = X_2 < X_3)$. The symbol exp (w) means e^w.

1.47. A *mode* of a distribution of one random variable X of the continuous or discrete type is a value of x that maximizes the p.d.f. $f(x)$. If there is only one such x, it is called the *mode of the distribution*. Find the mode of each of the following distributions:

    (a) $f(x) = (\frac{1}{2})^x$, $x = 1, 2, 3, \ldots$, zero elsewhere.

    (b) $f(x) = 12x^2(1 - x)$, $0 < x < 1$, zero elsewhere.

    (c) $f(x) = (\frac{1}{2})x^2e^{-x}$, $0 < x < \infty$, zero elsewhere.

1.48. A *median* of a distribution of one random variable X of the discrete or continuous type is a value of x such that Pr $(X < x) \leq \frac{1}{2}$ and Pr $(X \leq x) \geq \frac{1}{2}$. If there is only one such x, it is called the *median of the distribution*. Find the median of each of the following distributions:

    (a) $f(x) = \dfrac{4!}{x!\,(4 - x)!} \left(\dfrac{1}{4}\right)^x \left(\dfrac{3}{4}\right)^{4-x}$, $x = 0, 1, 2, 3, 4$, zero elsewhere.

    (b) $f(x) = 3x^2$, $0 < x < 1$, zero elsewhere.

    (c) $f(x) = \dfrac{1}{\pi(1 + x^2)}$, $-\infty < x < \infty$.

Hint. In parts (b) and (c), Pr $(X < x)$ = Pr $(X \leq x)$ and thus that common value must equal $\frac{1}{2}$ if x is to be the median of the distribution.

1.49. Let $0 < p < 1$. A $(100p)$th *percentile* (*quantile* of order p) of the distribution of a random variable X is a value ξ_p such that Pr $(X < \xi_p) \leq p$ and Pr $(X \leq \xi_p) \geq p$. Find the twentieth percentile of the distribution that has p.d.f. $f(x) = 4x^3$, $0 < x < 1$, zero elsewhere. *Hint.* With a continuous-type random variable X, Pr $(X < \xi_p)$ = Pr $(X \leq \xi_p)$ and hence that common value must equal p.

1.50. Show that

$$\int_0^\infty xe^{-x}\,dx = \int_0^\infty e^{-x}\,dx = 1,$$

and, for $k \geq 1$, that (by integrating by parts)

$$\int_0^\infty x^k e^{-x}\,dx = k \int_0^\infty x^{k-1} e^{-x}\,dx.$$

    (a) What is the value of $\int_0^\infty x^n e^{-x}\,dx$, where n is a nonnegative integer?

    (b) Formulate a reasonable definition of the now meaningless symbol $0!$.

(c) For what value of the constant c does the function $f(x) = cx^n e^{-x}$, $0 < x < \infty$, zero elsewhere, satisfy the properties of a p.d.f.?

1.51. Given that the nonnegative function $g(x)$ has the property that

$$\int_0^\infty g(x)\, dx = 1.$$

Show that $f(x_1, x_2) = [2g(\sqrt{x_1^2 + x_2^2})]/(\pi\sqrt{x_1^2 + x_2^2})$, $0 < x_1 < \infty$, $0 < x_2 < \infty$, zero elsewhere, satisfies the conditions of being a p.d.f. of two continuous-type random variables X_1 and X_2. *Hint.* Use polar coordinates.

1.7 The Distribution Function

Let the random variable X have the probability set function $P(A)$, where A is a one-dimensional set. Take x to be a real number and consider the set A which is an unbounded set from $-\infty$ to x, including the point x itself. For all such sets A we have $P(A) = \Pr(X \in A) = \Pr(X \leq x)$. This probability depends on the point x; that is, this probability is a function of the point x. This point function is denoted by the symbol $F(x) = \Pr(X \leq x)$. The function $F(x)$ is called the *distribution function* (sometimes, *cumulative distribution function*) of the random variable X. Since $F(x) = \Pr(X \leq x)$, then, with $f(x)$ the p.d.f., we have

$$F(x) = \sum_{w \leq x} f(w),$$

for the discrete type of random variable, and

$$F(x) = \int_{-\infty}^x f(w)\, dw,$$

for the continuous type of random variable. We speak of a distribution function $F(x)$ as being of the continuous or discrete type, depending on whether the random variable is of the continuous or discrete type.

Remark. If X is a random variable of the continuous type, the p.d.f. $f(x)$ has at most a finite number of discontinuities in every finite interval. This means (1) that the distribution function $F(x)$ is everywhere continuous and (2) that the derivative of $F(x)$ with respect to x exists and is equal to $f(x)$ at each point of continuity of $f(x)$. That is, $F'(x) = f(x)$ at each point of continuity of $f(x)$. If the random variable X is of the discrete type, most surely the p.d.f. $f(x)$ is *not* the derivative of $F(x)$ with respect to x (that is, with respect to Lebesgue measure); but $f(x)$ *is* the (Radon–Nikodym) derivative of $F(x)$ with respect to a counting measure. A derivative is often called a *density*. Accordingly, we call these derivatives *probability density functions*.

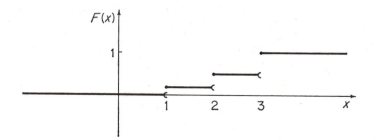

FIGURE I.3

Example 1. Let the random variable X of the discrete type have the p.d.f. $f(x) = x/6$, $x = 1, 2, 3$, zero elsewhere. The distribution function of X is

$$
\begin{aligned}
F(x) &= 0, & x &< 1,\\
&= \tfrac{1}{6}, & 1 &\le x < 2,\\
&= \tfrac{3}{6}, & 2 &\le x < 3,\\
&= 1, & 3 &\le x.
\end{aligned}
$$

Here, as depicted in Figure 1.3, $F(x)$ is a step function that is constant in every interval not containing 1, 2, or 3, but has steps of heights $\tfrac{1}{6}$, $\tfrac{2}{6}$, and $\tfrac{3}{6}$ at those respective points. It is also seen that $F(x)$ is everywhere continuous to the right.

Example 2. Let the random variable X of the continuous type have the p.d.f. $f(x) = 2/x^3$, $1 < x < \infty$, zero elsewhere. The distribution function of X is

$$
\begin{aligned}
F(x) &= \int_{-\infty}^{x} 0 \, dw = 0, & x &< 1,\\
&= \int_{1}^{x} \frac{2}{w^3} \, dw = 1 - \frac{1}{x^2}, & 1 &\le x.
\end{aligned}
$$

The graph of this distribution function is depicted in Figure 1.4. Here $F(x)$ is a continuous function for all real numbers x; in particular, $F(x)$ is everywhere continuous to the right. Moreover, the derivative of $F(x)$ with respect to x exists at all points except at $x = 1$. Thus the p.d.f. of X is defined by this derivative except at $x = 1$. Since the set $A = \{x; x = 1\}$ is a set of probability measure zero [that is, $P(A) = 0$], we are free to define the p.d.f. at $x = 1$ in any manner we please. One way to do this is to write $f(x) = 2/x^3$, $1 < x < \infty$, zero elsewhere.

There are several properties of a distribution function $F(x)$ that can be listed as a consequence of the properties of the probability set function. Some of these are the following. In listing these properties, we shall not restrict X to be a random variable of the discrete or continuous type. We shall use the symbols $F(\infty)$ and $F(-\infty)$ to mean $\lim_{x \to \infty} F(x)$

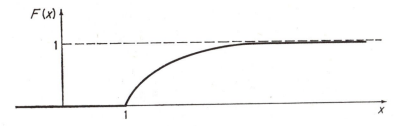

FIGURE 1.4

and $\lim\limits_{x \to -\infty} F(x)$, respectively. In like manner, the symbols $\{x; x \le \infty\}$ and $\{x; x \le -\infty\}$ represent, respectively, the limits of the sets $\{x; x \le b\}$ and $\{x; x \le -b\}$ as $b \to \infty$.

(a) $0 \le F(x) \le 1$ because $0 \le \Pr (X \le x) \le 1$.

(b) $F(x)$ is a nondecreasing function of x. For, if $x' < x''$, then

$$\{x; x \le x''\} = \{x; x \le x'\} \cup \{x; x' < x \le x''\}$$

and

$$\Pr (X \le x'') = \Pr (X \le x') + \Pr (x' < X \le x'').$$

That is,

$$F(x'') - F(x') = \Pr (x' < X \le x'') \ge 0.$$

(c) $F(\infty) = 1$ and $F(-\infty) = 0$ because the set $\{x; x \le \infty\}$ is the entire one-dimensional space and the set $\{x; x \le -\infty\}$ is the null set.

From the proof of (b), it is observed that, if $a < b$, then

$$\Pr (a < X \le b) = F(b) - F(a).$$

Suppose that we want to use $F(x)$ to compute the probability $\Pr (X = b)$. To do this, consider, with $h > 0$,

$$\lim_{h \to 0} \Pr (b - h < X \le b) = \lim_{h \to 0} [F(b) - F(b - h)].$$

Intuitively, it seems that $\lim\limits_{h \to 0} \Pr (b - h < X \le b)$ should exist and be equal to $\Pr (X = b)$ because, as h tends to zero, the limit of the set $\{x; b - h < x \le b\}$ is the set that contains the single point $x = b$. The fact that this limit is $\Pr (X = b)$ is a theorem that we accept without proof. Accordingly, we have

$$\Pr (X = b) = F(b) - F(b-),$$

where $F(b-)$ is the left-hand limit of $F(x)$ at $x = b$. That is, the probability that $X = b$ is the height of the step that $F(x)$ has at $x = b$. Hence, if the distribution function $F(x)$ is continuous at $x = b$, then $\Pr(X = b) = 0$.

There is a fourth property of $F(x)$ that is now listed.

(d) $F(x)$ is continuous to the right at each point x.

To prove this property, consider, with $h > 0$,

$$\lim_{h \to 0} \Pr(a < X \le a + h) = \lim_{h \to 0} [F(a + h) - F(a)].$$

We accept without proof a theorem which states, with $h > 0$, that

$$\lim_{h \to 0} \Pr(a < X \le a + h) = P(0) = 0.$$

Here also, the theorem is intuitively appealing because, as h tends to zero, the limit of the set $\{x; a < x \le a + h\}$ is the null set. Accordingly, we write

$$0 = F(a+) - F(a),$$

where $F(a+)$ is the right-hand limit of $F(x)$ at $x = a$. Hence $F(x)$ is continuous to the right at every point $x = a$.

The preceding discussion may be summarized in the following manner: A distribution function $F(x)$ is a nondecreasing function of x, which is everywhere continuous to the right and has $F(-\infty) = 0$, $F(\infty) = 1$. The probability $\Pr(a < X \le b)$ is equal to the difference $F(b) - F(a)$. If x is a discontinuity point of $F(x)$, then the probability $\Pr(X = x)$ is equal to the jump which the distribution function has at the point x. If x is a continuity point of $F(x)$, then $\Pr(X = x) = 0$.

Let X be a random variable of the continuous type that has p.d.f. $f(x)$, and let A be a set of probability measure zero; that is, $P(A) = \Pr(X \in A) = 0$. It has been observed that we may change the definition of $f(x)$ at any point in A without in any way altering the distribution of probability. The freedom to do this with the p.d.f. $f(x)$, of a continuous type of random variable does not extend to the distribution function $F(x)$; for, if $F(x)$ is changed at so much as one point x, the probability $\Pr(X \le x) = F(x)$ is changed, and we have a different distribution of probability. That is, the distribution function $F(x)$, not the p.d.f. $f(x)$, is really the fundamental concept.

Remark. The definition of the distribution function makes it clear that the probability set function P determines the distribution function F. It is true, although not so obvious, that a probability set function P can be found

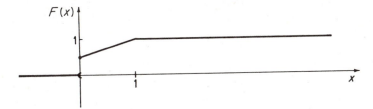

FIGURE I.5

from a distribution function F. That is, P and F give the same information about the distribution of probability, and which function is used is a matter of convenience.

We now give an illustrative example.

Example 3. Let a distribution function be given by

$$F(x) = 0, \qquad x < 0,$$

$$= \frac{x + 1}{2}, \qquad 0 \le x < 1,$$

$$= 1, \qquad 1 \le x.$$

Then, for instance,

$$\Pr\left(-3 < X \le \tfrac{1}{2}\right) = F(\tfrac{1}{2}) - F(-3) = \tfrac{3}{4} - 0 = \tfrac{3}{4}$$

and

$$\Pr\left(X = 0\right) = F(0) - F(0-) = \tfrac{1}{2} - 0 = \tfrac{1}{2}.$$

The graph of $F(x)$ is shown in Figure 1.5. We see that $F(x)$ is not always continuous, nor is it a step function. Accordingly, the corresponding distribution is neither of the continuous type nor of the discrete type. It may be described as a mixture of those types.

We shall now point out an important fact about a function of a random variable. Let X denote a random variable with space $\mathscr{A}$. Consider the function $Y = u(X)$ of the random variable X. Since X is a function defined on a sample space $\mathscr{C}$, then $Y = u(X)$ is a composite function defined on $\mathscr{C}$. That is, $Y = u(X)$ is itself a random variable which has its own space $\mathscr{B} = \{y; y = u(x), x \in \mathscr{A}\}$ and its own probability set function. If $y \in \mathscr{B}$, the event $Y = u(X) \le y$ occurs when, and only when, the event $X \in A \subset \mathscr{A}$ occurs, where $A = \{x; u(x) \le y\}$. That is, the distribution function of Y is

$$G(y) = \Pr\left(Y \le y\right) = \Pr\left[u(X) \le y\right] = P(A).$$

The following example illustrates a method of finding the distribution function and the p.d.f. of a function of a random variable.

Example 4. Let $f(x) = \frac{1}{2}$, $-1 < x < 1$, zero elsewhere, be the p.d.f. of the random variable X. Define the random variable Y by $Y = X^2$. We wish to find the p.d.f. of Y. If $y \geq 0$, the probability $\Pr(Y \leq y)$ is equivalent to

$$\Pr(X^2 \leq y) = \Pr(-\sqrt{y} \leq X \leq \sqrt{y}).$$

Accordingly, the distribution function of Y, $G(y) = \Pr(Y \leq y)$, is given by

$$G(y) = 0, \qquad y < 0,$$

$$= \int_{-\sqrt{y}}^{\sqrt{y}} \tfrac{1}{2}\, dx = \sqrt{y}, \qquad 0 \leq y < 1,$$

$$= 1, \qquad 1 \leq y.$$

Since Y is a random variable of the continuous type, the p.d.f. of Y is $g(y) = G'(y)$ at all points of continuity of $g(y)$. Thus we may write

$$g(y) = \frac{1}{2\sqrt{y}}, \qquad 0 < y < 1,$$

$$= 0 \text{ elsewhere.}$$

Let the random variables X and Y have the probability set function $P(A)$, where A is a two-dimensional set. If A is the unbounded set $\{(u, v); u \leq x, v \leq y\}$, where x and y are real numbers, we have

$$P(A) = \Pr[(X, Y) \in A] = \Pr(X \leq x, Y \leq y).$$

This function of the point (x, y) is called the *distribution function* of X and Y and is denoted by

$$F(x, y) = \Pr(X \leq x, Y \leq y).$$

If X and Y are random variables of the continuous type that have p.d.f. $f(x, y)$, then

$$F(x, y) = \int_{-\infty}^{y} \int_{-\infty}^{x} f(u, v)\, du\, dv.$$

Accordingly, at points of continuity of $f(x, y)$, we have

$$\frac{\partial^2 F(x, y)}{\partial x\, \partial y} = f(x, y).$$

It is left as an exercise to show, in every case, that

$$\Pr(a < X \leq b, c < Y \leq d) = F(b, d) - F(b, c) - F(a, d) + F(a, c),$$

for all real constants $a < b$, $c < d$.

The distribution function of the n random variables $X_1, X_2, \ldots, X_n$ is the point function

$$F(x_1, x_2, \ldots, x_n) = \Pr(X_1 \le x_1, X_2 \le x_2, \ldots, X_n \le x_n).$$

An illustrative example follows.

Example 5. Let $f(x, y, z) = e^{-(x+y+z)}$, $0 < x, y, z < \infty$, zero elsewhere, be the p.d.f. of the random variables X, Y, and Z. Then the distribution function of X, Y, and Z is given by

$$F(x, y, z) = \Pr(X \le x, Y \le y, Z \le z)$$

$$= \int_0^z \int_0^y \int_0^x e^{-u-v-w} \, du \, dv \, dw$$

$$= (1 - e^{-x})(1 - e^{-y})(1 - e^{-z}), \qquad 0 \le x, y, z < \infty,$$

and is equal to zero elsewhere. Incidentally, except for a set of probability measure zero, we have

$$\frac{\partial^3 F(x, y, z)}{\partial x \, \partial y \, \partial z} = f(x, y, z).$$

EXERCISES

1.52. Let $f(x)$ be the p.d.f. of a random variable X. Find the distribution function $F(x)$ of X and sketch its graph if:
(a) $f(x) = 1$, $x = 0$, zero elsewhere.
(b) $f(x) = \frac{1}{3}$, $x = -1, 0, 1$, zero elsewhere.
(c) $f(x) = x/15$, $x = 1, 2, 3, 4, 5$, zero elsewhere.
(d) $f(x) = 3(1 - x)^2$, $0 < x < 1$, zero elsewhere.
(e) $f(x) = 1/x^2$, $1 < x < \infty$, zero elsewhere.
(f) $f(x) = \frac{1}{3}$, $0 < x < 1$ or $2 < x < 4$, zero elsewhere.

1.53. Find the median of each of the distributions in Exercise 1.52.

1.54. Given the distribution function

$$F(x) = 0, \qquad x < -1,$$

$$= \frac{x+2}{4}, \qquad -1 \le x < 1,$$

$$= 1, \qquad 1 \le x.$$

Sketch the graph of $F(x)$ and then compute: (a) $\Pr(-\frac{1}{2} < X \le \frac{1}{2})$; (b) $\Pr(X = 0)$; (c) $\Pr(X = 1)$; (d) $\Pr(2 < X \le 3)$.

1.55. Let $F(x, y)$ be the distribution function of X and Y. Show that

$\Pr{(a < X \le b, c < Y \le d)} = F(b, d) - F(b, c) - F(a, d) + F(a, c)$, for all real constants $a < b, c < d$.

1.56. Let $f(x) = 1$, $0 < x < 1$, zero elsewhere, be the p.d.f. of X. Find the distribution function and the p.d.f. of $Y = \sqrt{X}$. *Hint.* $\Pr{(Y \le y)} = \Pr{(\sqrt{X} \le y)} = \Pr{(X \le y^2)}$, $0 < y < 1$.

1.57. Let $f(x) = x/6$, $x = 1, 2, 3$, zero elsewhere, be the p.d.f. of X. Find the distribution function and the p.d.f. of $Y = X^2$. *Hint.* Note that X is a random variable of the discrete type.

1.58. Let $f(x) = (4 - x)/16$, $-2 < x < 2$, zero elsewhere, be the p.d.f. of X.

(a) Sketch the distribution function and the p.d.f. of X on the same set of axes.

(b) If $Y = |X|$, compute $\Pr{(Y \le 1)}$.

(c) If $Z = X^2$, compute $\Pr{(Z \le \frac{1}{4})}$.

1.59. Let $f(x, y) = e^{-x-y}$, $0 < x < \infty$, $0 < y < \infty$, zero elsewhere, be the p.d.f. of X and Y. If $Z = X + Y$, compute $\Pr{(Z \le 0)}$, $\Pr{(Z \le 6)}$, and, more generally, $\Pr{(Z \le z)}$, for $0 < z < \infty$. What is the p.d.f. of Z?

1.60. Explain why, with $h > 0$, the two limits $\lim_{h \to 0} \Pr{(b - h < X \le b)}$ and $\lim_{h \to 0} F(b - h)$ exist. *Hint.* Note that $\Pr{(b - h < X \le b)}$ is bounded below by zero and $F(b - h)$ is bounded above by both $F(b)$ and 1.

1.61. Show that the function $F(x, y)$ that is equal to 1, provided $x + 2y \ge 1$, and that is equal to zero provided $x + 2y < 1$, cannot be a distribution function of two random variables. *Hint.* Find four numbers $a < b$, $c < d$, so that $F(b, d) - F(a, d) - F(b, c) + F(a, c)$ is less than zero.

1.62. Let $F(x)$ be the distribution function of the random variable X. If m is a number such that $F(m) = \frac{1}{2}$, show that m is a median of the distribution.

1.63. Let $f(x) = \frac{1}{3}$, $-1 < x < 2$, zero elsewhere, be the p.d.f. of X. Find the distribution function and the p.d.f. of $Y = X^2$. *Hint.* Consider $\Pr{(X^2 \le y)}$ for two cases: $0 \le y < 1$ and $1 \le y < 4$.

1.8 Certain Probability Models

Consider an experiment in which one chooses at random a point from the closed interval $[a, b]$ that is on the real line. Thus the sample space $\mathscr{C}$ is $[a, b]$. Let the random variable X be the identity function defined on $\mathscr{C}$. Thus the space $\mathscr{A}$ of X is $\mathscr{A} = \mathscr{C}$. Suppose that it is reasonable to *assume*, from the nature of the experiment, that if an interval A is a subset of $\mathscr{A}$, the probability of the event A is proportional to the length of A. Hence, if A is the interval $[a, x]$, $x \le b$, then

$$P(A) = \Pr\,(X \in A) = \Pr\,(a \le X \le x) = c(x - a),$$

where c is the constant of proportionality.

In the expression above, if we take $x = b$, we have

$$1 = \Pr\,(a \le X \le b) = c(b - a),$$

so $c = 1/(b - a)$. Thus we will have an appropriate probability model if we take the distribution function of X, $F(x) = \Pr\,(X \le x)$, to be

$$F(x) = 0, \qquad x < a,$$

$$= \frac{x - a}{b - a}, \qquad a \le x \le b,$$

$$= 1, \qquad b < x.$$

Accordingly, the p.d.f. of X, $f(x) = F'(x)$, may be written

$$f(x) = \frac{1}{b - a}, \qquad a \le x \le b,$$

$$= 0 \text{ elsewhere.}$$

The derivative of $F(x)$ does not exist at $x = a$ nor at $x = b$; but the set $\{x;\, x = a, b\}$ is a set of probability measure zero, and we elect to define $f(x)$ to be equal to $1/(b - a)$ at those two points, just as a matter of convenience. We observe that this p.d.f. is a constant on $\mathscr{A}$. If the p.d.f. of one or more variables of the continuous type or of the discrete type is a constant on the space $\mathscr{A}$, we say that the probability is distributed *uniformly* over $\mathscr{A}$. Thus, in the example above, we say that X has a *uniform distribution* over the interval $[a, b]$.

Consider next an experiment in which one chooses at random a point (X, Y) from the unit square $\mathscr{C} = \mathscr{A} = \{(x, y);\, 0 < x < 1,\, 0 < y < 1\}$. Suppose that our interest is not in X or in Y but in $Z = X + Y$. Once a suitable probability model has been adopted, we shall see how to find the p.d.f. of Z. To be specific, let the nature of the random experiment be such that it is reasonable to *assume* that the distribution of probability over the unit square is uniform. Then the p.d.f. of X and Y may be written

$$f(x, y) = 1, \qquad 0 < x < 1,\, 0 < y < 1,$$

$$= 0 \text{ elsewhere,}$$

and this describes the probability model. Now let the distribution function of Z be denoted by $G(z) = \text{Pr}\,(X + Y \le z)$. Then

$$G(z) = 0, \qquad z < 0,$$

$$= \int_0^z \int_0^{z-x} dy\,dx = \frac{z^2}{2}, \qquad 0 \le z < 1,$$

$$= 1 - \int_{z-1}^1 \int_{z-x}^1 dy\,dx = 1 - \frac{(2-z)^2}{2}, \qquad 1 \le z < 2,$$

$$= 1, \qquad 2 \le z.$$

Since $G'(z)$ exists for all values of z, the p.d.f. of Z may then be written

$$g(z) = z, \qquad 0 < z < 1,$$

$$= 2 - z, \qquad 1 \le z < 2,$$

$$= 0 \text{ elsewhere.}$$

It is clear that a different choice of the p.d.f. $f(x, y)$ that describes the probability model will, in general, lead to a different p.d.f. of Z.

We wish presently to extend and generalize some of the notions expressed in the next three sentences. Let the discrete type of random variable X have a uniform distribution of probability over the k points of the space $\mathscr{A} = \{x; x = 1, 2, \ldots, k\}$. The p.d.f. of X is then $f(x) = 1/k$, $x \in \mathscr{A}$, zero elsewhere. This type of p.d.f. is used to describe the probability model when each of the k points has the same probability, namely, $1/k$.

The probability model described in the preceding paragraph will now be adapted to a more general situation. Let a probability set function $P(C)$ be defined on a sample space $\mathscr{C}$. Here $\mathscr{C}$ may be a set in one, or two, or more dimensions. Let $\mathscr{C}$ be partitioned into k mutually disjoint subsets $C_1, C_2, \ldots, C_k$ in such a way that the union of these k mutually disjoint subsets is the sample space $\mathscr{C}$. Thus the events $C_1, C_2, \ldots, C_k$ are mutually exclusive and exhaustive. Suppose that the random experiment is of such a character that it may be *assumed* that each of the mutually exclusive and exhaustive events C_i, $i = 1, 2, \ldots, k$, has the same probability. Necessarily then, $P(C_i) = 1/k$, $i = 1, 2, \ldots, k$. Let the event E be the union of r of these mutually exclusive events, say

$$E = C_1 \cup C_2 \cup \cdots \cup C_r, \qquad r \le k.$$

Then

$$P(E) = P(C_1) + P(C_2) + \cdots + P(C_r) = \frac{r}{k}.$$

Frequently, the integer k is called the total number of ways (for this particular partition of $\mathscr{C}$) in which the random experiment can terminate and the integer r is called the number of ways that are favorable to the event E. So, in this terminology, $P(E)$ is equal to the number of ways favorable to the event E divided by the total number of ways in which the experiment can terminate. It should be emphasized that in order to assign, *in this manner*, the probability r/k to the event E, we must assume that each of the mutually exclusive and exhaustive events $C_1, C_2, \ldots, C_k$ has the same probability $1/k$. This assumption then becomes *part* of our probability model. Obviously, if this assumption is not realistic in an application, the probability of the event E cannot be computed in this way.

We next present two examples that are illustrative of this model.

Example 1. Let a card be drawn at random from an ordinary deck of 52 playing cards. The sample space $\mathscr{C}$ is the union of $k = 52$ outcomes, and it is reasonable to assume that each of these outcomes has the same probability $\frac{1}{52}$. Accordingly, if E_1 is the set of outcomes that are spades, $P(E_1) = \frac{13}{52} = \frac{1}{4}$ because there are $r_1 = 13$ spades in the deck; that is, $\frac{1}{4}$ is the probability of drawing a card that is a spade. If E_2 is the set of outcomes that are kings, $P(E_2) = \frac{4}{52} = \frac{1}{13}$ because there are $r_2 = 4$ kings in the deck; that is, $\frac{1}{13}$ is the probability of drawing a card that is a king. These computations are very easy because there are no difficulties in the determination of the appropriate values of r and k. However, instead of drawing only one card, suppose that five cards are taken, at random and without replacement, from this deck. We can think of each five-card hand as being an outcome in a sample space. It is reasonable to assume that each of these outcomes has the same probability. Now if E_1 is the set of outcomes in which each card of the hand is a spade, $P(E_1)$ is equal to the number r_1 of all spade hands divided by the total number, say k, of five-card hands. It is shown in many books on algebra that

$$r_1 = \binom{13}{5} = \frac{13!}{5!\,8!} \quad \text{and} \quad k = \binom{52}{5} = \frac{52!}{5!\,47!}.$$

In general, if n is a positive integer and if x is a nonnegative integer with $x \leq n$, then the binomial coefficient

$$\binom{n}{x} = \frac{n!}{x!\,(n-x)!}$$

is equal to the number of combinations of n things taken x at a time. Thus, here,

$$P(E_1) = \frac{\binom{13}{5}}{\binom{52}{5}} = \frac{(13)(12)(11)(10)(9)}{(52)(51)(50)(49)(48)} = 0.0005,$$

approximately. Next, let E_2 be the set of outcomes in which at least one card is a spade. Then E_2^* is the set of outcomes in which no card is a spade. There are $r_2^* = \binom{39}{5}$ such outcomes Hence

$$P(E_2^*) = \frac{\binom{39}{5}}{\binom{52}{5}} \quad \text{and} \quad P(E_2) = 1 - P(E_2^*).$$

Now suppose that E_3 is the set of outcomes in which exactly three cards are kings and exactly two cards are queens. We can select the three kings in any one of $\binom{4}{3}$ ways and the two queens in any one of $\binom{4}{2}$ ways By a well-known counting principle, the number of outcomes in E_3 is $r_3 = \binom{4}{3}\binom{4}{2}$. Thus $P(E_3) = \binom{4}{3}\binom{4}{2} / \binom{52}{5}$. Finally, let E_4 be the set of outcomes in which there are exactly two kings, two queens, and one jack. Then

$$P(E_4) = \frac{\binom{4}{2}\binom{4}{2}\binom{4}{1}}{\binom{52}{5}},$$

because the numerator of this fraction is the number of outcomes in E_4.

Example 2. A lot, consisting of 100 fuses, is inspected by the following procedure. Five of these fuses are chosen at random and tested; if all 5 "blow" at the correct amperage, the lot is accepted. If, in fact, there are 20 defective fuses in the lot, the probability of accepting the lot is, under appropriate assumptions,

$$\frac{\binom{80}{5}}{\binom{100}{5}} = 0.32,$$

approximately. More generally, let the random variable X be the number of defective fuses among the 5 that are inspected. The space of X is $\mathscr{A} = \{x; x = 0, 1, 2, 3, 4, 5\}$ and the p.d.f. of X is given by

$$f(x) = \Pr(X = x) = \frac{\binom{20}{x}\binom{80}{5-x}}{\binom{100}{5}}, \quad x = 0, 1, 2, 3, 4, 5,$$

$$= 0 \text{ elsewhere.}$$

This is an example of a discrete type of distribution called a *hypergeometric distribution*.

EXERCISES

(In order to solve some of these exercises, the reader must make certain assumptions.)

1.64. A bowl contains 16 chips, of which 6 are red, 7 are white, and 3 are blue. If 4 chips are taken at random and without replacement, find the probability that: (a) each of the 4 chips is red; (b) none of the 4 chips is red; (c) there is at least 1 chip of each color.

1.65. A person has purchased 10 of 1000 tickets sold in a certain raffle. To determine the five prize winners, 5 tickets are to be drawn at random and without replacement. Compute the probability that this person will win at least one prize. *Hint.* First compute the probability that the person does not win a prize.

1.66. Compute the probability of being dealt at random and without replacement a 13-card bridge hand consisting of: (a) 6 spades, 4 hearts, 2 diamonds, and 1 club; (b) 13 cards of the same suit.

1.67. Three distinct integers are chosen at random from the first 20 positive integers. Compute the probability that: (a) their sum is even; (b) their product is even.

1.68. There are five red chips and three blue chips in a bowl. The red chips are numbered 1, 2, 3, 4, 5, respectively, and the blue chips are numbered 1, 2, 3, respectively. If two chips are to be drawn at random and without replacement, find the probability that these chips have either the same number or the same color.

1.69. Let X have the uniform distribution given by the p.d.f. $f(x) = \frac{1}{5}$, $x = -2, -1, 0, 1, 2$, zero elsewhere. Find the p.d.f. of $Y = X^2$. *Hint.* Note that Y has a distribution of the discrete type.

1.70. Let X and Y have the p.d.f. $f(x, y) = 1$, $0 < x < 1$, $0 < y < 1$, zero elsewhere. Find the p.d.f. of the product $Z = XY$.

1.71. Let 13 cards be taken, at random and without replacement, from an ordinary deck of playing cards. If X is the number of spades in these 13 cards, find the p.d.f. of X. If, in addition, Y is the number of hearts in these 13 cards, find the probability $\Pr(X = 2, Y = 5)$. What is the p.d.f. of X and Y?

1.72. Four distinct integers are chosen at random and without replacement from the first 10 positive integers. Let the random variable X be the next to the smallest of these four numbers. Find the p.d.f. of X.

1.73. In a lot of 50 light bulbs, there are 2 bad bulbs. An inspector examines 5 bulbs, which are selected at random and without replacement.
(a) Find the probability of at least 1 defective bulb among the 5.

(b) How many bulbs should he examine so that the probability of finding at least 1 bad bulb exceeds $\frac{1}{2}$?

1.9 Mathematical Expectation

One of the more useful concepts in problems involving distributions of random variables is that of mathematical expectation. Let X be a random variable having a p.d.f. $f(x)$, and let $u(X)$ be a function of X such that

$$\int_{-\infty}^{\infty} u(x)f(x)\,dx$$

exists, if X is a continuous type of random variable, or such that

$$\sum_{x} u(x)f(x)$$

exists, if X is a discrete type of random variable. The integral, or the sum, as the case may be, is called the *mathematical expectation* (or expected value) of $u(X)$ and is denoted by $E[u(X)]$. That is,

$$E[u(X)] = \int_{-\infty}^{\infty} u(x)f(x)\,dx,$$

if X is a continuous type of random variable, or

$$E[u(X)] = \sum_{x} u(x)f(x),$$

if X is a discrete type of random variable.

Remarks. The usual definition of $E[u(X)]$ requires that the integral (or sum) converge absolutely. However, in this book, each $u(x)$ is of such a character that if the integral (or sum) exists, the convergence is absolute. Accordingly, we have not burdened the student with this additional provision. The terminology "mathematical expectation" or "expected value" has its origin in games of chance. This can be illustrated as follows: Three small similar discs, numbered 1, 2, and 2, respectively, are placed in a bowl and are mixed. A player is to be blindfolded and is to draw a disc from the bowl. If he draws the disc numbered 1, he will receive $9; if he draws either disc numbered 2, he will receive $3. It seems reasonable to assume that the player has a "$\frac{1}{3}$ claim" on the $9 and a "$\frac{2}{3}$ claim" on the $3. His "total claim" is $9(\frac{1}{3}) + 3(\frac{2}{3})$, or $5. If we take X to be a random variable having the p.d.f. $f(x) = x/3$, $x = 1$, 2, zero elsewhere, and $u(x) = 15 - 6x$, then $E[u(X)] = \sum_{x} u(x)f(x) = \sum_{x=1}^{2} (15 - 6x)(x/3) = 5$. That is, the mathematical expectation of $u(X)$ is precisely the player's "claim" or expectation.

The student may observe that $u(X)$ is a random variable Y with its own distribution of probability. Suppose the p.d.f. of Y is $g(y)$. Then $E(Y)$ is given by

$$\int_{-\infty}^{\infty} yg(y) \, dy \qquad \text{or} \qquad \sum_{y} yg(y),$$

according as Y is of the continuous type or of the discrete type. The question is: Does this have the same value as $E[u(X)]$, which was defined above? The answer to this question is in the affirmative, as will be shown in Chapter 4.

More generally, let $X_1, X_2, \ldots, X_n$ be random variables having p.d.f. $f(x_1, x_2, \ldots, x_n)$ and let $u(X_1, X_2, \ldots, X_n)$ be a function of these variables such that the n-fold integral

$$(1) \qquad \int_{-\infty}^{\infty} \cdots \int_{-\infty}^{\infty} u(x_1, x_2, \ldots, x_n) f(x_1, x_2, \ldots, x_n) \, dx_1 \, dx_2 \cdots dx_n$$

exists, if the random variables are of the continuous type, or such that the n-fold sum

$$(2) \qquad \sum_{x_n} \cdots \sum_{x_1} u(x_1, x_2, \ldots, x_n) f(x_1, x_2, \ldots, x_n)$$

exists if the random variables are of the discrete type. The n-fold integral (or the n-fold sum, as the case may be) is called the *mathematical expectation*, denoted by $E[u(X_1, X_2, \ldots, X_n)]$, of the function $u(X_1, X_2, \ldots, X_n)$.

Next, we shall point out some fairly obvious but useful facts about mathematical expectations when they exist.

(a) If k is a constant, then $E(k) = k$. This follows from expression (1) [or (2)] upon setting $u = k$ and recalling that an integral (or sum) of a constant times a function is the constant times the integral (or sum) of the function. Of course, the integral (or sum) of the function f is 1.

(b) If k is a constant and v is a function, then $E(kv) = kE(v)$. This follows from expression (1) [or (2)] upon setting $u = kv$ and rewriting expression (1) [or (2)] as k times the integral (or sum) of the product vf.

(c) If k_1 and k_2 are constants and v_1 and v_2 are functions, then $E(k_1 v_1 + k_2 v_2) = k_1 E(v_1) + k_2 E(v_2)$. This, too, follows from expression (1) [or (2)] upon setting $u = k_1 v_1 + k_2 v_2$ because the integral (or sum) of $(k_1 v_1 + k_2 v_2) f$ is equal to the integral (or sum) of $k_1 v_1 f$ plus the integral (or sum) of $k_2 v_2 f$. Repeated application of this property shows that if $k_1, k_2, \ldots, k_m$ are constants and $v_1, v_2, \ldots, v_m$ are functions, then

$$E(k_1 v_1 + k_2 v_2 + \cdots + k_m v_m) = k_1 E(v_1) + k_2 E(v_2) + \cdots + k_m E(v_m).$$

This property of mathematical expectation leads us to characterize the symbol E as a linear operator.

Example 1. Let X have the p.d.f.

$$f(x) = 2(1 - x), \quad 0 < x < 1,$$

$$= 0 \text{ elsewhere.}$$

Then

$$E(X) = \int_{-\infty}^{\infty} xf(x) \, dx = \int_0^1 (x)2(1 - x) \, dx = \tfrac{1}{3},$$

$$E(X^2) = \int_{-\infty}^{\infty} x^2 f(x) \, dx = \int_0^1 (x^2)2(1 - x) \, dx = \tfrac{1}{6},$$

and, of course,

$$E(6X + 3X^2) = 6(\tfrac{1}{3}) + 3(\tfrac{1}{6}) = \tfrac{5}{2}.$$

Example 2. Let X have the p.d.f.

$$f(x) = \frac{x}{6}, \quad x = 1, 2, 3,$$

$$= 0 \text{ elsewhere.}$$

Then

$$E(X^3) = \sum_x x^3 f(x) = \sum_{x=1}^3 x^3 \frac{x}{6}$$

$$= \tfrac{1}{6} + \tfrac{16}{6} + \tfrac{81}{6} = \tfrac{98}{6}.$$

Example 3. Let X and Y have the p.d.f.

$$f(x, y) = x + y, \quad 0 < x < 1, 0 < y < 1,$$

$$= 0 \text{ elsewhere.}$$

Accordingly,

$$E(XY^2) = \int_{-\infty}^{\infty} \int_{-\infty}^{\infty} xy^2 f(x, y) \, dx \, dy$$

$$= \int_0^1 \int_0^1 xy^2 (x + y) \, dx \, dy$$

$$= \tfrac{17}{72}.$$

Example 4. Let us divide, at random, a horizontal line segment of length 5 into two parts. If X is the length of the left-hand part, it is reasonable to assume that X has the p.d.f.

$$f(x) = \tfrac{1}{5}, \quad 0 < x < 5,$$

$$= 0 \text{ elsewhere.}$$

The expected value of the length X is $E(X) = \frac{5}{2}$ and the expected value of the length $5 - X$ is $E(5 - X) = \frac{5}{2}$. But the expected value of the product of the two lengths is equal to

$$E[X(5 - X)] = \int_0^5 x(5 - x)(\tfrac{1}{5})\, dx = \tfrac{25}{6} \neq (\tfrac{5}{2})^2.$$

That is, in general, the expected value of a product is not equal to the product of the expected values.

Example 5. A bowl contains five chips, which cannot be distinguished by a sense of touch alone. Three of the chips are marked \$1 each and the remaining two are marked \$4 each. A player is blindfolded and draws, at random and without replacement, two chips from the bowl. The player is paid an amount equal to the sum of the values of the two chips that he draws and the game is over. If it costs \$4.75 cents to play this game, would we care to participate for any protracted period of time? Because we are unable to distinguish the chips by sense of touch, we assume that each of the 10 pairs that can be drawn has the same probability of being drawn. Let the random variable X be the number of chips, of the two to be chosen, that are marked \$1. Then, under our assumption, X has the hypergeometric p.d.f.

$$f(x) = \frac{\binom{3}{x}\binom{2}{2-x}}{\binom{5}{2}}, \qquad x = 0, 1, 2,$$

$$= 0 \text{ elsewhere.}$$

If $X = x$, the player receives $u(x) = x + 4(2 - x) = 8 - 3x$ dollars. Hence his mathematical expectation is equal to

$$E[8 - 3X] = \sum_{x=0}^{2} (8 - 3x)f(x) = \tfrac{44}{10},$$

or \$4.40.

EXERCISES

1.74. Let X have the p.d.f. $f(x) = (x + 2)/18$, $-2 < x < 4$, zero elsewhere. Find $E(X)$, $E[(X + 2)^3]$, and $E[6X - 2(X + 2)^3]$.

1.75. Suppose that $f(x) = \frac{1}{5}$, $x = 1, 2, 3, 4, 5$, zero elsewhere, is the p.d.f. of the discrete type of random variable X. Compute $E(X)$ and $E(X^2)$. Use these two results to find $E[(X + 2)^2]$ by writing $(X + 2)^2 = X^2 + 4X + 4$.

1.76. If X and Y have the p.d.f. $f(x, y) = \frac{1}{3}$, $(x, y) = (0, 0), (0, 1), (1, 1)$, zero elsewhere, find $E[(X - \frac{1}{3})(Y - \frac{2}{3})]$.

1.77. Let the p.d.f. of X and Y be $f(x, y) = e^{-x-y}, 0 < x < \infty, 0 < y < \infty,$

zero elsewhere. Let $u(X, Y) = X$, $v(X, Y) = Y$, and $w(X, Y) = XY$. Show that $E[u(X, Y)] \cdot E[v(X, Y)] = E[w(X, Y)]$.

1.78. Let the p.d.f. of X and Y be $f(x, y) = 2$, $0 < x < y$, $0 < y < 1$, zero elsewhere. Let $u(X, Y) = X$, $v(X, Y) = Y$ and $w(X, Y) = XY$. Show that $E[u(X, Y)] \cdot E[v(X, Y)] \neq E[w(X, Y)]$.

1.79. Let X have a p.d.f. $f(x)$ that is positive at $x = -1, 0, 1$ and is zero elsewhere. (a) If $f(0) = \frac{1}{2}$, find $E(X^2)$. (b) If $f(0) = \frac{1}{2}$ and if $E(X) = \frac{1}{6}$, determine $f(-1)$ and $f(1)$.

1.80. A bowl contains 10 chips, of which 8 are marked $2 each and 2 are marked $5 each. Let a person choose, at random and without replacement, 3 chips from this bowl. If the person is to receive the sum of the resulting amounts, find his expectation.

1.81. Let X be a random variable of the continuous type that has p.d.f. $f(x)$. If m is the unique median of the distribution of X and b is a real constant, show that

$$E(|X - b|) = E(|X - m|) + 2 \int_m^b (b - x)f(x) \, dx,$$

provided that the expectations exist. For what value of b is $E(|X - b|)$ a minimum?

1.82. Let $f(x) = 2x$, $0 < x < 1$, zero elsewhere, be the p.d.f. of X. (a) Compute $E(\sqrt{X})$. (b) Find the distribution function and the p.d.f. of $Y = \sqrt{X}$. (c) Compute $E(Y)$ and compare this result with the answer obtained in part (a).

1.83. Two distinct integers are chosen at random and without replacement from the first six positive integers. Compute the expected value of the absolute value of the difference of these two numbers.

1.10 Some Special Mathematical Expectations

Certain mathematical expectations, if they exist, have special names and symbols to represent them. We shall mention now only those associated with one random variable. First, let $u(X) = X$, where X is a random variable of the discrete type having a p.d.f. $f(x)$. Then

$$E(X) = \sum_x xf(x).$$

If the discrete points of the space of positive probability density are $a_1, a_2, a_3, \ldots$, then

$$E(X) = a_1 f(a_1) + a_2 f(a_2) + a_3 f(a_3) + \cdots.$$

This sum of products is seen to be a "weighted average" of the values $a_1, a_2, a_3, \ldots$, the "weight" associated with each a_i being $f(a_i)$. This suggests that we call $E(X)$ the arithmetic mean of the values of X, or, more simply, the *mean value* of X (or the mean value of the distribution).

The mean value μ of a random variable X is defined, when it exists, to be $\mu = E(X)$, where X is a random variable of the discrete or of the continuous type.

Another special mathematical expectation is obtained by taking $u(X) = (X - \mu)^2$. If, initially, X is a random variable of the discrete type having a p.d.f. $f(x)$, then

$$E[(X - \mu)^2] = \sum_x (x - \mu)^2 f(x)$$
$$= (a_1 - \mu)^2 f(a_1) + (a_2 - \mu)^2 f(a_2) + \cdots,$$

if $a_1, a_2, \ldots$ are the discrete points of the space of positive probability density. This sum of products may be interpreted as a "weighted average" of the squares of the deviations of the numbers $a_1, a_2, \ldots$ from the mean value μ of those numbers where the "weight" associated with each $(a_i - \mu)^2$ is $f(a_i)$. This mean value of the square of the deviation of X from its mean value μ is called the *variance* of X (or the variance of the distribution).

The variance of X will be denoted by σ^2, and we define σ^2, if it exists, by $\sigma^2 = E[(X - \mu)^2]$, whether X is a discrete or a continuous type of random variable.

It is worthwhile to observe that

$$\sigma^2 = E[(X - \mu)^2] = E(X^2 - 2\mu X + \mu^2);$$

and since E is a linear operator,

$$\sigma^2 = E(X^2) - 2\mu E(X) + \mu^2$$
$$= E(X^2) - 2\mu^2 + \mu^2$$
$$= E(X^2) - \mu^2.$$

This frequency affords an easier way of computing the variance of X.

It is customary to call σ (the positive square root of the variance) the *standard deviation* of X (or the standard deviation of the distribution). The number σ is sometimes interpreted as a measure of the dispersion of the points of the space relative to the mean value μ. We note that if the space contains only one point x for which $f(x) > 0$, then $\sigma = 0$.

Remark. Let the random variable X of the continuous type have the p.d.f. $f(x) = 1/2a$, $-a < x < a$, zero elsewhere, so that $\sigma = a/\sqrt{3}$ is the standard deviation of the distribution of X. Next, let the random variable Y of the continuous type have the p.d.f. $g(y) = 1/4a$, $-2a < y < 2a$, zero elsewhere, so that $\sigma = 2a/\sqrt{3}$ is the standard deviation of the distribution of Y. Here the standard deviation of Y is greater than that of X; this reflects the fact that the probability for Y is more widely distributed (relative to the mean zero) than is the probability for X.

We next define a third special mathematical expectation, called the *moment-generating function* of a random variable X. Suppose that there is a positive number h such that for $-h < t < h$ the mathematical expectation $E(e^{tX})$ exists. Thus

$$E(e^{tX}) = \int_{-\infty}^{\infty} e^{tx} f(x)\, dx,$$

if X is a continuous type of random variable, or

$$E(e^{tX}) = \sum_{x} e^{tx} f(x),$$

if X is a discrete type of random variable. This expectation is called the moment-generating function of X (or of the distribution) and is denoted by $M(t)$. That is,

$$M(t) = E(e^{tX}).$$

It is evident that if we set $t = 0$, we have $M(0) = 1$. As will be seen by example, not every distribution has a moment-generating function, but it is difficult to overemphasize the importance of a moment-generating function when it does exist. This importance stems from the fact that the moment-generating function is unique and completely determines the distribution of the random variable; thus, if two random variables have the same moment-generating function, they have the same distribution. This property of a moment-generating function will be very useful in subsequent chapters. Proof of the uniqueness of the moment-generating function is based on the theory of transforms in analysis, and therefore we merely assert this uniqueness.

Although the fact that a moment-generating function (when it exists) completely determines a distribution of one random variable will not be proved, it does seem desirable to try to make the assertion plausible. This can be done if the random variable is of the discrete type. For example, let it be given that

$$M(t) = \tfrac{1}{10}e^{t} + \tfrac{2}{10}e^{2t} + \tfrac{3}{10}e^{3t} + \tfrac{4}{10}e^{4t}$$

is, for all real values of t, the moment-generating function of a random variable X of the discrete type. If we let $f(x)$ be the p.d.f. of X and let $a, b, c, d, \ldots$ be the discrete points in the space of X at which $f(x) > 0$, then

$$M(t) = \sum_x e^{tx} f(x),$$

or

$$\tfrac{1}{10}e^t + \tfrac{2}{10}e^{2t} + \tfrac{3}{10}e^{3t} + \tfrac{4}{10}e^{4t} = f(a)e^{at} + f(b)e^{bt} + \cdots.$$

Because this is an identity for all real values of t, it seems that the right-hand member should consist of but four terms and that each of the four should equal, respectively, one of those in the left-hand member; hence we may take $a = 1, f(a) = \tfrac{1}{10}; b = 2, f(b) = \tfrac{2}{10}; c = 3, f(c) = \tfrac{3}{10}; d = 4,$ $f(d) = \tfrac{4}{10}$. Or, more simply, the p.d.f. of X is

$$f(x) = \frac{x}{10}, \qquad x = 1, 2, 3, 4,$$

$$= 0 \text{ elsewhere.}$$

On the other hand, let X be a random variable of the continuous type and let it be given that

$$M(t) = \frac{1}{(1 - t)^2}, \qquad t < 1,$$

is the moment-generating function of X. That is, we are given

$$\frac{1}{(1 - t)^2} = \int_{-\infty}^{\infty} e^{tx} f(x) \, dx, \qquad t < 1.$$

It is not at all obvious how $f(x)$ is found. However, it is easy to see that a distribution with p.d.f.

$$f(x) = xe^{-x}, \qquad 0 < x < \infty,$$

$$= 0 \text{ elsewhere}$$

has the moment-generating function $M(t) = (1 - t)^{-2}, t < 1$. Thus the random variable X has a distribution with this p.d.f. in accordance with the assertion of the uniqueness of the moment-generating function.

Since a distribution that has a moment-generating function $M(t)$ is completely determined by $M(t)$, it would not be surprising if we could obtain some properties of the distribution directly from $M(t)$. For example, the existence of $M(t)$ for $-h < t < h$ implies that derivatives of all order exist at $t = 0$. Thus

$$\frac{dM(t)}{dt} = M'(t) = \int_{-\infty}^{\infty} xe^{tx} f(x) \, dx,$$

if X is of the continuous type, or

$$\frac{dM(t)}{dt} = M'(t) = \sum_x x e^{tx} f(x),$$

if X is of the discrete type. Upon setting $t = 0$, we have in either case

$$M'(0) = E(X) = \mu.$$

The second derivative of $M(t)$ is

$$M''(t) = \int_{-\infty}^{\infty} x^2 e^{tx} f(x)\, dx \qquad \text{or} \qquad \sum_x x^2 e^{tx} f(x),$$

so that $M''(0) = E(X^2)$. Accordingly,

$$\sigma^2 = E(X^2) - \mu^2 = M''(0) - [M'(0)]^2.$$

For example, if $M(t) = (1 - t)^{-2}$, $t < 1$, as in the illustration above, then

$$M'(t) = 2(1 - t)^{-3}$$

and

$$M''(t) = 6(1 - t)^{-4}.$$

Hence

$$\mu = M'(0) = 2$$

and

$$\sigma^2 = M''(0) - \mu^2 = 6 - 4 = 2.$$

Of course we could have computed μ and σ^2 from the p.d.f. by

$$\mu = \int_{-\infty}^{\infty} x f(x)\, dx \qquad \text{and} \qquad \sigma^2 = \int_{-\infty}^{\infty} x^2 f(x)\, dx - \mu^2,$$

respectively. Sometimes one way is easier than the other.

In general, if m is a positive integer and if $M^{(m)}(t)$ means the mth derivative of $M(t)$, we have, by repeated differentiation with respect to t,

$$M^{(m)}(0) = E(X^m).$$

Now

$$E(X^m) = \int_{-\infty}^{\infty} x^m f(x)\, dx \qquad \text{or} \qquad \sum_x x^m f(x),$$

and integrals (or sums) of this sort are, in mechanics, called *moments*. Since $M(t)$ generates the values of $E(X^m)$, $m = 1, 2, 3, \ldots$, it is called the moment-generating function. In fact, we shall sometimes call $E(X^m)$ the mth moment of the distribution, or the mth moment of X.

Example 1. Let X have the p.d.f.

$$f(x) = \tfrac{1}{2}(x + 1), \qquad -1 < x < 1,$$
$$= 0 \text{ elsewhere.}$$

Then the mean value of X is

$$\mu = \int_{-\infty}^{\infty} xf(x)\,dx = \int_{-1}^{1} x\,\frac{x+1}{2}\,dx = \frac{1}{3}$$

while the variance of X is

$$\sigma^2 = \int_{-\infty}^{\infty} x^2 f(x)\,dx - \mu^2 = \int_{-1}^{1} x^2\,\frac{x+1}{2}\,dx - (\tfrac{1}{3})^2 = \frac{2}{9}.$$

Example 2. If X has the p.d.f.

$$f(x) = \frac{1}{x^2}, \qquad 1 < x < \infty,$$
$$= 0 \text{ elsewhere,}$$

then the mean value of X does not exist, since

$$\int_{1}^{\infty} x\,\frac{1}{x^2}\,dx = \lim_{b \to \infty} \int_{1}^{b} \frac{1}{x}\,dx$$
$$= \lim_{b \to \infty} (\ln b - \ln 1)$$

does not exist.

Example 3. Given that the series

$$\frac{1}{1^2} + \frac{1}{2^2} + \frac{1}{3^2} + \cdots$$

converges to $\pi^2/6$. Then

$$f(x) = \frac{6}{\pi^2 x^2}, \qquad x = 1, 2, 3, \ldots,$$
$$= 0 \text{ elsewhere,}$$

is the p.d.f. of a discrete type of random variable X. The moment-generating function of this distribution, if it exists, is given by

$$M(t) = E(e^{tX}) = \sum_{x} e^{tx} f(x)$$
$$= \sum_{x=1}^{\infty} \frac{6e^{tx}}{\pi^2 x^2}.$$

The ratio test may be used to show that this series diverges if $t > 0$. Thus there does not exist a positive number h such that $M(t)$ exists for $-h < t < h$. Accordingly, the distribution having the p.d.f. $f(x)$ of this example does not have a moment-generating function.

Example 4. Let X have the moment-generating function $M(t) = e^{t^2/2}$, $-\infty < t < \infty$. We can differentiate $M(t)$ any number of times to find the moments of X. However, it is instructive to consider this alternative method. The function $M(t)$ is represented by the following MacLaurin's series.

$$e^{t^2/2} = 1 + \frac{1}{1!}\left(\frac{t^2}{2}\right) + \frac{1}{2!}\left(\frac{t^2}{2}\right)^2 + \cdots + \frac{1}{k!}\left(\frac{t^2}{2}\right)^k + \cdots$$

$$= 1 + \frac{1}{2!}t^2 + \frac{(3)(1)}{4!}t^4 + \cdots + \frac{(2k-1)\cdots(3)(1)}{(2k)!}t^{2k} + \cdots.$$

In general, the MacLaurin's series for $M(t)$ is

$$M(t) = M(0) + \frac{M'(0)}{1!}t + \frac{M''(0)}{2!}t^2 + \cdots + \frac{M^{(m)}(0)}{m!}t^m + \cdots$$

$$= 1 + \frac{E(X)}{1!}t + \frac{E(X^2)}{2!}t^2 + \cdots + \frac{E(X^m)}{m!}t^m + \cdots.$$

Thus the coefficient of $(t^m/m!)$ in the MacLaurin's series representation of $M(t)$ is $E(X^m)$. So, for our particular $M(t)$, we have

$$E(X^{2k}) = (2k-1)(2k-3)\cdots(3)(1) = \frac{(2k)!}{2^k k!},$$

$k = 1, 2, 3, \ldots$, and $E(X^{2k-1}) = 0$, $k = 1, 2, 3, \ldots$.

Remarks. In a more advanced course, we would not work with the moment-generating function because so many distributions do not have moment-generating functions. Instead, we would let i denote the imaginary unit, t an arbitrary real, and we would define $\varphi(t) = E(e^{itX})$. This expectation exists for *every* distribution and it is called the *characteristic function* of the distribution. To see why $\varphi(t)$ exists for all real t, we note, in the continuous case, that its absolute value

$$|\varphi(t)| = \left|\int_{-\infty}^{\infty} e^{itx} f(x)\, dx\right| \leq \int_{-\infty}^{\infty} |e^{itx} f(x)|\, dx.$$

However, $|f(x)| = f(x)$ since $f(x)$ is nonnegative and

$$|e^{itx}| = |\cos tx + i \sin tx| = \sqrt{\cos^2 tx + \sin^2 tx} = 1.$$

Thus

$$|\varphi(t)| \leq \int_{-\infty}^{\infty} f(x)\, dx = 1.$$

Accordingly, the integral for $\varphi(t)$ exists for all real values of t. In the discrete case, a summation would replace the integral.

Every distribution has a unique characteristic function; and to each characteristic function there corresponds a unique distribution of probability. If X has a distribution with characteristic function $\varphi(t)$, then, for instance, if $E(X)$ and $E(X^2)$ exist, they are given, respectively, by $iE(X) = \varphi'(0)$ and $i^2E(X^2) = \varphi''(0)$. Readers who are familiar with complex-valued functions may write $\varphi(t) = M(it)$ and, throughout this book, may prove certain theorems in complete generality.

Those who have studied Laplace and Fourier transforms will note a similarity between these transforms and $M(t)$ and $\varphi(t)$; it is the uniqueness of these transforms that allows us to assert the uniqueness of each of the moment-generating and characteristic functions.

EXERCISES

1.84. Find the mean and variance, if they exist, of each of the following distributions.

(a) $f(x) = \dfrac{3!}{x!\,(3-x)!}\left(\dfrac{1}{2}\right)^3$, $x = 0, 1, 2, 3$, zero elsewhere.

(b) $f(x) = 6x(1-x)$, $0 < x < 1$, zero elsewhere.

(c) $f(x) = 2/x^3$, $1 < x < \infty$, zero elsewhere.

1.85. Let $f(x) = (\tfrac{1}{2})^x$, $x = 1, 2, 3, \ldots$, zero elsewhere, be the p.d.f. of the random variable X. Find the moment-generating function, the mean, and the variance of X.

1.86. For each of the following probability density functions, compute $\Pr(\mu - 2\sigma < X < \mu + 2\sigma)$.

(a) $f(x) = 6x(1-x)$, $0 < x < 1$, zero elsewhere.

(b) $f(x) = (\tfrac{1}{2})^x$, $x = 1, 2, 3, \ldots$, zero elsewhere.

1.87. If the variance of the random variable X exists, show that $E(X^2) \geq [E(X)]^2$.

1.88. Let a random variable X of the continuous type have a p.d.f. $f(x)$ whose graph is symmetric with respect to $x = c$. If the mean value of X exists, show that $E(X) = c$. *Hint.* Show that $E(X - c)$ equals zero by writing $E(X - c)$ as the sum of two integrals: one from $-\infty$ to c and the other from c to ∞. In the first, let $y = c - x$; and, in the second, $z = x - c$. Finally, use the symmetry condition $f(c - y) = f(c + y)$ in the first.

1.89. Let the random variable X have mean μ, standard deviation σ, and moment-generating function $M(t)$, $-h < t < h$. Show that

$$E\left(\frac{X - \mu}{\sigma}\right) = 0, \qquad E\left[\left(\frac{X - \mu}{\sigma}\right)^2\right] = 1,$$

and

$$E\left\{\exp\left[t\left(\frac{X-\mu}{\sigma}\right)\right]\right\} = e^{-\mu t/\sigma}M\left(\frac{t}{\sigma}\right), \qquad -h\sigma < t < h\sigma.$$

1.90. Show that the moment-generating function of the random variable X having the p.d.f. $f(x) = \frac{1}{3}$, $-1 < x < 2$, zero elsewhere, is

$$M(t) = \frac{e^{2t} - e^{-t}}{3t}, \qquad t \neq 0,$$

$$= 1, \qquad t = 0.$$

1.91. Let X be a random variable such that $E[(X - b)^2]$ exists for all real b. Show that $E[(X - b)^2]$ is a minimum when $b = E(X)$.

1.92. Let $f(x_1, x_2) = 2x_1$, $0 < x_1 < 1$, $0 < x_2 < 1$, zero elsewhere, be the p.d.f. of X_1 and X_2. Compute $E(X_1 + X_2)$ and $E\{[X_1 + X_2 - E(X_1 + X_2)]^2\}$.

1.93. Let X denote a random variable for which $E[(X - a)^2]$ exists. Give an example of a distribution of a discrete type such that this expectation is zero. Such a distribution is called a *degenerate distribution*.

1.94. Let X be a random variable such that $K(t) = E(t^X)$ exists for all real values of t in a certain open interval that includes the point $t = 1$. Show that $K^{(m)}(1)$ is equal to the mth *factorial moment* $E[X(X - 1) \cdots (X - m + 1)]$.

1.95. Let X be a random variable. If m is a positive integer, the expectation $E[(X - b)^m]$, if it exists, is called the mth moment of the distribution about the point b. Let the first, second, and third moments of the distribution about the point 7 be 3, 11, and 15, respectively. Determine the mean μ of X, and then find the first, second, and third moments of the distribution about the point μ.

1.96. Let X be a random variable such that $R(t) = E(e^{t(X-b)})$ exists for $-h < t < h$. If m is a positive integer, show that $R^{(m)}(0)$ is equal to the mth moment of the distribution about the point b.

1.97. Let X be a random variable with mean μ and variance σ^2 such that the third moment $E[(X - \mu)^3]$ about the vertical line through μ exists. The value of the ratio $E[(X - \mu)^3]/\sigma^3$ is often used as a measure of *skewness*. Graph each of the following probability density functions and show that this measure is negative, zero, and positive for these respective distributions (said to be skewed to the left, not skewed, and skewed to the right, respectively).
  (a) $f(x) = (x + 1)/2$, $-1 < x < 1$, zero elsewhere.
  (b) $f(x) = \frac{1}{2}$, $-1 < x < 1$, zero elsewhere.
  (c) $f(x) = (1 - x)/2$, $-1 < x < 1$, zero elsewhere.

1.98. Let X be a random variable with mean μ and variance σ^2 such that the fourth moment $E[(X - \mu)^4]$ about the vertical line through μ exists. The value of the ratio $E[(X - \mu)^4]/\sigma^4$ is often used as a measure of *kurtosis*. Graph each of the following probability density functions and show that this measure is smaller for the first distribution.

(a) $f(x) = \frac{1}{2}$, $-1 < x < 1$, zero elsewhere.

(b) $f(x) = 3(1 - x^2)/4$, $-1 < x < 1$, zero elsewhere.

1.99. Let the random variable X have p.d.f.

$$f(x) = p, \qquad x = -1, 1,$$
$$= 1 - 2p, \qquad x = 0,$$
$$= 0 \text{ elsewhere,}$$

where $0 < p < \frac{1}{2}$. Find the measure of kurtosis as a function of p. Determine its value when $p = \frac{1}{3}$, $p = \frac{1}{5}$, $p = \frac{1}{10}$, and $p = \frac{1}{100}$. Note that the kurtosis increases as p decreases.

1.100. Let $\psi(t) = \ln M(t)$, where $M(t)$ is the moment-generating function of a distribution. Prove that $\psi'(0) = \mu$ and $\psi''(0) = \sigma^2$.

1.101. Find the mean and the variance of the distribution that has the distribution function

$$F(x) = 0, \qquad x < 0,$$
$$= \frac{x}{8}, \qquad 0 \le x < 2,$$
$$= \frac{x^2}{16}, \qquad 2 \le x < 4,$$
$$= 1, \qquad 4 \le x.$$

1.102. Find the moments of the distribution that has moment-generating function $M(t) = (1 - t)^{-3}$, $t < 1$. *Hint.* Differentiate twice the series

$$(1 - t)^{-1} = 1 + t + t^2 + t^3 + \cdots, \qquad -1 < t < 1.$$

1.103. Let X be a random variable of the continuous type with p.d.f. $f(x)$, which is positive provided $0 < x < b < \infty$, and is equal to zero elsewhere. Show that

$$E(X) = \int_0^b [1 - F(x)] \, dx,$$

where $F(x)$ is the distribution function of X.

1.11 Chebyshev's Inequality

In this section we shall prove a theorem that enables us to find upper (or lower) bounds for certain probabilities. These bounds, however, are not necessarily close to the exact probabilities and, accordingly, we ordinarily do not use the theorem to approximate a probability. The principal uses of the theorem and a special case of it are in theoretical discussions.

Theorem 6. *Let* $u(X)$ *be a nonnegative function of the random variable* X. *If* $E[u(X)]$ *exists, then, for every positive constant* c,

$$\Pr\left[u(X) \geq c\right] \leq \frac{E[u(X)]}{c}.$$

Proof. The proof is given when the random variable X is of the continuous type; but the proof can be adapted to the discrete case if we replace integrals by sums. Let $A = \{x; u(x) \geq c\}$ and let $f(x)$ denote the p.d.f. of X. Then

$$E[u(X)] = \int_{-\infty}^{\infty} u(x)f(x)\, dx = \int_{A} u(x)f(x)\, dx + \int_{A^*} u(x)f(x)\, dx.$$

Since each of the integrals in the extreme right-hand member of the preceding equation is nonnegative, the left-hand member is greater than or equal to either of them. In particular,

$$E[u(X)] \geq \int_{A} u(x)f(x)\, dx.$$

However, if $x \in A$, then $u(x) \geq c$; accordingly, the right-hand member of the preceding inequality is not increased if we replace $u(x)$ by c. Thus

$$E[u(X)] \geq c \int_{A} f(x)\, dx.$$

Since

$$\int_{A} f(x)\, dx = \Pr\,(X \in A) = \Pr\left[u(X) \geq c\right],$$

it follows that

$$E[u(X)] \geq c \Pr\left[u(X) \geq c\right],$$

which is the desired result.

The preceding theorem is a generalization of an inequality which is often called *Chebyshev's inequality*. This inequality will now be established.

Theorem 7. Chebyshev's Inequality. *Let the random variable X have a distribution of probability about which we assume only that there is a finite variance σ^2. This, of course, implies that there is a mean μ. Then for every $k > 0$,*

$$\Pr\left(|X - \mu| \geq k\sigma\right) \leq \frac{1}{k^2},$$

or, equivalently,

$$\Pr\left(|X - \mu| < k\sigma\right) \geq 1 - \frac{1}{k^2}.$$

Proof. In Theorem 6 take $u(X) = (X - \mu)^2$ and $c = k^2\sigma^2$. Then we have

$$\Pr\left[(X - \mu)^2 \geq k^2\sigma^2\right] \leq \frac{E[(X - \mu)^2]}{k^2\sigma^2}.$$

Since the numerator of the right-hand member of the preceding inequality is σ^2, the inequality may be written

$$\Pr\left(|X - \mu| \geq k\sigma\right) \leq \frac{1}{k^2},$$

which is the desired result. Naturally, we would take the positive number k to be greater than 1 to have an inequality of interest.

It is seen that the number $1/k^2$ is an upper bound for the probability $\Pr\left(|X - \mu| \geq k\sigma\right)$. In the following example this upper bound and the exact value of the probability are compared in special instances.

Example 1. Let X have the p.d.f.

$$f(x) = \frac{1}{2\sqrt{3}}, \qquad -\sqrt{3} < x < \sqrt{3},$$
$$= 0 \text{ elsewhere.}$$

Here $\mu = 0$ and $\sigma^2 = 1$. If $k = \frac{3}{2}$, we have the exact probability

$$\Pr\left(|X - \mu| \geq k\sigma\right) = \Pr\left(|X| \geq \frac{3}{2}\right) = 1 - \int_{-3/2}^{3/2} \frac{1}{2\sqrt{3}}\,dx = 1 - \frac{\sqrt{3}}{2}.$$

By Chebyshev's inequality, the preceding probability has the upper bound $1/k^2 = \frac{4}{9}$. Since $1 - \sqrt{3}/2 = 0.134$, approximately, the exact probability in this case is considerably less than the upper bound $\frac{4}{9}$. If we take $k = 2$, we have the exact probability $\Pr\left(|X - \mu| \geq 2\sigma\right) = \Pr\left(|X| \geq 2\right) = 0$. This again is considerably less than the upper bound $1/k^2 = \frac{1}{4}$ provided by Chebyshev's inequality.

In each instance in the preceding example, the probability $\Pr(|X - \mu| \geq k\sigma)$ and its upper bound $1/k^2$ differ considerably. This suggests that this inequality might be made sharper. However, if we want an inequality that holds for every $k > 0$ and holds for all random variables having finite variance, such an improvement is impossible, as is shown by the following example.

Example 2. Let the random variable X of the discrete type have probabilities $\frac{1}{8}, \frac{6}{8}, \frac{1}{8}$ at the points $x = -1, 0, 1$, respectively. Here $\mu = 0$ and $\sigma^2 = \frac{1}{4}$. If $k = 2$, then $1/k^2 = \frac{1}{4}$ and $\Pr(|X - \mu| \geq k\sigma) = \Pr(|X| \geq 1) = \frac{1}{4}$. That is, the probability $\Pr(|X - \mu| \geq k\sigma)$ here attains the upper bound $1/k^2 = \frac{1}{4}$. Hence the inequality cannot be improved without further assumptions about the distribution of X.

EXERCISES

1.104. Let X be a random variable with mean μ and let $E[(X - \mu)^{2k}]$ exist. Show, with $d > 0$, that $\Pr(|X - \mu| \geq d) \leq E[(X - \mu)^{2k}]/d^{2k}$.

1.105. Let X be a random variable such that $\Pr(X \leq 0) = 0$ and let $\mu = E(X)$ exist. Show that $\Pr(X \geq 2\mu) \leq \frac{1}{2}$.

1.106. If X is a random variable such that $E(X) = 3$ and $E(X^2) = 13$, use Chebyshev's inequality to determine a lower bound for the probability $\Pr(-2 < X < 8)$.

1.107. Let X be a random variable with moment-generating function $M(t), -h < t < h$. Prove that

$$\Pr(X \geq a) \leq e^{-at}M(t), \qquad 0 < t < h,$$

and that

$$\Pr(X \leq a) \leq e^{-at}M(t), \qquad -h < t < 0.$$

Hint. Let $u(x) = e^{tx}$ and $c = e^{ta}$ in Theorem 6. *Note.* These results imply that $\Pr(X \geq a)$ and $\Pr(X \leq a)$ are less than the respective greatest lower bounds of $e^{-at}M(t)$ when $0 < t < h$ and when $-h < t < 0$.

1.108. The moment-generating function of X exists for all real values of t and is given by

$$M(t) = \frac{e^t - e^{-t}}{2t}, \qquad t \neq 0, \quad M(0) = 1.$$

Use the results of the preceding exercise to show that $\Pr(X \geq 1) = 0$ and $\Pr(X \leq -1) = 0$. Note that here h is infinite.

Chapter 2

Conditional Probability and Stochastic Independence

2.1 Conditional Probability

In some random experiments, we are interested only in those outcomes that are elements of a subset C_1 of the sample space $\mathscr{C}$. This means, for our purposes, that the sample space is effectively the subset C_1. We are now confronted with the problem of defining a probability set function with C_1 as the "new" sample space.

Let the probability set function $P(C)$ be defined on the sample space $\mathscr{C}$ and let C_1 be a subset of $\mathscr{C}$ such that $P(C_1) > 0$. We agree to consider only those outcomes of the random experiment that are elements of C_1; in essence, then, we take C_1 to be a sample space. Let C_2 be another subset of $\mathscr{C}$. How, relative to the new sample space C_1, do we want to define the probability of the event C_2? Once defined, this probability is called the *conditional probability* of the event C_2, relative to the hypothesis of the event C_1; or, more briefly, the conditional probability of C_2, given C_1. Such a conditional probability is denoted by the symbol $P(C_2|C_1)$. We now return to the question that was raised about the definition of this symbol. Since C_1 is now the sample space, the only elements of C_2 that concern us are those, if any, that are also elements of C_1, that is, the elements of $C_1 \cap C_2$. It seems desirable, then, to define the symbol $P(C_2|C_1)$ in such a way that

$$P(C_1|C_1) = 1 \quad \text{and} \quad P(C_2|C_1) = P(C_1 \cap C_2|C_1).$$

Moreover, from a relative frequency point of view, it would seem logically inconsistent if we did not require that the ratio of the probabilities of the events $C_1 \cap C_2$ and C_1, relative to the space C_1, be the same as the ratio of the probabilities of these events relative to the space $\mathscr{C}$; that is, we should have

$$\frac{P(C_1 \cap C_2 | C_1)}{P(C_1 | C_1)} = \frac{P(C_1 \cap C_2)}{P(C_1)}.$$

These three desirable conditions imply that the relation

$$P(C_2 | C_1) = \frac{P(C_1 \cap C_2)}{P(C_1)}$$

is a suitable *definition* of the conditional probability of the event C_2, given the event C_1, provided $P(C_1) > 0$. Moreover, we have:

(a) $P(C_2 | C_1) \geq 0$.

(b) $P(C_2 \cup C_3 \cup \cdots | C_1) = P(C_2 | C_1) + P(C_3 | C_1) + \cdots$, provided $C_2, C_3, \ldots$ are mutually disjoint sets.

(c) $P(C_1 | C_1) = 1$.

Properties (a) and (c) are evident; proof of property (b) is left as an exercise. But these are precisely the conditions that a probability set function must satisfy. Accordingly, $P(C_2 | C_1)$ is a probability set function, defined for subsets of C_1. It may be called the conditional probability set function, relative to the hypothesis C_1; or the conditional probability set function, given C_1. It should be noted that this conditional probability set function, given C_1, is defined at this time only when $P(C_1) > 0$.

We have now defined the concept of conditional probability for subsets C of a sample space $\mathscr{C}$. We wish to do the same kind of thing for subsets A of $\mathscr{A}$, where $\mathscr{A}$ is the space of one or more random variables defined on $\mathscr{C}$. Let P denote the probability set function of the induced probability on $\mathscr{A}$. If A_1 and A_2 are subsets of $\mathscr{A}$, the conditional probability of the event A_2, given the event A_1, is

$$P(A_2 | A_1) = \frac{P(A_1 \cap A_2)}{P(A_1)}$$

provided $P(A_1) > 0$. This definition will apply to any space which has a probability set function assigned to it.

Example 1. A hand of 5 cards is to be dealt at random and without replacement from an ordinary deck of 52 playing cards. The conditional probability of an all-spade hand (C_2), relative to the hypothesis that there are at least 4 spades in the hand (C_1), is, since $C_1 \cap C_2 = C_2$,

$$P(C_2 | C_1) = \frac{P(C_2)}{P(C_1)} = \frac{\binom{13}{5} / \binom{52}{5}}{\left[\binom{13}{4} \binom{39}{1} + \binom{13}{5} \right] / \binom{52}{5}}.$$

It is worth noting, if we let the random variable X equal the number of spades in a 5-card hand, that a reasonable probability model for X is given by the hypergeometric p.d.f.

$$f(x) = \frac{\binom{13}{x}\binom{39}{5-x}}{\binom{52}{5}}, \qquad x = 0, 1, 2, 3, 4, 5,$$

$$= 0 \text{ elsewhere.}$$

Accordingly, we can write $P(C_2|C_1) = \text{Pr}(X = 5)/[\text{Pr}(X = 4, 5)] = f(5)/[f(4) + f(5)]$.

From the definition of the conditional probability set function, we observe that

$$P(C_1 \cap C_2) = P(C_1)P(C_2|C_1).$$

This relation is frequently called the *multiplication rule* for probabilities. Sometimes, after considering the nature of the random experiment, it is possible to make reasonable assumptions so that both $P(C_1)$ and $P(C_2|C_1)$ can be assigned. Then $P(C_1 \cap C_2)$ can be computed under these assumptions. This will be illustrated in Examples 2 and 3.

Example 2. A bowl contains eight chips. Three of the chips are red and the remaining five are blue. Two chips are to be drawn successively, at random and without replacement. We want to compute the probability that the first draw results in a red chip (C_1) and that the second draw results in a blue chip (C_2). It is reasonable to assign the following probabilities:

$$P(C_1) = \tfrac{3}{8} \qquad \text{and} \qquad P(C_2|C_1) = \tfrac{5}{7}.$$

Thus, under these assignments, we have $P(C_1 \cap C_2) = (\tfrac{3}{8})(\tfrac{5}{7}) = \tfrac{15}{56}$.

Example 3. From an ordinary deck of playing cards, cards are to be drawn successively, at random and without replacement. The probability that the third spade appears on the sixth draw is computed as follows. Let C_1 be the event of two spades in the first five draws and let C_2 be the event of a spade on the sixth draw. Thus the probability that we wish to compute is $P(C_1 \cap C_2)$. It is reasonable to take

$$P(C_1) = \frac{\binom{13}{2}\binom{39}{3}}{\binom{52}{5}} \qquad \text{and} \qquad P(C_2|C_1) = \tfrac{11}{47}.$$

The desired probability $P(C_1 \cap C_2)$ is then the product of these two numbers. More generally, if $X + 3$ is the number of draws necessary to produce exactly three spades, a reasonable probability model for the random variable X is given by the p.d.f.

$$f(x) = \left[\frac{\binom{13}{2}\binom{39}{x}}{\binom{52}{2+x}} \right] \left(\frac{11}{50-x} \right), \qquad x = 0, 1, 2, \ldots, 39,$$

$$= 0 \text{ elsewhere.}$$

Then the particular probability which we computed is $P(C_1 \cap C_2) = \Pr(X = 3) = f(3)$.

The multiplication rule can be extended to three or more events. In the case of three events, we have, by using the multiplication rule for two events,

$$\begin{aligned}
P(C_1 \cap C_2 \cap C_3) &= P[(C_1 \cap C_2) \cap C_3] \\
&= P(C_1 \cap C_2)P(C_3|C_1 \cap C_2).
\end{aligned}$$

But $P(C_1 \cap C_2) = P(C_1)P(C_2|C_1)$. Hence

$$P(C_1 \cap C_2 \cap C_3) = P(C_1)P(C_2|C_1)P(C_3|C_1 \cap C_2).$$

This procedure can be used to extend the multiplication rule to four or more events. The general formula for k events can be proved by mathematical induction.

Example 4. Four cards are to be dealt successively, at random and without replacement, from an ordinary deck of playing cards. The probability of receiving a spade, a heart, a diamond, and a club, in that order, is $\left(\frac{13}{52}\right)\left(\frac{13}{51}\right)\left(\frac{13}{50}\right)\left(\frac{13}{49}\right)$. This follows from the extension of the multiplication rule. In this computation, the assumptions that are involved seem clear.

EXERCISES

(In order to solve certain of these exercises, the student is required to make assumptions.)

2.1. If $P(C_1) > 0$ and if $C_2, C_3, C_4, \ldots$ are mutually disjoint sets, show that $P(C_2 \cup C_3 \cup \cdots |C_1) = P(C_2|C_1) + P(C_3|C_1) + \cdots$.

2.2. Prove that

$$P(C_1 \cap C_2 \cap C_3 \cap C_4) = P(C_1)P(C_2|C_1)P(C_3|C_1 \cap C_2)P(C_4|C_1 \cap C_2 \cap C_3).$$

2.3. A bowl contains eight chips. Three of the chips are red and five are blue. Four chips are to be drawn successively at random and without replacement. (a) Compute the probability that the colors alternate. (b) Compute the probability that the first blue chip appears on the third draw. (c) If

$X + 1$ is the number of draws needed to produce the first blue chip, determine the p.d.f. of X.

2.4. A hand of 13 cards is to be dealt at random and without replacement from an ordinary deck of playing cards. Find the conditional probability that there are at least three kings in the hand relative to the hypothesis that the hand contains at least two kings.

2.5. A drawer contains eight pairs of socks. If six socks are taken at random and without replacement, compute the probability that there is at least one matching pair among these six socks. *Hint.* Compute the probability that there is not a matching pair.

2.6. A bowl contains 10 chips. Four of the chips are red, 5 are white, and 1 is blue. If 3 chips are taken at random and without replacement, compute the conditional probability that there is 1 chip of each color relative to the hypothesis that there is exactly 1 red chip among the 3.

2.7. Let each of the mutually disjoint sets $C_1, \ldots, C_m$ have nonzero probability. If the set C is a subset of the union of $C_1, \ldots, C_m$, show that

$$P(C) = P(C_1)P(C|C_1) + \cdots + P(C_m)P(C|C_m).$$

If $P(C) > 0$, prove *Bayes' formula*:

$$P(C_i|C) = \frac{P(C_i)P(C|C_i)}{P(C_1)P(C|C_1) + \cdots + P(C_m)P(C|C_m)}, \qquad i = 1, \ldots, m.$$

Hint. $P(C)P(C_i|C) = P(C_i)P(C|C_i)$.

2.8. Bowl I contains 3 red chips and 7 blue chips. Bowl II contains 6 red chips and 4 blue chips. A bowl is selected at random and then 1 chip is drawn from this bowl. (a) Compute the probability that this chip is red. (b) Relative to the hypothesis that the chip is red, find the conditional probability that it is drawn from bowl II.

2.9. Bowl I contains 6 red chips and 4 blue chips. Five of these 10 chips are selected at random and without replacement and put in bowl II, which was originally empty. One chip is then drawn at random from bowl II. Relative to the hypothesis that this chip is blue, find the conditional probability that 2 red chips and 3 blue chips are transferred from bowl I to bowl II.

2.2 Marginal and Conditional Distributions

Let $f(x_1, x_2)$ be the p.d.f. of two random variables X_1 and X_2. From this point on, for emphasis and clarity, we shall call a p.d.f. or a distribution function a *joint* p.d.f. or a *joint* distribution function when more than one random variable is involved. Thus $f(x_1, x_2)$ is the joint p.d.f. of the random variables X_1 and X_2. Consider the event $a < X_1 < b$, $a < b$. This event can occur when and only when the event $a < X_1 < b$, $-\infty < X_2 < \infty$ occurs; that is, the two events are equivalent, so that they have the same probability. But the probability of the latter event has been defined and is given by

$$\Pr(a < X_1 < b, -\infty < X_2 < \infty) = \int_a^b \int_{-\infty}^{\infty} f(x_1, x_2) \, dx_2 \, dx_1$$

for the continuous case, and by

$$\Pr(a < X_1 < b, -\infty < X_2 < \infty) = \sum_{a < x_1 < b} \sum_{x_2} f(x_1, x_2)$$

for the discrete case. Now each of

$$\int_{-\infty}^{\infty} f(x_1, x_2) \, dx_2 \qquad \text{and} \qquad \sum_{x_2} f(x_1, x_2)$$

is a function of x_1 alone, say $f_1(x_1)$. Thus, for every $a < b$, we have

$$\Pr(a < X_1 < b) = \int_a^b f_1(x_1) \, dx_1 \qquad \text{(continuous case)},$$

$$= \sum_{a < x_1 < b} f_1(x_1) \qquad \text{(discrete case)},$$

so that $f_1(x_1)$ is the p.d.f. of X_1 alone. Since $f_1(x_1)$ is found by summing (or integrating) the joint p.d.f. $f(x_1, x_2)$ over all x_2 for a fixed x_1, we can think of recording this sum in the "margin" of the $x_1 x_2$-plane. Accordingly, $f_1(x_1)$ is called the marginal p.d.f. of X_1. In like manner

$$f_2(x_2) = \int_{-\infty}^{\infty} f(x_1, x_2) \, dx_1 \qquad \text{(continuous case)},$$

$$= \sum_{x_1} f(x_1, x_2) \qquad \text{(discrete case)},$$

is called the marginal p.d.f. of X_2.

Example 1. Let the joint p.d.f. of X_1 and X_2 be

$$f(x_1, x_2) = \frac{x_1 + x_2}{21}, \qquad x_1 = 1, 2, 3, x_2 = 1, 2,$$

$$= 0 \text{ elsewhere.}$$

Then, for instance,

$$\Pr(X_1 = 3) = f(3, 1) + f(3, 2) = \tfrac{3}{7}$$

and

$$\Pr(X_2 = 2) = f(1, 2) + f(2, 2) + f(3, 2) = \tfrac{4}{7}.$$

On the other hand, the marginal p.d.f. of X_1 is

$$f_1(x_1) = \sum_{x_2=1}^{2} \frac{x_1 + x_2}{21} = \frac{2x_1 + 3}{21}, \qquad x_1 = 1, 2, 3,$$

zero elsewhere, and the marginal p.d.f. of X_2 is

$$f_2(x_2) = \sum_{x_1=1}^{3} \frac{x_1 + x_2}{21} = \frac{6 + 3x_2}{21}, \qquad x_2 = 1, 2,$$

zero elsewhere. Thus the preceding probabilities may be computed as $\Pr(X_1 = 3) = f_1(3) = \frac{3}{7}$ and $\Pr(X_2 = 2) = f_2(2) = \frac{4}{7}$.

We shall now discuss the notion of a conditional p.d.f. Let X_1 and X_2 denote random variables of the discrete type which have the joint p.d.f. $f(x_1, x_2)$ which is positive on $\mathscr{A}$ and is zero elsewhere. Let $f_1(x_1)$ and $f_2(x_2)$ denote, respectively, the marginal probability density functions of X_1 and X_2. Take A_1 to be the set $A_1 = \{(x_1, x_2); x_1 = x_1',$ $-\infty < x_2 < \infty\}$, where x_1' is such that $P(A_1) = \Pr(X_1 = x_1') = f_1(x_1') > 0$, and take A_2 to be the set $A_2 = \{(x_1, x_2); -\infty < x_1 < \infty,$ $x_2 = x_2'\}$. Then, by definition, the conditional probability of the event A_2, given the event A_1, is

$$P(A_2|A_1) = \frac{P(A_1 \cap A_2)}{P(A_1)} = \frac{\Pr(X_1 = x_1', X_2 = x_2')}{\Pr(X_1 = x_1')} = \frac{f(x_1', x_2')}{f_1(x_1')}.$$

That is, if (x_1, x_2) is any point at which $f_1(x_1) > 0$, the conditional probability that $X_2 = x_2$, given that $X_1 = x_1$, is $f(x_1, x_2)/f_1(x_1)$. With x_1 held fast, and with $f_1(x_1) > 0$, this function of x_2 satisfies the conditions of being a p.d.f. of a discrete type of random variable X_2 because $f(x_1, x_2)/f_1(x_1)$ is not negative and

$$\sum_{x_2} \frac{f(x_1, x_2)}{f_1(x_1)} = \frac{1}{f_1(x_1)} \sum_{x_2} f(x_1, x_2) = \frac{f_1(x_1)}{f_1(x_1)} = 1.$$

We now define the symbol $f(x_2|x_1)$ by the relation

$$f(x_2|x_1) = \frac{f(x_1, x_2)}{f_1(x_1)}, \qquad f_1(x_1) > 0,$$

and we call $f(x_2|x_1)$ the *conditional p.d.f.* of the discrete type of random variable X_2, given that the discrete type of random variable $X_1 = x_1$. In a similar manner we define the symbol $f(x_1|x_2)$ by the relation

$$f(x_1|x_2) = \frac{f(x_1, x_2)}{f_2(x_2)}, \qquad f_2(x_2) > 0,$$

and we call $f(x_1|x_2)$ the conditional p.d.f. of the discrete type of random variable X_1, given that the discrete type of random variable $X_2 = x_2$.

Now let X_1 and X_2 denote random variables of the continuous type that have the joint p.d.f. $f(x_1, x_2)$ and the marginal probability density functions $f_1(x_1)$ and $f_2(x_2)$, respectively. We shall use the results of the preceding paragraph to motivate a definition of a conditional p.d.f. of a continuous type of random variable. When $f_1(x_1) > 0$, we define the symbol $f(x_2|x_1)$ by the relation

$$f(x_2|x_1) = \frac{f(x_1, x_2)}{f_1(x_1)}.$$

In this relation, x_1 is to be thought of as having a fixed (but any fixed) value for which $f_1(x_1) > 0$. It is evident that $f(x_2|x_1)$ is nonnegative and that

$$\int_{-\infty}^{\infty} f(x_2|x_1)\, dx_2 = \int_{-\infty}^{\infty} \frac{f(x_1, x_2)}{f_1(x_1)}\, dx_2$$

$$= \frac{1}{f_1(x_1)} \int_{-\infty}^{\infty} f(x_1, x_2)\, dx_2$$

$$= \frac{1}{f_1(x_1)} f_1(x_1) = 1.$$

That is, $f(x_2|x_1)$ has the properties of a p.d.f. of one continuous type of random variable. It is called the *conditional p.d.f.* of the continuous type of random variable X_2, given that the continuous type of random variable X_1 has the value x_1. When $f_2(x_2) > 0$, the conditional p.d.f. of the continuous type of random variable X_1, given that the continuous type of random variable X_2 has the value x_2, is defined by

$$f(x_1|x_2) = \frac{f(x_1, x_2)}{f_2(x_2)}, \qquad f_2(x_2) > 0.$$

Since each of $f(x_2|x_1)$ and $f(x_1|x_2)$ is a p.d.f. of one random variable (whether of the discrete or the continuous type), each has all the properties of such a p.d.f. Thus, we can compute probabilities and mathematical expectations. If the random variables are of the continuous type, the probability

$$\text{Pr}\,(a < X_2 < b | X_1 = x_1) = \int_a^b f(x_2|x_1)\, dx_2$$

is called "the conditional probability that $a < X_2 < b$, given that $X_1 = x_1$." If there is no ambiguity, this may be written in the form

$\Pr (a < X_2 < b|x_1)$. Similarly, the conditional probability that $c < X_1 < d$, given $X_2 = x_2$, is

$$\Pr (c < X_1 < d\,|X_2 = x_2) = \int_c^d f(x_1|x_2)\, dx_1.$$

If $u(X_2)$ is a function of X_2, the expectation

$$E[u(X_2)|x_1] = \int_{-\infty}^\infty u(x_2)f(x_2|x_1)\, dx_2$$

is called the conditional expectation of $u(X_2)$, given $X_1 = x_1$. In particular, if they exist, $E(X_2|x_1)$ is the mean and $E\{[X_2 - E(X_2|x_1)]^2|x_1\}$ is the variance of the conditional distribution of X_2, given $X_1 = x_1$. It is convenient to refer to these as the "conditional mean" and the "conditional variance" of X_2, given $X_1 = x_1$. Of course we have

$$E\{[X_2 - E(X_2|x_1)]^2|x_1\} = E(X_2^2|x_1) - [E(X_2|x_1)]^2$$

from an earlier result. In like manner, the conditional expectation of $u(X_1)$, given $X_2 = x_2$, is given by

$$E[u(X_1)|x_2] = \int_{-\infty}^\infty u(x_1)f(x_1|x_2)\, dx_1.$$

With random variables of the discrete type, these conditional probabilities and conditional expectations are computed by using summation instead of integration. An illustrative example follows.

Example 2. Let X_1 and X_2 have the joint p.d.f.

$$f(x_1, x_2) = 2, \qquad 0 < x_1 < x_2 < 1,$$
$$= 0 \text{ elsewhere.}$$

Then the marginal probability density functions are, respectively,

$$f_1(x_1) = \int_{x_1}^1 2\, dx_2 = 2(1 - x_1), \qquad 0 < x_1 < 1,$$
$$= 0 \text{ elsewhere,}$$

and

$$f_2(x_2) = \int_0^{x_2} 2\, dx_1 = 2x_2, \qquad 0 < x_2 < 1,$$
$$= 0 \text{ elsewhere.}$$

The conditional p.d.f. of X_1, given $X_2 = x_2$, is

$$f(x_1|x_2) = \frac{2}{2x_2} = \frac{1}{x_2}, \qquad 0 < x_1 < x_2, 0 < x_2 < 1,$$
$$= 0 \text{ elsewhere.}$$

Here the conditional mean and conditional variance of X_1, given $X_2 = x_2$, are, respectively,

$$E(X_1|x_2) = \int_{-\infty}^{\infty} x_1 f(x_1|x_2) \, dx_1$$

$$= \int_0^{x_2} x_1 \left(\frac{1}{x_2}\right) dx_1$$

$$= \frac{x_2}{2}, \qquad 0 < x_2 < 1,$$

and

$$E\{[X_1 - E(X_1|x_2)]^2|x_2\} = \int_0^{x_2} \left(x_1 - \frac{x_2}{2}\right)^2 \left(\frac{1}{x_2}\right) dx_1$$

$$= \frac{x_2^2}{12}, \qquad 0 < x_2 < 1.$$

Finally, we shall compare the values of $\Pr\left(0 < X_1 < \frac{1}{2}|X_2 = \frac{3}{4}\right)$ and $\Pr\left(0 < X_1 < \frac{1}{2}\right)$. We have

$$\Pr\left(0 < X_1 < \tfrac{1}{2}|X_2 = \tfrac{3}{4}\right) = \int_0^{1/2} f(x_1|\tfrac{3}{4}) \, dx_1 = \int_0^{1/2} (\tfrac{4}{3}) \, dx_1 = \tfrac{2}{3},$$

but

$$\Pr\left(0 < X_1 < \tfrac{1}{2}\right) = \int_0^{1/2} f_1(x_1) \, dx_1 = \int_0^{1/2} 2(1 - x_1) \, dx_1 = \tfrac{3}{4}.$$

We shall now discuss the notions of marginal and conditional probability density functions from the point of view of n random variables. All of the preceding definitions can be directly generalized to the case of n variables in the following manner. Let the random variables $X_1, X_2, \ldots, X_n$ have the joint p.d.f. $f(x_1, x_2, \ldots, x_n)$. If the random variables are of the continuous type, then by an argument similar to the two-variable case, we have for every $a < b$,

$$\Pr(a < X_1 < b) = \int_a^b f_1(x_1) \, dx_1,$$

where $f_1(x_1)$ is defined by the $(n-1)$-fold integral

$$f_1(x_1) = \int_{-\infty}^{\infty} \cdots \int_{-\infty}^{\infty} f(x_1, x_2, \ldots, x_n) \, dx_2 \cdots dx_n.$$

Accordingly, $f_1(x_1)$ is the p.d.f. of the one random variable X_1 and $f_1(x_1)$ is called the marginal p.d.f. of X_1. The marginal probability density functions $f_2(x_2), \ldots, f_n(x_n)$ of $X_2, \ldots, X_n$, respectively, are similar $(n-1)$-fold integrals.

Up to this point, each marginal p.d.f. has been a p.d.f. of one random variable. It is convenient to extend this terminology to joint probability density functions. We shall do this now. Let $f(x_1, x_2, \ldots, x_n)$ be the joint p.d.f. of the n random variables $X_1, X_2, \ldots, X_n$, just as before. Now, however, let us take any group of $k < n$ of these random variables and let us find the joint p.d.f. of them. This joint p.d.f. is called the marginal p.d.f. of this particular group of k variables. To fix the ideas, take $n = 6$, $k = 3$, and let us select the group X_2, X_4, X_5. Then the marginal p.d.f. of X_2, X_4, X_5 is the joint p.d.f. of this particular group of three variables, namely,

$$\int_{-\infty}^{\infty} \int_{-\infty}^{\infty} \int_{-\infty}^{\infty} f(x_1, x_2, x_3, x_4, x_5, x_6) \, dx_1 \, dx_3 \, dx_6,$$

if the random variables are of the continuous type.

We shall next extend the definition of a conditional p.d.f. If $f_1(x_1) > 0$, the symbol $f(x_2, \ldots, x_n | x_1)$ is defined by the relation

$$f(x_2, \ldots, x_n | x_1) = \frac{f(x_1, x_2, \ldots, x_n)}{f_1(x_1)},$$

and $f(x_2, \ldots, x_n | x_1)$ is called the *joint conditional p.d.f.* of $X_2, \ldots, X_n$, given $X_1 = x_1$. The joint conditional p.d.f. of any $n - 1$ random variables, say $X_1, \ldots, X_{i-1}, X_{i+1}, \ldots, X_n$, given $X_i = x_i$, is defined as the joint p.d.f. of $X_1, X_2, \ldots, X_n$ divided by marginal p.d.f. $f_i(x_i)$, provided $f_i(x_i) > 0$. More generally, the joint conditional p.d.f. of $n - k$ of the random variables, for given values of the remaining k variables, is defined as the joint p.d.f. of the n variables divided by the marginal p.d.f. of the particular group of k variables, provided the latter p.d.f. is positive. We remark that there are many other conditional probability density functions; for instance, see Exercise 2.17.

Because a conditional p.d.f. is a p.d.f. of a certain number of random variables, the mathematical expectation of a function of these random variables has been defined. To emphasize the fact that a conditional p.d.f. is under consideration, such expectations are called conditional expectations. For instance, the conditional expectation of $u(X_2, \ldots, X_n)$ given $X_1 = x_1$, is, for random variables of the continuous type, given by

$$E[u(X_2, \ldots, X_n) | x_1]$$
$$= \int_{-\infty}^{\infty} \cdots \int_{-\infty}^{\infty} u(x_2, \ldots, x_n) f(x_2, \ldots, x_n | x_1) \, dx_2 \cdots dx_n,$$

provided $f_1(x_1) > 0$ and the integral converges (absolutely). If the random variables are of the discrete type, conditional mathematical expectations are, of course, computed by using sums instead of integrals.

EXERCISES

2.10. Let X_1 and X_2 have the joint p.d.f. $f(x_1, x_2) = x_1 + x_2, 0 < x_1 < 1,$ $0 < x_2 < 1$, zero elsewhere. Find the conditional mean and variance of X_2, given $X_1 = x_1, 0 < x_1 < 1$.

2.11. Let $f(x_1|x_2) = c_1 x_1/x_2^2$, $0 < x_1 < x_2$, $0 < x_2 < 1$, zero elsewhere, and $f_2(x_2) = c_2 x_2^4$, $0 < x_2 < 1$, zero elsewhere, denote, respectively, the conditional p.d.f. of X_1, given $X_2 = x_2$, and the marginal p.d.f. of X_2. Determine: (a) the constants c_1 and c_2; (b) the joint p.d.f. of X_1 and X_2; (c) $\Pr\left(\frac{1}{4} < X_1 < \frac{1}{2}|X_2 = \frac{5}{8}\right)$; and (d) $\Pr\left(\frac{1}{4} < X_1 < \frac{1}{2}\right)$.

2.12. Let $f(x_1, x_2) = 21x_1^2 x_2^3$, $0 < x_1 < x_2 < 1$, zero elsewhere, be the joint p.d.f. of X_1 and X_2. Find the conditional mean and variance of X_1, given $X_2 = x_2, 0 < x_2 < 1$.

2.13. If X_1 and X_2 are random variables of the discrete type having p.d.f. $f(x_1, x_2) = (x_1 + 2x_2)/18$, $(x_1, x_2) = (1, 1)$, $(1, 2)$, $(2, 1)$, $(2, 2)$, zero elsewhere, determine the conditional mean and variance of X_2, given $X_1 = x_1$, $x_1 = 1$ or 2.

2.14. Five cards are drawn at random and without replacement from a bridge deck. Let the random variables X_1, X_2, and X_3 denote, respectively, the number of spades, the number of hearts, and the number of diamonds that appear among the five cards. (a) Determine the joint p.d.f. of X_1, X_2, and X_3. (b) Find the marginal probability density functions of X_1, X_2, and X_3. (c) What is the joint conditional p.d.f. of X_2 and X_3, given that $X_1 = 3$?

2.15. Let X_1 and X_2 have the joint p.d.f. $f(x_1, x_2)$ described as follows:

(x_1, x_2)	$(0, 0)$	$(0, 1)$	$(1, 0)$	$(1, 1)$	$(2, 0)$	$(2, 1)$
$f(x_1, x_2)$	$\frac{1}{18}$	$\frac{3}{18}$	$\frac{4}{18}$	$\frac{3}{18}$	$\frac{6}{18}$	$\frac{1}{18}$

and $f(x_1, x_2)$ is equal to zero elsewhere. Find the two marginal probability density functions and the two conditional means.

2.16. Let us choose at random a point from the interval $(0, 1)$ and let the random variable X_1 be equal to the number which corresponds to that point. Then choose a point at random from the interval $(0, x_1)$, where x_1 is the experimental value of X_1; and let the random variable X_2 be equal to the number which corresponds to this point. (a) Make assumptions about the marginal p.d.f. $f_1(x_1)$ and the conditional p.d.f. $f(x_2|x_1)$. (b) Compute $\Pr(X_1 + X_2 \geq 1)$. (c) Find the conditional mean $E(X_1|x_2)$.

2.17. Let $f(x)$ and $F(x)$ denote, respectively, the p.d.f. and the distribution function of the random variable X. The conditional p.d.f. of X, given $X > x_0$, x_0 a fixed number, is defined by $f(x|X > x_0) = f(x)/[1 - F(x_0)]$, $x_0 < x$, zero elsewhere. This kind of conditional p.d.f. finds application in a problem of time until death, given survival until time x_0. (a) Show that $f(x|X > x_0)$ is a p.d.f. (b) Let $f(x) = e^{-x}$, $0 < x < \infty$, zero elsewhere. Compute $\Pr(X > 2| X > 1)$.

2.3 The Correlation Coefficient

Let X, Y, and Z denote random variables that have joint p.d.f. $f(x, y, z)$. If $u(x, y, z)$ is a function of x, y, and z, then $E[u(X, Y, Z)]$ was defined, subject to its existence, on p. 45. The existence of all mathematical expectations will be assumed in this discussion. The means of X, Y, and Z, say μ_1, μ_2, and μ_3, are obtained by taking $u(x, y, z)$ to be x, y, and z, respectively; and the variances of X, Y, and Z, say σ_1^2, σ_2^2, and σ_3^2, are obtained by setting the function $u(x, y, z)$ equal to $(x - \mu_1)^2$, $(y - \mu_2)^2$, and $(z - \mu_3)^2$, respectively. Consider the mathematical expectation

$$E[(X - \mu_1)(Y - \mu_2)] = E(XY - \mu_2 X - \mu_1 Y + \mu_1\mu_2)$$
$$= E(XY) - \mu_2 E(X) - \mu_1 E(Y) + \mu_1\mu_2$$
$$= E(XY) - \mu_1\mu_2.$$

This number is called the *covariance* of X and Y. The covariance of X and Z is given by $E[(X - \mu_1)(Z - \mu_3)]$, and the covariance of Y and Z is $E[(Y - \mu_2)(Z - \mu_3)]$. If each of σ_1 and σ_2 is positive, the number

$$\rho_{12} = \frac{E[(X - \mu_1)(Y - \mu_2)]}{\sigma_1\sigma_2}$$

is called the *correlation coefficient* of X and Y. If the standard deviations are positive, the correlation coefficient of any two random variables is defined to be the covariance of the two random variables divided by the product of the standard deviations of the two random variables. It should be noted that the expected value of the product of two random variables is equal to the product of their expectations plus their covariance.

Example 1. Let the random variables X and Y have the joint p.d.f.

$$f(x, y) = x + y, \qquad 0 < x < 1, 0 < y < 1,$$
$$= 0 \text{ elsewhere.}$$

We shall compute the correlation coefficient of X and Y. When only two variables are under consideration, we shall denote the correlation coefficient by ρ. Now

$$\mu_1 = E(X) = \int_0^1 \int_0^1 x(x + y)\, dx\, dy = \tfrac{7}{12}$$

$$E(X \cdot Y) = E(X) \cdot E(Y) + E[(X - \mu_1) \cdot (Y - \mu_2)]$$

and

$$\sigma_1^2 = E(X^2) - \mu_1^2 = \int_0^1 \int_0^1 x^2(x+y)\,dx\,dy - (\tfrac{7}{12})^2 = \tfrac{11}{144}.$$

Similarly,

$$\mu_2 = E(Y) = \tfrac{7}{12} \quad \text{and} \quad \sigma_2^2 = E(Y^2) - \mu_2^2 = \tfrac{11}{144}.$$

The covariance of X and Y is

$$E(XY) - \mu_1\mu_2 = \int_0^1 \int_0^1 xy(x+y)\,dx\,dy - (\tfrac{7}{12})^2 = -\tfrac{1}{144}.$$

Accordingly, the correlation coefficient of X and Y is

$$\rho = \frac{-\tfrac{1}{144}}{\sqrt{(\tfrac{11}{144})(\tfrac{11}{144})}} = -\frac{1}{11}.$$

Remark. For certain kinds of distributions of two random variables, say X and Y, the correlation coefficient ρ proves to be a very useful characteristic of the distribution. Unfortunately, the formal definition of ρ does not reveal this fact. At this time we make some observations about ρ, some of which will be explored more fully at a later stage. It will soon be seen that if a joint distribution of two variables has a correlation coefficient (that is, if both of the variances are positive), then ρ satisfies $-1 \le \rho \le 1$. If $\rho = 1$, there is a line with equation $y = a + bx$, $b > 0$, the graph of which contains all of the probability for the distribution of X and Y. In this extreme case, we have $\Pr(Y = a + bX) = 1$. If $\rho = -1$, we have the same state of affairs except that $b < 0$. This suggests the following interesting question: When ρ does not have one of its extreme values, is there a line in the xy-plane such that the probability for X and Y tends to be concentrated in a band about this line? Under certain restrictive conditions this is in fact the case, and under those conditions we can look upon ρ as a measure of the intensity of the concentration of the probability for X and Y about that line.

Next, let $f(x, y)$ denote the joint p.d.f. of two random variables X and Y and let $f_1(x)$ denote the marginal p.d.f. of X. The conditional p.d.f. of Y, given $X = x$, is

$$f(y|x) = \frac{f(x, y)}{f_1(x)}$$

at points where $f_1(x) > 0$. Then the conditional mean of Y, given $X = x$, is given by

$$E(Y|x) = \int_{-\infty}^{\infty} yf(y|x)\,dy = \frac{\int_{-\infty}^{\infty} yf(x, y)\,dy}{f_1(x)},$$

when dealing with random variables of the continuous type. This conditional mean of Y, given $X = x$, is, of course, a function of x alone, say $\varphi(x)$. In like vein, the conditional mean of X, given $Y = y$, is a function of y alone, say $\psi(y)$.

In case $\varphi(x)$ is a linear function of x, say $\varphi(x) = a + bx$, we say the conditional mean of Y is linear in x; or that Y has a linear conditional mean. When $\varphi(x) = a + bx$, the constants a and b have simple values which will now be determined.

It will be assumed that neither σ_1^2 nor σ_2^2, the variances of X and Y, is zero. From

$$E(Y|x) = \frac{\int_{-\infty}^{\infty} yf(x, y)\, dy}{f_1(x)} = a + bx,$$

we have

(1) $$\int_{-\infty}^{\infty} yf(x, y)\, dy = (a + bx)f_1(x).$$

If both members of Equation (1) are integrated on x, it is seen that

$$E(Y) = a + bE(X),$$

or

(2) $$\mu_2 = a + b\mu_1,$$

where $\mu_1 = E(X)$ and $\mu_2 = E(Y)$. If both members of Equation (1) are first multiplied by x and then integrated on x, we have

$$E(XY) = aE(X) + bE(X^2),$$

or

(3) $$\rho\sigma_1\sigma_2 + \mu_1\mu_2 = a\mu_1 + b(\sigma_1^2 + \mu_1^2),$$

where $\rho\sigma_1\sigma_2$ is the covariance of X and Y. The simultaneous solution of Equations (2) and (3) yields

$$a = \mu_2 - \rho\frac{\sigma_2}{\sigma_1}\mu_1 \qquad \text{and} \qquad b = \rho\frac{\sigma_2}{\sigma_1}.$$

That is,

$$\varphi(x) = E(Y|x) = \mu_2 + \rho\frac{\sigma_2}{\sigma_1}(x - \mu_1)$$

is the conditional mean of Y, given $X = x$, when the conditional mean of Y is linear in x. If the conditional mean of X, given $Y = y$, is linear in y, then that conditional mean is given by

$$\psi(y) = E(X|y) = \mu_1 + \rho\frac{\sigma_1}{\sigma_2}(y - \mu_2).$$

We shall next investigate the variance of a conditional distribution under the assumption that the conditional mean is linear. The conditional variance of Y is given by

$$(4) \quad E\{[Y - E(Y|x)]^2|x\} = \int_{-\infty}^{\infty} \left[y - \mu_2 - \rho\frac{\sigma_2}{\sigma_1}(x - \mu_1)\right]^2 f(y|x)\, dy$$

$$= \frac{\int_{-\infty}^{\infty} \left[(y - \mu_2) - \rho\frac{\sigma_2}{\sigma_1}(x - \mu_1)\right]^2 f(x, y)\, dy}{f_1(x)}$$

when the random variables are of the continuous type. This variance is nonnegative and is at most a function of x alone. If then, it is multiplied by $f_1(x)$ and integrated on x, the result obtained will be nonnegative. This result is

$$\int_{-\infty}^{\infty}\int_{-\infty}^{\infty} \left[(y - \mu_2) - \rho\frac{\sigma_2}{\sigma_1}(x - \mu_1)\right]^2 f(x, y)\, dy\, dx$$

$$= \int_{-\infty}^{\infty}\int_{-\infty}^{\infty} \left[(y - \mu_2)^2 - 2\rho\frac{\sigma_2}{\sigma_1}(y - \mu_2)(x - \mu_1) + \rho^2\frac{\sigma_2^2}{\sigma_1^2}(x - \mu_1)^2\right]$$
$$\times f(x, y)\, dy\, dx$$

$$= E[(Y - \mu_2)^2] - 2\rho\frac{\sigma_2}{\sigma_1} E[(X - \mu_1)(Y - \mu_2)] + \rho^2\frac{\sigma_2^2}{\sigma_1^2} E[(X - \mu_1)^2]$$

$$= \sigma_2^2 - 2\rho\frac{\sigma_2}{\sigma_1}\rho\sigma_1\sigma_2 + \rho^2\frac{\sigma_2^2}{\sigma_1^2}\sigma_1^2$$

$$= \sigma_2^2 - 2\rho^2\sigma_2^2 + \rho^2\sigma_2^2 = \sigma_2^2(1 - \rho^2) \geq 0.$$

That is, if the variance, Equation (4), is denoted by $k(x)$, then $E[k(X)] = \sigma_2^2(1 - \rho^2) \geq 0$. Accordingly, $\rho^2 \leq 1$, or $-1 \leq \rho \leq 1$. It is left as an exercise to prove that $-1 \leq \rho \leq 1$ whether the conditional mean is or is not linear.

Suppose that the variance, Equation (4), is positive but not a function of x; that is, the variance is a constant $k > 0$. Now if k is multiplied by $f_1(x)$ and integrated on x, the result is k, so that $k = \sigma_2^2(1 - \rho^2)$. Thus, in this case, the variance of each conditional distribution of Y, given $X = x$, is $\sigma_2^2(1 - \rho^2)$. If $\rho = 0$, the variance of each conditional distribution of Y, given $X = x$, is σ_2^2, the variance of the marginal distribution of Y. On the other hand, if ρ^2 is near one, the variance of each conditional distribution of Y, given $X = x$, is relatively small, and there is a high concentration of the probability for this conditional distribution near the mean $E(Y|x) = \mu_2 + \rho(\sigma_2/\sigma_1)(x - \mu_1)$.

It should be pointed out that if the random variables X and Y in the preceding discussion are taken to be of the discrete type, the results just obtained are valid.

Example 2. Let the random variables X and Y have the linear conditional means $E(Y|x) = 4x + 3$ and $E(X|y) = \frac{1}{16}y - 3$. In accordance with the general formulas for the linear conditional means, we see that $E(Y|x) = \mu_2$ if $x = \mu_1$ and $E(X|y) = \mu_1$ if $y = \mu_2$. Accordingly, in this special case, we have $\mu_2 = 4\mu_1 + 3$ and $\mu_1 = \frac{1}{16}\mu_2 - 3$ so that $\mu_1 = -\frac{15}{4}$ and $\mu_2 = -12$. The general formulas for the linear conditional means also show that the product of the coefficients of x and y, respectively, is equal to ρ^2 and that the quotient of these coefficients is equal to σ_2^2/σ_1^2. Here $\rho^2 = 4(\frac{1}{16}) = \frac{1}{4}$ with $\rho = \frac{1}{2}$ (not $-\frac{1}{2}$), and $\sigma_2^2/\sigma_1^2 = 64$. Thus, from the two linear conditional means, we are able to find the values of μ_1, μ_2, ρ, and σ_2/σ_1, but not the values of σ_1 and σ_2.

This section will conclude with a definition and an illustrative example. Let $f(x, y)$ denote the joint p.d.f. of the two random variables X and Y. If $E(e^{t_1 X + t_2 Y})$ exists for $-h_1 < t_1 < h_1$, $-h_2 < t_2 < h_2$, where h_1 and h_2 are positive, it is denoted by $M(t_1, t_2)$ and is called the *moment-generating function* of the joint distribution of X and Y. As in the case of one random variable, the moment-generating function $M(t_1, t_2)$ completely determines the joint distribution of X and Y, and hence the marginal distributions of X and Y. In fact,

$$M(t_1, 0) = E(e^{t_1 X}) = M(t_1)$$

and

$$M(0, t_2) = E(e^{t_2 Y}) = M(t_2).$$

In addition, in the case of random variables of the continuous type,

$$\frac{\partial^{k+m} M(t_1, t_2)}{\partial t_1^k \partial t_2^m} = \int_{-\infty}^{\infty} \int_{-\infty}^{\infty} x^k y^m e^{t_1 x + t_2 y} f(x, y)\, dx\, dy,$$

so that

$$\frac{\partial^{k+m} M(t_1, t_2)}{\partial t_1^k \partial t_2^m}\bigg|_{t_1 = t_2 = 0} = \int_{-\infty}^{\infty} \int_{-\infty}^{\infty} x^k y^m f(x, y)\, dx\, dy = E(X^k Y^m).$$

For instance, in a simplified notation which appears to be clear,

$$\mu_1 = E(X) = \frac{\partial M(0, 0)}{\partial t_1}, \qquad \mu_2 = E(Y) = \frac{\partial M(0, 0)}{\partial t_2},$$

$$\sigma_1^2 = E(X^2) - \mu_1^2 = \frac{\partial^2 M(0, 0)}{\partial t_1^2} - \mu_1^2,$$

(5)

$$\sigma_2^2 = E(Y^2) - \mu_2^2 = \frac{\partial^2 M(0, 0)}{\partial t_2^2} - \mu_2^2,$$

$$E[(X - \mu_1)(Y - \mu_2)] = \frac{\partial^2 M(0, 0)}{\partial t_1\, \partial t_2} - \mu_1 \mu_2.$$

It is fairly obvious that the results of Equations (5) hold if X and Y are random variables of the discrete type. Thus the correlation coefficients may be computed by using the moment-generating function of the joint distribution if that function is readily available. An illustrative example follows.

Example 3. Let the continuous-type random variables X and Y have the joint p.d.f.

$$f(x, y) = e^{-y}, \qquad 0 < x < y < \infty,$$

$$= 0 \text{ elsewhere.}$$

The moment-generating function of this joint distribution is

$$M(t_1, t_2) = \int_0^\infty \int_x^\infty \exp(t_1 x + t_2 y - y) \, dy \, dx$$

$$= \frac{1}{(1 - t_1 - t_2)(1 - t_2)},$$

provided $t_1 + t_2 < 1$ and $t_2 < 1$. For this distribution, Equations (5) become

$$\mu_1 = 1, \qquad \mu_2 = 2,$$

(6) $$\sigma_1^2 = 1, \qquad \sigma_2^2 = 2,$$

$$E[(X - \mu_1)(Y - \mu_2)] = 1.$$

Verification of results of Equations (6) is left as an exercise. If, momentarily, we accept these results, the correlation coefficient of X and Y is $\rho = 1/\sqrt{2}$. Furthermore, the moment-generating functions of the marginal distributions of X and Y are, respectively,

$$M(t_1, 0) = \frac{1}{1 - t_1}, \qquad t_1 < 1,$$

$$M(0, t_2) = \frac{1}{(1 - t_2)^2}, \qquad t_2 < 1.$$

These moment-generating functions are, of course, respectively, those of the marginal probability density functions,

$$f_1(x) = \int_x^\infty e^{-y} \, dy = e^{-x}, \qquad 0 < x < \infty,$$

zero elsewhere, and

$$f_2(y) = e^{-y} \int_0^y dx = y e^{-y}, \qquad 0 < y < \infty,$$

zero elsewhere.

EXERCISES

2.18. Let the random variables X and Y have the joint p.d.f.
(a) $f(x, y) = \frac{1}{3}$, $(x, y) = (0, 0)$, $(1, 1)$, $(2, 2)$, zero elsewhere.
(b) $f(x, y) = \frac{1}{3}$, $(x, y) = (0, 2)$, $(1, 1)$, $(2, 0)$, zero elsewhere.
(c) $f(x, y) = \frac{1}{3}$, $(x, y) = (0, 0)$, $(1, 1)$, $(2, 0)$, zero elsewhere.
In each case compute the correlation coefficient of X and Y.

2.19. Let X and Y have the joint p.d.f. described as follows:

(x, y)	$(1, 1)$	$(1, 2)$	$(1, 3)$	$(2, 1)$	$(2, 2)$	$(2, 3)$
$f(x, y)$	$\frac{2}{15}$	$\frac{4}{15}$	$\frac{3}{15}$	$\frac{1}{15}$	$\frac{1}{15}$	$\frac{4}{15}$

and $f(x, y)$ is equal to zero elsewhere. Find the correlation coefficient ρ.

2.20. Let $f(x, y) = 2, 0 < x < y, 0 < y < 1$, zero elsewhere, be the joint p.d.f. of X and Y. Show that the conditional means are, respectively, $(1 + x)/2$, $0 < x < 1$, and $y/2$, $0 < y < 1$. Show that the correlation coefficient of X and Y is $\rho = \frac{1}{2}$.

2.21. Show that the variance of the conditional distribution of Y, given $X = x$, in Exercise 2.20, is $(1 - x)^2/12$, $0 < x < 1$, and that the variance of the conditional distribution of X, given $Y = y$, is $y^2/12$, $0 < y < 1$.

2.22. Verify the results of Equations (6) of this section.

2.23. Let X and Y have the joint p.d.f. $f(x, y) = 1$, $-x < y < x$, $0 < x < 1$, zero elsewhere. Show that, on the set of positive probability density, the graph of $E(Y|x)$ is a straight line, whereas that of $E(X|y)$ is not a straight line.

2.24. If the correlation coefficient ρ of X and Y exists, show that $-1 \le \rho \le 1$. *Hint.* Consider the discriminant of the nonnegative quadratic function $h(v) = E\{[(X - \mu_1) + v(Y - \mu_2)]^2\}$, where v is real and is not a function of X nor of Y.

2.25. Let $\psi(t_1, t_2) = \ln M(t_1, t_2)$, where $M(t_1, t_2)$ is the moment-generating function of X and Y. Show that

$$\frac{\partial \psi(0, 0)}{\partial t_i}, \qquad \frac{\partial^2 \psi(0, 0)}{\partial t_i^2}, \qquad i = 1, 2,$$

and

$$\frac{\partial^2 \psi(0, 0)}{\partial t_1 \, \partial t_2}$$

yield the means, the variances, and the covariance of the two random variables.

2.26. Let X_1, X_2, and X_3 be three random variables with means, variances, and correlation coefficients, denoted by μ_1, μ_2, μ_3; σ_1^2, σ_2^2, σ_3^2; and

ρ_{12}, ρ_{13}, ρ_{23}, respectively. If $E(X_1 - \mu_1 | x_2, x_3) = b_2(x_2 - \mu_2) + b_3(x_3 - \mu_3)$, where b_2 and b_3 are constants, determine b_2 and b_3 in terms of the variances and the correlation coefficients.

2.4 Stochastic Independence

Let X_1 and X_2 denote random variables of either the continuous or the discrete type which have the joint p.d.f. $f(x_1, x_2)$ and marginal probability density functions $f_1(x_1)$ and $f_2(x_2)$, respectively. In accordance with the definition of the conditional p.d.f. $f(x_2|x_1)$, we may write the joint p.d.f. $f(x_1, x_2)$ as

$$f(x_1, x_2) = f(x_2|x_1) f_1(x_1).$$

Suppose we have an instance where $f(x_2|x_1)$ does not depend upon x_1. Then the marginal p.d.f. of X_2 is, for random variables of the continuous type,

$$f_2(x_2) = \int_{-\infty}^{\infty} f(x_2|x_1) f_1(x_1)\,dx_1$$
$$= f(x_2|x_1) \int_{-\infty}^{\infty} f_1(x_1)\,dx_1$$
$$= f(x_2|x_1).$$

Accordingly,

$$f_2(x_2) = f(x_2|x_1) \qquad \text{and} \qquad f(x_1, x_2) = f_1(x_1) f_2(x_2),$$

when $f(x_2|x_1)$ does not depend upon x_1. That is, if the conditional distribution of X_2, given $X_1 = x_1$, is independent of any assumption about x_1, then $f(x_1, x_2) = f_1(x_1) f_2(x_2)$. These considerations motivate the following definition.

Definition 1. Let the random variables X_1 and X_2 have the joint p.d.f. $f(x_1, x_2)$ and the marginal probability density functions $f_1(x_1)$ and $f_2(x_2)$, respectively. The random variables X_1 and X_2 are said to be stochastically independent if, and only if, $f(x_1, x_2) \equiv f_1(x_1) f_2(x_2)$. Random variables that are not stochastically independent are said to be stochastically dependent.

Remarks. Two comments should be made about the preceding definition. First the product of two nonnegative functions $f_1(x_1) f_2(x_2)$ means a function that is positive on a product space. That is, if $f_1(x_1)$ and $f_2(x_2)$ are positive on, and only on, the respective spaces $\mathscr{A}_1$ and $\mathscr{A}_2$, then the product of $f_1(x_1)$ and $f_2(x_2)$ is positive on, and only on, the product space $\mathscr{A} = \{(x_1, x_2); x_1 \in \mathscr{A}_1, x_2 \in \mathscr{A}_2\}$. For instance, if $\mathscr{A}_1 = \{x_1; 0 < x_1 < 1\}$ and $\mathscr{A}_2 = \{x_2; 0 < x_2 < 3\}$, then $\mathscr{A} = \{(x_1, x_2); 0 < x_1 < 1, 0 < x_2 < 3\}$. The second remark pertains to the identity. The identity in Definition 1 should be interpreted as follows. There may be certain points $(x_1, x_2) \in \mathscr{A}$ at which $f(x_1, x_2) \neq f_1(x_1)f_2(x_2)$. However, if A is the set of points (x_1, x_2) at which the equality does not hold, then $P(A) = 0$. In the subsequent theorems and the subsequent generalizations, a product of nonnegative functions and an identity should be interpreted in an analogous manner.

Example 1. Let the joint p.d.f. of X_1 and X_2 be

$$f(x_1, x_2) = x_1 + x_2, \qquad 0 < x_1 < 1, \ 0 < x_2 < 1,$$
$$= 0 \text{ elsewhere.}$$

It will be shown that X_1 and X_2 are stochastically dependent. Here the marginal probability density functions are

$$f_1(x_1) = \int_{-\infty}^{\infty} f(x_1, x_2)\, dx_2 = \int_0^1 (x_1 + x_2)\, dx_2 = x_1 + \tfrac{1}{2}, \qquad 0 < x_1 < 1,$$
$$= 0 \text{ elsewhere,}$$

and

$$f_2(x_2) = \int_{-\infty}^{\infty} f(x_1, x_2)\, dx_1 = \int_0^1 (x_1 + x_2)\, dx_1 = \tfrac{1}{2} + x_2, \qquad 0 < x_2 < 1,$$
$$= 0 \text{ elsewhere.}$$

Since $f(x_1, x_2) \not\equiv f_1(x_1)f_2(x_2)$, the random variables X_1 and X_2 are stochastically dependent.

The following theorem makes it possible to assert, without computing the marginal probability density functions, that the random variables X_1 and X_2 of Example 1 are stochastically dependent.

Theorem 1. *Let the random variables X_1 and X_2 have the joint p.d.f. $f(x_1, x_2)$. Then X_1 and X_2 are stochastically independent if and only if $f(x_1, x_2)$ can be written as a product of a nonnegative function of x_1 alone and a nonnegative function of x_2 alone. That is,*

$$f(x_1, x_2) \equiv g(x_1)h(x_2),$$

where $g(x_1) > 0$, $x_1 \in \mathscr{A}_1$, zero elsewhere, and $h(x_2) > 0$, $x_2 \in \mathscr{A}_2$, zero elsewhere.

Proof. If X_1 and X_2 are stochastically independent, then $f(x_1, x_2) \equiv f_1(x_1)f_2(x_2)$, where $f_1(x_1)$ and $f_2(x_2)$ are the marginal probability density functions of X_1 and X_2, respectively. Thus, the condition $f(x_1, x_2) \equiv g(x_1)h(x_2)$ is fulfilled.

Conversely, if $f(x_1, x_2) \equiv g(x_1)h(x_2)$, then, for random variables of the continuous type, we have

$$f_1(x_1) = \int_{-\infty}^{\infty} g(x_1)h(x_2)\, dx_2 = g(x_1) \int_{-\infty}^{\infty} h(x_2)\, dx_2 = c_1 g(x_1)$$

and

$$f_2(x_2) = \int_{-\infty}^{\infty} g(x_1)h(x_2)\, dx_1 = h(x_2) \int_{-\infty}^{\infty} g(x_1)\, dx_1 = c_2 h(x_2),$$

where c_1 and c_2 are constants, not functions of x_1 or x_2. Moreover, $c_1 c_2 = 1$ because

$$1 = \int_{-\infty}^{\infty} \int_{-\infty}^{\infty} g(x_1)h(x_2)\, dx_1\, dx_2 = \left[\int_{-\infty}^{\infty} g(x_1)\, dx_1 \right]\left[\int_{-\infty}^{\infty} h(x_2)\, dx_2 \right] = c_2 c_1.$$

These results imply that

$$f(x_1, x_2) \equiv g(x_1)h(x_2) \equiv c_1 g(x_1)c_2 h(x_2) \equiv f_1(x_1)f_2(x_2).$$

Accordingly, X_1 and X_2 are stochastically independent.

If we now refer to Example 1, we see that the joint p.d.f.

$$f(x_1, x_2) = x_1 + x_2, \qquad 0 < x_1 < 1, 0 < x_2 < 1,$$
$$= 0 \text{ elsewhere},$$

cannot be written as the product of a nonnegative function of x_1 alone and a nonnegative function of x_2 alone. Accordingly, X_1 and X_2 are stochastically dependent.

Example 2. Let the p.d.f. of the random variables X_1 and X_2 be $f(x_1, x_2) = 8x_1 x_2, 0 < x_1 < x_2 < 1$, zero elsewhere. The formula $8x_1 x_2$ might suggest to some that X_1 and X_2 are stochastically independent. However, if we consider the space $\mathscr{A} = \{(x_1, x_2); 0 < x_1 < x_2 < 1\}$, we see that it is not a product space. This should make it clear that, in general, X_1 and X_2 must be stochastically dependent if the space of positive probability density of X_1 and X_2 is bounded by a curve that is neither a horizontal nor a vertical line.

We now give a theorem that frequently simplifies the calculations of probabilities of events which involve stochastically independent variables.

Thereom 2. *If X_1 and X_2 are stochastically independent random variables with marginal probability density functions $f_1(x_1)$ and $f_2(x_2)$, respectively, then*

$$\Pr\,(a < X_1 < b, c < X_2 < d) = \Pr\,(a < X_1 < b)\,\Pr\,(c < X_2 < d)$$

for every $a < b$ and $c < d$, where $a, b, c,$ and d are constants.

Proof. From the stochastic independence of X_1 and X_2, the joint p.d.f. of X_1 and X_2 is $f_1(x_1)f_2(x_2)$. Accordingly, in the continuous case,

$$\Pr\,(a < X_1 < b, c < X_2 < d) = \int_a^b \int_c^d f_1(x_1)f_2(x_2)\,dx_2\,dx_1$$

$$= \left[\int_a^b f_1(x_1)\,dx_1\right]\left[\int_c^d f_2(x_2)\,dx_2\right]$$

$$= \Pr\,(a < X_1 < b)\,\Pr\,(c < X_2 < d);$$

or, in the discrete case,

$$\Pr\,(a < X_1 < b, c < X_2 < d) = \sum_{a<x_1<b}\ \sum_{c<x_2<d} f_1(x_1)f_2(x_2)$$

$$= \left[\sum_{a<x_1<b} f_1(x_1)\right]\left[\sum_{c<x_2<d} f_2(x_2)\right]$$

$$= \Pr\,(a < X_1 < b)\,\Pr\,(c < X_2 < d),$$

as was to be shown.

Example 3. In Example 1, X_1 and X_2 were found to be stochastically dependent. There, in general,

$$\Pr\,(a < X_1 < b, c < X_2 < d) \neq \Pr\,(a < X_1 < b)\,\Pr\,(c < X_2 < d).$$

For instance,

$$\Pr\,(0 < X_1 < \tfrac{1}{2}, 0 < X_2 < \tfrac{1}{2}) = \int_0^{1/2} \int_0^{1/2} (x_1 + x_2)\,dx_1\,dx_2 = \tfrac{1}{8},$$

whereas

$$\Pr\,(0 < X_1 < \tfrac{1}{2}) = \int_0^{1/2} (x_1 + \tfrac{1}{2})\,dx_1 = \tfrac{3}{8}$$

and

$$\Pr\,(0 < X_2 < \tfrac{1}{2}) = \int_0^{1/2} (\tfrac{1}{2} + x_2)\,dx_2 = \tfrac{3}{8}.$$

Not merely are calculations of some probabilities usually simpler when we have stochastically independent random variables, but many mathematical expectations, including certain moment-generating functions, have comparably simpler computations. The following result will prove so useful that we state it in form of a theorem.

Theorem 3. *Let the stochastically independent random variables X_1 and X_2 have the marginal probability density functions $f_1(x_1)$ and $f_2(x_2)$, respectively. The expected value of the product of a function $u(X_1)$ of X_1 alone and a function $v(X_2)$ of X_2 alone is, subject to their existence, equal to the product of the expected value of $u(X_1)$ and the expected value of $v(X_2)$; that is,*

$$E[u(X_1)v(X_2)] = E[u(X_1)]E[v(X_2)].$$

Proof. The stochastic independence of X_1 and X_2 implies that the joint p.d.f. of X_1 and X_2 is $f_1(x_1)f_2(x_2)$. Thus, we have, by definition of mathematical expectation, in the continuous case,

$$
\begin{aligned}
E[u(X_1)v(X_2)] &= \int_{-\infty}^{\infty}\int_{-\infty}^{\infty} u(x_1)v(x_2)f_1(x_1)f_2(x_2)\,dx_1\,dx_2 \\
&= \left[\int_{-\infty}^{\infty} u(x_1)f_1(x_1)\,dx_1\right]\left[\int_{-\infty}^{\infty} v(x_2)f_2(x_2)\,dx_2\right] \\
&= E[u(X_1)]E[v(X_2)];
\end{aligned}
$$

or, in the discrete case,

$$
\begin{aligned}
E[u(X_1)v(X_2)] &= \sum_{x_2}\sum_{x_1} u(x_1)v(x_2)f_1(x_1)f_2(x_2) \\
&= \left[\sum_{x_1} u(x_1)f_1(x_1)\right]\left[\sum_{x_2} v(x_2)f_2(x_2)\right] \\
&= E[u(X_1)]E[v(X_2)],
\end{aligned}
$$

as stated in the theorem.

Example 4. Let X and Y be two stochastically independent random variables with means μ_1 and μ_2 and positive variances σ_1^2 and σ_2^2, respectively. We shall show that the stochastic independence of X and Y implies that the correlation coefficient of X and Y is zero. This is true because the covariance of X and Y is equal to

$$
E[(X - \mu_1)(Y - \mu_2)] = E(X - \mu_1)E(Y - \mu_2) = 0.
$$

We shall now prove a very useful theorem about stochastically independent random variables. The proof of the theorem relies heavily upon our assertion that a moment-generating function, when it exists, is unique and that it uniquely determines the distribution of probability.

Thereom 4. *Let X_1 and X_2 denote random variables that have the joint p.d.f. $f(x_1, x_2)$ and the marginal probability density functions $f_1(x_1)$ and $f_2(x_2)$, respectively. Furthermore, let $M(t_1, t_2)$ denote the moment-generating function of the distribution. Then X_1 and X_2 are stochastically independent if and only if $M(t_1, t_2) = M(t_1, 0)M(0, t_2)$.*

Proof. If X_1 and X_2 are stochastically independent, then

$$
\begin{aligned}
M(t_1, t_2) &= E(e^{t_1 X_1 + t_2 X_2}) \\
&= E(e^{t_1 X_1}e^{t_2 X_2}) \\
&= E(e^{t_1 X_1})E(e^{t_2 X_2}) \\
&= M(t_1, 0)M(0, t_2).
\end{aligned}
$$

Thus the stochastic independence of X_1 and X_2 implies that the moment-generating function of the joint distribution factors into the product of the moment-generating functions of the two marginal distributions.

Suppose next that the moment-generating function of the joint distribution of X_1 and X_2 is given by $M(t_1, t_2) = M(t_1, 0)M(0, t_2)$. Now X_1 has the unique moment-generating function which, in the continuous case, is given by

$$M(t_1, 0) = \int_{-\infty}^{\infty} e^{t_1 x_1} f_1(x_1) \, dx_1.$$

Similarly, the unique moment-generating function of X_2, in the continuous case, is given by

$$M(0, t_2) = \int_{-\infty}^{\infty} e^{t_2 x_2} f_2(x_2) \, dx_2.$$

Thus we have

$$M(t_1, 0)M(0, t_2) = \left[\int_{-\infty}^{\infty} e^{t_1 x_1} f_1(x_1) \, dx_1 \right]\left[\int_{-\infty}^{\infty} e^{t_2 x_2} f_2(x_2) \, dx_2 \right]$$

$$= \int_{-\infty}^{\infty} \int_{-\infty}^{\infty} e^{t_1 x_1 + t_2 x_2} f_1(x_1) f_2(x_2) \, dx_1 \, dx_2.$$

We are given that $M(t_1, t_2) = M(t_1, 0)M(0, t_2)$; so

$$M(t_1, t_2) = \int_{-\infty}^{\infty} \int_{-\infty}^{\infty} e^{t_1 x_1 + t_2 x_2} f_1(x_1) f_2(x_2) \, dx_1 \, dx_2.$$

But $M(t_1, t_2)$ is the moment-generating function of X_1 and X_2. Thus also

$$M(t_1, t_2) = \int_{-\infty}^{\infty} \int_{-\infty}^{\infty} e^{t_1 x_1 + t_2 x_2} f(x_1, x_2) \, dx_1 \, dx_2.$$

The uniqueness of the moment-generating function implies that the two distributions of probability that are described by $f_1(x_1)f_2(x_2)$ and $f(x_1, x_2)$ are the same. Thus

$$f(x_1, x_2) \equiv f_1(x_1)f_2(x_2).$$

That is, if $M(t_1, t_2) = M(t_1, 0)M(0, t_2)$, then X_1 and X_2 are stochastically independent. This completes the proof when the random variables are of the continuous type. With random variables of the discrete type, the proof is made by using summation instead of integration.

Let the random variables $X_1, X_2, \ldots, X_n$ have the joint p.d.f. $f(x_1, x_2, \ldots, x_n)$ and the marginal probability density functions $f_1(x_1), f_2(x_2), \ldots, f_n(x_n)$, respectively. The definition of the stochastic independence of X_1 and X_2 is generalized to the mutual stochastic independence of $X_1, X_2, \ldots, X_n$ as follows: The random variables

$X_1, X_2, \ldots, X_n$ are said to be mutually stochastically independent if and only if $f(x_1, x_2, \ldots, x_n) \equiv f_1(x_1)f_2(x_2) \cdots f_n(x_n)$. It follows immediately from this definition of the mutual stochastic independence of $X_1, X_2, \ldots, X_n$ that

$$\Pr(a_1 < X_1 < b_1, a_2 < X_2 < b_2, \ldots, a_n < X_n < b_n)$$
$$= \Pr(a_1 < X_1 < b_1)\Pr(a_2 < X_2 < b_2) \cdots \Pr(a_n < X_n < b_n)$$
$$= \prod_{i=1}^{n} \Pr(a_i < X_i < b_i),$$

where the symbol $\prod_{i=1}^{n} \varphi(i)$ is defined to be

$$\prod_{i=1}^{n} \varphi(i) = \varphi(1)\varphi(2) \cdots \varphi(n).$$

The theorem that $E[u(X_1)v(X_2)] = E[u(X_1)]E[v(X_2)]$ for stochastically independent random variables X_1 and X_2 becomes, for mutually stochastically independent random variables $X_1, X_2, \ldots, X_n$,

$$E[u_1(X_1)u_2(X_2) \cdots u_n(X_n)] = E[u_1(X_1)]E[u_2(X_2)] \cdots E[u_n(X_n)],$$

or

$$E\left[\prod_{i=1}^{n} u_i(X_i)\right] = \prod_{i=1}^{n} E[u_i(X_i)].$$

The moment-generating function of the joint distribution of n random variables $X_1, X_2, \ldots, X_n$ is defined as follows. Let

$$E[\exp(t_1 X_1 + t_2 X_2 + \cdots + t_n X_n)]$$

exist for $-h_i < t_i < h_i$, $i = 1, 2, \ldots, n$, where each h_i is positive. This expectation is denoted by $M(t_1, t_2, \ldots, t_n)$ and it is called the *moment-generating* function of the joint distribution of $X_1, \ldots, X_n$ (or simply the moment-generating function of $X_1, \ldots, X_n$). As in the cases of one and two variables, this moment-generating function is unique and uniquely determines the joint distribution of the n variables (and hence all marginal distributions). For example, the moment-generating function of the marginal distribution of X_i is $M(0, \ldots, 0, t_i, 0, \ldots, 0)$, $i = 1, 2, \ldots, n$; that of the marginal distribution of X_i and X_j is $M(0, \ldots, 0, t_i, 0, \ldots, 0, t_j, 0, \ldots, 0)$; and so on. Theorem 4 of this chapter can be generalized, and the factorization

$$M(t_1, t_2, \ldots, t_n) = \prod_{i=1}^{n} M(0, \ldots, 0, t_i, 0, \ldots, 0)$$

is a necessary and sufficient condition for the mutual stochastic independence of $X_1, X_2, \ldots, X_n$.

Remark. If X_1, X_2, and X_3 are mutually stochastically independent, they are _pairwise stochastically independent_ (that is, X_i and X_j, $i \neq j$, where $i, j = 1$, 2, 3 are stochastically independent). However, the following example, due to S. Bernstein, shows that pairwise independence does not necessarily imply mutual independence. Let X_1, X_2, and X_3 have the joint p.d.f.

$$f(x_1, x_2, x_3) = \tfrac{1}{4}, \qquad (x_1, x_2, x_3) \in \{(1, 0, 0),\ (0, 1, 0),\ (0, 0, 1),\ (1, 1, 1)\},$$
$$= 0 \text{ elsewhere.}$$

The joint p.d.f. of X_i and X_j, $i \neq j$, is

$$f_{ij}(x_i, x_j) = \tfrac{1}{4}, \qquad (x_i, x_j) \in \{(0, 0),\ (1, 0),\ (0, 1),\ (1, 1)\},$$
$$= 0 \text{ elsewhere,}$$

whereas the marginal p.d.f. of X_i is

$$f_i(x_i) = \tfrac{1}{2}, \qquad x_i = 0, 1,$$
$$= 0 \text{ elsewhere.}$$

Obviously, if $i \neq j$, we have

$$f_{ij}(x_i, x_j) \equiv f_i(x_i) f_j(x_j),$$

and thus X_i and X_j are stochastically independent. However,

$$f(x_1, x_2, x_3) \not\equiv f_1(x_1) f_2(x_2) f_3(x_3).$$

Thus X_1, X_2, and X_3 are not mutually stochastically independent.

Example 5. Let X_1, X_2, and X_3 be three mutually stochastically independent random variables and let each have the p.d.f. $f(x) = 2x$, $0 < x < 1$, zero elsewhere. The joint p.d.f. of X_1, X_2, X_3 is $f(x_1) f(x_2) f(x_3) = 8x_1 x_2 x_3$, $0 < x_i < 1$, $i = 1$, 2, 3, zero elsewhere. Let Y be the maximum of X_1, X_2, and X_3. Then, for instance, we have

$$\Pr\left(Y \leq \tfrac{1}{2}\right) \doteq \Pr\left(X_1 \leq \tfrac{1}{2}, X_2 \leq \tfrac{1}{2}, X_3 \leq \tfrac{1}{2}\right)$$
$$= \int_0^{1/2} \int_0^{1/2} \int_0^{1/2} 8x_1 x_2 x_3 \, dx_1 \, dx_2 \, dx_3$$
$$= \left(\tfrac{1}{2}\right)^6 = \tfrac{1}{64}.$$

In a similar manner, we find that the distribution function of Y is

$$G(y) = \Pr(Y \leq y) = 0, \qquad y < 0,$$
$$= y^6, \qquad 0 \leq y < 1,$$
$$= 1, \qquad 1 \leq y.$$

Accordingly, the p.d.f. of Y is

$$g(y) = 6y^5, \qquad 0 < y < 1,$$
$$= 0 \text{ elsewhere.}$$

Example 6. Let a fair coin be tossed at random on successive independent trials. Let the random variable $X_i = 1$ or $X_i = 0$ according to whether the outcome on the ith toss is a head or a tail, $i = 1, 2, 3, \ldots$. Let the p.d.f. of each X_i be $f(x) = \frac{1}{2}, x = 0, 1$, zero elsewhere. Since the trials are independent, we say that the random variables $X_1, X_2, X_3, \ldots$ are mutually stochastically independent. Then, for example, the probability that the first head appears on the third trial is

Pr $(X_1 = 0, X_2 = 0, X_3 = 1)$

$$= \text{Pr } (X_1 = 0) \text{ Pr } (X_2 = 0) \text{ Pr } (X_3 = 1) = (\tfrac{1}{2})^3 = \tfrac{1}{8}.$$

In general, if Y is the number of the trial on which the first head appears, then the p.d.f. of Y is

$$g(y) = (\tfrac{1}{2})^y, \qquad y = 1, 2, 3, \ldots,$$
$$= 0 \text{ elsewhere.}$$

In particular, Pr $(Y = 3) = g(3) = \tfrac{1}{8}$.

EXERCISES

2.27. Show that the random variables X_1 and X_2 with joint p.d.f. $f(x_1, x_2) = 12x_1x_2(1 - x_2), \ 0 < x_1 < 1, \ 0 < x_2 < 1$, zero elsewhere, are stochastically independent.

2.28. If the random variables X_1 and X_2 have the joint p.d.f. $f(x_1, x_2) = 2e^{-x_1-x_2}, 0 < x_1 < x_2, 0 < x_2 < \infty$, zero elsewhere, show that X_1 and X_2 are stochastically dependent.

2.29. Let $f(x_1, x_2) = \frac{1}{16}, \ x_1 = 1, 2, 3, 4$, and $x_2 = 1, 2, 3, 4$, zero elsewhere, be the joint p.d.f. of X_1 and X_2. Show that X_1 and X_2 are stochastically independent.

2.30. Find Pr $(0 < X_1 < \frac{1}{3}, 0 < X_2 < \frac{1}{3})$ if the random variables X_1 and X_2 have the joint p.d.f. $f(x_1, x_2) = 4x_1(1 - x_2), 0 < x_1 < 1, 0 < x_2 < 1$, zero elsewhere.

2.31. Find the probability of the union of the events $a < X_1 < b$, $-\infty < X_2 < \infty$ and $-\infty < X_1 < \infty$, $c < X_2 < d$ if X_1 and X_2 are two stochastically independent variables with Pr $(a < X_1 < b) = \frac{2}{3}$ and Pr $(c < X_2 < d) = \frac{5}{8}$.

2.32. If $f(x_1, x_2) = e^{-x_1-x_2}, 0 < x_1 < \infty, 0 < x_2 < \infty$, zero elsewhere, is the joint p.d.f. of the random variables X_1 and X_2, show that X_1 and X_2 are stochastically independent and that

$$E(e^{t(X_1 + X_2)}) = (1 - t)^{-2}, \qquad t < 1.$$

2.33. Let X_1, X_2, X_3, and X_4 be four mutually stochastically independent random variables, each with p.d.f. $f(x) = 3(1 - x)^2, \ 0 < x < 1$, zero elsewhere. If Y is the minimum of these four variables, find the distribution function and the p.d.f. of Y.

2.34. A fair die is cast at random three independent times. Let the random variable X_i be equal to the number of spots which appear on the ith trial, $i = 1, 2, 3$. Let the random variable Y be equal to max (X_i). Find the distribution function and the p.d.f. of Y. *Hint.* $\Pr(Y \leq y) = \Pr(X_i \leq y, i = 1, 2, 3)$.

2.35. Suppose a man leaves for work between 8:00 A.M. and 8:30 A.M. and takes between 40 and 50 minutes to get to the office. Let X denote the time of departure and let Y denote the time of travel. If we assume that these random variables are stochastically independent and uniformly distributed, find the probability that he arrives at the office before 9:00 A.M.

2.36. Let $M(t_1, t_2, t_3)$ be the moment-generating function of the random variables X_1, X_2, and X_3 of Bernstein's example, described in the final remark of this section. Show that $M(t_1, t_2, 0) = M(t_1, 0, 0)M(0, t_2, 0)$, $M(t_1, 0, t_3) = M(t_1, 0, 0)M(0, 0, t_3)$, $M(0, t_2, t_3) = M(0, t_2, 0)M(0, 0, t_3)$, but $M(t_1, t_2, t_3) \neq M(t_1, 0, 0)M(0, t_2, 0)M(0, 0, t_3)$. Thus X_1, X_2, X_3 are pairwise stochastically independent but not mutually stochastically independent.

2.37. Generalize Theorem 1 of this chapter to the case of n mutually stochastically independent random variables.

2.38. Generalize Theorem 4 of this chapter to the case of n mutually stochastically independent random variables.

Chapter 3
Some Special Distributions

3.1 The Binomial, Trinomial, and Multinomial Distributions

In Chapter 1 we introduced the *uniform distribution* and the *hypergeometric distribution*. In this chapter we shall discuss some other important distributions of random variables frequently used in statistics. We begin with the binomial distribution.

Recall, if n is a positive integer, that

$$(a + b)^n = \sum_{x=0}^{n} \binom{n}{x} b^x a^{n-x}.$$

Consider the function defined by

$$f(x) = \binom{n}{x} p^x (1 - p)^{n-x}, \qquad x = 0, 1, 2, \ldots, n,$$

$$= 0 \text{ elsewhere,}$$

where n is a positive integer and $0 < p < 1$. Under these conditions it is clear that $f(x) \geq 0$ and that

$$\sum_{x} f(x) = \sum_{x=0}^{n} \binom{n}{x} p^x (1 - p)^{n-x}$$

$$= [(1 - p) + p]^n = 1.$$

That is, $f(x)$ satisfies the conditions of being a p.d.f. of a random variable X of the discrete type. A random variable X that has a p.d.f. of the form of $f(x)$ is said to have a *binomial distribution*, and any such $f(x)$ is called a *binomial p.d.f.* A binomial distribution will be denoted by the symbol $b(n, p)$. The constants n and p are called the *parameters* of the binomial distribution. Thus, if we say that X is $b(5, \frac{1}{3})$, we mean that X has the binomial p.d.f.

$$f(x) = \binom{5}{x}\left(\frac{1}{3}\right)^x\left(\frac{2}{3}\right)^{5-x}, \qquad x = 0, 1, \ldots, 5,$$
$$= 0 \text{ elsewhere.}$$

Remark. The binomial distribution serves as an excellent mathematical model in a number of experimental situations. Consider a random experiment, the outcome of which can be classified in but one of two mutually exclusive and exhaustive ways, say, success or failure (for example, head or tail, life or death, effective or noneffective, etc.). Let the random experiment be repeated n independent times. Assume further that the probability of success, say p, is the same on each repetition; thus the probability of failure on each repetition is $1 - p$. Define the random variable X_i, $i = 1, 2, \ldots, n$, to be zero, if the outcome of the ith performance is a failure, and to be 1 if that outcome is a success. We then have $\Pr(X_i = 0) = 1 - p$ and $\Pr(X_i = 1) = p$, $i = 1, 2, \ldots, n$. Since it has been assumed that the experiment is to be repeated n independent times, the random variables $X_1, X_2, \ldots, X_n$ are mutually stochastically independent. According to the definition of X_i, the sum $Y = X_1 + X_2 + \cdots + X_n$ is the number of successes throughout the n repetitions of the random experiment. The following argument shows that Y has a binomial distribution. Let y be an element of $\{y; y = 0, 1, 2, \ldots, n\}$. Then $Y = y$ if and only if exactly y of the variables $X_1, X_2, \ldots, X_n$ have the value 1, and each of the remaining $n - y$ variables is equal to zero. There are $\binom{n}{y}$ ways in which exactly y *ones* can be assigned to y of the variables $X_1, X_2, \ldots, X_n$. Since $X_1, X_2, \ldots, X_n$ are mutually stochastically independent, the probability of each of these ways is $p^y(1 - p)^{n-y}$. Now $\Pr(Y = y)$ is the sum of the probabilities of these $\binom{n}{y}$ mutually exclusive events; that is,

$$\Pr(Y = y) = \binom{n}{y}p^y(1 - p)^{n-y}, \qquad y = 0, 1, 2, \ldots, n,$$

zero elsewhere. This is the p.d.f. of a binomial distribution.

The moment-generating function of a binomial distribution is easily found. It is

$$M(t) = \sum_x e^{tx}f(x) = \sum_{x=0}^{n} e^{tx}\binom{n}{x}p^x(1 - p)^{n-x}$$
$$= \sum_{x=0}^{n} \binom{n}{x}(pe^t)^x(1 - p)^{n-x}$$
$$= [(1 - p) + pe^t]^n$$

for all real values of t. The mean μ and the variance σ^2 of X may be computed from $M(t)$. Since

$$M'(t) = n[(1 - p) + pe^t]^{n-1}(pe^t)$$

and

$$M''(t) = n[(1 - p) + pe^t]^{n-1}(pe^t) + n(n - 1)[(1 - p) + pe^t]^{n-2}(pe^t)^2,$$

it follows that

$$\mu = M'(0) = np$$

and

$$\sigma^2 = M''(0) - \mu^2 = np + n(n - 1)p^2 - (np)^2 = np(1 - p).$$

Example 1. The binomial distribution with p.d.f.

$$f(x) = \binom{7}{x}\left(\frac{1}{2}\right)^x\left(1 - \frac{1}{2}\right)^{7-x}, \qquad x = 0, 1, 2, \ldots, 7,$$

$$= 0 \text{ elsewhere},$$

has the moment-generating function

$$M(t) = (\tfrac{1}{2} + \tfrac{1}{2}e^t)^7,$$

has mean $\mu = np = \frac{7}{2}$, and has variance $\sigma^2 = np(1 - p) = \frac{7}{4}$. Furthermore, if X is the random variable with this distribution, we have

$$\Pr(0 \le X \le 1) = \sum_{x=0}^{1} f(x) = \frac{1}{128} + \frac{7}{128} = \frac{8}{128}$$

and

$$\Pr(X = 5) = f(5)$$

$$= \frac{7!}{5!\,2!}\left(\frac{1}{2}\right)^5\left(\frac{1}{2}\right)^2 = \frac{21}{128}.$$

Example 2. If the moment-generating function of a random variable X is

$$M(t) = (\tfrac{2}{3} + \tfrac{1}{3}e^t)^5,$$

then X has a binomial distribution with $n = 5$ and $p = \frac{1}{3}$; that is, the p.d.f. of X is

$$f(x) = \binom{5}{x}\left(\frac{1}{3}\right)^x\left(\frac{2}{3}\right)^{5-x}, \qquad x = 0, 1, 2, \ldots, 5,$$

$$= 0 \text{ elsewhere}.$$

Here $\mu = np = \frac{5}{3}$ and $\sigma^2 = np(1 - p) = \frac{10}{9}$.

Example 3. If Y is $b(n, \frac{1}{3})$, then $\Pr(Y \geq 1) = 1 - \Pr(Y = 0) = 1 - (\frac{2}{3})^n$. Suppose we wish to find the smallest value of n that yields $\Pr(Y \geq 1) > 0.80$. We have $1 - (\frac{2}{3})^n > 0.80$ and $0.20 > (\frac{2}{3})^n$. Either by inspection or by use of logarithms, we see that $n = 4$ is the solution. That is, the probability of at least one success throughout $n = 4$ independent repetitions of a random experiment with probability of success $p = \frac{1}{3}$ is greater than 0.80.

Example 4. Let the random variable Y be equal to the number of successes throughout n independent repetitions of a random experiment with probability p of success. That is, Y is $b(n, p)$. The ratio Y/n is called the relative frequency of success. For every $\epsilon > 0$, we have

$$\Pr\left(\left|\frac{Y}{n} - p\right| \geq \epsilon\right) = \Pr\left(|Y - np| \geq \epsilon n\right)$$

$$= \Pr\left(|Y - \mu| \geq \epsilon \sqrt{\frac{n}{p(1-p)}}\, \sigma\right),$$

where $\mu = np$ and $\sigma^2 = np(1-p)$. In accordance with Chebyshev's inequality with $k = \epsilon\sqrt{n/p(1-p)}$, we have

$$\Pr\left(|Y - \mu| \geq \epsilon \sqrt{\frac{n}{p(1-p)}}\, \sigma\right) \leq \frac{p(1-p)}{n\epsilon^2}$$

and hence

$$\Pr\left(\left|\frac{Y}{n} - p\right| \geq \epsilon\right) \leq \frac{p(1-p)}{n\epsilon^2}.$$

Now, for every fixed $\epsilon > 0$, the right-hand member of the preceding inequality is close to zero for sufficiently large n. That is,

$$\lim_{n \to \infty} \Pr\left(\left|\frac{Y}{n} - p\right| \geq \epsilon\right) = 0$$

and

$$\lim_{n \to \infty} \Pr\left(\left|\frac{Y}{n} - p\right| < \epsilon\right) = 1.$$

Since this is true for every fixed $\epsilon > 0$, we see, in a certain sense, that the relative frequency of success is for large values of n, close to the probability p of success. This result is one form of the *law of large numbers*. It was alluded to in the initial discussion of probability in Chapter 1 and will be considered again, along with related concepts, in Chapter 5.

Example 5. Let the mutually stochastically independent random variables X_1, X_2, X_3 have the same distribution function $F(x)$. Let Y be the middle value of X_1, X_2, X_3. To determine the distribution function of Y, say

$G(y) = \Pr (Y \le y)$, we note that $Y \le y$ if and only if at least two of the random variables X_1, X_2, X_3 are less than or equal to y. Let us say that the ith "trial" is a success if $X_i \le y$, $i = 1, 2, 3$; here each "trial" has the probability of success $F(y)$. In this terminology, $G(y) = \Pr (Y \le y)$ is then the probability of at least two successes in three independent trials. Thus

$$G(y) = \binom{3}{2}[F(y)]^2[1 - F(y)] + [F(y)]^3.$$

If $F(x)$ is a continuous type of distribution function so that the p.d.f. of X is $F'(x) = f(x)$, then the p.d.f. of Y is

$$g(y) = G'(y) = 6[F(y)][1 - F(y)]f(y).$$

Example 6. Consider a sequence of independent repetitions of a random experiment with constant probability p of success. Let the random variable Y denote the total number of failures in this sequence before the rth success; that is, $Y + r$ is equal to the number of trials necessary to produce exactly r successes. Here r is a fixed positive integer. To determine the p.d.f. of Y, let y be an element of $\{y; y = 0, 1, 2, \ldots\}$. Then, by the multiplication rule of probabilities, $\Pr (Y = y) = g(y)$ is equal to the product of the probability

$$\binom{y + r - 1}{r - 1} p^{r-1}(1 - p)^y$$

of obtaining exactly $r - 1$ successes in the first $y + r - 1$ trials and the probability p of a success on the $(y + r)$th trial. Thus the p.d.f. $g(y)$ of Y is given by

$$g(y) = \binom{y + r - 1}{r - 1} p^r(1 - p)^y, \qquad y = 0, 1, 2, \ldots,$$

$$= 0 \text{ elsewhere.}$$

A distribution with a p.d.f. of the form $g(y)$ is called a *negative binomial distribution*; and any such $g(y)$ is called a negative binomial p.d.f. The distribution derives its name from the fact that $g(y)$ is a general term in the expansion of $p^r[1 - (1 - p)]^{-r}$. It is left as an exercise to show that the moment-generating function of this distribution is $M(t) = p^r[1 - (1 - p)e^t]^{-r}$, for $t < -\ln (1 - p)$. If $r = 1$, then Y has the p.d.f.

$$g(y) = p(1 - p)^y, \qquad y = 0, 1, 2, \ldots,$$

zero elsewhere, and the moment-generating function $M(t) = p[1 - (1 - p)e^t]^{-1}$. In this special case, $r = 1$, we say that Y has a *geometric distribution*.

The binomial distribution can be generalized to the trinomial distribution. If n is a positive integer and a_1, a_2, a_3 are fixed constants, we have

$$(1) \quad \sum_{x=0}^{n} \sum_{y=0}^{n-x} \frac{n!}{x!\, y!\, (n-x-y)!}\, a_1^x a_2^y a_3^{n-x-y}$$

$$= \sum_{x=0}^{n} \frac{n!\, a_1^x}{x!\, (n-x)!} \sum_{y=0}^{n-x} \frac{(n-x)!}{y!\, (n-x-y)!}\, a_2^y a_3^{n-x-y}$$

$$= \sum_{x=0}^{n} \frac{n!}{x!\, (n-x)!}\, a_1^x (a_2 + a_3)^{n-x}$$

$$= (a_1 + a_2 + a_3)^n.$$

Let the function $f(x, y)$ be given by

$$f(x, y) = \frac{n!}{x!\, y!\, (n-x-y)!}\, p_1^x p_2^y p_3^{n-x-y},$$

where x and y are nonnegative integers with $x + y \le n$, and p_1, p_2, and p_3 are positive proper fractions with $p_1 + p_2 + p_3 = 1$; and let $f(x, y) = 0$ elsewhere. Accordingly, $f(x, y)$ satisfies the conditions of being a joint p.d.f. of two random variables X and Y of the discrete type; that is, $f(x, y)$ is nonnegative and its sum over all points (x, y) at which $f(x, y)$ is positive is equal to $(p_1 + p_2 + p_3)^n = 1$. The random variables X and Y which have a joint p.d.f. of the form $f(x, y)$ are said to have a *trinomial distribution*, and any such $f(x, y)$ is called a *trinomial p.d.f.* The moment-generating function of a trinomial distribution, in accordance with Equation (1), is given by

$$M(t_1, t_2) = \sum_{x=0}^{n} \sum_{y=0}^{n-x} \frac{n!}{x!\, y!\, (n-x-y)!}\, (p_1 e^{t_1})^x (p_2 e^{t_2})^y p_3^{n-x-y}$$

$$= (p_1 e^{t_1} + p_2 e^{t_2} + p_3)^n$$

for all real values of t_1 and t_2. The moment-generating functions of the marginal distributions of X and Y are, respectively,

$$M(t_1, 0) = (p_1 e^{t_1} + p_2 + p_3)^n = [(1 - p_1) + p_1 e^{t_1}]^n$$

and

$$M(0, t_2) = (p_1 + p_2 e^{t_2} + p_3)^n = [(1 - p_2) + p_2 e^{t_2}]^n.$$

We see immediately, from Theorem 4, Section 2.4, that X and Y are stochastically dependent. In addition, X is $b(n, p_1)$ and Y is $b(n, p_2)$. Accordingly, the means and the variances of X and Y are, respectively, $\mu_1 = np_1$, $\mu_2 = np_2$, $\sigma_1^2 = np_1(1 - p_1)$, and $\sigma_2^2 = np_2(1 - p_2)$.

Consider next the conditional p.d.f. of Y, given $X = x$. We have

$$f(y|x) = \frac{(n-x)!}{y!\,(n-x-y)!} \left(\frac{p_2}{1-p_1}\right)^y \left(\frac{p_3}{1-p_1}\right)^{n-x-y}, \qquad y = 0, 1, \ldots, n-x,$$

$$= 0 \text{ elsewhere.}$$

Thus the conditional distribution of Y, given $X = x$, is $b[n - x, p_2/(1 - p_1)]$. Hence the conditional mean of Y, given $X = x$, is the linear function

$$E(Y|x) = (n-x)\left(\frac{p_2}{1-p_1}\right).$$

Likewise, we find that the conditional distribution of X, given $Y = y$, is $b[n - y, p_1/(1 - p_2)]$ and thus

$$E(X|y) = (n-y)\left(\frac{p_1}{1-p_2}\right).$$

Now recall (Example 2, Section 2.3) that the square of the correlation coefficient, say ρ^2, is equal to the product of $-p_2/(1 - p_1)$ and $-p_1/(1 - p_2)$, the coefficients of x and y in the respective conditional means. Since both of these coefficients are negative (and thus ρ is negative), we have

$$\rho = -\sqrt{\frac{p_1 p_2}{(1-p_1)(1-p_2)}}.$$

The trinomial distribution is generalized to the multinomial distribution as follows. Let a random experiment be repeated n independent times. On each repetition the experiment terminates in but one of k mutually exclusive and exhaustive ways, say $C_1, C_2, \ldots, C_k$. Let p_i be the probability that the outcome is an element of C_i and let p_i remain constant throughout the n independent repetitions, $i = 1, 2, \ldots, k$. Define the random variable X_i to be equal to the number of outcomes which are elements of C_i, $i = 1, 2, \ldots, k - 1$. Furthermore, let x_1, $x_2, \ldots, x_{k-1}$ be nonnegative integers so that $x_1 + x_2 + \cdots + x_{k-1} \leq n$. Then the probability that exactly x_1 terminations of the experiment are in $C_1, \ldots$, exactly x_{k-1} terminations are in C_{k-1}, and hence exactly $n - (x_1 + \cdots + x_{k-1})$ terminations are in C_k is

$$\frac{n!}{x_1! \cdots x_{k-1}!\, x_k!}\, p_1{}^{x_1} \cdots p_k{}^{x_{k-1}} p_k{}^{x_k},$$

where x_k is merely an abbreviation for $n - (x_1 + \cdots + x_{k-1})$. This is the *multinomial p.d.f.* of $k - 1$ random variables $X_1, X_2, \ldots, X_{k-1}$ of the discrete type. The moment-generating function of a multinomial distribution is given by

$$M(t_1, \ldots, t_{k-1}) = (p_1 e^{t_1} + \cdots + p_{k-1} e^{t_{k-1}} + p_k)^n$$

for all real values of $t_1, t_2, \ldots, t_{k-1}$. Thus each one-variable marginal p.d.f. is binomial, each two-variable marginal p.d.f. is trinomial, and so on.

EXERCISES

3.1. If the moment-generating function of a random variable X is $(\frac{1}{3} + \frac{2}{3} e^t)^5$, find $\Pr (X = 2 \text{ or } 3)$.

3.2. The moment-generating function of a random variable X is $(\frac{2}{3} + \frac{1}{3} e^t)^9$. Show that

$$\Pr (\mu - 2\sigma < X < \mu + 2\sigma) = \sum_{x=1}^{5} \binom{9}{x} \left(\frac{1}{3}\right)^x \left(\frac{2}{3}\right)^{9-x}.$$

3.3. If X is $b(n, p)$, show that

$$E\left(\frac{X}{n}\right) = p \quad \text{and} \quad E\left[\left(\frac{X}{n} - p\right)^2\right] = \frac{p(1 - p)}{n}.$$

3.4. Let the mutually stochastically independent random variables X_1, X_2, X_3 have the same p.d.f. $f(x) = 3x^2$, $0 < x < 1$, zero elsewhere. Find the probability that exactly two of these three variables exceed $\frac{1}{2}$.

3.5. Let Y be the number of successes in n independent repetitions of a random experiment having the probability of success $p = \frac{2}{3}$. If $n = 3$, compute $\Pr (2 \le Y)$; if $n = 5$, compute $\Pr (3 \le Y)$.

3.6. Let Y be the number of successes throughout n independent repetitions of a random experiment having probability of success $p = \frac{1}{4}$. Determine the smallest value of n so that $\Pr (1 \le Y) \ge 0.70$.

3.7. Let the stochastically independent random variables X_1 and X_2 have binomial distributions with parameters $n_1 = 3$, $p_1 = \frac{2}{3}$ and $n_2 = 4$, $p_2 = \frac{1}{2}$, respectively. Compute $\Pr (X_1 = X_2)$. *Hint.* List the four mutually exclusive ways that $X_1 = X_2$ and compute the probability of each.

3.8. Let $X_1, X_2, \ldots, X_{k-1}$ have a multinomial distribution. (a) Find the moment-generating function of $X_2, X_3, \ldots, X_{k-1}$. (b) What is the p.d.f. of $X_2, X_3, \ldots, X_{k-1}$? (c) Determine the conditional p.d.f. of X_1, given that $X_2 = x_2, \ldots, X_{k-1} = x_{k-1}$. (d) What is the conditional expectation $E(X_1 \mid x_2, \ldots, x_{k-1})$?

3.9. Let X be $b(2, p)$ and let Y be $b(4, p)$. If $\Pr(X \geq 1) = \frac{5}{9}$, find $\Pr(Y \geq 1)$.

3.10. If $x = r$ is the unique mode of a distribution that is $b(n, p)$, show that

$$(n + 1)p - 1 < r < (n + 1)p.$$

Hint. Determine the values of x for which the ratio $f(x + 1)/f(x) > 1$.

3.11. One of the numbers $1, 2, \ldots, 6$ is to be chosen by casting an unbiased die. Let this random experiment be repeated five independent times. Let the random variable X_1 be the number of terminations in the set $\{x; x = 1, 2, 3\}$ and let the random variable X_2 be the number of terminations in the set $\{x; x = 4, 5\}$. Compute $\Pr(X_1 = 2, X_2 = 1)$.

3.12. Show that the moment-generating function of the negative binomial distribution is $M(t) = p^r[1 - (1 - p)e^t]^{-r}$. Find the mean and the variance of this distribution. *Hint.* In the summation representing $M(t)$, make use of the MacLaurin's series for $(1 - w)^{-r}$.

3.13. Let X_1 and X_2 have a trinomial distribution. Differentiate the moment-generating function to show that their covariance is $-np_1p_2$.

3.14. If a fair coin is tossed at random five independent times, find the conditional probability of five heads relative to the hypothesis that there are at least four heads.

3.15. Let an unbiased die be cast at random seven independent times. Compute the conditional probability that each side appears at least once relative to the hypothesis that side 1 appears exactly twice.

3.16. Compute the measures of skewness and kurtosis of the binomial distribution $b(n, p)$.

3.17. Let

$$f(x_1, x_2) = \binom{x_1}{x_2}\left(\frac{1}{2}\right)^{x_1}\left(\frac{x_1}{15}\right), \qquad \begin{matrix} x_2 = 0, 1, \ldots, x_1, \\ x_1 = 1, 2, 3, 4, 5, \end{matrix}$$

zero elsewhere, be the joint p.d.f. of X_1 and X_2. Determine: (a) $E(X_2)$, (b) $u(x_1) = E(X_2|x_1)$, and (c) $E[u(X_1)]$. Compare the answers to parts (a) and (c). *Hint.* Note that $E(X_2) = \sum\limits_{x_1=1}^{5} \sum\limits_{x_2=0}^{x_1} x_2 f(x_1, x_2)$ and use the fact that $\sum\limits_{y=0}^{n} y\binom{n}{y}(\frac{1}{2})^n = n/2$. Why?

3.18. Three fair dice are cast. In 10 independent casts, let X be the number of times all three faces are alike and let Y be the number of times only two faces are alike. Find the joint p.d.f. of X and Y and compute $E(6XY)$.

3.2 The Poisson Distribution

Recall that the series

$$1 + m + \frac{m^2}{2!} + \frac{m^3}{3!} + \cdots = \sum_{x=0}^{\infty} \frac{m^x}{x!}$$

converge, for all values of m, to e^m. Consider the function $f(x)$ defined by

$$f(x) = \frac{m^x e^{-m}}{x!}, \qquad x = 0, 1, 2, \ldots,$$

$$= 0 \text{ elsewhere,}$$

where $m > 0$. Since $m > 0$, then $f(x) \geq 0$ and

$$\sum_x f(x) = \sum_{x=0}^{\infty} \frac{m^x e^{-m}}{x!} = e^{-m} \sum_{x=0}^{\infty} \frac{m^x}{x!} = e^{-m} e^m = 1;$$

that is, $f(x)$ satisfies the conditions of being a p.d.f. of a discrete type of random variable. A random variable that has a p.d.f. of the form $f(x)$ is said to have a *Poisson distribution,* and any such $f(x)$ is called a *Poisson p.d.f.*

Remarks. Experience indicates that the Poisson p.d.f. may be used in a number of applications with quite satisfactory results. For example, let the random variable X denote the number of alpha particles emitted by a radioactive substance that enter a prescribed region during a prescribed interval of time. With a suitable value of m, it is found that X may be assumed to have a Poisson distribution. Again let the random variable X denote the number of defects on a manufactured article, such as a refrigerator door. Upon examining many of these doors, it is found, with an appropriate value of m, that X may be said to have a Poisson distribution. The number of automobile accidents in some unit of time (or the number of insurance claims in some unit of time) is often assumed to be a random variable which has a Poisson distribution. Each of these instances can be thought of as a process that generates a number of changes (accidents, claims, etc.) in a fixed interval (of time or space and so on). If a process leads to a Poisson distribution, that process is called a *Poisson process.* Some assumptions that ensure a Poisson process will now be enumerated.

Let $g(x, w)$ denote the probability of x changes in each interval of length w. Furthermore, let the symbol $o(h)$ represent any function such that $\lim_{h \to 0} [o(h)/h] = 0$; for example, $h^2 = o(h)$ and $o(h) + o(h) = o(h)$. The Poisson postulates are the following:

(a) $g(1, h) = \lambda h + o(h)$, where λ is a positive constant and $h > 0$.

(b) $\sum\limits_{x=2}^{\infty} g(x, h) = o(h)$.

(c) The numbers of changes in nonoverlapping intervals are stochastically independent.

Postulates (a) and (c) state, in effect, that the probability of one change in a short interval h is independent of changes in other nonoverlapping intervals and is approximately proportional to the length of the interval. The substance of (b) is that the probability of two or more changes in the same short interval h is essentially equal to zero. If $x = 0$, we take $g(0, 0) = 1$. In accordance with postulates (a) and (b), the probability of at least one change in an interval of length h is $\lambda h + o(h) + o(h) = \lambda h + o(h)$. Hence the probability of zero changes in this interval of length h is $1 - \lambda h - o(h)$. Thus the probability $g(0, w + h)$ of zero changes in an interval of length $w + h$ is, in accordance with postulate (c), equal to the product of the probability $g(0, w)$ of zero changes in an interval of length w and the probability $[1 - \lambda h - o(h)]$ of zero changes in a nonoverlapping interval of length h. That is,

$$g(0, w + h) = g(0, w)[1 - \lambda h - o(h)].$$

Then

$$\frac{g(0, w + h) - g(0, w)}{h} = -\lambda g(0, w) - \frac{o(h)g(0, w)}{h}.$$

If we take the limit as $h \to 0$, we have

$$D_w[g(0, w)] = -\lambda g(0, w).$$

The solution of this differential equation is

$$g(0, w) = ce^{-\lambda w}.$$

The condition $g(0, 0) = 1$ implies that $c = 1$; so

$$g(0, w) = e^{-\lambda w}.$$

If x is a positive integer, we take $g(x, 0) = 0$. The postulates imply that

$$g(x, w + h) = [g(x, w)][1 - \lambda h - o(h)] + [g(x - 1, w)][\lambda h + o(h)] + o(h).$$

Accordingly, we have

$$\frac{g(x, w + h) - g(x, w)}{h} = -\lambda g(x, w) + \lambda g(x - 1, w) + \frac{o(h)}{h}$$

and

$$D_w[g(x, w)] = -\lambda g(x, w) + \lambda g(x - 1, w),$$

for $x = 1, 2, 3, \ldots$. It can be shown, by mathematical induction, that the solutions to these differential equations, with boundary conditions $g(x, 0) = 0$ for $x = 1, 2, 3, \ldots$, are, respectively,

$$g(x, w) = \frac{(\lambda w)^x e^{-\lambda w}}{x!}, \qquad x = 1, 2, 3, \ldots.$$

Hence the number of changes X in an interval of length w has a Poisson distribution with parameter $m = \lambda w$.

The moment-generating function of a Poisson distribution is given by

$$M(t) = \sum_x e^{tx} f(x) = \sum_{x=0}^{\infty} e^{tx} \frac{m^x e^{-m}}{x!}$$

$$= e^{-m} \sum_{x=0}^{\infty} \frac{(me^t)^x}{x!}$$

$$= e^{-m} e^{me^t} = e^{m(e^t - 1)}$$

for all real values of t. Since

$$M'(t) = e^{m(e^t - 1)}(me^t)$$

and

$$M''(t) = e^{m(e^t - 1)}(me^t) + e^{m(e^t - 1)}(me^t)^2,$$

then

$$\mu = M'(0) = m$$

and

$$\sigma^2 = M''(0) - \mu^2 = m + m^2 - m^2 = m.$$

That is, a Poisson distribution has $\mu = \sigma^2 = m > 0$. On this account, a Poisson p.d.f. is frequently written

$$f(x) = \frac{\mu^x e^{-\mu}}{x!}, \qquad x = 0, 1, 2, \ldots,$$

$$= 0 \text{ elsewhere.}$$

Thus the parameter m in a Poisson p.d.f. is the mean μ. Table I in Appendix B gives approximately the distribution function of the Poisson distribution for various values of the parameter $m = \mu$.

Example 1. Suppose that X has a Poisson distribution with $\mu = 2$. Then the p.d.f. of X is

$$f(x) = \frac{2^x e^{-2}}{x!}, \qquad x = 0, 1, 2, \ldots,$$

$$= 0 \text{ elsewhere.}$$

The variance of this distribution is $\sigma^2 = \mu = 2$. If we wish to compute $\Pr(1 \le X)$, we have

$$\Pr(1 \le X) = 1 - \Pr(X = 0)$$
$$= 1 - f(0) = 1 - e^{-2} = 0.865,$$

approximately, by Table I of Appendix B.

Example 2. If the moment-generating function of a random variable X is

$$M(t) = e^{4(e^t - 1)},$$

then X has a Poisson distribution with $\mu = 4$. Accordingly, by way of example,

$$\Pr(X = 3) = \frac{4^3 e^{-4}}{3!} = \frac{32}{3} e^{-4};$$

or, by Table I,

$$\Pr(X = 3) = \Pr(X \le 3) - \Pr(X \le 2) = 0.433 - 0.238 = 0.195.$$

Example 3. Let the probability of exactly one blemish in 1 foot of wire be about $\frac{1}{1000}$ and let the probability of two or more blemishes in that length be, for all practical purposes, zero. Let the random variable X be the number of blemishes in 3000 feet of wire. If we assume the stochastic independence of the numbers of blemishes in nonoverlapping intervals, then the postulates of the Poisson process are approximated, with $\lambda = \frac{1}{1000}$ and $w = 3000$. Thus X has an approximate Poisson distribution with mean $3000(\frac{1}{1000}) = 3$. For example, the probability that there are exactly five blemishes in 3000 feet of wire is

$$\Pr(X = 5) = \frac{3^5 e^{-3}}{5!}$$

and, by Table I,

$$\Pr(X = 5) = \Pr(X \le 5) - \Pr(X \le 4) = 0.101,$$

approximately.

EXERCISES

3.19. If the random variable X has a Poisson distribution such that $\Pr(X = 1) = \Pr(X = 2)$, find $\Pr(X = 4)$.

3.20. The moment-generating function of a random variable X is $e^{4(e^t - 1)}$. Show that $\Pr(\mu - 2\sigma < X < \mu + 2\sigma) = 0.931$.

3.21. In a lengthy manuscript, it is discovered that only 13.5 per cent of the pages contain no typing errors. If we assume that the number of errors per page is a random variable with a Poisson distribution, find the percentage of pages that have exactly one error.

3.22. Let the p.d.f. $f(x)$ be positive on and only on the nonnegative integers. Given that $f(x) = (4/x)f(x - 1)$, $x = 1, 2, 3, \ldots$. Find $f(x)$. *Hint.* Note that $f(1) = 4f(0), f(2) = (4^2/2!)f(0)$, and so on. That is, find each $f(x)$ in terms of $f(0)$ and then determine $f(0)$ from $1 = f(0) + f(1) + f(2) + \cdots$.

3.23. Let X have a Poisson distribution with $\mu = 100$. Use Chebyshev's inequality to determine a lower bound for $\Pr(75 < X < 125)$.

3.24. Given that $g(x, 0) = 0$ and that

$$D_w[g(x, w)] = -\lambda g(x, w) + \lambda g(x - 1, w)$$

for $x = 1, 2, 3, \ldots$. If $g(0, w) = e^{-\lambda w}$, show, by mathematical induction, that

$$g(x, w) = \frac{(\lambda w)^x e^{-\lambda w}}{x!}, \qquad x = 1, 2, 3, \ldots.$$

3.25. Let the number of chocolate drops in a certain type of cookie have a Poisson distribution. We want the probability that a cookie of this type contains at least two chocolate drops to be greater than 0.99. Find the smallest value that the mean of the distribution can take.

3.26. Compute the measures of skewness and kurtosis of the Poisson distribution with mean μ.

3.27. Let X and Y have the joint p.d.f. $f(x, y) = e^{-2}/[x! \, (y - x)!]$, $y = 0, 1, 2, \ldots; x = 0, 1, \ldots, y$, zero elsewhere.

(a) Find the moment-generating function $M(t_1, t_2)$ of this joint distribution.

(b) Compute the means, the variances, and the correlation coefficient of X and Y.

(c) Determine the conditional mean $E(X|y)$. *Hint.* Note that

$$\sum_{x=0}^{y} [\exp(t_1 x)]y!/[x! \, (y - x)!] = [1 + \exp(t_1)]^y.$$

Why?

3.3 The Gamma and Chi-Square Distributions

In this section we introduce the gamma and chi-square distributions. It is proved in books on advanced calculus that the integral

$$\int_0^\infty y^{\alpha - 1} e^{-y} \, dy$$

exists for $\alpha > 0$ and that the value of the integral is a positive number. The integral is called the gamma function of α, and we write

$$\Gamma(\alpha) = \int_0^\infty y^{\alpha-1}e^{-y}\,dy.$$

If $\alpha = 1$, clearly

$$\Gamma(1) = \int_0^\infty e^{-y}\,dy = 1.$$

If $\alpha > 1$, an integration by parts shows that

$$\Gamma(\alpha) = (\alpha - 1)\int_0^\infty y^{\alpha-2}e^{-y}\,dy = (\alpha - 1)\Gamma(\alpha - 1).$$

Accordingly, if α is a positive integer greater than 1,

$$\Gamma(\alpha) = (\alpha - 1)(\alpha - 2)\cdots(3)(2)(1)\Gamma(1) = (\alpha - 1)!.$$

Since $\Gamma(1) = 1$, this suggests that we take $0! = 1$, as we have done.

In the integral that defines $\Gamma(\alpha)$, let us introduce a new variable x by writing $y = x/\beta$, where $\beta > 0$. Then

$$\Gamma(\alpha) = \int_0^\infty \left(\frac{x}{\beta}\right)^{\alpha-1} e^{-x/\beta}\left(\frac{1}{\beta}\right)dx,$$

or, equivalently,

$$1 = \int_0^\infty \frac{1}{\Gamma(\alpha)\beta^\alpha}\,x^{\alpha-1}e^{-x/\beta}\,dx.$$

Since $\alpha > 0$, $\beta > 0$, and $\Gamma(\alpha) > 0$, we see that

$$f(x) = \frac{1}{\Gamma(\alpha)\beta^\alpha}\,x^{\alpha-1}e^{-x/\beta}, \qquad 0 < x < \infty,$$

$$= 0 \text{ elsewhere,}$$

is a p.d.f. of a random variable of the continuous type. A random variable X that has a p.d.f. of this form is said to have a *gamma distribution* with parameters α and β; and any such $f(x)$ is called a *gamma-type p.d.f.*

Remark. The gamma distribution is frequently the probability model for waiting times; for instances, in life testing, the waiting time until "death" is the random variable which frequently has a gamma distribution. To see this, let us assume the postulates of a Poisson process and let the interval of length w be a time interval. Specifically, let the random variable W be the time that is needed to obtain exactly k changes (possibly deaths), where k is a fixed positive integer. Then the distribution function of W is

$$G(w) = \Pr\,(W \le w) = 1 - \Pr\,(W > w).$$

However, the event $W > w$, for $w > 0$, is equivalent to the event in which there are less than k changes in a time interval of length w. That is, if the random variable X is the number of changes in an interval of length w, then

$$\Pr(W > w) = \sum_{x=0}^{k-1} \Pr(X = x) = \sum_{x=0}^{k-1} \frac{(\lambda w)^x e^{-\lambda w}}{x!}.$$

It is left as an exercise to verify that

$$\int_{\lambda w}^{\infty} \frac{z^{k-1} e^{-z}}{(k-1)!} \, dz = \sum_{x=0}^{k-1} \frac{(\lambda w)^x e^{-\lambda w}}{x!}.$$

If, momentarily, we accept this result, we have, for $w > 0$,

$$G(w) = 1 - \int_{\lambda w}^{\infty} \frac{z^{k-1} e^{-z}}{\Gamma(k)} \, dz = \int_{0}^{\lambda w} \frac{z^{k-1} e^{-z}}{\Gamma(k)} \, dz,$$

and for $w \le 0$, $G(w) = 0$. If we change the variable of integration in the integral that defines $G(w)$ by writing $z = \lambda y$, then

$$G(w) = \int_{0}^{w} \frac{\lambda^k y^{k-1} e^{-\lambda y}}{\Gamma(k)} \, dy, \qquad w > 0,$$

and $G(w) = 0$, $w \le 0$. Accordingly, the p.d.f. of W is

$$g(w) = G'(w) = \frac{\lambda^k w^{k-1} e^{-\lambda w}}{\Gamma(k)}, \qquad 0 < w < \infty,$$

$$= 0 \text{ elsewhere.}$$

That is, W has a gamma distribution with $\alpha = k$ and $\beta = 1/\lambda$. If W is the waiting time until the first change, that is, if $k = 1$, the p.d.f. of W is

$$g(w) = \lambda e^{-\lambda w}, \qquad 0 < w < \infty,$$

$$= 0 \text{ elsewhere,}$$

and W is said to have an *exponential distribution*.

We now find the moment-generating function of a gamma distribution. Since

$$M(t) = \int_{0}^{\infty} e^{tx} \frac{1}{\Gamma(\alpha)\beta^\alpha} x^{\alpha-1} e^{-x/\beta} \, dx$$

$$= \int_{0}^{\infty} \frac{1}{\Gamma(\alpha)\beta^\alpha} x^{\alpha-1} e^{-x(1-\beta t)/\beta} \, dx,$$

we may set $y = x(1 - \beta t)/\beta$, $t < 1/\beta$, or $x = \beta y/(1 - \beta t)$, to obtain

$$M(t) = \int_0^\infty \frac{\beta/(1 - \beta t)}{\Gamma(\alpha)\beta^\alpha} \left(\frac{\beta y}{1 - \beta t}\right)^{\alpha - 1} e^{-y}\, dy.$$

That is,

$$M(t) = \left(\frac{1}{1 - \beta t}\right)^\alpha \int_0^\infty \frac{1}{\Gamma(\alpha)} y^{\alpha - 1} e^{-y}\, dy$$

$$= \frac{1}{(1 - \beta t)^\alpha}, \qquad t < \frac{1}{\beta}.$$

Now

$$M'(t) = (-\alpha)(1 - \beta t)^{-\alpha - 1}(-\beta)$$

and

$$M''(t) = (-\alpha)(-\alpha - 1)(1 - \beta t)^{-\alpha - 2}(-\beta)^2.$$

Hence, for a gamma distribution, we have

$$\mu = M'(0) = \alpha\beta$$

and

$$\sigma^2 = M''(0) - \mu^2 = \alpha(\alpha + 1)\beta^2 - \alpha^2\beta^2 = \alpha\beta^2.$$

Example 1. Let the waiting time W have a gamma p.d.f. with $\alpha = k$ and $\beta = 1/\lambda$. Accordingly, $E(W) = k/\lambda$. If $k = 1$, then $E(W) = 1/\lambda$; that is, the expected waiting time for $k = 1$ changes is equal to the reciprocal of λ.

Example 2. Let X be a random variable such that

$$E(X^m) = \frac{(m + 3)!}{3!} 3^m, \qquad m = 1, 2, 3, \ldots.$$

Then the moment-generating function of X is given by the series

$$M(t) = 1 + \frac{4!\ 3}{3!\ 1!} t + \frac{5!\ 3^2}{3!\ 2!} t^2 + \frac{6!\ 3^3}{3!\ 3!} t^3 + \cdots.$$

This, however, is the Maclaurin's series for $(1 - 3t)^{-4}$, provided that $-1 < 3t < 1$. Accordingly, X has a gamma distribution with $\alpha = 4$ and $\beta = 3$.

Remark. The gamma distribution is not only a good model for waiting times, but one for many nonnegative random variables of the continuous type. For illustrations, the distribution of certain incomes could be modeled satisfactorily by the gamma distribution, since the two parameters α and β provide a great deal of flexibility.

Let us now consider the special case of the gamma distribution in which $\alpha = r/2$, where r is a positive integer, and $\beta = 2$. A random variable X of the continuous type that has the p.d.f.

$$f(x) = \frac{1}{\Gamma(r/2)2^{r/2}} x^{r/2-1} e^{-x/2}, \qquad 0 < x < \infty,$$

$$= 0 \text{ elsewhere,}$$

and the moment-generating function

$$M(t) = (1 - 2t)^{-r/2}, \qquad t < \tfrac{1}{2},$$

is said to have a *chi-square distribution*, and any $f(x)$ of this form is called a *chi-square p.d.f.* The mean and the variance of a chi-square distribution are $\mu = \alpha\beta = (r/2)2 = r$ and $\sigma^2 = \alpha\beta^2 = (r/2)2^2 = 2r$, respectively. For no obvious reason, we call the parameter r the number of degrees of freedom of the chi-square distribution (or of the chi-square p.d.f.). Because the chi-square distribution has an important role in statistics and occurs so frequently, we write, for brevity, that X is $\chi^2(r)$ to mean that the random variable X has a chi-square distribution with r degrees of freedom.

Example 3. If X has the p.d.f.

$$f(x) = \tfrac{1}{4}xe^{-x/2}, \qquad 0 < x < \infty,$$

$$= 0 \text{ elsewhere,}$$

then X is $\chi^2(4)$. Hence $\mu = 4$, $\sigma^2 = 8$, and $M(t) = (1 - 2t)^{-2}$, $t < \tfrac{1}{2}$.

Example 4. If X has the moment-generating function $M(t) = (1 - 2t)^{-8}$, $t < \tfrac{1}{2}$, then X is $\chi^2(16)$.

If the random variable X is $\chi^2(r)$, then, with $c_1 < c_2$, we have

$$\Pr(c_1 \le X \le c_2) = \Pr(X \le c_2) - \Pr(X \le c_1),$$

since $\Pr(X = c_1) = 0$. To compute such a probability, we need the value of an integral like

$$\Pr(X \le x) = \int_0^x \frac{1}{\Gamma(r/2)2^{r/2}} w^{r/2-1} e^{-w/2} \, dw.$$

Tables of this integral for selected values of r and x have been prepared and are partially reproduced in Table II in Appendix B.

Example 5. Let X be $\chi^2(10)$. Then, by Table II of Appendix B, with $r = 10$,

$$\Pr(3.25 \le X \le 20.5) = \Pr(X \le 20.5) - \Pr(X \le 3.25)$$

$$= 0.975 - 0.025 = 0.95.$$

Again, by way of example, if $\Pr(a < X) = 0.05$, then $\Pr(X \le a) = 0.95$, and thus $a = 18.3$ from Table II with $r = 10$.

Example 6. Let X have a gamma distribution with $\alpha = r/2$, where r is a positive integer, and $\beta > 0$. Define the random variable $Y = 2X/\beta$. We seek the p.d.f. of Y. Now the distribution function of Y is

$$G(y) = \Pr(Y \le y) = \Pr\left(X \le \frac{\beta y}{2}\right).$$

If $y \le 0$, then $G(y) = 0$; but if $y > 0$, then

$$G(y) = \int_0^{\beta y/2} \frac{1}{\Gamma(r/2)\beta^{r/2}} x^{r/2-1} e^{-x/\beta}\, dx.$$

Accordingly, the p.d.f. of Y is

$$g(y) = G'(y) = \frac{\beta/2}{\Gamma(r/2)\beta^{r/2}} (\beta y/2)^{r/2-1} e^{-y/2}$$

$$= \frac{1}{\Gamma(r/2)2^{r/2}} y^{r/2-1} e^{-y/2}$$

if $y > 0$. That is, Y is $\chi^2(r)$.

EXERCISES

3.28. If $(1 - 2t)^{-6}$, $t < \frac{1}{2}$, is the moment-generating function of the random variable X, find $\Pr(X < 5.23)$.

3.29. If X is $\chi^2(5)$, determine the constants c and d so that $\Pr(c < X < d) = 0.95$ and $\Pr(X < c) = 0.025$.

3.30. If X has a gamma distribution with $\alpha = 3$ and $\beta = 4$, find $\Pr(3.28 < X < 25.2)$. *Hint.* Consider the probability of the equivalent event $1.64 < Y < 12.6$, where $Y = 2X/4 = X/2$.

3.31. Let X be a random variable such that $E(X^m) = (m + 1)! \, 2^m$, $m = 1, 2, 3, \ldots$. Determine the distribution of X.

3.32. Show that

$$\int_\mu^\infty \frac{1}{\Gamma(k)} z^{k-1} e^{-z}\, dz = \sum_{x=0}^{k-1} \frac{\mu^x e^{-\mu}}{x!}, \qquad k = 1, 2, 3, \ldots.$$

This demonstrates the relationship between the distribution functions of the gamma and Poisson distributions. *Hint.* Either integrate by parts $k - 1$ times or simply note that the "antiderivative" of $z^{k-1}e^{-z}$ is $-z^{k-1}e^{-z} - (k-1)z^{k-2}e^{-z} - \cdots - (k-1)! \, e^{-z}$ by differentiating the latter expression.

3.33. Let X_1, X_2, and X_3 be mutually stochastically independent random variables, each with p.d.f. $f(x) = e^{-x}$, $0 < x < \infty$, zero elsewhere. Find the distribution of Y = minimum (X_1, X_2, X_3). *Hint.* Pr $(Y \leq y) = 1 -$ Pr $(Y > y) = 1 - $ Pr $(X_i > y, i = 1, 2, 3)$.

3.34. Let X have a gamma distribution with p.d.f.

$$f(x) = \frac{1}{\beta^2} x e^{-x/\beta}, \qquad 0 < x < \infty,$$

zero elsewhere. If $x = 2$ is the unique mode of the distribution, find the parameter β and Pr $(X < 9.49)$.

3.35. Compute the measures of skewness and kurtosis of a gamma distribution with parameters α and β.

3.36. Let X have a gamma distribution with parameters α and β. Show that Pr $(X \geq 2\alpha\beta) \leq (2/e)^\alpha$. *Hint.* Use the result of Exercise 1.107.

3.37. Give a reasonable definition of a chi-square distribution with zero degrees of freedom. *Hint.* Work with the moment-generating function of a distribution that is $\chi^2(r)$ and let $r = 0$.

3.38. In the Poisson postulates on page 99, let λ be a nonnegative function of w, say $\lambda(w)$, such that $D_w[g(0, w)] = -\lambda(w)g(0, w)$. Suppose that $\lambda(w) = krw^{r-1}$, $r \geq 1$. (a) Find $g(0, w)$ noting that $g(0, 0) = 1$. (b) Let W be the time that is needed to obtain exactly one change. Find the distribution function of W, namely $G(w) = $ Pr $(W \leq w) = 1 - $ Pr $(W > w) = 1 - g(0, w)$, $0 \leq w$, and then find the p.d.f. of W. This p.d.f. is that of the *Weibull distribution*, which is used in the study of breaking strengths of materials.

3.39. Let X have a Poisson distribution with parameter m. If m is an experimental value of a random variable having a gamma distribution with $\alpha = 2$ and $\beta = 1$, compute Pr $(X = 0, 1, 2)$.

3.40. Let X have the uniform distribution with p.d.f. $f(x) = 1, 0 < x < 1$, zero elsewhere. Find the distribution function of $Y = -2 \ln X$. What is the p.d.f. of Y?

3.4 The Normal Distribution

Consider the integral

$$I = \int_{-\infty}^{\infty} \exp(-y^2/2) \, dy.$$

This integral exists because the integrand is a positive continuous function which is bounded by an integrable function; that is,

$$0 < \exp(-y^2/2) < \exp(-|y| + 1), \qquad -\infty < y < \infty,$$

and

$$\int_{-\infty}^{\infty} \exp(-|y| + 1) \, dy = 2e.$$

To evaluate the integral I, we note that $I > 0$ and that I^2 may be written

$$I^2 = \int_{-\infty}^{\infty} \int_{-\infty}^{\infty} \exp\left(-\frac{y^2 + z^2}{2}\right) dy\, dz.$$

This iterated integral can be evaluated by changing to polar coordinates. If we set $y = r \cos \theta$ and $z = r \sin \theta$, we have

$$I^2 = \int_0^{2\pi} \int_0^{\infty} e^{-r^2/2} r\, dr\, d\theta$$

$$= \int_0^{2\pi} d\theta = 2\pi.$$

Accordingly, $I = \sqrt{2\pi}$ and

$$\int_{-\infty}^{\infty} \frac{1}{\sqrt{2\pi}} e^{-y^2/2}\, dy = 1.$$

If we introduce a new variable of integration, say x, by writing

$$y = \frac{x - a}{b}, \qquad b > 0,$$

the preceding integral becomes

$$\int_{-\infty}^{\infty} \frac{1}{b\sqrt{2\pi}} \exp\left[-\frac{(x - a)^2}{2b^2}\right] dx = 1.$$

Since $b > 0$, this implies that

$$f(x) = \frac{1}{b\sqrt{2\pi}} \exp\left[-\frac{(x - a)^2}{2b^2}\right], \qquad -\infty < x < \infty$$

satisfies the conditions of being a p.d.f. of a continuous type of random variable. A random variable of the continuous type that has a p.d.f. of the form of $f(x)$ is said to have a *normal distribution*, and any $f(x)$ of this form is called a normal p.d.f.

We can find the moment-generating function of a normal distribution as follows. In

$$M(t) = \int_{-\infty}^{\infty} e^{tx} \frac{1}{b\sqrt{2\pi}} \exp\left[-\frac{(x - a)^2}{2b^2}\right] dx$$

$$= \int_{-\infty}^{\infty} \frac{1}{b\sqrt{2\pi}} \exp\left(-\frac{-2b^2tx + x^2 - 2ax + a^2}{2b^2}\right) dx$$

we complete the square in the exponent. Thus $M(t)$ becomes

$$M(t) = \exp\left[-\frac{a^2 - (a + b^2t)^2}{2b^2}\right] \int_{-\infty}^{\infty} \frac{1}{b\sqrt{2\pi}} \exp\left[-\frac{(x - a - b^2t)^2}{2b^2}\right] dx$$

$$= \exp\left(at + \frac{b^2t^2}{2}\right)$$

because the integrand of the last integral can be thought of as a normal p.d.f. with a replaced by $a + b^2t$, and hence it is equal to 1.

The mean μ and variance σ^2 of a normal distribution will be calculated from $M(t)$. Now

$$M'(t) = M(t)(a + b^2t)$$

and

$$M''(t) = M(t)(b^2) + M(t)(a + b^2t)^2.$$

Thus

$$\mu = M'(0) = a$$

and

$$\sigma^2 = M''(0) - \mu^2 = b^2 + a^2 - a^2 = b^2.$$

This permits us to write a normal p.d.f. in the form of

$$f(x) = \frac{1}{\sigma\sqrt{2\pi}} \exp\left[-\frac{(x - \mu)^2}{2\sigma^2}\right], \qquad -\infty < x < \infty,$$

a form that shows explicitly the values of μ and σ^2. The moment-generating function $M(t)$ can be written

$$M(t) = \exp\left(\mu t + \frac{\sigma^2 t^2}{2}\right).$$

Example 1. If X has the moment-generating function

$$M(t) = e^{2t + 32t^2},$$

then X has a normal distribution with $\mu = 2$, $\sigma^2 = 64$.

The normal p.d.f. occurs so frequently in certain parts of statistics that we denote it, for brevity, by $n(\mu, \sigma^2)$. Thus, if we say that the random variable X is $n(0, 1)$, we mean that X has a normal distribution with mean $\mu = 0$ and variance $\sigma^2 = 1$, so that the p.d.f. of X is

$$f(x) = \frac{1}{\sqrt{2\pi}} e^{-x^2/2}, \qquad -\infty < x < \infty.$$

If we say that X is $n(5, 4)$, we mean that X has a normal distribution with mean $\mu = 5$ and variance $\sigma^2 = 4$, so that the p.d.f. of X is

$$f(x) = \frac{1}{2\sqrt{2\pi}} \exp\left[-\frac{(x-5)^2}{2(4)}\right], \qquad -\infty < x < \infty.$$

Moreover, if

$$M(t) = e^{t^2/2},$$

then X is $n(0, 1)$.

The graph of

$$f(x) = \frac{1}{\sigma\sqrt{2\pi}} \exp\left[-\frac{(x-\mu)^2}{2\sigma^2}\right], \qquad -\infty < x < \infty,$$

is seen (1) to be symmetric about a vertical axis through $x = \mu$, (2) to have its maximum of $1/\sigma\sqrt{2\pi}$ at $x = \mu$, and (3) to have the x-axis as a horizontal asymptote. It should be verified that (4) there are points of inflection at $x = \mu \pm \sigma$.

Remark. Each of the special distributions considered thus far has been "justified" by some derivation that is based upon certain concepts found in elementary probability theory. Such a motivation for the normal distribution is not given at this time; a motivation is presented in Chapter 5. However, the normal distribution is one of the more widely used distributions in applications of statistical methods. Variables that are often assumed to be random variables having normal distributions (with appropriate values of μ and σ) are the diameter of a hole made by a drill press, the score on a test, the yield of a grain on a plot of ground, and the length of a newborn child.

We now prove a very useful theorem.

Theorem 1. *If the random variable X is $n(\mu, \sigma^2)$, $\sigma^2 > 0$, then the random variable $W = (X - \mu)/\sigma$ is $n(0, 1)$.*

Proof. The distribution function $G(w)$ of W is, since $\sigma > 0$,

$$G(w) = \Pr\left(\frac{X-\mu}{\sigma} \leq w\right) = \Pr\left(X \leq w\sigma + \mu\right).$$

That is,

$$G(w) = \int_{-\infty}^{w\sigma+\mu} \frac{1}{\sigma\sqrt{2\pi}} \exp\left[-\frac{(x-\mu)^2}{2\sigma^2}\right] dx.$$

If we change the variable of integration by writing $y = (x - \mu)/\sigma$, then

$$G(w) = \int_{-\infty}^{w} \frac{1}{\sqrt{2\pi}} e^{-y^2/2} \, dy.$$

Accordingly, the p.d.f. $g(w) = G'(w)$ of the continuous-type random variable W is

$$g(w) = \frac{1}{\sqrt{2\pi}} e^{-w^2/2}, \qquad -\infty < w < \infty.$$

Thus W is $n(0, 1)$, which is the desired result.

This fact considerably simplifies calculations of probabilities concerning normally distributed variables, as will be seen presently. Suppose that X is $n(\mu, \sigma^2)$. Then, with $c_1 < c_2$ we have, since $\Pr(X = c_1) = 0$,

$$\Pr(c_1 < X < c_2) = \Pr(X < c_2) - \Pr(X < c_1)$$

$$= \Pr\left(\frac{X - \mu}{\sigma} < \frac{c_2 - \mu}{\sigma}\right) - \Pr\left(\frac{X - \mu}{\sigma} < \frac{c_1 - \mu}{\sigma}\right)$$

$$= \int_{-\infty}^{(c_2 - \mu)/\sigma} \frac{1}{\sqrt{2\pi}} e^{-w^2/2} \, dw - \int_{-\infty}^{(c_1 - \mu)/\sigma} \frac{1}{\sqrt{2\pi}} e^{-w^2/2} \, dw$$

because $W = (X - \mu)/\sigma$ is $n(0, 1)$. That is, probabilities concerning X, which is $n(\mu, \sigma^2)$, can be expressed in terms of probabilities concerning W, which is $n(0, 1)$. However, an integral such as

$$\int_{-\infty}^{k} \frac{1}{\sqrt{2\pi}} e^{-w^2/2} \, dw$$

cannot be evaluated by the fundamental theorem of calculus because an "antiderivative" of $e^{-w^2/2}$ is not expressible as an elementary function. Instead, tables of the approximate value of this integral for various values of k have been prepared and are partially reproduced in Table III in Appendix B. We use the notation (for normal)

$$N(x) = \int_{-\infty}^{x} \frac{1}{\sqrt{2\pi}} e^{-w^2/2} \, dw;$$

thus, if X is $n(\mu, \sigma^2)$, then

$$\Pr(c_1 < X < c_2) = \Pr\left(\frac{X - \mu}{\sigma} < \frac{c_2 - \mu}{\sigma}\right) - \Pr\left(\frac{X - \mu}{\sigma} < \frac{c_1 - \mu}{\sigma}\right)$$

$$= N\left(\frac{c_2 - \mu}{\sigma}\right) - N\left(\frac{c_1 - \mu}{\sigma}\right).$$

It is left as an exercise to show that $N(-x) = 1 - N(x)$.

Example 2. Let X be $n(2, 25)$. Then, by Table III,

$$\Pr(0 < X < 10) = N\left(\frac{10 - 2}{5}\right) - N\left(\frac{0 - 2}{5}\right)$$

$$= N(1.6) - N(-0.4)$$

$$= 0.945 - (1 - 0.655) = 0.600$$

and

$$\Pr(-8 < X < 1) = N\left(\frac{1 - 2}{5}\right) - N\left(\frac{-8 - 2}{5}\right)$$

$$= N(-0.2) - N(-2)$$

$$= (1 - 0.579) - (1 - 0.977) = 0.398.$$

Example 3. Let X be $n(\mu, \sigma^2)$. Then, by Table III,

$$\Pr(\mu - 2\sigma < X < \mu + 2\sigma) = N\left(\frac{\mu + 2\sigma - \mu}{\sigma}\right) - N\left(\frac{\mu - 2\sigma - \mu}{\sigma}\right)$$

$$= N(2) - N(-2)$$

$$= 0.977 - (1 - 0.977) = 0.954.$$

Example 4. Suppose that 10 per cent of the probability for a certain distribution that is $n(\mu, \sigma^2)$ is below 60 and that 5 per cent is above 90. What are the values of μ and σ? We are given that the random variable X is $n(\mu, \sigma^2)$ and that $\Pr(X \le 60) = 0.10$ and $\Pr(X \le 90) = 0.95$. Thus $N[(60 - \mu)/\sigma] = 0.10$ and $N[(90 - \mu)/\sigma] = 0.95$. From Table III we have

$$\frac{60 - \mu}{\sigma} = -1.282, \qquad \frac{90 - \mu}{\sigma} = 1.645.$$

These conditions require that $\mu = 73.1$ and $\sigma = 10.2$ approximately.

We close this section with an important theorem.

Theorem 2. *If the random variable X is $n(\mu, \sigma^2)$, $\sigma^2 > 0$, then the random variable $V = (X - \mu)^2/\sigma^2$ is $\chi^2(1)$.*

Proof. Because $V = W^2$, where $W = (X - \mu)/\sigma$ is $n(0, 1)$, the distribution function $G(v)$ of V is, for $v \ge 0$,

$$G(v) = \Pr(W^2 \le v) = \Pr(-\sqrt{v} \le W \le \sqrt{v}).$$

That is,

$$G(v) = 2 \int_0^{\sqrt{v}} \frac{1}{\sqrt{2\pi}} e^{-w^2/2} \, dw, \qquad 0 \le v,$$

and

$$G(v) = 0, \qquad v < 0.$$

If we change the variable of integration by writing $w = \sqrt{y}$, then

$$G(v) = \int_0^v \frac{1}{\sqrt{2\pi}\sqrt{y}} e^{-y/2} \, dy, \qquad 0 \le v.$$

Hence the p.d.f. $g(v) = G'(v)$ of the continuous-type random variable V is

$$g(v) = \frac{1}{\sqrt{\pi}\sqrt{2}} v^{1/2 - 1} e^{-v/2}, \qquad 0 < v < \infty,$$

$$= 0 \text{ elsewhere.}$$

Since $g(v)$ is a p.d.f. and hence

$$\int_0^\infty g(v) \, dv = 1,$$

it must be that $\Gamma(\tfrac{1}{2}) = \sqrt{\pi}$ and thus V is $\chi^2(1)$.

EXERCISES

3.41. If

$$N(x) = \int_{-\infty}^x \frac{1}{\sqrt{2\pi}} e^{-w^2/2} \, dw,$$

show that $N(-x) = 1 - N(x)$.

3.42. If X is $n(75, 100)$, find Pr $(X < 60)$ and Pr $(70 < X < 100)$.

3.43. If X is $n(\mu, \sigma^2)$, find b so that Pr $[-b < (X - \mu)/\sigma < b] = 0.90$.

3.44. Let X be $n(\mu, \sigma^2)$ so that Pr $(X < 89) = 0.90$ and Pr $(X < 94) = 0.95$. Find μ and σ^2.

3.45. Show that the constant c can be selected so that $f(x) = c2^{-x^2}$, $-\infty < x < \infty$, satisfies the conditions of a normal p.d.f. *Hint.* Write $2 = e^{\ln 2}$.

3.46. If X is $n(\mu, \sigma^2)$, show that $E(|X - \mu|) = \sigma\sqrt{2/\pi}$.

3.47. Show that the graph of a p.d.f. $n(\mu, \sigma^2)$ has points of inflection at $x = \mu - \sigma$ and $x = \mu + \sigma$.

3.48. Determine the ninetieth percentile of the distribution, which is $n(65, 25)$.

3.49. If $e^{3t + 8t^2}$ is the moment-generating function of the random variable X, find $\Pr(-1 < X < 9)$.

3.50. Let the random variable X have the p.d.f.

$$f(x) = \frac{2}{\sqrt{2\pi}} e^{-x^2/2}, \qquad 0 < x < \infty, \text{ zero elsewhere.}$$

Find the mean and variance of X. *Hint.* Compute $E(X)$ directly and $E(X^2)$ by comparing that integral with the integral representing the variance of a variable that is $n(0, 1)$.

3.51. Let X be $n(5, 10)$. Find $\Pr[0.04 < (X - 5)^2 < 38.4]$.

3.52. If X is $n(1, 4)$, compute the probability $\Pr(1 < X^2 < 9)$.

3.53. If X is $n(75, 25)$, find the conditional probability that X is greater than 80 relative to the hypothesis that X is greater than 77. See Exercise 2.17.

3.54. Let X be a random variable such that $E(X^{2m}) = (2m)!/(2^m m!)$, $m = 1, 2, 3, \ldots$ and $E(X^{2m-1}) = 0$, $m = 1, 2, 3, \ldots$. Find the moment-generating function and the p.d.f. of X.

3.55. Let the mutually stochastically independent random variables X_1, X_2, and X_3 be $n(0, 1)$, $n(2, 4)$, and $n(-1, 1)$, respectively. Compute the probability that exactly two of these three variables are less than zero.

3.56. Compute the measures of skewness and kurtosis of a distribution which is $n(\mu, \sigma^2)$.

3.57. Let the random variable X have a distribution that is $n(\mu, \sigma^2)$.
(a) Does the random variable $Y = X^2$ also have a normal distribution?
(b) Would the random variable $Y = aX + b$, a and b nonzero constants, have a normal distribution? *Hint.* In each case, first determine $\Pr(Y \leq y)$.

3.58. Let the random variable X be $n(\mu, \sigma^2)$. What would this distribution be if $\sigma^2 = 0$? *Hint.* Look at the moment-generating function of X for $\sigma^2 > 0$ and investigate its limit as $\sigma^2 \to 0$.

3.59. Let $n(x)$ and $N(x)$ be the p.d.f. and distribution function of a distribution that is $n(0, 1)$. Let Y have a *truncated* distribution with p.d.f. $g(y) = n(y)/[N(b) - N(a)]$, $a < y < b$, zero elsewhere. Show that $E(Y)$ is equal to $[n(a) - n(b)]/[N(b) - N(a)]$.

3.60. Let $f(x)$ and $F(x)$ be the p.d.f. and the distribution function of a distribution of the continuous type such that $f'(x)$ exists for all x. Let the mean of the truncated distribution that has p.d.f. $g(y) = f(y)/F(b)$, $-\infty < y < b$, zero elsewhere, be equal to $-f(b)/F(b)$ for all real b. Prove that $f(x)$ is $n(0, 1)$.

3.61. Let X and Y be stochastically independent random variables, each with a distribution that is $n(0, 1)$. Let $Z = X + Y$. Find the integral that represents the distribution function $G(z) = \Pr(X + Y \leq z)$ of Z. Determine the p.d.f. of Z. *Hint.* We have that $G(z) = \int_{-\infty}^{\infty} H(x, z)\, dx$, where

$$H(x, z) = \int_{-\infty}^{z-x} \frac{1}{2\pi} \exp\left[-(x^2 + y^2)/2\right] dy.$$

Find $G'(z)$ by evaluating $\int_{-\infty}^{\infty} [\partial H(x, z)/\partial z]\, dx$.

3.5 The Bivariate Normal Distribution

Let us investigate the function

$$f(x, y) = \frac{1}{2\pi\sigma_1\sigma_2\sqrt{1 - \rho^2}}\, e^{-q/2}, \qquad -\infty < x < \infty, \ -\infty < y < \infty,$$

where, with $\sigma_1 > 0$, $\sigma_2 > 0$, and $-1 < \rho < 1$,

$$q = \frac{1}{1 - \rho^2}\left[\left(\frac{x - \mu_1}{\sigma_1}\right)^2 - 2\rho\left(\frac{x - \mu_1}{\sigma_1}\right)\left(\frac{y - \mu_2}{\sigma_2}\right) + \left(\frac{y - \mu_2}{\sigma_2}\right)^2\right].$$

At this point we do not know that the constants μ_1, μ_2, σ_1^2, σ_2^2, and ρ represent parameters of a distribution. As a matter of fact, we do not know that $f(x, y)$ has the properties of a joint p.d.f. It will now be shown that:

(a) $f(x, y)$ is a joint p.d.f.
(b) X is $n(\mu_1, \sigma_1^2)$ and Y is $n(\mu_2, \sigma_2^2)$.
(c) ρ is the correlation coefficient of X and Y.

A joint p.d.f. of this form is called a *bivariate normal p.d.f.*, and the random variables X and Y are said to have a *bivariate normal distribution*.

That the nonnegative function $f(x, y)$ is actually a joint p.d.f. can be seen as follows. Define $f_1(x)$ by

$$f_1(x) = \int_{-\infty}^{\infty} f(x, y)\, dy.$$

Now

$$(1 - \rho^2)q = \left[\left(\frac{y - \mu_2}{\sigma_2}\right) - \rho\left(\frac{x - \mu_1}{\sigma_1}\right)\right]^2 + (1 - \rho^2)\left(\frac{x - \mu_1}{\sigma_1}\right)^2$$

$$= \left(\frac{y - b}{\sigma_2}\right)^2 + (1 - \rho^2)\left(\frac{x - \mu_1}{\sigma_1}\right)^2,$$

where $b = \mu_2 + \rho(\sigma_2/\sigma_1)(x - \mu_1)$. Thus

$$f_1(x) = \frac{\exp\left[-(x - \mu_1)^2/2\sigma_1^2\right]}{\sigma_1\sqrt{2\pi}} \int_{-\infty}^{\infty} \frac{\exp\left\{-(y - b)^2/[2\sigma_2^2(1 - \rho^2)]\right\}}{\sigma_2\sqrt{1 - \rho^2}\sqrt{2\pi}}\, dy.$$

For the purpose of integration, the integrand of the integral in this expression for $f_1(x)$ may be considered a normal p.d.f. with mean b and variance $\sigma_2^2(1 - \rho^2)$. Thus this integral is equal to 1 and

$$f_1(x) = \frac{1}{\sigma_1\sqrt{2\pi}}\exp\left[-\frac{(x - \mu_1)^2}{2\sigma_1^2}\right], \qquad -\infty < x < \infty.$$

Since

$$\int_{-\infty}^{\infty}\int_{-\infty}^{\infty} f(x, y)\, dy\, dx = \int_{-\infty}^{\infty} f_1(x)\, dx = 1,$$

the nonnegative function $f(x, y)$ is a joint p.d.f. of two continuous-type random variables X and Y. Accordingly, the function $f_1(x)$ is the marginal p.d.f. of X, and X is seen to be $n(\mu_1, \sigma_1^2)$. In like manner, we see that Y is $n(\mu_2, \sigma_2^2)$.

Moreover, from the development above, we note that

$$f(x, y) = f_1(x)\left(\frac{1}{\sigma_2\sqrt{1 - \rho^2}\sqrt{2\pi}}\exp\left[-\frac{(y - b)^2}{2\sigma_2^2(1 - \rho^2)}\right]\right),$$

where $b = \mu_2 + \rho(\sigma_2/\sigma_1)(x - \mu_1)$. Accordingly, the second factor in the right-hand member of the equation above is the conditional p.d.f of Y, given that $X = x$. That is, the conditional p.d.f of Y, given $X = x$, is itself normal with mean $\mu_2 + \rho(\sigma_2/\sigma_1)(x - \mu_1)$ and variance $\sigma_2^2(1 - \rho^2)$. Thus, with a bivariate normal distribution, the conditional mean of Y, given that $X = x$, is linear in x and is given by

$$E(Y|x) = \mu_2 + \rho\frac{\sigma_2}{\sigma_1}(x - \mu_1).$$

Since the coefficient of x in this linear conditional mean $E(Y|x)$ is $\rho\sigma_2/\sigma_1$, and since σ_1 and σ_2 represent the respective standard deviations, the number ρ is, in fact, the correlation coefficient of X and Y. This follows from the result, established in Section 2.3, that the coefficient of x in a general linear conditional mean $E(Y|x)$ is the product of the correlation coefficient and the ratio σ_2/σ_1.

Although the mean of the conditional distribution of Y, given $X = x$, depends upon x (unless $\rho = 0$), the variance $\sigma_2^2(1 - \rho^2)$ is the same for all real values of x. Thus, by way of example, given that $X = x$, the conditional probability that Y is within $(2.576)\sigma_2\sqrt{1 - \rho^2}$ units of the conditional mean is 0.99, whatever the value of x. In this sense, most of the probability for the distribution of X and Y lies in the band

$$\mu_2 + \rho\frac{\sigma_2}{\sigma_1}(x - \mu_1) \pm (2.576)\sigma_2\sqrt{1 - \rho^2}$$

about the graph of the linear conditional mean. For every fixed positive σ_2, the width of this band depends upon ρ. Because the band is narrow when ρ^2 is nearly 1, we see that ρ does measure the intensity of the concentration of the probability for X and Y about the linear conditional mean. This is the fact to which we alluded in the remark of Section 2.3.

In a similar manner we can show that the conditional distribution of X, given $Y = y$, is the normal distribution

$$n\left[\mu_1 + \rho\frac{\sigma_1}{\sigma_2}(y - \mu_2), \sigma_1^2(1 - \rho^2)\right].$$

Example 1. Let us assume that in a certain population of married couples the height X_1 of the husband and the height X_2 of the wife have a bivariate normal distribution with parameters $\mu_1 = 5.8$ feet, $\mu_2 = 5.3$ feet, $\sigma_1 = \sigma_2 = 0.2$ foot, and $\rho = 0.6$. The conditional p.d.f. of X_2, given $x_1 = 6.3$, is normal with mean $5.3 + (0.6)(6.3 - 5.8) = 5.6$ and standard deviation $(0.2)\sqrt{(1 - 0.36)} = 0.16$. Accordingly, given that the height of the husband is 6.3 feet, the probability that his wife has a height between 5.28 and 5.92 feet is

$$\Pr(5.28 < X_2 < 5.92 | x_1 = 6.3) = N(2) - N(-2) = 0.954.$$

The moment-generating function of a bivariate normal distribution can be determined as follows. We have

$$M(t_1, t_2) = \int_{-\infty}^{\infty}\int_{-\infty}^{\infty} e^{t_1 x + t_2 y} f(x, y)\, dx\, dy$$

$$= \int_{-\infty}^{\infty} e^{t_1 x} f_1(x)\left[\int_{-\infty}^{\infty} e^{t_2 y} f(y|x)\, dy\right] dx$$

for all real values of t_1 and t_2. The integral within the brackets is the moment-generating function of the conditional p.d.f. $f(y|x)$. Since $f(y|x)$ is a normal p.d.f. with mean $\mu_2 + \rho(\sigma_2/\sigma_1)(x - \mu_1)$ and variance $\sigma_2^2(1 - \rho^2)$, then

$$\int_{-\infty}^{\infty} e^{t_2 y} f(y|x)\, dy = \exp\left\{t_2\left[\mu_2 + \rho\frac{\sigma_2}{\sigma_1}(x - \mu_1)\right] + \frac{t_2^2\sigma_2^2(1 - \rho^2)}{2}\right\}.$$

Accordingly, $M(t_1, t_2)$ can be written in the form

$$\exp\left\{t_2\mu_2 - t_2\rho\frac{\sigma_2}{\sigma_1}\mu_1 + \frac{t_2^2\sigma_2^2(1 - \rho^2)}{2}\right\}\int_{-\infty}^{\infty}\exp\left[\left(t_1 + t_2\rho\frac{\sigma_2}{\sigma_1}\right)x\right]f_1(x)\, dx.$$

But $E(e^{tX}) = \exp\left[\mu_1 t + (\sigma_1^2 t^2)/2\right]$ for all real values of t. Accordingly, if we set $t = t_1 + t_2\rho(\sigma_2/\sigma_1)$, we see that $M(t_1, t_2)$ is given by

$$\exp\left\{t_2\mu_2 - t_2\rho\frac{\sigma_2}{\sigma_1}\mu_1 + \frac{t_2^2\sigma_2^2(1 - \rho^2)}{2}\right.$$
$$\left. + \mu_1\left(t_1 + t_2\rho\frac{\sigma_2}{\sigma_1}\right) + \sigma_1^2\frac{\left(t_1 + t_2\rho\dfrac{\sigma_2}{\sigma_1}\right)^2}{2}\right\}$$

or, equivalently,

$$M(t_1, t_2) = \exp\left(\mu_1 t_1 + \mu_2 t_2 + \frac{\sigma_1^2 t_1^2 + 2\rho\sigma_1\sigma_2 t_1 t_2 + \sigma_2^2 t_2^2}{2}\right).$$

It is interesting to note that if, in this moment-generating function $M(t_1, t_2)$, the correlation coefficient ρ is set equal to zero, then

$$M(t_1, t_2) = M(t_1, 0)M(0, t_2).$$

Thus X and Y are stochastically independent when $\rho = 0$, If, conversely, $M(t_1, t_2) \equiv M(t_1, 0)M(0, t_2)$, we have $e^{\rho\sigma_1\sigma_2 t_1 t_2} = 1$. Since each of σ_1 and σ_2 is positive, then $\rho = 0$. Accordingly, we have the following theorem.

Theorem 3. *Let X and Y have a bivariate normal distribution with means μ_1 and μ_2, positive variances σ_1^2 and σ_2^2, and correlation coefficient ρ. Then X and Y are stochastically independent if and only if $\rho = 0$.*

As a matter of fact, if any two random variables are stochastically independent and have positive standard deviations, we have noted in Example 4 of Section 2.4 that $\rho = 0$. However, $\rho = 0$ does not in general imply that two variables are stochastically independent; this can be seen in Exercises 2.18(c) and 2.23. The importance of Theorem 3 lies in the fact that we now know when and only when two random variables that have a bivariate normal distribution are stochastically independent.

EXERCISES

3.62. Let X and Y have a bivariate normal distribution with parameters $\mu_1 = 3$, $\mu_2 = 1$, $\sigma_1^2 = 16$, $\sigma_2^2 = 25$, and $\rho = \frac{3}{5}$. Determine the following probabilities:

(a) $\Pr(3 < Y < 8)$.

(b) $\Pr(3 < Y < 8 | x = 7)$.

(c) $\Pr(-3 < X < 3)$.

(d) $\Pr(-3 < X < 3 | y = -4)$.

3.63. If $M(t_1, t_2)$ is the moment-generating function of a bivariate normal distribution, compute the covariance by using the formula

$$\frac{\partial^2 M(0, 0)}{\partial t_1 \, \partial t_2} - \frac{\partial M(0, 0)}{\partial t_1} \frac{\partial M(0, 0)}{\partial t_2}.$$

Now let $\psi(t_1, t_2) = \ln M(t_1, t_2)$. Show that $\partial^2 \psi(0, 0)/\partial t_1 \, \partial t_2$ gives this covariance directly.

3.64. Let X and Y have a bivariate normal distribution with parameters $\mu_1 = 5$, $\mu_2 = 10$, $\sigma_1^2 = 1$, $\sigma_2^2 = 25$, and $\rho > 0$. If $\Pr(4 < Y < 16 | x = 5) = 0.954$, determine ρ.

3.65. Let X and Y have a bivariate normal distribution with parameters $\mu_1 = 20$, $\mu_2 = 40$, $\sigma_1^2 = 9$, $\sigma_2^2 = 4$, and $\rho = 0.6$. Find the shortest interval for which 0.90 is the conditional probability that Y is in this interval, given that $X = 22$.

3.66. Let $f(x, y) = (1/2\pi) \exp[-\frac{1}{2}(x^2 + y^2)]\{1 + xy \exp[-\frac{1}{2}(x^2 + y^2 - 2)]\}$, where $-\infty < x < \infty$, $-\infty < y < \infty$. If $f(x, y)$ is a joint p.d.f., it is not a normal bivariate p.d.f. Show that $f(x, y)$ actually is a joint p.d.f. and that each marginal p.d.f. is normal. Thus the fact that each marginal p.d.f. is normal does not imply that the joint p.d.f. is bivariate normal.

3.67. Let X, Y, and Z have the joint p.d.f.

$$(1/2\pi)^{3/2} \exp[-(x^2 + y^2 + z^2)/2]\{1 + xyz \exp[-(x^2 + y^2 + z^2)/2]\},$$

where $-\infty < x < \infty$, $-\infty < y < \infty$, and $-\infty < z < \infty$. While X, Y, and Z are obviously stochastically dependent, show that X, Y, and Z are pairwise stochastically independent and that each pair has a bivariate normal distribution.

3.68. Let X and Y have a bivariate normal distribution with parameters $\mu_1 = \mu_2 = 0$, $\sigma_1^2 = \sigma_2^2 = 1$, and correlation coefficient ρ. Find the distribution of the random variable $Z = aX + bY$ in which a and b are nonzero constants. *Hint.* Write $G(z) = \Pr(Z \le z)$ as an iterated integral and compute $G'(z) = g(z)$ by differentiating under the first integral sign and then evaluating the resulting integral by completing the square in the exponent.

Chapter *4*

Distributions of Functions of Random Variables

4.1 Sampling Theory

Let $X_1, X_2, \ldots, X_n$ denote n random variables that have the joint p.d.f. $f(x_1, x_2, \ldots, x_n)$. These variables may or may not be stochastically independent. Problems such as the following are very interesting in themselves; but more importantly, their solutions often provide the basis for making statistical inferences. Let Y be a random variable that is defined by a function of $X_1, X_2, \ldots, X_n$, say $Y = u(X_1, X_2, \ldots, X_n)$. Once the p.d.f. $f(x_1, x_2, \ldots, x_n)$ is given, can we find the p.d.f. of Y? In some of the preceding chapters, we have solved a few of these problems. Among them are the following two. If $n = 1$ and if X_1 is $n(\mu, \sigma^2)$, then $Y = (X_1 - \mu)/\sigma$ is $n(0, 1)$. If n is a positive integer, if the random variables $X_i, i = 1, 2, \ldots, n$, are mutually stochastically independent, and each X_i has the same p.d.f. $f(x) = p^x(1 - p)^{1-x}$, $x = 0, 1$, and zero elsewhere, and if $Y = \sum_1^n X_i$, then Y is $b(n, p)$. It should be observed that $Y = u(X_1) = (X_1 - \mu)/\sigma$ is a function of X_1 that depends upon the two parameters of the normal distribution; whereas $Y = u(X_1, X_2, \ldots, X_n) = \sum_1^n X_i$ does not depend upon p, the parameter of the common p.d.f. of the $X_i, i = 1, 2, \ldots, n$. The distinction that we make between these functions is brought out in the following definition.

Definition 1. A function of one or more random variables that does not depend upon any *unknown* parameter is called a *statistic*.

In accordance with this definition, the random variable $Y = \sum_{1}^{n} X_i$ discussed above is a statistic. But the random variable $Y = (X_1 - \mu)/\sigma$ is not a statistic unless μ and σ are known numbers. It should be noted that, although a statistic does not depend upon any unknown parameter, the *distribution* of that statistic may very well depend upon unknown parameters.

Remark. We remark, for the benefit of the more advanced reader, that a statistic is usually defined to be a measurable function of the random variables. In this book, however, we wish to minimize the use of measure theoretic terminology so we have suppressed the modifier "measurable." It is quite clear that a statistic is a random variable. In fact, some probabilists avoid the use of the word "statistic" altogether, and they refer to a measureable function of random variables as a random variable. We decided to use the word "statistic" because the reader will encounter it so frequently in books and journals.

We can motivate the study of the distribution of a statistic in the following way. Let a random variable X be defined on a sample space $\mathscr{C}$ and let the space of X be denoted by $\mathscr{A}$. In many situations confronting us, the distribution of X is not completely known. For instance, we may know the distribution except for the value of an unknown parameter. To obtain more information about this distribution (or the unknown parameter), we shall repeat under identical conditions the random experiment n independent times. Let the random variable X_i be a function of the ith outcome, $i = 1, 2, \ldots, n$. Then we call X_1, $X_2, \ldots, X_n$ the items of a random sample from the distribution under consideration. Suppose that we can define a statistic $Y = u(X_1, X_2, \ldots, X_n)$ whose p.d.f. is found to be $g(y)$. Perhaps this p.d.f. shows that there is a great probability that Y has a value close to the unknown parameter. Once the experiment has been repeated in the manner indicated and we have $X_1 = x_1, \ldots, X_n = x_n$, then $y = u(x_1, x_2, \ldots, x_n)$ is a known number. It is to be hoped that this known number can in some manner be used to elicit information about the unknown parameter. Thus a statistic may prove to be useful.

Remarks. Let the random variable X be defined as the diameter of a hole to be drilled by a certain drill press and let it be assumed that X has a normal distribution. Past experience with many drill presses makes this assumption plausible; but the assumption does not specify the mean μ nor the variance σ^2 of this normal distribution. The only way to obtain information about μ and σ^2 is to have recourse to experimentation. Thus we shall drill a number, say $n = 20$, of these holes whose diameters will be $X_1, X_2, \ldots,$ X_{20}. Then $X_1, X_2, \ldots, X_{20}$ is a random sample from the normal distribution under consideration. Once the holes have been drilled and the diameters measured, the 20 numbers may be used, as will be seen later, to elicit information about μ and σ^2.

The term "random sample" is now defined in a more formal manner.

Definition 2. Let $X_1, X_2, \ldots, X_n$ denote n mutually stochastically independent random variables, each of which has the same but possibly unknown p.d.f. $f(x)$; that is, the probability density functions of $X_1, X_2, \ldots, X_n$ are, respectively, $f_1(x_1) = f(x_1), f_2(x_2) = f(x_2), \ldots, f_n(x_n)$ $= f(x_n)$, so that the joint p.d.f. is $f(x_1)f(x_2)\cdots f(x_n)$. The random variables $X_1, X_2, \ldots, X_n$ are then said to constitute a *random sample* from a distribution that has p.d.f. $f(x)$.

Later we shall define what we mean by a random sample from a distribution of more than one random variable.

Sometimes it is convenient to refer to a random sample of size n from a given distribution and, as has been remarked, to refer to $X_1, X_2, \ldots, X_n$ as the items of the random sample. A reexamination of Example 5 of Section 2.4 reveals that we found the p.d.f. of the statistic, which is the maximum of the items of a random sample of size $n = 3$, from a distribution with p.d.f. $f(x) = 2x, 0 < x < 1$, zero elsewhere. In the first Remark of Section 3.1 (and referred to in this section), we found the p.d.f. of the statistic, which is the sum of the items of a random sample of size n from a distribution that has p.d.f. $f(x) = p^x(1 - p)^{1-x}, x = 0, 1$, zero elsewhere.

In this book, most of the statistics that we shall encounter will be functions of the items of a random sample from a given distribution. Next, we define two important statistics of this type.

Definition 3. Let $X_1, X_2, \ldots, X_n$ denote a random sample of size n from a given distribution. The statistic

$$\bar{X} = \frac{X_1 + X_2 + \cdots + X_n}{n} = \sum_{i=1}^{n} \frac{X_i}{n}$$

is called the *mean of the random* sample, and the statistic

$$S^2 = \sum_{i=1}^{n} \frac{(X_i - \bar{X})^2}{n} = \sum_{i=1}^{n} \frac{X_i^2}{n} - \bar{X}^2$$

is called the *variance* of the random sample.

Remark. Many writers do not define the variance of a random sample as we have done but, instead, they take $S^2 = \sum_{1}^{n} (X_i - \bar{X})^2/(n-1)$. There are good reasons for doing this. But a certain price has to be paid, as we shall indicate. Let $x_1, x_2, \ldots, x_n$ denote experimental values of the random variable X that has the p.d.f. $f(x)$ and the distribution function $F(x)$. Thus we may look upon $x_1, x_2, \ldots, x_n$ as the experimental values of a random sample of size n from the given distribution. The *distribution of the sample* is then defined to be the distribution obtained by assigning a probability of $1/n$ to each of the points $x_1, x_2, \ldots, x_n$. This is a distribution of the discrete type. The corresponding distribution function will be denoted by $F_n(x)$ and it is a step function. If we let f_x denote the number of sample values that are less than or equal to x, then $F_n(x) = f_x/n$, so that $F_n(x)$ gives the relative frequency of the event $X \le x$ in the set of n observations. The function $F_n(x)$ is often called the "empirical distribution function" and it has a number of uses.

Because the distribution of the sample is a discrete distribution, the mean and the variance have been defined and are, respectively, $\sum_{1}^{n} x_i/n = \bar{x}$ and $\sum_{1}^{n} (x_i - \bar{x})^2/n = s^2$. Thus, if one finds the distribution of the sample and the associated empirical distribution function to be useful concepts, it would seem logically inconsistent to define the variance of a random sample in any way other than we have.

Random sampling distribution theory means the general problem of finding distributions of functions of the items of a random sample. Up to this point, the only method, other than direct probabilistic arguments, of finding the distribution of a function of one or more random variables is the *distribution function technique*. That is, if $X_1, X_2, \ldots, X_n$ are random variables, the distribution of $Y = u(X_1, X_2, \ldots, X_n)$ is determined by computing the distribution function of Y,

$$G(y) = \Pr\left[u(X_1, X_2, \ldots, X_n) \le y\right].$$

Even in what superficially appears to be a very simple problem, this can be quite tedious. This fact is illustrated in the next paragraph.

Let X_1, X_2, X_3 denote a random sample of size 3 from a distribution that is $n(0, 1)$. Let Y denote the statistic that is the sum of the squares of the sample items. The distribution function of Y is given by

$$G(y) = \Pr\left(X_1^2 + X_2^2 + X_3^2 \le y\right).$$

If $y < 0$, then $G(y) = 0$. However, if $y \ge 0$, then

$$G(y) = \iiint_A \frac{1}{(2\pi)^{3/2}} \exp\left[-\frac{1}{2}(x_1^2 + x_2^2 + x_3^2)\right] dx_1\, dx_2\, dx_3,$$

where A is the set of points (x_1, x_2, x_3) interior to, or on the surface of, a sphere with center at $(0, 0, 0)$ and radius equal to $\sqrt{y}$. This is not a simple integral. We might hope to make progress by changing to spherical coordinates:

$$x_1 = \rho \cos \theta \sin \varphi, \qquad x_2 = \rho \sin \theta \sin \varphi, \qquad x_3 = \rho \cos \varphi,$$

where $\rho \geq 0$, $0 \leq \theta < 2\pi$, $0 \leq \varphi \leq \pi$. Then, for $y \geq 0$,

$$G(y) = \int_0^{\sqrt{y}} \int_0^{2\pi} \int_0^{\pi} \frac{1}{(2\pi)^{3/2}} e^{-\rho^2/2} \rho^2 \sin \varphi \, d\varphi \, d\theta \, d\rho$$

$$= \sqrt{\frac{2}{\pi}} \int_0^{\sqrt{y}} \rho^2 e^{-\rho^2/2} \, d\rho.$$

If we change the variable of integration by setting $\rho = \sqrt{w}$, we have

$$G(y) = \sqrt{\frac{2}{\pi}} \int_0^{y} \frac{\sqrt{w}}{2} e^{-w/2} \, dw,$$

for $y \geq 0$. Since Y is a random variable of the continuous type, the p.d.f. of Y is $g(y) = G'(y)$. Thus

$$g(y) = \frac{1}{\sqrt{2\pi}} y^{3/2-1} e^{-y/2}, \qquad 0 < y < \infty,$$

$$= 0 \text{ elsewhere.}$$

Because $\Gamma(\frac{3}{2}) = (\frac{1}{2})\Gamma(\frac{1}{2}) = (\frac{1}{2})\sqrt{\pi}$, and thus $\sqrt{2\pi} = \Gamma(\frac{3}{2})2^{3/2}$, we see that Y is $\chi^2(3)$.

The problem that we have just solved points up the desirability of having, if possible, various methods of determining the distribution of a function of random variables. We shall find that other techniques are available and that often a particular technique is vastly superior to the others in a given situation. These techniques will be discussed in subsequent sections.

Example 1. Let the random variable Y be distributed uniformly over the unit interval $0 < y < 1$; that is, the distribution function of Y is

$$G(y) = 0, y \leq 0$$

$$= y, 0 < y < 1,$$

$$= 1, 1 \leq y.$$

Suppose that $F(x)$ is a distribution function of the continuous type which is strictly increasing when $0 < F(x) < 1$. If we define the random variable X by the relationship $Y = F(X)$, we now show that X has a distribution which corresponds to $F(x)$. If $0 < F(x) < 1$, the inequalities $X \leq x$ and $F(X) \leq F(x)$ are equivalent. Thus, with $0 < F(x) < 1$, the distribution function of X is

$$\Pr (X \leq x) = \Pr [F(X) \leq F(x)] = \Pr [Y \leq F(x)]$$

because $Y = F(X)$. However, $\Pr (Y \leq y) = G(y)$, so we have

$$\Pr (X \leq x) = G[F(x)] = F(x), \qquad 0 < F(x) < 1.$$

That is, the distribution function of X is $F(x)$.

This result permits us to *simulate* random variables of different types. This is done by simply determining values of the uniform variable Y, usually with a computer. Then, after determining the observed value $Y = y$, solve the equation $y = F(x)$, either explicitly or by numerical methods. This yields the inverse function $x = F^{-1}(y)$. By the preceding result, this number x will be an observed value of X that has distribution function $F(x)$.

It is also interesting to note that the converse of this result is true. If X has distribution function $F(x)$ of the continuous type, then $Y = F(X)$ is uniformly distributed over $0 < y < 1$. The reason for this is, for $0 < y < 1$, that

$$\Pr (Y \leq y) = \Pr [F(X) \leq y] = \Pr [X \leq F^{-1}(y)].$$

However, it is given that $\Pr (X \leq x) = F(x)$, so

$$\Pr (Y \leq y) = F[F^{-1}(y)] = y, \qquad 0 < y < 1.$$

This is the distribution function of a random variable that is distributed uniformly on the interval $(0, 1)$.

EXERCISES

4.1. Show that

$$S^2 = \frac{1}{n} \sum_{1}^{n} (X_i - \bar{X})^2 = \frac{1}{n} \sum_{1}^{n} X_i^2 - \bar{X}^2,$$

where $\bar{X} = \sum_{1}^{n} X_i/n$.

4.2. Find the probability that exactly four items of a random sample of size 5 from the distribution having p.d.f. $f(x) = (x + 1)/2$, $-1 < x < 1$, zero elsewhere, exceed zero.

4.3. Let X_1, X_2, X_3 be a random sample of size 3 from a distribution that is $n(6, 4)$. Determine the probability that the largest sample item exceeds 8.

4.4. Let X_1, X_2 be a random sample from the distribution having p.d.f. $f(x) = 2x$, $0 < x < 1$, zero elsewhere. Find $\Pr(X_1/X_2 \le \frac{1}{2})$.

4.5. If the sample size is $n = 2$, find the constant c so that $S^2 = c(X_1 - X_2)^2$.

4.6. If $x_i = i$, $i = 1, 2, \ldots, n$, compute the values of $\bar{x} = \sum x_i/n$ and $s^2 = \sum (x_i - \bar{x})^2/n$.

4.7. Let $y_i = a + bx_i$, $i = 1, 2, \ldots, n$, where a and b are constants. Find $\bar{y} = \sum y_i/n$ and $s_y^2 = \sum (y_i - \bar{y})^2/n$ in terms of a, b, $\bar{x} = \sum x_i/n$, and $s_x^2 = \sum (x_i - \bar{x})^2/n$.

4.8. Let X_1 and X_2 denote a random sample of size 2 from a distribution that is $n(0, 1)$. Find the p.d.f. of $Y = X_1^2 + X_2^2$. *Hint.* In the double integral representing $\Pr(Y \le y)$, use polar coordinates.

4.9. The four values $y_1 = 0.42$, $y_2 = 0.31$, $y_3 = 0.87$, and $y_4 = 0.65$ represent the observed values of a random sample of size $n = 4$ from the uniform distribution over $0 < y < 1$. Using these four values, find a corresponding observed random sample from a distribution that has p.d.f. $f(x) = e^{-x}$, $0 < x < \infty$, zero elsewhere.

4.10. Let X_1, X_2 denote a random sample of size 2 from a distribution with p.d.f. $f(x) = \frac{1}{2}$, $0 < x < 2$, zero elsewhere. Find the joint p.d.f. of X_1 and X_2. Let $Y = X_1 + X_2$. Find the distribution function and the p.d.f. of Y.

4.11. Let X_1 and X_2 denote a random sample of size 2 from a distribution with p.d.f. $f(x) = 1$, $0 < x < 1$, zero elsewhere. Find the distribution function and the p.d.f. of $Y = X_1/X_2$.

4.12. Let X_1, X_2, X_3 be a random sample of size 3 from a distribution having p.d.f. $f(x) = 5x^4$, $0 < x < 1$, zero elsewhere. Let Y be the largest item in the sample. Find the distribution function and p.d.f. of Y.

4.13. Let X_1 and X_2 be items of a random sample from a distribution with p.d.f. $f(x) = 2x$, $0 < x < 1$, zero elsewhere. Evaluate the conditional probability $\Pr(X_1 < X_2 | X_1 < 2X_2)$.

4.2 Transformations of Variables of the Discrete Type

An alternative method of finding the distribution of a function of one or more random variables is called the *change of variable technique*. There are some delicate questions (with particular reference to random variables of the continuous type) involved in this technique, and these make it desirable for us first to consider special cases.

Let X have the Poisson p.d.f.

$$f(x) = \frac{\mu^x e^{-\mu}}{x!}, \qquad x = 0, 1, 2, \ldots,$$

$$= 0 \text{ elsewhere.}$$

As we have done before, let $\mathscr{A}$ denote the space $\mathscr{A} = \{x; x = 0, 1, 2, \ldots\}$, so that $\mathscr{A}$ is the set where $f(x) > 0$. Define a new random variable Y by $Y = 4X$. We wish to find the p.d.f. of Y by the change-of-variable technique. Let $y = 4x$. We call $y = 4x$ a transformation from x to y, and we say that the transformation maps the space $\mathscr{A}$ onto the space $\mathscr{B} = \{y; y = 0, 4, 8, 12, \ldots\}$. The space $\mathscr{B}$ is obtained by transforming each point in $\mathscr{A}$ in accordance with $y = 4x$. We note two things about this transformation. It is such that to each point in $\mathscr{A}$ there corresponds one, and only one, point in $\mathscr{B}$; and conversely, to each point in $\mathscr{B}$ there corresponds one, and only one, point in $\mathscr{A}$. That is, the transformation $y = 4x$ sets up a one-to-one correspondence between the points of $\mathscr{A}$ and those of $\mathscr{B}$. Any function $y = u(x)$ (not merely $y = 4x$) that maps a space $\mathscr{A}$ (not merely our $\mathscr{A}$) onto a space $\mathscr{B}$ (not merely our $\mathscr{B}$) such that there is a one-to-one correspondence between the points of $\mathscr{A}$ and those of $\mathscr{B}$ is called a *one-to-one transformation*. It is important to note that a one-to-one transformation, $y = u(x)$, implies that y is a single-valued function of x, and that x is a single-valued function of y. In our case this is obviously true, since $y = 4x$ and $x = (\frac{1}{4})y$.

Our problem is that of finding the p.d.f. $g(y)$ of the discrete type of random variable $Y = 4X$. Now $g(y) = \text{Pr}\,(Y = y)$. Because there is a one-to-one correspondence between the points of $\mathscr{A}$ and those of $\mathscr{B}$, the event $Y = y$ or $4X = y$ can occur when, and only when, the event $X = (\frac{1}{4})y$ occurs. That is, the two events are equivalent and have the same probability. Hence

$$g(y) = \text{Pr}\,(Y = y) = \text{Pr}\left(X = \frac{y}{4}\right) = \frac{\mu^{y/4} e^{-\mu}}{(y/4)!}, \qquad y = 0, 4, 8, \ldots,$$

$$0 = \text{elsewhere.}$$

The foregoing detailed discussion should make the subsequent text easier to read. Let X be a random variable of the discrete type, having p.d.f. $f(x)$. Let $\mathscr{A}$ denote the set of discrete points, at each of which $f(x) > 0$, and let $y = u(x)$ define a one-to-one transformation that maps $\mathscr{A}$ onto $\mathscr{B}$. If we solve $y = u(x)$ for x in terms of y, say, $x = w(y)$, then for each $y \in \mathscr{B}$, we have $x = w(y) \in \mathscr{A}$. Consider the random variable $Y = u(X)$. If $y \in \mathscr{B}$, then $x = w(y) \in \mathscr{A}$, and the events $Y = y$ [or

$u(X) = y]$ and $X = w(y)$ are equivalent. Accordingly, the p.d.f. of Y is

$$g(y) = \Pr(Y = y) = \Pr[X = w(y)] = f[w(y)], \qquad y \in \mathscr{B},$$
$$= 0 \text{ elsewhere.}$$

Example 1. Let X have the binomial p.d.f.

$$f(x) = \frac{3!}{x!\,(3-x)!} \left(\frac{2}{3}\right)^x \left(\frac{1}{3}\right)^{3-x}, \qquad x = 0, 1, 2, 3,$$
$$= 0 \text{ elsewhere.}$$

We seek the p.d.f. $g(y)$ of the random variable $Y = X^2$. The transformation $y = u(x) = x^2$ maps $\mathscr{A} = \{x; x = 0, 1, 2, 3\}$ onto $\mathscr{B} = \{y; y = 0, 1, 4, 9\}$. In general, $y = x^2$ does not define a one-to-one transformation; here, however, it does, for there are no negative values of x in $\mathscr{A} = \{x; x = 0, 1, 2, 3\}$. That is, we have the single-valued inverse function $x = w(y) = \sqrt{y}$ (not $-\sqrt{y}$), and so

$$g(y) = f(\sqrt{y}) = \frac{3!}{(\sqrt{y})!\,(3 - \sqrt{y})!} \left(\frac{2}{3}\right)^{\sqrt{y}} \left(\frac{1}{3}\right)^{3 - \sqrt{y}}, \qquad y = 0, 1, 4, 9,$$
$$= 0 \text{ elsewhere.}$$

There are no essential difficulties involved in a problem like the following. Let $f(x_1, x_2)$ be the joint p.d.f. of two discrete-type random variables X_1 and X_2 with $\mathscr{A}$ the (two-dimensional) set of points at which $f(x_1, x_2) > 0$. Let $y_1 = u_1(x_1, x_2)$ and $y_2 = u_2(x_1, x_2)$ define a one-to-one transformation that maps $\mathscr{A}$ onto $\mathscr{B}$. The joint p.d.f. of the two new random variables $Y_1 = u_1(X_1, X_2)$ and $Y_2 = u_2(X_1, X_2)$ is given by

$$g(y_1, y_2) = f[w_1(y_1, y_2), w_2(y_1, y_2)], \qquad (y_1, y_2) \in \mathscr{B},$$
$$= 0 \text{ elsewhere,}$$

where $x_1 = w_1(y_1, y_2)$ and $x_2 = w_2(y_1, y_2)$ are the single-valued inverses of $y_1 = u_1(x_1, x_2)$ and $y_2 = u_2(x_1, x_2)$. From this joint p.d.f. $g(y_1, y_2)$ we may obtain the marginal p.d.f. of Y_1 by summing on y_2 or the marginal p.d.f. of Y_2 by summing on y_1.

Perhaps it should be emphasized that the technique of change of variables involves the introduction of as many "new" variables as there were "old" variables. That is, suppose that $f(x_1, x_2, x_3)$ is the joint p.d.f. of X_1, X_2, and X_3, with $\mathscr{A}$ the set where $f(x_1, x_2, x_3) > 0$. Let us say we seek the p.d.f. of $Y_1 = u_1(X_1, X_2, X_3)$. We would then define (if possible) $Y_2 = u_2(X_1, X_2, X_3)$ and $Y_3 = u_3(X_1, X_2, X_3)$, so that $y_1 = u_1(x_1, x_2, x_3)$, $y_2 = u_2(x_1, x_2, x_3)$, $y_3 = u_3(x_1, x_2, x_3)$ define a one-to-one transformation of $\mathscr{A}$ onto $\mathscr{B}$. This would enable us to find the joint p.d.f. of Y_1, Y_2, and Y_3 from which we would get the marginal p.d.f. of Y_1 by summing on y_2 and y_3.

Example 2. Let X_1 and X_2 be two stochastically independent random variables that have Poisson distributions with means μ_1 and μ_2, respectively. The joint p.d.f. of X_1 and X_2 is

$$\frac{\mu_1^{x_1}\mu_2^{x_2}e^{-\mu_1-\mu_2}}{x_1!\, x_2!}, \qquad x_1 = 0, 1, 2, 3, \ldots, \quad x_2 = 0, 1, 2, 3, \ldots,$$

and is zero elsewhere. Thus the space $\mathscr{A}$ is the set of points (x_1, x_2), where each of x_1 and x_2 is a nonnegative integer. We wish to find the p.d.f. of $Y_1 = X_1 + X_2$. If we use the change of variable technique, we need to define a second random variable Y_2. Because Y_2 is of no interest to us, let us choose it in such a way that we have a simple one-to-one transformation. For example, take $Y_2 = X_2$. Then $y_1 = x_1 + x_2$ and $y_2 = x_2$ represent a one-to-one transformation that maps $\mathscr{A}$ onto

$$\mathscr{B} = \{(y_1, y_2); y_2 = 0, 1, \ldots, y_1 \quad \text{and} \quad y_1 = 0, 1, 2, \ldots\}.$$

Note that, if $(y_1, y_2) \in \mathscr{B}$, then $0 \le y_2 \le y_1$. The inverse functions are given by $x_1 = y_1 - y_2$ and $x_2 = y_2$. Thus the joint p.d.f. of Y_1 and Y_2 is

$$g(y_1, y_2) = \frac{\mu_1^{y_1-y_2}\mu_2^{y_2}e^{-\mu_1-\mu_2}}{(y_1 - y_2)!\, y_2!}, \qquad (y_1, y_2) \in \mathscr{B},$$

and is zero elsewhere. Consequently, the marginal p.d.f. of Y_1 is given by

$$
\begin{aligned}
g_1(y_1) &= \sum_{y_2=0}^{y_1} g(y_1, y_2) \\
&= \frac{e^{-\mu_1-\mu_2}}{y_1!} \sum_{y_2=0}^{y_1} \frac{y_1!}{(y_1 - y_2)!\, y_2!}\, \mu_1^{y_1-y_2}\mu_2^{y_2} \\
&= \frac{(\mu_1 + \mu_2)^{y_1}e^{-\mu_1-\mu_2}}{y_1!}, \qquad y_1 = 0, 1, 2, \ldots,
\end{aligned}
$$

and is zero elsewhere. That is, $Y_1 = X_1 + X_2$ has a Poisson distribution with parameter $\mu_1 + \mu_2$.

EXERCISES

4.14. Let X have a p.d.f. $f(x) = \frac{1}{3}$, $x = 1, 2, 3$, zero elsewhere. Find the p.d.f. of $Y = 2X + 1$.

4.15. If $f(x_1, x_2) = (\frac{2}{3})^{x_1+x_2}(\frac{1}{3})^{2-x_1-x_2}$, $(x_1, x_2) = (0, 0), (0, 1), (1, 0), (1, 1)$. zero elsewhere, is the joint p.d.f. of X_1 and X_2, find the joint p.d.f. of $Y_1 = X_1 - X_2$ and $Y_2 = X_1 + X_2$.

4.16. Let X have the p.d.f. $f(x) = (\frac{1}{2})^x$, $x = 1, 2, 3, \ldots$, zero elsewhere. Find the p.d.f. of $Y = X^3$.

4.17. Let X_1 and X_2 have the joint p.d.f. $f(x_1, x_2) = x_1 x_2/36$, $x_1 = 1, 2, 3$ and $x_2 = 1, 2, 3$, zero elsewhere. Find first the joint p.d.f. of $Y_1 = X_1 X_2$ and $Y_2 = X_2$, and then find the marginal p.d.f. of Y_1.

4.18. Let the stochastically independent random variables X_1 and X_2 be $b(n_1, p)$ and $b(n_2, p)$, respectively. Find the joint p.d.f. of $Y_1 = X_1 + X_2$ and $Y_2 = X_2$, and then find the marginal p.d.f. of Y_1. *Hint.* ·Use the fact that

$$\sum_{w=0}^{k} \binom{n_1}{w}\binom{n_2}{k-w} = \binom{n_1+n_2}{k}.$$

This can be proved by comparing the coefficients of x^k in each member of the identity $(1 + x)^{n_1}(1 + x)^{n_2} \equiv (1 + x)^{n_1+n_2}$.

4.19. Let X_1 and X_2 be stochastically independent random variables of the discrete type with joint p.d.f. $f_1(x_1)f_2(x_2)$, $(x_1, x_2) \in \mathscr{A}$. Let $y_1 = u_1(x_1)$ and $y_2 = u_2(x_2)$ denote a one-to-one transformation that maps $\mathscr{A}$ onto $\mathscr{B}$. Show that $Y_1 = u_1(X_1)$ and $Y_2 = u_2(X_2)$ are stochastically independent.

4.3 Transformations of Variables of the Continuous Type

In the preceding section we introduced the notion of a one-to-one transformation and the mapping of a set $\mathscr{A}$ onto a set $\mathscr{B}$ under that transformation. Those ideas were sufficient to enable us to find the distribution of a function of several random variables of the discrete type. In this section we shall examine the same problem when the random variables are of the continuous type. It is again helpful to begin with a special problem.

Example 1. Let X be a random variable of the continuous type, having p.d.f.

$$f(x) = 2x, \qquad 0 < x < 1,$$

$$= 0 \text{ elsewhere.}$$

Here $\mathscr{A}$ is the space $\{x; 0 < x < 1\}$, where $f(x) > 0$. Define the random variable Y by $Y = 8X^3$ and consider the transformation $y = 8x^2$. Under the transformation $y = 8x^2$, the set $\mathscr{A}$ is mapped onto the set $\mathscr{B} = \{y; 0 < y < 8\}$, and, moreover, the transformation is one-to-one. For every $0 < a < b < 8$, the event $a < Y < b$ will occur when, and only when, the event $\frac{1}{2}\sqrt[3]{a} < X <$

$\frac{1}{2}\sqrt[3]{b}$ occurs because there is a one-to-one correspondence between the points of $\mathcal{A}$ and $\mathcal{B}$. Thus

$$\Pr\ (a < Y < b) = \Pr\ (\tfrac{1}{2}\sqrt[3]{a} < X < \tfrac{1}{2}\sqrt[3]{b})$$

$$= \int_{\sqrt[3]{a}/2}^{\sqrt[3]{b}/2} 2x\ dx.$$

Let us rewrite this integral by changing the variable of integration from x to y by writing $y = 8x^3$ or $x = \tfrac{1}{2}\sqrt[3]{y}$. Now

$$\frac{dx}{dy} = \frac{1}{6y^{2/3}},$$

and, accordingly, we have

$$\Pr\ (a < Y < b) = \int_a^b 2\left(\frac{\sqrt[3]{y}}{2}\right)\left(\frac{1}{6y^{2/3}}\right) dy$$

$$= \int_a^b \frac{1}{6y^{1/3}}\ dy.$$

Since this is true for every $0 < a < b < 8$, the p.d.f. $g(y)$ of Y is the integrand; that is,

$$g(y) = \frac{1}{6y^{1/3}}, \qquad 0 < y < 8,$$

$$= 0 \text{ elsewhere.}$$

It is worth noting that we found the p.d.f. of the random variable $Y = 8X^3$ by using a theorem on the change of variable in a definite integral. However, to obtain $g(y)$ we actually need only two things: (1) the set $\mathcal{B}$ of points y where $g(y) > 0$ and (2) the integrand of the integral on y to which $\Pr\ (a < Y < b)$ is equal. These can be found by two simple rules:

(a) Verify that the transformation $y = 8x^3$ maps $\mathcal{A} = \{x; 0 < x < 1\}$ onto $\mathcal{B} = \{y; 0 < y < 8\}$ and that the transformation is one-to-one.

(b) Determine $g(y)$ on this set $\mathcal{B}$ by substituting $\tfrac{1}{2}\sqrt[3]{y}$ for x in $f(x)$ and then multiplying this result by the derivative of $\tfrac{1}{2}\sqrt[3]{y}$. That is,

$$g(y) = f\left(\frac{\sqrt[3]{y}}{2}\right)\frac{d[(\tfrac{1}{2})\sqrt[3]{y}]}{dy} = \frac{1}{6y^{1/3}}, \qquad 0 < y < 8,$$

$$= 0 \text{ elsewhere.}$$

We shall accept a theorem in analysis on the change of variable in a definite integral to enable us to state a more general result. Let X be a random variable of the continuous type having p.d.f. $f(x)$. Let $\mathcal{A}$ be the one-dimensional space where $f(x) > 0$. Consider the random variable $Y = u(X)$, where $y = u(x)$ defines a one-to-one transformation that maps the set $\mathscr{A}$ onto the set $\mathscr{B}$. Let the inverse of $y = u(x)$ be denoted by $x = w(y)$, and let the derivative $dx/dy = w'(y)$ be continuous and not vanish for all points y in $\mathscr{B}$. Then the p.d.f. of the random variable $Y = u(X)$ is given by

$$g(y) = f[w(y)]|w'(y)|, \qquad y \in \mathscr{B},$$
$$= 0 \text{ elsewhere,}$$

where $|w'(y)|$ represents the absolute value of $w'(y)$. This is precisely what we did in Example 1 of this section, except there we deliberately chose $y = 8x^3$ to be an increasing function so that

$$\frac{dx}{dy} = w'(y) = \frac{1}{6y^{2/3}}, \qquad 0 < y < 8,$$

is positive, and hence

$$\left| \frac{1}{6y^{2/3}} \right| = \frac{1}{6y^{2/3}}, \qquad 0 < y < 8.$$

Henceforth we shall refer to $dx/dy = w'(y)$ as the Jacobian (denoted by J) of the transformation. In most mathematical areas, $J = w'(y)$ is referred to as the Jacobian of the inverse transformation $x = w(y)$, but in this book it will be called the Jacobian of the transformation, simply for convenience.

Example 2. Let X have the p.d.f.

$$f(x) = 1, \qquad 0 < x < 1,$$
$$= 0 \text{ elsewhere.}$$

We are to show that the random variable $Y = -2 \ln X$ has a chi-square distribution with 2 degrees of freedom. Here the transformation is $y = u(x) = -2 \ln x$, so that $x = w(y) = e^{-y/2}$. The space $\mathscr{A}$ is $\mathscr{A} = \{x; 0 < x < 1\}$, which the one-to-one transformation $y = -2 \ln x$ maps onto $\mathscr{B} = \{y; 0 < y < \infty\}$. The Jacobian of the transformation is

$$J = \frac{dx}{dy} = w'(y) = -\frac{1}{2}e^{-y/2}.$$

Accordingly, the p.d.f. $g(y)$ of $Y = -2 \ln X$ is

$$g(y) = f(e^{-y/2})|J| = \tfrac{1}{2}e^{-y/2}, \qquad 0 < y < \infty,$$
$$= 0 \text{ elsewhere,}$$

a p.d.f. that is chi-square with 2 degrees of freedom. Note that this problem was first proposed in Exercise 3.40.

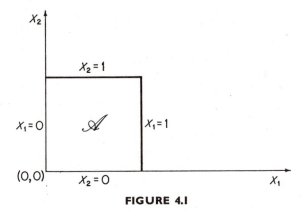

FIGURE 4.I

This method of finding the p.d.f. of a function of one random variable of the continuous type will now be extended to functions of two random variables of this type. Again, only functions that define a one-to-one transformation will be considered at this time. Let $y_1 = u_1(x_1, x_2)$ and $y_2 = u_2(x_1, x_2)$ define a one-to-one transformation that maps a (two-dimensional) set $\mathscr{A}$ in the x_1x_2-plane onto a (two-dimensional) set $\mathscr{B}$ in the y_1y_2-plane. If we express each of x_1 and x_2 in terms of y_1 and y_2, we can write $x_1 = w_1(y_1, y_2)$, $x_2 = w_2(y_1, y_2)$. The determinant of order 2,

$$\begin{vmatrix} \dfrac{\partial x_1}{\partial y_1} & \dfrac{\partial x_1}{\partial y_2} \\ \dfrac{\partial x_2}{\partial y_1} & \dfrac{\partial x_2}{\partial y_2} \end{vmatrix},$$

is called the *Jacobian* of the transformation and will be denoted by the symbol J. It will be assumed that these first-order partial derivatives are continuous and that the Jacobian J is not identically equal to zero in $\mathscr{B}$. An illustrative example may be desirable before we proceed with the extension of the change of variable technique to two random variables of the continuous type.

Example 3. Let $\mathscr{A}$ be the set $\mathscr{A} = \{(x_1, x_2); 0 < x_1 < 1, 0 < x_2 < 1\}$, depicted in Figure 4.1. We wish to determine the set $\mathscr{B}$ in the y_1y_2-plane that is the mapping of $\mathscr{A}$ under the one-to-one transformation

$$y_1 = u_1(x_1, x_2) = x_1 + x_2,$$
$$y_2 = u_2(x_1, x_2) = x_1 - x_2,$$

and we wish to compute the Jacobian of the transformation. Now

$$x_1 = w_1(y_1, y_2) = \tfrac{1}{2}(y_1 + y_2),$$
$$x_2 = w_2(y_1, y_2) = \tfrac{1}{2}(y_1 - y_2).$$

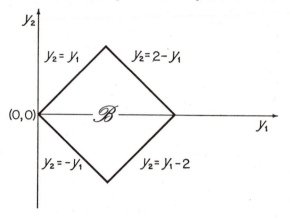

FIGURE 4.2

To determine the set $\mathscr{B}$ in the y_1y_2-plane onto which $\mathscr{A}$ is mapped under the transformation, note that the boundaries of $\mathscr{A}$ are transformed as follows into the boundaries of $\mathscr{B}$;

$$
\begin{aligned}
x_1 &= 0 \quad &\text{into} \quad& 0 = \tfrac{1}{2}(y_1 + y_2),\\
x_1 &= 1 \quad &\text{into} \quad& 1 = \tfrac{1}{2}(y_1 + y_2),\\
x_2 &= 0 \quad &\text{into} \quad& 0 = \tfrac{1}{2}(y_1 - y_2),\\
x_2 &= 1 \quad &\text{into} \quad& 1 = \tfrac{1}{2}(y_1 - y_2).
\end{aligned}
$$

Accordingly, $\mathscr{B}$ is as shown in Figure 4.2. Finally,

$$
J = \begin{vmatrix} \dfrac{\partial x_1}{\partial y_1} & \dfrac{\partial x_1}{\partial y_2} \\[2mm] \dfrac{\partial x_2}{\partial y_1} & \dfrac{\partial x_2}{\partial y_2} \end{vmatrix} = \begin{vmatrix} \dfrac{1}{2} & \dfrac{1}{2} \\[2mm] \dfrac{1}{2} & -\dfrac{1}{2} \end{vmatrix} = -\dfrac{1}{2}.
$$

We now proceed with the problem of finding the joint p.d.f. of two functions of two continuous-type random variables. Let X_1 and X_2 be random variables of the continuous type, having joint p.d.f. $\varphi(x_1, x_2)$. Let $\mathscr{A}$ be the two-dimensional set in the x_1x_2-plane where $\varphi(x_1, x_2) > 0$. Let $Y_1 = u_1(X_1, X_2)$ be a random variable whose p.d.f. is to be found. If $y_1 = u_1(x_1, x_2)$ and $y_2 = u_2(x_1, x_2)$ define a one-to-one transformation of $\mathscr{A}$ onto a set $\mathscr{B}$ in the y_1y_2-plane (with nonidentically vanishing Jacobian), we can find, by use of a theorem in analysis, the joint p.d.f. of $Y_1 = u_1(X_1, X_2)$ and $Y_2 = u_2(X_1, X_2)$. Let A be a subset of $\mathscr{A}$, and let B denote the mapping of A under the one-to-one transformation (see Figure 4.3). The events $(X_1, X_2) \in A$ and $(Y_1, Y_2) \in B$ are equivalent. Hence

$$
\begin{aligned}
\Pr\left[(Y_1, Y_2) \in B\right] &= \Pr\left[(X_1, X_2) \in A\right]\\
&= \int_A\!\!\int \varphi(x_1, x_2)\, dx_1\, dx_2.
\end{aligned}
$$

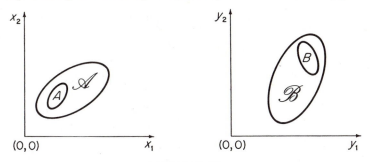

FIGURE 4.3

We wish now to change variables of integration by writing $y_1 = u_1(x_1, x_2)$, $y_2 = u_2(x_1, x_2)$, or $x_1 = w_1(y_1, y_2)$, $x_2 = w_2(y_1, y_2)$. It has been proved in analysis that this change of variables requires

$$\int_A \int \varphi(x_1, x_2)\, dx_1\, dx_2 = \int_B \int \varphi[w_1(y_1, y_2), w_2(y_1, y_2)]|J|\, dy_1\, dy_2.$$

Thus for every set B in $\mathscr{B}$,

$$\Pr\left[(Y_1, Y_2) \in B\right] = \int_B \int \varphi[w_1(y_1, y_2), w_2(y_1, y_2)]|J|\, dy_1\, dy_2,$$

which implies that the joint p.d.f. $g(y_1, y_2)$ of Y_1 and Y_2 is

$$g(y_1, y_2) = \varphi[w_1(y_1, y_2), w_2(y_1, y_2)]|J|, \qquad (y_1, y_2) \in \mathscr{B},$$
$$= 0 \text{ elsewhere.}$$

Accordingly, the marginal p.d.f. $g_1(y_1)$ of Y_1 can be obtained from the joint p.d.f. $g(y_1, y_2)$ in the usual manner by integrating on y_2. Five examples of this result will be given.

Example 4. Let the random variable X have the p.d.f.

$$f(x) = 1, \qquad 0 < x < 1,$$
$$= 0 \text{ elsewhere,}$$

and let X_1, X_2 denote a random sample from this distribution. The joint p.d.f. of X_1 and X_2 is then

$$\varphi(x_1, x_2) = f(x_1)f(x_2) = 1, \qquad 0 < x_1 < 1, 0 < x_2 < 1,$$
$$= 0 \text{ elsewhere.}$$

Consider the two random variables $Y_1 = X_1 + X_2$ and $Y_2 = X_1 - X_2$. We wish to find the joint p.d.f of Y_1 and Y_2. Here the two-dimensional space $\mathscr{A}$ in the $x_1 x_2$-plane is that of Example 3 of this section. The one-to-one transformation $y_1 = x_1 + x_2$, $y_2 = x_1 - x_2$ maps $\mathscr{A}$ onto the space $\mathscr{B}$ of that example. Moreover, the Jacobian of that transformation has been shown to be $J = -\frac{1}{2}$. Thus

$$
\begin{aligned}
g(y_1, y_2) &= \varphi[\tfrac{1}{2}(y_1 + y_2), \tfrac{1}{2}(y_1 - y_2)]|J| \\
&= f[\tfrac{1}{2}(y_1 + y_2)]f[\tfrac{1}{2}(y_1 - y_2)]|J| = \tfrac{1}{2}, \qquad (y_1, y_2) \in \mathcal{B}, \\
&= 0 \text{ elsewhere.}
\end{aligned}
$$

Because $\mathcal{B}$ is not a product space, the random variables Y_1 and Y_2 are stochastically dependent. The marginal p.d.f. of Y_1 is given by

$$
g_1(y_1) = \int_{-\infty}^{\infty} g(y_1, y_2)\, dy_2.
$$

If we refer to Figure 4.2, it is seen that

$$
\begin{aligned}
g_1(y_1) &= \int_{-y_1}^{y_1} \tfrac{1}{2}\, dy_2 = y_1, \qquad 0 < y_1 \le 1, \\
&= \int_{y_1 - 2}^{2 - y_1} \tfrac{1}{2}\, dy_2 = 2 - y_1, \qquad 1 < y_1 < 2, \\
&= 0 \text{ elsewhere.}
\end{aligned}
$$

In a similar manner, the marginal p.d.f. $g_2(y_2)$ is given by

$$
\begin{aligned}
g_2(y_2) &= \int_{-y_2}^{y_2 + 2} \tfrac{1}{2}\, dy_1 = y_2 + 1, \qquad -1 < y_2 \le 0, \\
&= \int_{y_2}^{2 - y_2} \tfrac{1}{2}\, dy_1 = 1 - y_2, \qquad 0 < y_2 < 1, \\
&= 0 \text{ elsewhere.}
\end{aligned}
$$

Example 5. Let X_1 and X_2 be two stochastically independent random variables that have gamma distributions and joint p.d.f.

$$
f(x_1, x_2) = \frac{1}{\Gamma(\alpha)\Gamma(\beta)} x_1^{\alpha - 1} x_2^{\beta - 1} e^{-x_1 - x_2}, \qquad 0 < x_1 < \infty, 0 < x_2 < \infty,
$$

zero elsewhere, where $\alpha > 0$, $\beta > 0$. Let $Y_1 = X_1 + X_2$ and $Y_2 = X_1/(X_1 + X_2)$. We shall show that Y_1 and Y_2 are stochastically independent.

The space $\mathcal{A}$ is, exclusive of the points on the coordinate axes, the first quadrant of the $x_1 x_2$-plane. Now

$$
y_1 = u_1(x_1, x_2) = x_1 + x_2,
$$

$$
y_2 = u_2(x_1, x_2) = \frac{x_1}{x_1 + x_2}
$$

may be written $x_1 = y_1 y_2$, $x_2 = y_1(1 - y_2)$, so

$$
J = \begin{vmatrix} y_2 & y_1 \\ 1 - y_2 & -y_1 \end{vmatrix} = -y_1 \not\equiv 0.
$$

The transformation is one-to-one, and it maps $\mathcal{A}$ onto $\mathcal{B} = \{(y_1, y_2);$

$0 < y_1 < \infty$, $0 < y_2 < 1\}$ in the $y_1 y_2$-plane. The joint p.d.f. of Y_1 and Y_2 is then

$$g(y_1, y_2) = (y_1) \frac{1}{\Gamma(\alpha)\Gamma(\beta)} (y_1 y_2)^{\alpha - 1}[y_1(1 - y_2)]^{\beta - 1} e^{-y_1}$$

$$= \frac{y_2^{\alpha - 1}(1 - y_2)^{\beta - 1}}{\Gamma(\alpha)\Gamma(\beta)} y_1^{\alpha + \beta - 1} e^{-y_1}, \qquad 0 < y_1 < \infty, 0 < y_2 < 1,$$

$$= 0 \text{ elsewhere.}$$

In accordance with Theorem 1, Section 2.4, the random variables are stochastically independent. The marginal p.d.f. of Y_2 is

$$g_2(y_2) = \frac{y_2^{\alpha - 1}(1 - y_2)^{\beta - 1}}{\Gamma(\alpha)\Gamma(\beta)} \int_0^\infty y_1^{\alpha + \beta - 1} e^{-y_1} \, dy_1,$$

$$= \frac{\Gamma(\alpha + \beta)}{\Gamma(\alpha)\Gamma(\beta)} y_2^{\alpha - 1}(1 - y_2)^{\beta - 1}, \qquad 0 < y_2 < 1,$$

$$= 0 \text{ elsewhere.}$$

This p.d.f. is that of the *beta distribution* with parameters α and β. Since $g(y_1, y_2) \equiv g_1(y_1)g_2(y_2)$, it must be that the p.d.f. of Y_1 is

$$g_1(y_1) = \frac{1}{\Gamma(\alpha + \beta)} y_1^{\alpha + \beta - 1} e^{-y_1}, \qquad 0 < y_1 < \infty$$

$$= 0 \text{ elsewhere,}$$

which is that of a gamma distribution with parameter values of $\alpha + \beta$ and 1.

It is an easy exercise to show that the mean and the variance of Y_2, which has a beta distribution with parameters α and β, are, respectively,

$$\mu = \frac{\alpha}{\alpha + \beta}, \qquad \sigma^2 = \frac{\alpha\beta}{(\alpha + \beta + 1)(\alpha + \beta)^2}.$$

Example 6. Let $Y_1 = \frac{1}{2}(X_1 - X_2)$, where X_1 and X_2 are stochastically independent random variables, each being $\chi^2(2)$. The joint p.d.f. of X_1 and X_2 is

$$f(x_1)f(x_2) = \frac{1}{4} \exp\left(-\frac{x_1 + x_2}{2}\right), \qquad 0 < x_1 < \infty, 0 < x_2 < \infty,$$

$$= 0 \text{ elsewhere.}$$

Let $Y_2 = X_2$ so that $y_1 = \frac{1}{2}(x_1 - x_2)$, $y_2 = x_2$, or $x_1 = 2y_1 + y_2$, $x_2 = y_2$ define a one-to-one transformation from $\mathscr{A} = \{(x_1, x_2); 0 < x_1 < \infty, 0 < x_2 < \infty\}$ onto $\mathscr{B} = \{(y_1, y_2); -2y_1 < y_2 \text{ and } 0 < y_2, -\infty < y_1 < \infty\}$. The Jacobian of the transformation is

$$J = \begin{vmatrix} 2 & 1 \\ 0 & 1 \end{vmatrix} = 2;$$

hence the joint p.d.f. of Y_1 and Y_2 is

$$g(y_1, y_2) = \frac{|2|}{4} e^{-y_1 - y_2}, \qquad (y_1, y_2) \in \mathscr{B},$$

$$= 0 \text{ elsewhere.}$$

Thus the p.d.f. of Y_1 is given by

$$g_1(y_1) = \int_{-2y_1}^{\infty} \tfrac{1}{2} e^{-y_1 - y_2} \, dy_2 = \tfrac{1}{2} e^{y_1}, \qquad -\infty < y_1 < 0,$$

$$= \int_{0}^{\infty} \tfrac{1}{2} e^{-y_1 - y_2} \, dy_2 = \tfrac{1}{2} e^{-y_1}, \qquad 0 \le y_1 < \infty,$$

or

$$g_1(y_1) = \tfrac{1}{2} e^{-|y_1|}, \qquad -\infty < y_1 < \infty.$$

This p.d.f. is now frequently called the *double exponential p.d.f.*

Example 7. In this example a rather important result is established. Let X_1 and X_2 be stochastically independent random variables of the continuous type with joint p.d.f. $f_1(x_1)f_2(x_2)$ that is positive on the two-dimensional space $\mathscr{A}$. Let $Y_1 = u_1(X_1)$, a function of X_1 alone, and $Y_2 = u_2(X_2)$, a function of X_2 alone. We assume for the present that $y_1 = u_1(x_1)$, $y_2 = u_2(x_2)$ define a one-to-one transformation from $\mathscr{A}$ onto a two-dimensional set $\mathscr{B}$ in the $y_1 y_2$-plane. Solving for x_1 and x_2 in terms of y_1 and y_2, we have $x_1 = w_1(y_1)$ and $x_2 = w_2(y_2)$, so

$$J = \begin{vmatrix} w_1'(y_1) & 0 \\ 0 & w_2'(y_2) \end{vmatrix} = w_1'(y_1) w_2'(y_2) \ne 0.$$

Hence the joint p.d.f. of Y_1 and Y_2 is

$$g(y_1, y_2) = f_1[w_1(y_1)] f_2[w_2(y_2)] |w_1'(y_1) w_2'(y_2)|, \qquad (y_1, y_2) \in \mathscr{B},$$

$$= 0 \text{ elsewhere.}$$

However, from the procedure for changing variables in the case of one random variable, we see that the marginal probability density functions of Y_1 and Y_2 are, respectively, $g_1(y_1) = f_1[w_1(y_1)]|w_1'(y_1)|$ and $g_2(y_2) = f_2[w_2(y_2)]|w_2'(y_2)|$ for y_1 and y_2 in some appropriate sets. Consequently,

$$g(y_1, y_2) \equiv g_1(y_1)g_2(y_2).$$

Thus, summarizing, we note that, if X_1 and X_2 are stochastically independent random variables, then $Y_1 = u_1(X_1)$ and $Y_2 = u_2(X_2)$ are also stochastically independent random variables. It has been seen that the result holds if X_1 and X_2 are of the discrete type; see Exercise 4.19.

Example 8. In the *simulation* of random variables using uniform random variables, it is frequently difficult to solve $y = F(x)$ for x. Thus other methods are necessary. For instance, consider the important normal case in which we desire to determine X so that it is $n(0, 1)$. Of course, once X is determined, other normal variables can then be obtained through X by the transformation $Z = \sigma X + \mu$.

To simulate normal variables, Box and Muller suggested the following scheme. Let Y_1, Y_2 be a random sample from the uniform distribution over $0 < y < 1$. Define X_1 and X_2 by

$$X_1 = (-2 \ln Y_1)^{1/2} \cos (2\pi Y_2),$$

$$X_2 = (-2 \ln Y_1)^{1/2} \sin (2\pi Y_2).$$

The corresponding transformation is one-to-one and maps $\{(y_1, y_2);$ $0 < y_1 < 1, 0 < y_2 < 1\}$ onto $\{(x_1, x_2); -\infty < x_1 < \infty, -\infty < x_2 < \infty\}$ except for sets involving $x_1 = 0$ and $x_2 = 0$, which have probability zero. The inverse transformation is given by

$$y_1 = \exp\left(-\frac{x_1^2 + x_2^2}{2}\right),$$

$$y_2 = \frac{1}{2\pi} \arctan \frac{x_2}{x_1}.$$

This has the Jacobian

$$J = \begin{vmatrix} (-x_1) \exp\left(-\dfrac{x_1^2 + x_2^2}{2}\right) & (-x_2) \exp\left(-\dfrac{x_1^2 + x_2^2}{2}\right) \\[2ex] \dfrac{-x_2/x_1^2}{(2\pi)(1 + x_2^2/x_1^2)} & \dfrac{1/x_1}{(2\pi)(1 + x_2^2/x_1^2)} \end{vmatrix}$$

$$= \frac{-(1 + x_2^2/x_1^2) \exp\left(-\dfrac{x_1^2 + x_2^2}{2}\right)}{(2\pi)(1 + x_2^2/x_1^2)} = \frac{-\exp\left(-\dfrac{x_1^2 + x_2^2}{2}\right)}{2\pi}.$$

Since the joint p.d.f. of Y_1 and Y_2 is 1 on $0 < y_1 < 1, 0 < y_2 < 1$, and zero elsewhere, the joint p.d.f. of X_1 and X_2 is

$$\frac{\exp\left(-\dfrac{x_1^2 + x_2^2}{2}\right)}{2\pi}, \qquad -\infty < x_1 < \infty, \quad -\infty < x_2 < \infty.$$

That is, X_1 and X_2 are stochastically independent random variables, each being $n(0, 1)$.

EXERCISES

4.20. Let X have the p.d.f. $f(x) = x^2/9$, $0 < x < 3$, zero elsewhere. Find the p.d.f. of $Y = X^3$.

4.21. If the p.d.f. of X is $f(x) = 2xe^{-x^2}$, $0 < x < \infty$, zero elsewhere, determine the p.d.f. of $Y = X^2$.

4.22. Let X_1, X_2 be a random sample from the normal distribution $n(0, 1)$. Show that the marginal p.d.f. of $Y_1 = X_1/X_2$ is the *Cauchy p.d.f.*

$$g_1(y_1) = \frac{1}{\pi(1 + y_1^2)}, \qquad -\infty < y_1 < \infty.$$

Hint. Let $Y_2 = X_2$ and take the p.d.f. of X_2 to be equal to zero at $x_2 = 0$. Then determine the joint p.d.f. of Y_1 and Y_2. Be sure to multiply by the absolute value of the Jacobian.

4.23. Find the mean and variance of the beta distribution considered in Example 5. *Hint.* From that example, we know that

$$\int_0^1 y^{\alpha-1}(1 - y)^{\beta-1}\, dy = \frac{\Gamma(\alpha)\Gamma(\beta)}{\Gamma(\alpha + \beta)}$$

for all $\alpha > 0$, $\beta > 0$.

4.24. Determine the constant c in each of the following so that each $f(x)$ is a *beta p.d.f.*
(a) $f(x) = cx(1 - x)^3$, $0 < x < 1$, zero elsewhere.
(b) $f(x) = cx^4(1 - x)^5$, $0 < x < 1$, zero elsewhere.
(c) $f(x) = cx^2(1 - x)^8$, $0 < x < 1$, zero elsewhere.

4.25. Determine the constant c so that $f(x) = cx(3 - x)^4$, $0 < x < 3$, zero elsewhere, is a p.d.f.

4.26. Show that the graph of the beta p.d.f. is symmetric about the vertical line through $x = \frac{1}{2}$ if $\alpha = \beta$.

4.27. Show, for $k = 1, 2, \ldots, n$, that

$$\int_p^1 \frac{n!}{(k-1)!\,(n-k)!}\, z^{k-1}(1 - z)^{n-k}\, dz = \sum_{x=0}^{k-1} \binom{n}{x} p^x (1 - p)^{n-x}.$$

This demonstrates the relationship between the distribution functions of the beta and binomial distributions.

4.28. Let X have the *logistic p.d.f.* $f(x) = e^{-x}/(1 + e^{-x})^2$, $-\infty < x < \infty$.
(a) Show that the graph of $f(x)$ is symmetric about the vertical axis through $x = 0$.
(b) Find the distribution function of X.
(c) Show that the moment-generating function $M(t)$ of X is

$\Gamma(1 - t)\Gamma(1 + t)$, $-1 < t < 1$. *Hint.* In the integral representing $M(t)$, let $y = (1 + e^{-x})^{-1}$.

4.29. Let X have the uniform distribution over the interval $(-\pi/2, \pi/2)$. Show that $Y = \tan X$ has a Cauchy distribution.

4.30. Let X_1 and X_2 be two stochastically independent random variables of the continuous type with probability density functions $f(x_1)$ and $g(x_2)$, respectively. Show that the p.d.f. $h(y)$ of $Y = X_1 + X_2$ can be found by the *convolution formula,*

$$h(y) = \int_{-\infty}^{\infty} f(y - w)g(w)\ dw.$$

4.31. Let X_1 and X_2 be two stochastically independent normal random variables, each with mean zero and variance one (possibly resulting from a Box–Muller transformation). Show that

$$Z_1 = \mu_1 + \sigma_1 X_1,$$
$$Z_2 = \mu_2 + \rho\sigma_2 X_1 + \sigma_2\sqrt{1 - \rho^2}X_2,$$

where $0 < \sigma_1$, $0 < \sigma_2$, and $0 < \rho < 1$, have a bivariate normal distribution with respective parameters μ_1, μ_2, σ_1^2, σ_2^2, and ρ.

4.32. Let X_1 and X_2 denote a random sample of size 2 from a distribution that is $n(\mu, \sigma^2)$. Let $Y_1 = X_1 + X_2$ and $Y_2 = X_1 - X_2$. Find the joint p.d.f. of Y_1 and Y_2 and show that these random variables are stochastically independent.

4.33. Let X_1 and X_2 denote a random sample of size 2 from a distribution that is $n(\mu, \sigma^2)$. Let $Y_1 = X_1 + X_2$ and $Y_2 = X_1 + 2X_2$. Show that the joint p.d.f. of Y_1 and Y_2 is bivariate normal with correlation coefficient $3/\sqrt{10}$.

4.4 The *t* and *F* Distributions

It is the purpose of this section to define two additional distributions quite useful in certain problems of statistical inference. These are called, respectively, the (Student's) t distribution and the F distribution.

Let W denote a random variable that is $n(0, 1)$; let V denote a random variable that is $\chi^2(r)$; and let W and V be stochastically independent. Then the joint p.d.f. of W and V, say $\varphi(w, v)$, is the product of the p.d.f. of W and that of V or

$$\varphi(w, v) = \frac{1}{\sqrt{2\pi}}e^{-w^2/2}\frac{1}{\Gamma(r/2)2^{r/2}}v^{r/2-1}e^{-v/2}, \quad -\infty < w < \infty, 0 < v < \infty,$$

$$= 0 \text{ elsewhere.}$$

Define a new random variable T by writing

$$T = \frac{W}{\sqrt{V/r}}.$$

The change-of-variable technique will be used to obtain the p.d.f. $g_1(t)$ of T. The equations

$$t = \frac{w}{\sqrt{v/r}} \quad \text{and} \quad u = v$$

define a one-to-one transformation that maps $\mathcal{A} = \{(w, v);\ -\infty < w < \infty,\ 0 < v < \infty\}$ onto $\mathcal{B} = \{(t, u);\ -\infty < t < \infty,\ 0 < u < \infty\}$. Since $w = t\sqrt{u}/\sqrt{r}$, $v = u$, the absolute value of the Jacobian of the transformation is $|J| = \sqrt{u}/\sqrt{r}$. Accordingly, the joint p.d.f. of T and $U = V$ is given by

$$g(t, u) = \varphi\left(\frac{t\sqrt{u}}{\sqrt{r}}, u\right)|J|$$

$$= \frac{1}{\sqrt{2\pi}\,\Gamma(r/2)2^{r/2}}\, u^{r/2-1}\exp\left[-\frac{u}{2}\left(1 + \frac{t^2}{r}\right)\right]\frac{\sqrt{u}}{\sqrt{r}},$$

$$-\infty < t < \infty, 0 < u < \infty,$$

$$= 0 \text{ elsewhere.}$$

The marginal p.d.f. of T is then

$$g_1(t) = \int_{-\infty}^{\infty} g(t, u)\, du$$

$$= \int_0^{\infty} \frac{1}{\sqrt{2\pi r}\,\Gamma(r/2)2^{r/2}}\, u^{(r+1)/2-1}\exp\left[-\frac{u}{2}\left(1 + \frac{t^2}{r}\right)\right] du.$$

In this integral let $z = u[1 + (t^2/r)]/2$, and it is seen that

$$g_1(t) = \int_0^{\infty} \frac{1}{\sqrt{2\pi r}\,\Gamma(r/2)2^{r/2}}\left(\frac{2z}{1 + t^2/r}\right)^{(r+1)/2-1} e^{-z}\left(\frac{2}{1 + t^2/r}\right) dz$$

$$= \frac{\Gamma[(r + 1)/2]}{\sqrt{\pi r}\,\Gamma(r/2)}\frac{1}{(1 + t^2/r)^{(r+1)/2}}, \quad -\infty < t < \infty.$$

Thus, if W is $n(0, 1)$, if V is $\chi^2(r)$, and if W and V are stochastically independent, then

$$T = \frac{W}{\sqrt{V/r}}$$

has the immediately preceding p.d.f. $g_1(t)$. The distribution of the random variable T is usually called a *t distribution*. It should be observed that a t distribution is completely determined by the parameter r, the number of degrees of freedom of the random variable that has the chi-square distribution. Some approximate values of

$$\Pr(T \le t) = \int_{-\infty}^{t} g_1(w)\, dw$$

for selected values of r and t, can be found in Table IV in Appendix B.

Next consider two stochastically independent chi-square random variables U and V having r_1 and r_2 degrees of freedom, respectively. The joint p.d.f. $\varphi(u, v)$ of U and V is then

$$\varphi(u, v) = \frac{1}{\Gamma(r_1/2)\Gamma(r_2/2)2^{(r_1+r_2)/2}} u^{r_1/2-1}v^{r_2/2-1}e^{-(u+v)/2},$$

$$0 < u < \infty, 0 < v < \infty,$$

$$= 0 \text{ elsewhere.}$$

We define the new random variable

$$F = \frac{U/r_1}{V/r_2}$$

and we propose finding the p.d.f. $g_1(f)$ of F. The equations

$$f = \frac{u/r_1}{v/r_2}, \qquad z = v,$$

define a one-to-one transformation that maps the set $\mathcal{A} = \{(u, v); 0 < u < \infty, 0 < v < \infty\}$ onto the set $\mathcal{B} = \{(f, z); 0 < f < \infty, 0 < z < \infty\}$. Since $u = (r_1/r_2)zf$, $v = z$, the absolute value of the Jacobian of the transformation is $|J| = (r_1/r_2)z$. The joint p.d.f. $g(f, z)$ of the random variables F and $Z = V$ is then

$$g(f, z) = \frac{1}{\Gamma(r_1/2)\Gamma(r_2/2)2^{(r_1+r_2)/2}} \left(\frac{r_1 zf}{r_2}\right)^{r_1/2-1} z^{r_2/2-1}$$

$$\times \exp\left[-\frac{z}{2}\left(\frac{r_1 f}{r_2} + 1\right)\right]\frac{r_1 z}{r_2},$$

provided that $(f, z) \in \mathcal{B}$, and zero elsewhere. The marginal p.d.f. $g_1(f)$ of F is then

$$g_1(f) = \int_{-\infty}^{\infty} g(f, z)\, dz$$

$$= \int_{0}^{\infty} \frac{(r_1/r_2)^{r_1/2}(f)^{r_1/2-1}}{\Gamma(r_1/2)\Gamma(r_2/2)2^{(r_1+r_2)/2}} z^{(r_1+r_2)/2-1} \exp\left[-\frac{z}{2}\left(\frac{r_1 f}{r_2} + 1\right)\right] dz.$$

If we change the variable of integration by writing

$$y = \frac{z}{2}\left(\frac{r_1 f}{r_2} + 1\right),$$

it can be seen that

$$g_1(f) = \int_0^\infty \frac{(r_1/r_2)^{r_1/2}(f)^{r_1/2-1}}{\Gamma(r_1/2)\Gamma(r_2/2)2^{(r_1+r_2)/2}} \left(\frac{2y}{r_1 f/r_2 + 1}\right)^{(r_1+r_2)/2-1} e^{-y}$$

$$\times \left(\frac{2}{r_1 f/r_2 + 1}\right) dy$$

$$= \frac{\Gamma[(r_1 + r_2)/2](r_1/r_2)^{r_1/2}}{\Gamma(r_1/2)\Gamma(r_2/2)} \frac{(f)^{r_1/2-1}}{(1 + r_1 f/r_2)^{(r_1+r_2)/2}}, \qquad 0 < f < \infty,$$

$$= 0 \text{ elsewhere.}$$

Accordingly, if U and V are stochastically independent chi-square variables with r_1 and r_2 degrees of freedom, respectively, then

$$F = \frac{U/r_1}{V/r_2}$$

has the immediately preceding p.d.f. $g_1(f)$. The distribution of this random variable is usually called an F distribution. It should be observed that an F distribution is completely determined by the two parameters r_1 and r_2. Table V in Appendix B gives some approximate values of

$$\Pr(F \le f) = \int_0^f g_1(w)\, dw$$

for selected values of r_1, r_2, and f.

EXERCISES

4.34. Let T have a t distribution with 10 degrees of freedom. Find $\Pr(|T| > 2.228)$ from Table IV.

4.35. Let T have a t distribution with 14 degrees of freedom. Determine b so that $\Pr(-b < T < b) = 0.90$.

4.36. Let F have an F distribution with parameters r_1 and r_2. Prove that $1/F$ has an F distribution with parameters r_2 and r_1.

4.37. If F has an F distribution with parameters $r_1 = 5$ and $r_2 = 10$, find a and b so that $\Pr(F \le a) = 0.05$ and $\Pr(F \le b) = 0.95$, and, accordingly, $\Pr(a < F < b) = 0.90$. *Hint.* Write $\Pr(F \le a) = \Pr(1/F \ge 1/a) = 1 - \Pr(1/F \le 1/a)$, and use the result of Exercise 4.36 and Table V.

4.38. Let $T = W/\sqrt{V/r}$, where the stochastically independent variables W and V are, respectively, normal with mean zero and variance 1 and chi-square with r degrees of freedom. Show that T^2 has an F distribution with parameters $r_1 = 1$ and $r_2 = r$. *Hint.* What is the distribution of the numerator of T^2?

4.39. Show that the t distribution with $r = 1$ degree of freedom and the Cauchy distribution are the same.

4.40. Show that

$$Y = \frac{1}{1 + (r_1/r_2)F},$$

where F has an F distribution with parameters r_1 and r_2, has a beta distribution.

4.41. Let X_1, X_2 be a random sample from a distribution having the p.d.f. $f(x) = e^{-x}$, $0 < x < \infty$, zero elsewhere. Show that $Z = X_1/X_2$ has an F distribution.

4.5 Extensions of the Change-of-Variable Technique

In Section 4.3 it was seen that the determination of the joint p.d.f. of two functions of two random variables of the continuous type was essentially a corollary to a theorem in analysis having to do with the change of variables in a twofold integral. This theorem has a natural extension to n-fold integrals. This extension is as follows. Consider an integral of the form

$$\int \cdots \int_A \varphi(x_1, x_2, \ldots, x_n) \, dx_1 \, dx_2 \cdots dx_n$$

taken over a subset A of an n-dimensional space $\mathscr{A}$. Let

$$y_1 = u_1(x_1, x_2, \ldots, x_n), \qquad y_2 = u_2(x_1, x_2, \ldots, x_n), \ldots,$$
$$y_n = u_n(x_1, \ldots, x_n),$$

together with the inverse functions

$$x_1 = w_1(y_1, y_2, \ldots, y_n), \qquad x_2 = w_2(y_1, y_2, \ldots, y_n), \ldots,$$
$$x_n = w_n(y_1, y_2, \ldots, y_n)$$

define a one-to-one transformation that maps $\mathscr{A}$ onto $\mathscr{B}$ in the $y_1, y_2, \ldots, y_n$ space (and hence maps the subset A of $\mathscr{A}$ onto a subset B

of $\mathscr{B}$). Let the first partial derivatives of the inverse functions be continuous and let the n by n determinant (called the Jacobian)

$$J = \begin{vmatrix} \dfrac{\partial x_1}{\partial y_1} & \dfrac{\partial x_1}{\partial y_2} & \cdots & \dfrac{\partial x_1}{\partial y_n} \\[2mm] \dfrac{\partial x_2}{\partial y_1} & \dfrac{\partial x_2}{\partial y_2} & \cdots & \dfrac{\partial x_2}{\partial y_n} \\[2mm] \vdots & \vdots & & \vdots \\[2mm] \dfrac{\partial x_n}{\partial y_1} & \dfrac{\partial x_n}{\partial y_2} & \cdots & \dfrac{\partial x_n}{\partial y_n} \end{vmatrix}$$

not vanish identically in $\mathscr{B}$. Then

$$\int \cdots_A \int \varphi(x_1, x_2, \ldots, x_n)\, dx_1\, dx_2 \cdots dx_n$$

$$= \int \cdots_B \int \varphi[w_1(y_1, \ldots, y_n), w_2(y_1, \ldots, y_n), \ldots, w_n(y_1, \ldots, y_n)]$$

$$\times |J|\, dy_1\, dy_2 \cdots dy_n.$$

Whenever the conditions of this theorem are satisfied, we can determine the joint p.d.f. of n functions of n random variables. Appropriate changes of notation in Section 4.3 (to indicate n-space as opposed to 2-space) is all that is needed to show that the joint p.d.f. of the random variables $Y_1 = u_1(X_1, X_2, \ldots, X_n)$, $Y_2 = u_2(X_1, X_2, \ldots, X_n)$, $\ldots$, $Y_n = u_n(X_1, X_2, \ldots, X_n)$—where the joint p.d.f. of $X_1, X_2, \ldots, X_n$ is $\varphi(x_1, \ldots, x_n)$—is given by

$$g(y_1, y_2, \ldots, y_n) = |J|\varphi[w_1(y_1, \ldots, y_n), \ldots, w_n(y_1, \ldots, y_n)],$$

when $(y_1, y_2, \ldots, y_n) \in \mathscr{B}$, and is zero elsewhere.

Example 1. Let $X_1, X_2, \ldots, X_{k+1}$ be mutually stochastically independent random variables, each having a gamma distribution with $\beta = 1$. The joint p.d.f. of these variables may be written as

$$\varphi(x_1, x_2, \ldots, x_{k+1}) = \prod_{i=1}^{k+1} \frac{1}{\Gamma(\alpha_i)} x_i^{\alpha_i - 1} e^{-x_i}, \qquad 0 < x_i < \infty,$$

$$= 0 \text{ elsewhere.}$$

Let

$$Y_i = \frac{X_i}{X_1 + X_2 + \cdots + X_{k+1}}, \qquad i = 1, 2, \ldots, k,$$

and $Y_{k+1} = X_1 + X_2 + \cdots + X_{k+1}$ denote $k + 1$ new random variables.

The associated transformation maps $\mathscr{A} = \{(x_1, \ldots, x_{k+1}); \; 0 < x_i < \infty,$ $i = 1, \ldots, k + 1\}$ onto the space

$$\mathscr{B} = \{(y_1, \ldots, y_k, y_{k+1}); \; 0 < y_i, \; i = 1, \ldots, k,$$
$$y_1 + \cdots + y_k < 1, \; 0 < y_{k+1} < \infty\}.$$

The single-valued inverse functions are $x_1 = y_1 y_{k+1}, \ldots, x_k = y_k y_{k+1},$ $x_{k+1} = y_{k+1}(1 - y_1 - \cdots - y_k)$, so that the Jacobian is

$$J = \begin{vmatrix} y_{k+1} & 0 & \cdots & 0 & y_1 \\ 0 & y_{k+1} & \cdots & 0 & y_2 \\ \vdots & \vdots & & \vdots & \vdots \\ 0 & 0 & \cdots & y_{k+1} & y_k \\ -y_{k+1} & -y_{k+1} & \cdots & -y_{k+1} & (1 - y_1 - \cdots - y_k) \end{vmatrix} = y_{k+1}^k.$$

Hence the joint p.d.f. of $Y_1, \ldots, Y_k, Y_{k+1}$ is given by

$$\frac{y_{k+1}^{\alpha_1 + \cdots + \alpha_{k+1} - 1} y_1^{\alpha_1 - 1} \cdots y_k^{\alpha_k - 1}(1 - y_1 - \cdots - y_k)^{\alpha_{k+1} - 1} e^{-y_{k+1}}}{\Gamma(\alpha_1) \cdots \Gamma(\alpha_k) \Gamma(\alpha_{k+1})},$$

provided that $(y_1, \ldots, y_k, y_{k+1}) \in \mathscr{B}$ and is equal to zero elsewhere. The joint p.d.f. of $Y_1, \ldots, Y_k$ is seen by inspection to be given by

$$g(y_1, \ldots, y_k) = \frac{\Gamma(\alpha_1 + \cdots + \alpha_{k+1})}{\Gamma(\alpha_1) \cdots \Gamma(\alpha_{k+1})} y_1^{\alpha_1 - 1} \cdots y_k^{\alpha_k - 1}(1 - y_1 - \cdots - y_k)^{\alpha_{k+1} - 1},$$

when $0 < y_i, \; i = 1, \ldots, k, \; y_1 + \cdots + y_k < 1$, while the function g is equal to zero elsewhere. Random variables $Y_1, \ldots, Y_k$ that have a joint p.d.f. of this form are said to have a *Dirichlet distribution* with parameters $\alpha_1, \ldots,$ α_k, α_{k+1}, and any such $g(y_1, \ldots, y_k)$ is called a *Dirichlet p.d.f.* It is seen, in the special case of $k = 1$, that the Dirichlet p.d.f. becomes a beta p.d.f. Moreover, it is also clear from the joint p.d.f. of $Y_1, \ldots, Y_k, Y_{k+1}$ that Y_{k+1} has a gamma distribution with parameters $\alpha_1 + \cdots + \alpha_k + \alpha_{k+1}$ and $\beta = 1$ and that Y_{k+1} is stochastically independent of $Y_1, Y_2, \ldots, Y_k$.

We now consider some other problems that are encountered when transforming variables. Let X have the Cauchy p.d.f.

$$f(x) = \frac{1}{\pi(1 + x^2)}, \qquad -\infty < x < \infty,$$

and let $Y = X^2$. We seek the p.d.f. $g(y)$ of Y. Consider the transformation $y = x^2$. This transformation maps the space of X, $\mathscr{A} = \{x; -\infty < x < \infty\}$, onto $\mathscr{B} = \{y; 0 \le y < \infty\}$. However, the transformation is not one-to-one. To each $y \in \mathscr{B}$, with the exception of $y = 0$, there correspond two points $x \in \mathscr{A}$. For example, if $y = 4$, we may have either $x = 2$ or $x = -2$. In such an instance, we represent $\mathscr{A}$ as the union of two disjoint sets A_1 and A_2 such that $y = x^2$ defines a one-to-one transformation that maps each of A_1 and A_2 onto $\mathscr{B}$. If we take A_1 to be $\{x; -\infty < x < 0\}$ and A_2 to be $\{x; 0 \le x < \infty\}$, we see that A_1 is mapped onto $\{y; 0 < y < \infty\}$, whereas A_2 is mapped onto $\{y; 0 \le y < \infty\}$, and these sets are not the same. Our difficulty is caused by the fact that $x = 0$ is an element of $\mathscr{A}$. Why, then, do we not return to the Cauchy p.d.f. and take $f(0) = 0$? Then our new $\mathscr{A}$ is $\mathscr{A} = \{-\infty < x < \infty$ but $x \ne 0\}$. We then take $A_1 = \{x; -\infty < x < 0\}$ and $A_2 = \{x; 0 < x < \infty\}$. Thus $y = x^2$, with the inverse $x = -\sqrt{y}$, maps A_1 onto $\mathscr{B} = \{y; 0 < y < \infty\}$ and the transformation is one-to-one. Moreover, the transformation $y = x^2$, with inverse $x = \sqrt{y}$, maps A_2 onto $\mathscr{B} = \{y; 0 < y < \infty\}$ and the transformation is one-to-one. Consider the probability $\Pr(Y \in B)$, where $B \subset \mathscr{B}$. Let $A_3 = \{x; x = -\sqrt{y}, y \in B\} \subset A_1$ and let $A_4 = \{x; x = \sqrt{y}, y \in B\} \subset A_2$. Then $Y \in B$ when and only when $X \in A_3$ or $X \in A_4$. Thus we have

$$\Pr(Y \in B) = \Pr(X \in A_3) + \Pr(X \in A_4)$$
$$= \int_{A_3} f(x)\,dx + \int_{A_4} f(x)\,dx.$$

In the first of these integrals, let $x = -\sqrt{y}$. Thus the Jacobian, say J_1, is $-1/2\sqrt{y}$; moreover, the set A_3 is mapped onto B. In the second integral let $x = \sqrt{y}$. Thus the Jacobian, say J_2, is $1/2\sqrt{y}$; moreover, the set A_4 is also mapped onto B. Finally,

$$\Pr(Y \in B) = \int_B f(-\sqrt{y})\left| -\frac{1}{2\sqrt{y}}\right|dy + \int_B f(\sqrt{y})\frac{1}{2\sqrt{y}}\,dy$$
$$= \int_B [f(-\sqrt{y}) + f(\sqrt{y})]\frac{1}{2\sqrt{y}}\,dy.$$

Hence the p.d.f. of Y is given by

$$g(y) = \frac{1}{2\sqrt{y}}[f(-\sqrt{y}) + f(\sqrt{y})], \qquad y \in B.$$

With $f(x)$ the Cauchy p.d.f. we have

$$g(y) = \frac{1}{\pi(1 + y)\sqrt{y}}, \qquad 0 < y < \infty,$$
$$= 0 \text{ elsewhere.}$$

In the preceding discussion of a random variable of the continuous type, we had two inverse functions, $x = -\sqrt{y}$ and $x = \sqrt{y}$. That is why we sought to partition $\mathscr{A}$ (or a modification of $\mathscr{A}$) into two disjoint subsets such that the transformation $y = x^2$ maps each onto the same $\mathscr{B}$. Had there been three inverse functions, we would have sought to partition $\mathscr{A}$ (or a modified form of $\mathscr{A}$) into three disjoint subsets, and so on. It is hoped that this detailed discussion will make the following paragraph easier to read.

Let $\varphi(x_1, x_2, \ldots, x_n)$ be the joint p.d.f. of $X_1, X_2, \ldots, X_n$, which are random variables of the continuous type. Let $\mathscr{A}$ be the n-dimensional space where $\varphi(x_1, x_2, \ldots, x_n) > 0$, and consider the transformation $y_1 = u_1(x_1, x_2, \ldots, x_n), y_2 = u_2(x_1, x_2, \ldots, x_n), \ldots, y_n = u_n(x_1, x_2, \ldots, x_n)$, which maps $\mathscr{A}$ onto $\mathscr{B}$ in the $y_1, y_2, \ldots, y_n$ space. To each point of $\mathscr{A}$ there will correspond, of course, but one point in $\mathscr{B}$; but to a point in $\mathscr{B}$ there may correspond more than one point in $\mathscr{A}$. That is, the transformation may not be one-to-one. Suppose, however, that we can represent $\mathscr{A}$ as the union of a finite number, say k, of mutually disjoint sets $A_1, A_2, \ldots, A_k$ so that

$$y_1 = u_1(x_1, x_2, \ldots, x_n), \ldots, \qquad y_n = u_n(x_1, x_2, \ldots, x_n)$$

define a one-to-one transformation of each A_i onto $\mathscr{B}$. Thus, to each point in $\mathscr{B}$ there will correspond exactly one point in each of $A_1, A_2, \ldots, A_k$. Let

$$x_1 = w_{1i}(y_1, y_2, \ldots, y_n),$$
$$x_2 = w_{2i}(y_1, y_2, \ldots, y_n), \qquad i = 1, 2, \ldots, k,$$
$$\vdots$$
$$x_n = w_{ni}(y_1, y_2, \ldots, y_n),$$

denote the k groups of n inverse functions, one group for each of these k transformations. Let the first partial derivatives be continuous and let each

$$J_i = \begin{vmatrix} \dfrac{\partial w_{1i}}{\partial y_1} & \dfrac{\partial w_{1i}}{\partial y_2} & \cdots & \dfrac{\partial w_{1i}}{\partial y_n} \\[2ex] \dfrac{\partial w_{2i}}{\partial y_1} & \dfrac{\partial w_{2i}}{\partial y_2} & \cdots & \dfrac{\partial w_{2i}}{\partial y_n} \\[1ex] \vdots & \vdots & & \vdots \\[1ex] \dfrac{\partial w_{ni}}{\partial y_1} & \dfrac{\partial w_{ni}}{\partial y_2} & \cdots & \dfrac{\partial w_{ni}}{\partial y_n} \end{vmatrix}, \qquad i = 1, 2, \ldots, k,$$

be not identically equal to zero in $\mathscr{B}$. From a consideration of the probability of the union of k mutually exclusive events and by applying the change of variable technique to the probability of each of these events, it can be seen that the joint p.d.f. of $Y_1 = u_1(X_1, X_2, \ldots, X_n)$, $Y_2 = u_2(X_1, X_2, \ldots, X_n), \ldots, Y_n = u_n(X_1, X_2, \ldots, X_n)$, is given by

$$g(y_1, y_2, \ldots, y_n) = \sum_{i=1}^{k} |J_i| \varphi[w_{1i}(y_1, \ldots, y_n), \ldots, w_{ni}(y_1, \ldots, y_n)],$$

provided that $(y_1, y_2, \ldots, y_n) \in \mathscr{B}$, and equals zero elsewhere. The p.d.f. of any Y_i, say Y_1, is then

$$g_1(y_1) = \int_{-\infty}^{\infty} \cdots \int_{-\infty}^{\infty} g(y_1, y_2, \ldots, y_n) \, dy_2 \cdots dy_n.$$

An illustrative example follows.

Example 2. To illustrate the result just obtained, take $n = 2$ and let X_1, X_2 denote a random sample of size 2 from a distribution that is $n(0, 1)$. The joint p.d.f. of X_1 and X_2 is

$$f(x_1, x_2) = \frac{1}{2\pi} \exp\left(-\frac{x_1^2 + x_2^2}{2}\right), \qquad -\infty < x_1 < \infty, \ -\infty < x_2 < \infty.$$

Let Y_1 denote the mean and let Y_2 denote twice the variance of the random sample. The associated transformation is

$$y_1 = \frac{x_1 + x_2}{2},$$

$$y_2 = \frac{(x_1 - x_2)^2}{2}.$$

This transformation maps $\mathscr{A} = \{(x_1, x_2); -\infty < x_1 < \infty, -\infty < x_2 < \infty\}$ onto $\mathscr{B} = \{(y_1, y_2); -\infty < y_1 < \infty, 0 \le y_2 < \infty\}$. But the transformation is not one-to-one because, to each point in $\mathscr{B}$, exclusive of points where $y_2 = 0$, there correspond two points in $\mathscr{A}$. In fact, the two groups of inverse functions are

$$x_1 = y_1 - \sqrt{\frac{y_2}{2}} \qquad\qquad x_2 = y_1 + \sqrt{\frac{y_2}{2}},$$

and

$$x_1 = y_1 + \sqrt{\frac{y_2}{2}} \qquad\qquad x_2 = y_1 - \sqrt{\frac{y_2}{2}}.$$

Moreover, the set $\mathscr{A}$ cannot be represented as the union of two disjoint sets, each of which under our transformation maps onto $\mathscr{B}$. Our difficulty is caused by those points of $\mathscr{A}$ that lie on the line whose equation is $x_2 = x_1$. At each of these points we have $y_2 = 0$. However, we can define $f(x_1, x_2)$ to be zero at each point where $x_1 = x_2$. We can do this without altering the distribution of probability, because the probability measure of this set is zero. Thus we have a new $\mathscr{A} = \{(x_1, x_2); -\infty < x_1 < \infty, -\infty < x_2 < \infty,$ but

$x_1 \neq x_2$}. This space is the union of the two disjoint sets $A_1 = \{(x_1, x_2);$ $x_2 > x_1\}$ and $A_2 = \{(x_1, x_2); x_2 < x_1\}$. Moreover, our transformation now defines a one-to-one transformation of each A_i, $i = 1, 2$, onto the new $\mathscr{B} = \{(y_1, y_2); -\infty < y_1 < \infty, 0 < y_2 < \infty\}$. We can now find the joint p.d.f., say $g(y_1, y_2)$, of the mean Y_1 and twice the variance Y_2 of our random sample. An easy computation shows that $|J_1| = |J_2| = 1/\sqrt{2y_2}$. Thus

$$g(y_1, y_2) = \frac{1}{2\pi} \exp\left[-\frac{(y_1 - \sqrt{y_2/2})^2}{2} - \frac{(y_1 + \sqrt{y_2/2})^2}{2}\right] \frac{1}{\sqrt{2y_2}}$$

$$+ \frac{1}{2\pi} \exp\left[-\frac{(y_1 + \sqrt{y_2/2})^2}{2} - \frac{(y_1 - \sqrt{y_2/2})^2}{2}\right] \frac{1}{\sqrt{2y_2}}$$

$$= \sqrt{\frac{2}{2\pi}}\, e^{-y_1^2} \frac{1}{\sqrt{2}\,\Gamma(\frac{1}{2})}\, y_2^{1/2-1} e^{-y_2/2}, \qquad -\infty < y_1 < \infty, 0 < y_2 < \infty.$$

We can make three interesting observations. The mean Y_1 of our random sample is $n(0, \frac{1}{2})$; Y_2, which is twice the variance of our sample, is $\chi^2(1)$; and the two are stochastically independent. Thus the mean and the variance of our sample are stochastically independent.

EXERCISES

4.42. Let X_1, X_2, X_3 denote a random sample from a normal distribution $n(0, 1)$. Let the random variables Y_1, Y_2, Y_3 be defined by

$$X_1 = Y_1 \cos Y_2 \sin Y_3, \qquad X_2 = Y_1 \sin Y_2 \sin Y_3, \qquad X_3 = Y_1 \cos Y_3,$$

where $0 \leq Y_1 < \infty, 0 \leq Y_2 < 2\pi, 0 \leq Y_3 \leq \pi$. Show that Y_1, Y_2, Y_3 are mutually stochastically independent.

4.43. Let X_1, X_2, X_3 denote a random sample from the distribution having p.d.f. $f(x) = e^{-x}$, $0 < x < \infty$, zero elsewhere. Show that

$$Y_1 = \frac{X_1}{X_1 + X_2}, \qquad Y_2 = \frac{X_1 + X_2}{X_1 + X_2 + X_3}, \qquad Y_3 = X_1 + X_2 + X_3$$

are mutually stochastically independent.

4.44. Let $X_1, X_2, \ldots, X_r$ be r mutually stochastically independent gamma variables with parameters $\alpha = \alpha_i$ and $\beta = 1$, $i = 1, 2, \ldots, r$, respectively. Show that $Y_1 = X_1 + X_2 + \cdots + X_r$ has a gamma distribution with parameters $\alpha = \alpha_1 + \cdots + \alpha_r$ and $\beta = 1$. *Hint.* Let $Y_2 = X_2 + \cdots + X_r, Y_3 = X_3 + \cdots + X_r, \ldots, Y_r = X_r$.

4.45. Let $Y_1, \ldots, Y_k$ have a Dirichlet distribution with parameters $\alpha_1, \ldots, \alpha_k, \alpha_{k+1}$.

(a) Show that Y_1 has a beta distribution with parameters $\alpha = \alpha_1$ and $\beta = \alpha_2 + \cdots + \alpha_{k+1}$.

(b) Show that $Y_1 + \cdots + Y_r$, $r \leq k$, has a beta distribution with parameters $\alpha = \alpha_1 + \cdots + \alpha_r$ and $\beta = \alpha_{r+1} + \cdots + \alpha_{k+1}$.

(c) Show that $Y_1 + Y_2$, $Y_3 + Y_4$, $Y_5, \ldots, Y_k$, $k \geq 5$, have a Dirichlet distribution with parameters $\alpha_1 + \alpha_2$, $\alpha_3 + \alpha_4$, $\alpha_5, \ldots, \alpha_k$, α_{k+1}. *Hint.* Recall the definition of Y_i in Example 1 and use the fact that the sum of several stochastically independent gamma variables with $\beta = 1$ is a gamma variable (Exercise 4.44).

4.46. Let X_1, X_2, and X_3 be three mutually stochastically independent chi-square variables with r_1, r_2, and r_3 degrees of freedom, respectively.

(a) Show that $Y_1 = X_1/X_2$ and $Y_2 = X_1 + X_2$ are stochastically independent and that Y_2 is $\chi^2(r_1 + r_2)$.

(b) Deduce that

$$\frac{X_1/r_1}{X_2/r_2} \quad \text{and} \quad \frac{X_3/r_3}{(X_1 + X_2)/(r_1 + r_2)}$$

are stochastically independent F variables.

4.47. If $f(x) = \frac{1}{2}$, $-1 < x < 1$, zero elsewhere, is the p.d.f. of the random variable X, find the p.d.f. of $Y = X^2$.

4.48. If X_1, X_2 is a random sample from a distribution that is $n(0, 1)$, find the joint p.d.f. of $Y_1 = X_1^2 + X_2^2$ and $Y_2 = X_2$ and the marginal p.d.f. of Y_1. *Hint.* Note that the space of Y_1 and Y_2 is given by $-\sqrt{y_1} < y_2 < \sqrt{y_1}$, $0 < y_1 < \infty$.

4.49. If X has the p.d.f. $f(x) = \frac{1}{4}$, $-1 < x < 3$, zero elsewhere, find the p.d.f. of $Y = X^2$. *Hint.* Here $\mathscr{B} = \{y; 0 \leq y < 9\}$ and the event $Y \in B$ is the union of two mutually exclusive events if $B = \{y; 0 < y < 1\}$.

4.6 Distributions of Order Statistics

In this section the notion of an order statistic will be defined and we shall investigate some of the simpler properties of such a statistic. These statistics have in recent times come to play an important role in statistical inference partly because some of their properties do not depend upon the distribution from which the random sample is obtained.

Let $X_1, X_2, \ldots, X_n$ denote a random sample from a distribution of the *continuous type* having a p.d.f. $f(x)$ that is positive, provided that $a < x < b$. Let Y_1 be the smallest of these X_i, Y_2 the next X_i in order of magnitude, $\ldots$, and Y_n the largest X_i. That is, $Y_1 < Y_2 < \cdots < Y_n$ represent $X_1, X_2, \ldots, X_n$ when the latter are arranged in ascending order of magnitude. Then Y_i, $i = 1, 2, \ldots, n$, is called the ith order statistic of the random sample $X_1, X_2, \ldots, X_n$. It will be shown that the joint p.d.f. of $Y_1, Y_2, \ldots, Y_n$ is given by

$$(1) \qquad g(y_1, y_2, \ldots, y_n) = (n!)f(y_1)f(y_2)\cdots f(y_n),$$
$$a < y_1 < y_2 < \cdots < y_n < b,$$

$$= 0 \text{ elsewhere.}$$

We shall prove this only for the case $n = 3$, but the argument is seen to be entirely general. With $n = 3$, the joint p.d.f. of X_1, X_2, X_3 is $f(x_1)f(x_2)f(x_3)$. Consider a probability such as $\Pr(a < X_1 = X_2 < b, a < X_3 < b)$. This probability is given by

$$\int_a^b \int_a^b \int_{x_2}^{x_2} f(x_1)f(x_2)f(x_3)\, dx_1\, dx_2\, dx_3 = 0,$$

since

$$\int_{x_2}^{x_2} f(x_1)\, dx_1$$

is defined in calculus to be zero. As has been pointed out, we may, without altering the distribution of X_1, X_2, X_3, define the joint p.d.f. $f(x_1)f(x_2)f(x_3)$ to be zero at all points (x_1, x_2, x_3) that have at least two of their coordinates equal. Then the set $\mathscr{A}$, where $f(x_1)f(x_2)f(x_3) > 0$, is the union of the six mutually disjoint sets:

$$A_1 = \{(x_1, x_2, x_3); a < x_1 < x_2 < x_3 < b\},$$
$$A_2 = \{(x_1, x_2, x_3); a < x_2 < x_1 < x_3 < b\},$$
$$A_3 = \{(x_1, x_2, x_3); a < x_1 < x_3 < x_2 < b\},$$
$$A_4 = \{(x_1, x_2, x_3); a < x_2 < x_3 < x_1 < b\},$$
$$A_5 = \{(x_1, x_2, x_3); a < x_3 < x_1 < x_2 < b\},$$
$$A_6 = \{(x_1, x_2, x_3); a < x_3 < x_2 < x_1 < b\}.$$

There are six of these sets because we can arrange x_1, x_2, x_3 in precisely $3! = 6$ ways. Consider the functions $y_1 = $ minimum of x_1, x_2, x_3; $y_2 = $ middle in magnitude of x_1, x_2, x_3; and $y_3 = $ maximum of x_1, x_2, x_3. These functions define one-to-one transformations that map each of $A_1, A_2, \ldots, A_6$ onto the same set $\mathscr{B} = \{(y_1, y_2, y_3); a < y_1 < y_2 < y_3 < b\}$. The inverse functions are, for points in A_1, $x_1 = y_1$, $x_2 = y_2$, $x_3 = y_3$; for points in A_2, they are $x_1 = y_2$, $x_2 = y_1$, $x_3 = y_3$; and so on, for each of the remaining four sets. Then we have that

$$J_1 = \begin{vmatrix} 1 & 0 & 0 \\ 0 & 1 & 0 \\ 0 & 0 & 1 \end{vmatrix} = 1$$

and

$$J_2 = \begin{vmatrix} 0 & 1 & 0 \\ 1 & 0 & 0 \\ 0 & 0 & 1 \end{vmatrix} = -1.$$

It is easily verified that the absolute value of each of the $3! = 6$ Jacobians is $+1$. Thus the joint p.d.f. of the three order statistics $Y_1 = $ minimum of X_1, X_2, X_3; $Y_2 = $ middle in magnitude of X_1, X_2, X_3; $Y_3 = $ maximum of X_1, X_2, X_3 is

$$g(y_1, y_2, y_3) = |J_1| f(y_1) f(y_2) f(y_3) + |J_2| f(y_2) f(y_1) f(y_3) + \cdots$$
$$+ |J_6| f(y_3) f(y_2) f(y_1), \qquad a < y_1 < y_2 < y_3 < b,$$
$$= (3!) f(y_1) f(y_2) f(y_3), \qquad a < y_1 < y_2 < y_3 < b,$$
$$= 0 \text{ elsewhere.}$$

This is Equation (1) with $n = 3$.

In accordance with the natural extension of Theorem 1, Section 2.4, to distributions of more than two random variables, it is seen that the order statistics, unlike the items of the random sample, are stochastically dependent.

Example 1. Let X denote a random variable of the continuous type with a p.d.f. $f(x)$ that is positive and continuous provided that $a < x < b$, and is zero elsewhere. The distribution function $F(x)$ of X may be written

$$F(x) = \int_a^x f(w) \, dw, \qquad a < x < b.$$

If $x \le a$, $F(x) = 0$; and if $b \le x$, $F(x) = 1$. Thus there is a unique median m of the distribution with $F(m) = \frac{1}{2}$. Let X_1, X_2, X_3 denote a random sample from this distribution and let $Y_1 < Y_2 < Y_3$ denote the order statistics of the sample. We shall compute the probability that $Y_2 \le m$. The joint p.d.f. of the three order statistics is

$$g(y_1, y_2, y_3) = 6 f(y_1) f(y_2) f(y_3), \qquad a < y_1 < y_2 < y_3 < b,$$
$$= 0 \text{ elsewhere.}$$

The p.d.f. of Y_2 is then

$$h(y_2) = 6 f(y_2) \int_{y_2}^b \int_a^{y_2} f(y_1) f(y_3) \, dy_1 \, dy_3,$$
$$= 6 f(y_2) F(y_2) [1 - F(y_2)], \qquad a < y_2 < b,$$
$$= 0 \text{ elsewhere.}$$

Accordingly,

$$\Pr\,(Y_2 \le m) = 6 \int_a^m \{F(y_2)f(y_2) - [F(y_2)]^2 f(y_2)\}\, dy_2$$

$$= 6 \left\{ \frac{[F(y_2)]^2}{2} - \frac{[F(y_2)]^3}{3} \right\}_a^m = \frac{1}{2}.$$

The procedure used in Example 1 can be used to obtain general formulas for the marginal probability density functions of the order statistics. We shall do this now. Let X denote a random variable of the continuous type having a p.d.f. $f(x)$ that is positive and continuous, provided that $a < x < b$, and is zero elsewhere. Then the distribution function $F(x)$ may be written

$$F(x) = 0, \qquad x \le a,$$

$$= \int_a^x f(w)\, dw, \qquad a < x < b,$$

$$= 1, \qquad b \le x.$$

Accordingly, $F'(x) = f(x)$, $a < x < b$. Moreover, if $a < x < b$,

$$1 - F(x) = F(b) - F(x)$$

$$= \int_a^b f(w)\, dw - \int_a^x f(w)\, dw$$

$$= \int_x^b f(w)\, dw.$$

Let $X_1, X_2, \ldots, X_n$ denote a random sample of size n from this distribution, and let $Y_1, Y_2, \ldots, Y_n$ denote the order statistics of this random sample. Then the joint p.d.f. of $Y_1, Y_2, \ldots, Y_n$ is

$$g(y_1, y_2, \ldots, y_n) = n! f(y_1)f(y_2)\cdots f(y_n), \qquad a < y_1 < y_2 < \cdots < y_n < b,$$

$$= 0 \text{ elsewhere.}$$

It will first be shown how the marginal p.d.f. of Y_n may be expressed in terms of the distribution function $F(x)$ and the p.d.f. $f(x)$ of the random variable X. If $a < y_n < b$, the marginal p.d.f. of y_n is given by

$$g_n(y_n) = \int_a^{y_n} \cdots \int_a^{y_4} \int_a^{y_3} \int_a^{y_2} n! f(y_1)f(y_2)\cdots f(y_n)\, dy_1\, dy_2\, dy_3 \cdots dy_{n-1}$$

$$= \int_a^{y_n} \cdots \int_a^{y_4} \int_a^{y_3} n! \left(\int_a^{y_2} f(y_1)\, dy_1 \right) f(y_2)\cdots f(y_n)\, dy_2 \cdots dy_{n-1}$$

$$= \int_a^{y_n} \cdots \int_a^{y_4} \int_a^{y_3} n!\, F(y_2)f(y_2)\cdots f(y_n)\, dy_2 \cdots dy_{n-1},$$

since $F(x) = \int_a^x f(w)\, dw$. Now

$$\int_a^{y_3} F(y_2)f(y_2)\, dy_2 = \left.\frac{[F(y_2)]^2}{2}\right|_a^{y_3}$$

$$= \frac{[F(y_3)]^2}{2},$$

since $F(a) = 0$. Thus

$$g_n(y_n) = \int_a^{y_n}\cdots\int_a^{y_4} n!\, \frac{[F(y_3)]^2}{2} f(y_3)\cdots f(y_n)\, dy_3\cdots dy_{n-1}.$$

But

$$\int_a^{y_4} \frac{[F(y_3)]^2}{2} f(y_3)\, dy_3 = \left.\frac{[F(y_3)]^3}{2\cdot 3}\right|_a^{y_4} = \frac{[F(y_4)]^3}{2\cdot 3},$$

so

$$g_n(y_n) = \int_a^{y_n}\cdots\int_a^{y_5} n!\, \frac{[F(y_4)]^3}{3!} f(y_4)\cdots f(y_n)\, dy_4\cdots dy_{n-1}.$$

If the successive integrations on $y_4, \ldots, y_{n-1}$ are carried out, it is seen that

$$g_n(y_n) = n!\, \frac{[F(y_n)]^{n-1}}{(n-1)!} f(y_n)$$

$$= n[F(y_n)]^{n-1}f(y_n), \qquad a < y_n < b,$$

$$= 0 \text{ elsewhere.}$$

It will next be shown how to express the marginal p.d.f. of Y_1 in terms of $F(x)$ and $f(x)$. We have, for $a < y_1 < b$,

$$g_1(y_1) = \int_{y_1}^b\cdots\int_{y_{n-3}}^b \int_{y_{n-2}}^b \int_{y_{n-1}}^b n!\, f(y_1)f(y_2)\cdots f(y_n)\, dy_n\, dy_{n-1}\cdots dy_2$$

$$= \int_{y_1}^b\cdots\int_{y_{n-3}}^b \int_{y_{n-2}}^b n!\, f(y_1)f(y_2)\cdots$$

$$f(y_{n-1})[1 - F(y_{n-1})]\, dy_{n-1}\cdots dy_2.$$

But

$$\int_{y_{n-2}}^b [1 - F(y_{n-1})]f(y_{n-1})\, dy_{n-1} = \left.-\frac{[1 - F(y_{n-1})]^2}{2}\right|_{y_{n-2}}^b$$

$$= \frac{[1 - F(y_{n-2})]^2}{2},$$

so that

$$g_1(y_1) = \int_{y_1}^{b} \cdots \int_{y_{n-3}}^{b} n! \, f(y_1) \cdots f(y_{n-2}) \frac{[1 - F(y_{n-2})]^2}{2} \, dy_{n-2} \cdots dy_2.$$

Upon completing the integrations, it is found that

$$g_1(y_1) = n[1 - F(y_1)]^{n-1} f(y_1), \qquad a < y_1 < b,$$
$$= 0 \text{ elsewhere.}$$

Once it is observed that

$$\int_{a}^{x} [F(w)]^{\alpha-1} f(w) \, dw = \frac{[F(x)]^{\alpha}}{\alpha}, \qquad \alpha > 0$$

and that

$$\int_{y}^{b} [1 - F(w)]^{\beta-1} f(w) \, dw = \frac{[1 - F(y)]^{\beta}}{\beta}, \qquad \beta > 0,$$

it is easy to express the marginal p.d.f. of any order statistic, say Y_k, in terms of $F(x)$ and $f(x)$. This is done by evaluating the integral

$$g_k(y_k) = \int_{a}^{y_k} \cdots \int_{a}^{y_2} \int_{y_k}^{b} \cdots \int_{y_{n-1}}^{b} n! \, f(y_1) f(y_2) \cdots f(y_n) \, dy_n \cdots$$
$$dy_{k+1} \, dy_1 \cdots dy_{k-1}.$$

The result is

$$(2) \quad g_k(y_k) = \frac{n!}{(k-1)! \, (n-k)!} [F(y_k)]^{k-1} [1 - F(y_k)]^{n-k} f(y_k),$$
$$a < y_k < b,$$
$$= 0 \text{ elsewhere.}$$

Example 2. Let $Y_1 < Y_2 < Y_3 < Y_4$ denote the order statistics of a random sample of size 4 from a distribution having p.d.f.

$$f(x) = 2x, \qquad 0 < x < 1,$$
$$= 0 \text{ elsewhere.}$$

We shall express the p.d.f. of Y_3 in terms of $f(x)$ and $F(x)$ and then compute $\Pr\left(\frac{1}{2} < Y_3\right)$. Here $F(x) = x^2$, provided that $0 < x < 1$, so that

$$g_3(y_3) = \frac{4!}{2! \, 1!} (y_3^2)^2 (1 - y_3^2)(2y_3), \qquad 0 < y_3 < 1,$$
$$= 0 \text{ elsewhere.}$$

Thus

$$\Pr\left(\tfrac{1}{2} < Y_3\right) = \int_{1/2}^{\infty} g_3(y_3)\, dy_3$$

$$= \int_{1/2}^{1} 24(y_3^5 - y_3^7)\, dy_3 = \tfrac{243}{256}.$$

Finally, the joint p.d.f. of any two order statistics, say $Y_i < Y_j$, is as easily expressed in terms of $F(x)$ and $f(x)$. We have

$$g_{ij}(y_i, y_j) = \int_a^{y_i} \cdots \int_a^{y_2} \int_{y_i}^{y_j} \cdots \int_{y_{j-2}}^{y_j} \int_{y_j}^{b} \cdots \int_{y_{n-1}}^{b} n!\, f(y_1) \cdots$$

$$f(y_n)\, dy_n \cdots dy_{j+1}\, dy_{j-1} \cdots dy_{i+1}\, dy_1 \cdots dy_{i-1}.$$

Since, for $\gamma > 0$,

$$\int_x^y [F(y) - F(w)]^{\gamma - 1} f(w)\, dw = -\frac{[F(y) - F(w)]^\gamma}{\gamma}\bigg|_x^y$$

$$= \frac{[F(y) - F(x)]^\gamma}{\gamma},$$

it is found that

$$(3) \quad g_{ij}(y_i, y_j) = \frac{n!}{(i - 1)!\,(j - i - 1)!\,(n - j)!}$$

$$\times [F(y_i)]^{i-1}[F(y_j) - F(y_i)]^{j-i-1}[1 - F(y_j)]^{n-j} f(y_i) f(y_j)$$

for $a < y_i < y_j < b$, and zero elsewhere.

Remark. There is an easy method of remembering a p.d.f. like that given in Formula (3). The probability $\Pr(y_i < Y_i < y_i + \Delta_i, y_j < Y_j < y_j + \Delta_j)$, where Δ_i and Δ_j are small, can be approximated by the following multinomial probability. In n independent trials, $i - 1$ outcomes must be less than y_i (an event that has probability $p_1 = F(y_i)$ on each trial); $j - i - 1$ outcomes must be between $y_i + \Delta_i$ and y_j [an event with approximate probability $p_2 = F(y_j) - F(y_i)$ on each trial]; $n - j$ outcomes must be greater than $y_j + \Delta_j$ (an event with approximate probability $p_3 = 1 - F(y_j)$ on each trial); one outcome must be between y_i and $y_i + \Delta_i$ (an event with approximate probability $p_4 = f(y_i)\Delta_i$ on each trial); and finally one outcome must be between y_j and $y_j + \Delta_j$ [an event with approximate probability $p_5 = f(y_j)\Delta_j$ on each trial]. This multinomial probability is

$$\frac{n!}{(i - 1)!\,(j - i - 1)!\,(n - j)!\,1!\,1!}\, p_1^{i-1} p_2^{j-i-1} p_3^{n-j} p_4 p_5,$$

which is $g(y_i, y_j)\Delta_i\Delta_j$.

Certain functions of the order statistics $Y_1, Y_2, \ldots, Y_n$ are important statistics themselves. A few of these are: (a) $Y_n - Y_1$, which is called the range of the random sample; (b) $(Y_1 + Y_n)/2$, which is called the midrange of the random sample; and (c) if n is odd, $Y_{(n+1)/2}$, which is called the median of the random sample.

Example 3. Let Y_1, Y_2, Y_3 be the order statistics of a random sample of size 3 from a distribution having p.d.f.

$$f(x) = 1, \quad 0 < x < 1,$$
$$= 0 \text{ elsewhere.}$$

We seek the p.d.f. of the sample range $Z_1 = Y_3 - Y_1$. Since $F(x) = x$, $0 < x < 1$, the joint p.d.f. of Y_1 and Y_3 is

$$g_{13}(y_1, y_3) = 6(y_3 - y_1), \quad 0 < y_1 < y_3 < 1,$$
$$= 0 \text{ elsewhere.}$$

In addition to $Z_1 = Y_3 - Y_1$, let $Z_2 = Y_3$. Consider the functions $z_1 = y_3 - y_1$, $z_2 = y_3$, and their inverses $y_1 = z_2 - z_1$, $y_3 = z_2$, so that the corresponding Jacobian of the one-to-one-transformation is

$$J = \begin{vmatrix} \dfrac{\partial y_1}{\partial z_1} & \dfrac{\partial y_1}{\partial z_2} \\[2mm] \dfrac{\partial y_3}{\partial z_1} & \dfrac{\partial y_3}{\partial z_2} \end{vmatrix} = \begin{vmatrix} -1 & 1 \\ 0 & 1 \end{vmatrix} = -1.$$

Thus the joint p.d.f. of Z_1 and Z_2 is

$$h(z_1, z_2) = |-1|6z_1 = 6z_1, \quad 0 < z_1 < z_2 < 1.$$
$$= 0 \text{ elsewhere.}$$

Accordingly, the p.d.f. of the range $Z_1 = Y_3 - Y_1$ of the random sample of size 3 is

$$h_1(z_1) = \int_{z_1}^{1} 6z_1 \, dz_2 = 6z_1(1 - z_1), \quad 0 < z_1 < 1,$$
$$= 0 \text{ elsewhere.}$$

EXERCISES

4.50. Let $Y_1 < Y_2 < Y_3 < Y_4$ be the order statistics of a random sample of size 4 from the distribution having p.d.f. $f(x) = e^{-x}$, $0 < x < \infty$, zero elsewhere. Find $\Pr(3 \le Y_4)$.

4.51. Let X_1, X_2, X_3 be a random sample from a distribution of the continuous type having p.d.f. $f(x) = 2x, 0 < x < 1$, zero elsewhere. Compute the probability that the smallest of these X_i exceeds the median of the distribution.

4.52. Let $f(x) = \frac{1}{6}$, $x = 1, 2, 3, 4, 5, 6$, zero elsewhere, be the p.d.f. of a distribution of the discrete type. Show that the p.d.f. of the smallest item of a random sample of size 5 from this distribution is

$$g_1(y_1) = \left(\frac{7 - y_1}{6}\right)^5 - \left(\frac{6 - y_1}{6}\right)^5, \qquad y_1 = 1, 2, \ldots, 6,$$

zero elsewhere. Note that in this exercise the random sample is from a distribution of the discrete type. All formulas in the text were derived under the assumption that the random sample is from a distribution of the continuous type and are not applicable. Why?

4.53. Let $Y_1 < Y_2 < Y_3 < Y_4 < Y_5$ denote the order statistics of a random sample of size 5 from a distribution having p.d.f. $f(x) = e^{-x}$, $0 < x < \infty$, zero elsewhere. Show that $Z_1 = Y_2$ and $Z_2 = Y_4 - Y_2$ are stochastically independent. *Hint.* First find the joint p.d.f. of Y_2 and Y_4.

4.54. Let $Y_1 < Y_2 < \cdots < Y_n$ be the order statistics of a random sample of size n from a distribution with p.d.f. $f(x) = 1$, $0 < x < 1$, zero elsewhere. Show that the kth order statistic Y_k has a beta p.d.f. with parameters $\alpha = k$ and $\beta = n - k + 1$.

4.55. Let $Y_1 < Y_2 < \cdots < Y_n$ be the order statistics from a Weibull distribution, Exercise 3.38, Section 3.3. Find the distribution function and p.d.f. of Y_1.

4.56. Find the probability that the range of a random sample of size 4 from the uniform distribution having the p.d.f. $f(x) = 1$, $0 < x < 1$, zero elsewhere, is less than $\frac{1}{2}$.

4.57. Let $Y_1 < Y_2 < Y_3$ be the order statistics of a random sample of size 3 from a distribution having the p.d.f. $f(x) = 2x$, $0 < x < 1$, zero elsewhere. Show that $Z_1 = Y_1/Y_2$, $Z_2 = Y_2/Y_3$, and $Z_3 = Y_3$ are mutually stochastically independent.

4.58. If a random sample of size 2 is taken from a distribution having p.d.f. $f(x) = 2(1 - x)$, $0 < x < 1$, zero elsewhere, compute the probability that one sample item is at least twice as large as the other.

4.59. Let $Y_1 < Y_2 < Y_3$ denote the order statistics of a random sample of size 3 from a distribution with p.d.f. $f(x) = 1$, $0 < x < 1$, zero elsewhere. Let $Z = (Y_1 + Y_3)/2$ be the midrange of the sample. Find the p.d.f. of Z.

4.60. Let $Y_1 < Y_2$ denote the order statistics of a random sample of size 2 from $n(0, \sigma^2)$. Show that $E(Y_1) = -\sigma/\sqrt{\pi}$. *Hint.* Evaluate $E(Y_1)$ by using the joint p.d.f. of Y_1 and Y_2, and first integrating on y_1.

4.61. Let $Y_1 < Y_2$ be the order statistics of a random sample of size 2 from a distribution of the continuous type which has p.d.f. $f(x)$ such that $f(x) > 0$, provided $x \geq 0$, and $f(x) = 0$ elsewhere. Show that the stochastic independence of $Z_1 = Y_1$ and $Z_2 = Y_2 - Y_1$ characterizes the gamma p.d.f. $f(x)$, which has parameters $\alpha = 1$ and $\beta > 0$. *Hint.* Use the change-of-variable technique to find the joint p.d.f. of Z_1 and Z_2 from that of Y_1 and Y_2. Accept the fact that the functional equation $h(0)h(x + y) \equiv h(x)h(y)$ has the solution $h(x) = c_1 e^{c_2 x}$, where c_1 and c_2 are constants.

4.62. Let Y denote the median of a random sample of size $n = 2k + 1$, k a positive integer, from a distribution which is $n(\mu, \sigma^2)$. Prove that the graph of the p.d.f. of Y is symmetric with respect to the vertical axis through $y = \mu$ and deduce that $E(Y) = \mu$.

4.63. Let X and Y denote stochastically independent random variables with respective probability density functions $f(x) = 2x$, $0 < x < 1$, zero elsewhere, and $g(y) = 3y^2$, $0 < y < 1$, zero elsewhere. Let $U = \min(X, Y)$ and $V = \max(X, Y)$. Find the joint p.d.f. of U and V. *Hint.* Here the two inverse transformations are given by $x = u$, $y = v$ and $x = v$, $y = u$.

4.64. Let the joint p.d.f. of X and Y be $f(x, y) = \frac{12}{7}x(x + y)$, $0 < x < 1$, $0 < y < 1$, zero elsewhere. Let $U = \min(X, Y)$ and $V = \max(X, Y)$. Find the joint p.d.f. of U and V.

4.65. Let $X_1, X_2, \ldots, X_n$ be a random sample from a distribution of either type. A measure of spread is *Gini's mean difference*

$$G = \sum_{j=2}^{n} \sum_{i=1}^{j-1} |X_i - X_j| \Big/ \binom{n}{2}.$$

(a) If $n = 10$, find $a_1, a_2, \ldots, a_{10}$ so that $G = \sum_{i=1}^{10} a_i Y_i$, where $Y_1, Y_2, \ldots,$ Y_{10} are the order statistics of the sample.

(b) Show that $E(G) = 2\sigma/\sqrt{\pi}$ if the sample arises from the normal distribution $n(\mu, \sigma^2)$.

4.66. Let $Y_1 < Y_2 < \cdots < Y_n$ be the order statistics of a random sample of size n from the exponential distribution with p.d.f. $f(x) = e^{-x}$, $0 < x < \infty$, zero elsewhere.

(a) Show that $Z_1 = nY_1$, $Z_2 = (n - 1)(Y_2 - Y_1)$, $Z_3 = (n - 2)$ $(Y_3 - Y_2), \ldots, Z_n = Y_n - Y_{n-1}$ are stochastically independent and that each Z_i has the exponential distribution.

(b) Demonstrate that all linear functions of $Y_1, Y_2, \ldots, Y_n$, such as $\sum_{1}^{n} a_i Y_i$, can be expressed as linear functions of stochastically independent random variables.

4.67. In the Program Evaluation and Review Technique (PERT), we are interested in the total time to complete a project that is comprised of a large number of subprojects. For illustration, let X_1, X_2, X_3 be three stochastically independent random times for three subprojects. If these subprojects are in series (the first one must be completed before the second starts, etc.), then we are interested in the sum $Y = X_1 + X_2 + X_3$. If these are in parallel (can be worked on simultaneously), then we are interested in $Z = \max(X_1, X_2, X_3)$. In the case each of these random variables has the uniform distribution with p.d.f. $f(x) = 1$, $0 < x < 1$, zero elsewhere, find (a) the p.d.f. of Y and (b) the p.d.f. of Z.

4.7 The Moment-Generating-Function Technique

The change-of-variable procedure has been seen, in certain cases, to be an effective method of finding the distribution of a function of several random variables. An alternative procedure, built around the concept of the moment-generating function of a distribution, will be presented in this section. This procedure is particularly effective in certain instances. We should recall that a moment-generating function, when it exists, is unique and that it uniquely determines the distribution of a probability.

Let $\varphi(x_1, x_2, \ldots, x_n)$ denote the joint p.d.f. of the n random variables $X_1, X_2, \ldots, X_n$. These random variables may or may not be the items of a random sample from some distribution that has a given p.d.f. $f(x)$. Let $Y_1 = u_1(X_1, X_2, \ldots, X_n)$. We seek $g(y_1)$, the p.d.f. of the random variable Y_1. Consider the moment-generating function of Y_1. If it exists, it is given by

$$M(t) = E(e^{tY_1}) = \int_{-\infty}^{\infty} e^{ty_1} g(y_1)\, dy_1$$

in the continuous case. It would seem that we need to know $g(y_1)$ before we can compute $M(t)$. That this is not the case is a fundamental fact. To see this consider

(1) $\qquad \int_{-\infty}^{\infty} \cdots \int_{-\infty}^{\infty} \exp\left[tu_1(x_1, \ldots, x_n)\right]\varphi(x_1, \ldots, x_n)\, dx_1 \cdots dx_n,$

which we assume to exist for $-h < t < h$. We shall introduce n new variables of integration. They are $y_1 = u_1(x_1, x_2, \ldots, x_n), \ldots, y_n = u_n(x_1, x_2, \ldots, x_n)$. Momentarily, we assume that these functions define a one-to-one transformation. Let $x_i = w_i(y_1, y_2, \ldots, y_n)$, $i = 1, 2, \ldots, n$, denote the inverse functions and let J denote the Jacobian. Under this transformation, display (1) becomes

(2) $\qquad \int_{-\infty}^{\infty} \cdots \int_{-\infty}^{\infty} e^{ty_1} |J| \varphi(w_1, \ldots, w_n)\, dy_2 \cdots dy_n\, dy_1.$

In accordance with Section 4.5,

$$|J|\varphi[w_1(y_1, y_2, \ldots, y_n), \ldots, w_n(y_1, y_2, \ldots, y_n)]$$

is the joint p.d.f. of $Y_1, Y_2, \ldots, Y_n$. The marginal p.d.f. $g(y_1)$ of Y_1 is obtained by integrating this joint p.d.f. on $y_2, \ldots, y_n$. Since the factor e^{ty_1} does not involve the variables $y_2, \ldots, y_n$, display (2) may be written as

(3)
$$\int_{-\infty}^{\infty} e^{ty_1} g(y_1) \, dy_1.$$

But this is by definition the moment-generating function $M(t)$ of the distribution of Y_1. That is, we can compute $E[\exp(tu_1(X_1, \ldots, X_n))]$ and have the value of $E(e^{tY_1})$, where $Y_1 = u_1(X_1, \ldots, X_n)$. This fact provides another technique to help us find the p.d.f. of a function of several random variables. For if the moment-generating function of Y_1 is seen to be that of a certain kind of distribution, the uniqueness property makes it certain that Y_1 has that kind of distribution. When the p.d.f. of Y_1 is obtained in this manner, we say that we use the *moment-generating-function technique*.

The reader will observe that we have assumed the transformation to be one-to-one. We did this for simplicity of presentation. If the transformation is not one-to-one, let

$$x_j = w_{ji}(y_1, \ldots, y_n), \qquad j = 1, 2, \ldots, n, \quad i = 1, 2, \ldots, k,$$

denote the k groups of n inverse functions each. Let J_i, $i = 1, 2, \ldots, k$, denote the k Jacobians. Then

(4)
$$\sum_{i=1}^{k} |J_i| \varphi[w_{1i}(y_1, \ldots, y_n), \ldots, w_{ni}(y_1, \ldots, y_n)]$$

is the joint p.d.f. of $Y_1, \ldots, Y_n$. Then display (1) becomes display (2) with $|J|\varphi(w_1, \ldots, w_n)$ replaced by display (4). Hence our result is valid if the transformation is not one-to-one. It seems evident that we can treat the discrete case in an analogous manner with the same result.

It should be noted that the expectation, subject to its existence, of any function of Y_1 can be computed in like manner. That is, if $w(y_1)$ is a function of y_1, then

$$E[w(Y_1)] = \int_{-\infty}^{\infty} w(y_1) g(y_1) \, dy_1$$

$$= \int_{-\infty}^{\infty} \cdots \int_{-\infty}^{\infty} w[u_1(x_1, \ldots, x_n)] \varphi(x_1, \ldots, x_n) \, dx_1 \cdots dx_n.$$

We shall now give some examples and prove some theorems where we use the moment-generating-function technique. In the first example, to emphasize the nature of the problem, we find the distribution of a rather simple statistic both by a direct probabilistic argument and by the moment-generating-function technique.

Example 1. Let the stochastically independent random variables X_1 and X_2 have the same p.d.f.

$$f(x) = \frac{x}{6}, \qquad x = 1, 2, 3,$$

$$= 0 \text{ elsewhere};$$

that is, the p.d.f. of X_1 is $f(x_1)$ and that of X_2 is $f(x_2)$; and so the joint p.d.f. of X_1 and X_2 is

$$f(x_1)f(x_2) = \frac{x_1 x_2}{36}, \qquad x_1 = 1, 2, 3, x_2 = 1, 2, 3,$$

$$= 0 \text{ elsewhere}.$$

A probability, such as $\Pr(X_1 = 2, X_2 = 3)$, can be seen immediately to be $(2)(3)/36 = \frac{1}{6}$. However, consider a probability such as $\Pr(X_1 + X_2 = 3)$. The computation can be made by first observing that the event $X_1 + X_2 = 3$ is the union, exclusive of the events with probability zero, of the two mutually exclusive events $(X_1 = 1, X_2 = 2)$ and $(X_1 = 2, X_2 = 1)$. Thus

$$\Pr(X_1 + X_2 = 3) = \Pr(X_1 = 1, X_2 = 2) + \Pr(X_1 = 2, X_2 = 1)$$

$$= \frac{(1)(2)}{36} + \frac{(2)(1)}{36} = \frac{4}{36}.$$

More generally, let y represent any of the numbers 2, 3, 4, 5, 6. The probability of each of the events $X_1 + X_2 = y$, $y = 2, 3, 4, 5, 6$, can be computed as in the case $y = 3$. Let $g(y) = \Pr(X_1 + X_2 = y)$. Then the table

y	2	3	4	5	6
$g(y)$	$\frac{1}{36}$	$\frac{4}{36}$	$\frac{10}{36}$	$\frac{12}{36}$	$\frac{9}{36}$

gives the values of $g(y)$ for $y = 2, 3, 4, 5, 6$. For all other values of y, $g(y) = 0$. What we have actually done is to define a new random variable Y by $Y = X_1 + X_2$, and we have found the p.d.f. $g(y)$ of this random variable Y. We shall now solve the same problem, and by the moment-generating-function technique.

Now the moment-generating function of Y is

$$M(t) = E(e^{t(X_1 + X_2)})$$

$$= E(e^{tX_1} e^{tX_2})$$

$$= E(e^{tX_1})E(e^{tX_2}),$$

since X_1 and X_2 are stochastically independent. In this example X_1 and X_2 have the same distribution, so they have the same moment-generating function; that is,

$$E(e^{tX_1}) = E(e^{tX_2}) = \tfrac{1}{6}e^t + \tfrac{2}{6}e^{2t} + \tfrac{3}{6}e^{3t}.$$

Thus

$$M(t) = (\tfrac{1}{6}e^t + \tfrac{2}{6}e^{2t} + \tfrac{3}{6}e^{3t})^2$$

$$= \tfrac{1}{36}e^{2t} + \tfrac{4}{36}e^{3t} + \tfrac{10}{36}e^{4t} + \tfrac{12}{36}e^{5t} + \tfrac{9}{36}e^{6t}.$$

This form of $M(t)$ tells us immediately that the p.d.f. $g(y)$ of Y is zero except at $y = 2, 3, 4, 5, 6$, and that $g(y)$ assumes the values $\tfrac{1}{36}, \tfrac{4}{36}, \tfrac{10}{36}, \tfrac{12}{36}, \tfrac{9}{36}$, respectively, at these points where $g(y) > 0$. This is, of course, the same result that was obtained in the first solution. There appears here to be little, if any, preference for one solution over the other. But in more complicated situations, and particularly with random variables of the continuous type, the moment-generating-function technique can prove very powerful.

Example 2. Let X_1 and X_2 be stochastically independent with normal distributions $n(\mu_1, \sigma_1^2)$ and $n(\mu_2, \sigma_2^2)$, respectively. Define the random variable Y by $Y = X_1 - X_2$. The problem is to find $g(y)$, the p.d.f. of Y. This will be done by first finding the moment-generating function of Y. It is

$$\begin{aligned} M(t) &= E(e^{t(X_1 - X_2)}) \\ &= E(e^{tX_1}e^{-tX_2}) \\ &= E(e^{tX_1})E(e^{-tX_2}), \end{aligned}$$

since X_1 and X_2 are stochastically independent. It is known that

$$E(e^{tX_1}) = \exp\left(\mu_1 t + \frac{\sigma_1^2 t^2}{2}\right)$$

and that

$$E(e^{tX_2}) = \exp\left(\mu_2 t + \frac{\sigma_2^2 t^2}{2}\right)$$

for all real t. Then $E(e^{-tX_2})$ can be obtained from $E(e^{tX_2})$ by replacing t by $-t$. That is,

$$E(e^{-tX_2}) = \exp\left(-\mu_2 t + \frac{\sigma_2^2 t^2}{2}\right).$$

Finally, then,

$$\begin{aligned} M(t) &= \exp\left(\mu_1 t + \frac{\sigma_1^2 t^2}{2}\right)\exp\left(-\mu_2 t + \frac{\sigma_2^2 t^2}{2}\right) \\ &= \exp\left((\mu_1 - \mu_2)t + \frac{(\sigma_1^2 + \sigma_2^2)t^2}{2}\right). \end{aligned}$$

The distribution of Y is completely determined by its moment-generating function $M(t)$, and it is seen that Y has the p.d.f. $g(y)$, which is $n(\mu_1 - \mu_2, \sigma_1^2 + \sigma_2^2)$. That is, the difference between two stochastically independent, normally distributed, random variables is itself a random variable which is normally distributed with mean equal to the difference of the means (in the order indicated) and the variance equal to the sum of the variances.

The following theorem, which is a generalization of Example 2, is very important in distribution theory.

Theorem 1. *Let $X_1, X_2, \ldots, X_n$ be mutually stochastically independent random variables having, respectively, the normal distributions $n(\mu_1, \sigma_1^2), n(\mu_2, \sigma_2^2), \ldots,$ and $n(\mu_n, \sigma_n^2)$. The random variable $Y = k_1X_1 + k_2X_2 + \cdots + k_nX_n$, where $k_1, k_2, \ldots, k_n$ are real constants, is normally distributed with mean $k_1\mu_1 + \cdots + k_n\mu_n$ and variance $k_1^2\sigma_1^2 + \cdots + k_n^2\sigma_n^2$. That is, Y is $n\left(\sum_1^n k_i\mu_i, \sum_1^n k_i^2\sigma_i^2\right)$.*

Proof. Because $X_1, X_2, \ldots, X_n$ are mutually stochastically independent, the moment-generating function of Y is given by

$$M(t) = E\{\exp\left[t(k_1X_1 + k_2X_2 + \cdots + k_nX_n)\right]\}$$
$$= E(e^{tk_1X_1})E(e^{tk_2X_2}) \cdots E(e^{tk_nX_n}).$$

Now

$$E(e^{tX_i}) = \exp\left(\mu_i t + \frac{\sigma_i^2 t^2}{2}\right),$$

for all real $t, i = 1, 2, \ldots, n$. Hence we have

$$E(e^{tk_iX_i}) = \exp\left[\mu_i(k_i t) + \frac{\sigma_i^2(k_i t)^2}{2}\right].$$

That is, the moment-generating function of Y is

$$M(t) = \prod_{i=1}^n \exp\left[(k_i\mu_i)t + \frac{(k_i^2\sigma_i^2)t^2}{2}\right]$$

$$= \exp\left[\left(\sum_1^n k_i\mu_i\right)t + \frac{\left(\sum_1^n k_i^2\sigma_i^2\right)t^2}{2}\right].$$

But this is the moment-generating function of a distribution that is $n\left(\sum_1^n k_i\mu_i, \sum_1^n k_i^2\sigma_i^2\right)$. This is the desired result.

If, in Theorem 1, we set each $k_i = 1$, we see that the sum of n mutually stochastically independent normally distributed variables has a normal distribution. The next theorem proves a similar result for chi-square variables.

Theorem 2. *Let $X_1, X_2, \ldots, X_n$ be mutually stochastically independent variables that have, respectively, the chi-square distributions $\chi^2(r_1)$, $\chi^2(r_2), \ldots$, and $\chi^2(r_n)$. Then the random variable $Y = X_1 + X_2 + \cdots + X_n$ has a chi-square distribution with $r_1 + \cdots + r_n$ degrees of freedom; that is, Y is $\chi^2(r_1 + \cdots + r_n)$.*

Proof. The moment-generating function of Y is

$$M(t) = E\{\exp[t(X_1 + X_2 + \cdots + X_n)]\}$$
$$= E(e^{tX_1})E(e^{tX_2}) \cdots E(e^{tX_n})$$

because $X_1, X_2, \ldots, X_n$ are mutually stochastically independent. Since

$$E(e^{tX_i}) = (1 - 2t)^{-r_i/2}, \qquad t < \tfrac{1}{2}, i = 1, 2, \ldots, n,$$

we have

$$M(t) = (1 - 2t)^{-(r_1 + r_2 + \cdots + r_n)/2}, \qquad t < \tfrac{1}{2}.$$

But this is the moment-generating function of a distribution that is $\chi^2(r_1 + r_2 + \cdots + r_n)$. Accordingly, Y has this chi-square distribution.

Next, let $X_1, X_2, \ldots, X_n$ be a random sample of size n from a distribution that is $n(\mu, \sigma^2)$. In accordance with Theorem 2 of Section 3.4, each of the random variables $(X_i - \mu)^2/\sigma^2, i = 1, 2, \ldots, n$, is $\chi^2(1)$. Moreover, these n random variables are mutually stochastically independent. Accordingly, by Theorem 2, the random variable $Y = \sum_1^n [(X_i - \mu)/\sigma]^2$ is $\chi^2(n)$. This proves the following theorem.

Theorem 3. *Let $X_1, X_2, \ldots, X_n$ denote a random sample of size n from a distribution that is $n(\mu, \sigma^2)$. The random variable*

$$Y = \sum_1^n \left(\frac{X_i - \mu}{\sigma}\right)^2$$

has a chi-square distribution with n degrees of freedom.

Not always do we sample from a distribution of one random variable. Let the random variables X and Y have the joint p.d.f. $f(x, y)$ and let the $2n$ random variables $(X_1, Y_1), (X_2, Y_2), \ldots, (X_n, Y_n)$ have the joint p.d.f.

$$f(x_1, y_1)f(x_2, y_2) \cdots f(x_n, y_n).$$

The n random pairs $(X_1, Y_1), (X_2, Y_2), \ldots, (X_n, Y_n)$ are then mutually stochastically independent and are said to constitute a random sample of size n from the distribution of X and Y. In the next paragraph we shall take $f(x, y)$ to be the normal bivariate p.d.f., and we shall solve a problem in sampling theory when we are sampling from this two-variable distribution.

Let $(X_1, Y_1), (X_2, Y_2), \ldots, (X_n, Y_n)$ denote a random sample of size n from a bivariate normal distribution with p.d.f. $f(x, y)$ and parameters $\mu_1, \mu_2, \sigma_1^2, \sigma_2^2$, and ρ. We wish to find the joint p.d.f. of the two statistics $\bar{X} = \sum_1^n X_i/n$ and $\bar{Y} = \sum_1^n Y_i/n$. We call $\bar{X}$ the mean of $X_1, \ldots, X_n$ and $\bar{Y}$ the mean of $Y_1, \ldots, Y_n$. Since the joint p.d.f. of the $2n$ random variables (X_i, Y_i), $i = 1, 2, \ldots, n$, is given by

$$\varphi = f(x_1, y_1)f(x_2, y_2) \cdots f(x_n, y_n),$$

the moment-generating function of the two means $\bar{X}$ and $\bar{Y}$ is given by

$$M(t_1, t_2) = \int_{-\infty}^{\infty} \cdots \int_{-\infty}^{\infty} \exp\left(\frac{t_1 \sum_1^n x_i}{n} + \frac{t_2 \sum_1^n y_i}{n}\right) \varphi \, dx_1 \cdots dy_n$$

$$= \prod_{i=1}^{n} \left[\int_{-\infty}^{\infty} \int_{-\infty}^{\infty} \exp\left(\frac{t_1 x_i}{n} + \frac{t_2 y_i}{n}\right) f(x_i, y_i) \, dx_i \, dy_i\right].$$

The justification of the form of the right-hand member of the second equality is that each pair (X_i, Y_i) has the same p.d.f., and that these n pairs are mutually stochastically independent. The twofold integral in the brackets in the last equality is the moment-generating function of X_i and Y_i (see Section 3.5) with t_1 replaced by t_1/n and t_2 replaced by t_2/n. Accordingly,

$$M(t_1, t_2) = \prod_{i=1}^{n} \exp\left[\frac{t_1 \mu_1}{n} + \frac{t_2 \mu_2}{n}\right.$$

$$\left. + \frac{\sigma_1^2(t_1/n)^2 + 2\rho\sigma_1\sigma_2(t_1/n)(t_2/n) + \sigma_2^2(t_2/n)^2}{2}\right]$$

$$= \exp\left[t_1 \mu_1 + t_2 \mu_2 + \frac{(\sigma_1^2/n)t_1^2 + 2\rho(\sigma_1\sigma_2/n)t_1 t_2 + (\sigma_2^2/n)t_2^2}{2}\right].$$

But this is the moment-generating function of a bivariate normal distribution with means μ_1 and μ_2, variances σ_1^2/n and σ_2^2/n, and correlation coefficient ρ; therefore, $\bar{X}$ and $\bar{Y}$ have this joint distribution.

EXERCISES

4.68. Let the stochastically independent random variables X_1 and X_2 have the same p.d.f. $f(x) = \frac{1}{6}$, $x = 1, 2, 3, 4, 5, 6$, zero elsewhere. Find the p.d.f. of $Y = X_1 + X_2$. Note, under appropriate assumptions, that Y may be interpreted as the sum of the spots that appear when two dice are cast.

4.69. Let X_1 and X_2 be stochastically independent with normal distributions $n(6, 1)$ and $n(7, 1)$, respectively. Find $\Pr(X_1 > X_2)$. *Hint.* Write $\Pr(X_1 > X_2) = \Pr(X_1 - X_2 > 0)$ and determine the distribution of $X_1 - X_2$.

4.70. Let X_1 and X_2 be stochastically independent random variables. Let X_1 and $Y = X_1 + X_2$ have chi-square distributions with r_1 and r degrees of freedom, respectively. Here $r_1 < r$. Show that X_2 has a chi-square distribution with $r - r_1$ degrees of freedom. *Hint.* Write $M(t) = E(e^{t(X_1+X_2)})$ and make use of the stochastic independence of X_1 and X_2.

4.71. Let the stochastically independent random variables X_1 and X_2 have binomial distributions with parameters n_1, $p_1 = \frac{1}{2}$ and n_2, $p_2 = \frac{1}{2}$, respectively. Show that $Y = X_1 - X_2 + n_2$ has a binomial distribution with parameters $n = n_1 + n_2$, $p = \frac{1}{2}$.

4.72. Let X be $n(0, 1)$. Use the moment-generating-function technique to show that $Y = X^2$ is $\chi^2(1)$. *Hint.* Evaluate the integral that represents $E(e^{tX^2})$ by writing $w = x\sqrt{1 - 2t}$, $t < \frac{1}{2}$.

4.73. Let $X_1, X_2, \ldots, X_n$ denote n mutually stochastically independent random variables with the moment-generating functions $M_1(t), M_2(t), \ldots, M_n(t)$, respectively.

(a) Show that $Y = k_1 X_1 + k_2 X_2 + \cdots + k_n X_n$, where $k_1, k_2, \ldots, k_n$ are real constants, has the moment-generating function $M(t) = \prod_1^n M_i(k_i t)$.

(b) If each $k_i = 1$ and if X_i is Poisson with mean μ_i, $i = 1, 2, \ldots, n$, prove that Y is Poisson with mean $\mu_1 + \cdots + \mu_n$.

4.74. If $X_1, X_2, \ldots, X_n$ is a random sample from a distribution with moment-generating function $M(t)$, show that the moment-generating functions of $\sum_1^n X_i$ and $\sum_1^n X_i/n$ are, respectively, $[M(t)]^n$ and $[M(t/n)]^n$.

4.75. In Exercise 4.67 concerning PERT, find: (a) the p.d.f. of Y; (b) the p.d.f. of Z in case each of the three stochastically independent variables has the p.d.f. $f(x) = e^{-x}$, $0 < x < \infty$, zero elsewhere.

4.76. If X and Y have a bivariate normal distribution with parameters μ_1, μ_2, σ_1^2, σ_2^2, and ρ, show that $Z = aX + bY + c$ is $n(a\mu_1 + b\mu_2 + c,$ $a^2\sigma_1^2 + 2ab\rho\sigma_1\sigma_2 + b^2\sigma_2^2)$, where a, b, and c are constants. *Hint.* Use the moment-generating function $M(t_1, t_2)$ of X and Y to find the moment-generating function of Z.

4.77. Let X and Y have a bivariate normal distribution with parameters $\mu_1 = 25$, $\mu_2 = 35$, $\sigma_1^2 = 4$, $\sigma_2^2 = 16$, and $\rho = \frac{17}{32}$. If $Z = 3X - 2Y$, find $\Pr(-2 < Z < 19)$.

4.78. Let U and V be stochastically independent random variables, each having a normal distribution with mean zero and variance 1. Show that the moment-generating function $E(e^{t(UV)})$ of the product UV is $(1 - t^2)^{-1/2}$, $-1 < t < 1$. *Hint.* Compare $E(e^{tUV})$ with the integral of a bivariate normal p.d.f. that has means equal to zero.

4.79. Let X and Y have a bivariate normal distribution with the parameters μ_1, μ_2, σ_1^2, σ_2^2, and ρ. Show that $W = X - \mu_1$ and $Z = (Y - \mu_2) - \rho(\sigma_2/\sigma_1)(X - \mu_1)$ are stochastically independent normal variables.

4.80. Let X_1, X_2, X_3 be a random sample of size $n = 3$ from the normal distribution $n(0, 1)$.

(a) Show that $Y_1 = X_1 + \delta X_3$, $Y_2 = X_2 + \delta X_3$ has a bivariate normal distribution.

(b) Find the value of δ so that the correlation coefficient $\rho = \frac{1}{2}$.

(c) What additional transformation involving Y_1 and Y_2 would produce a bivariate normal distribution with means μ_1 and μ_2, variances σ_1^2 and σ_2^2, and the same correlation coefficient ρ?

4.81. Let X_1, X_2, ..., X_n be a random sample of size n from the normal distribution $n(\mu, \sigma^2)$. Find the joint distribution of $Y = \sum_1^n a_i X_i$ and $Z = \sum_1^n b_i X_i$, where the a_i and b_i are real constants. When, and only when, are Y and Z stochastically independent? *Hint.* Note that the joint moment-generating function $E\left[\exp\left(t_1 \sum_1^n a_i X_i + t_2 \sum_1^n b_i X_i\right)\right]$ is that of a bivariate normal distribution.

4.82. Let X_1, X_2 be a random sample of size 2 from a distribution with positive variance and moment-generating function $M(t)$. If $Y = X_1 + X_2$ and $Z = X_1 - X_2$ are stochastically independent, prove that the distribution from which the sample is taken is a normal distribution. *Hint.* Show that $m(t_1, t_2) = E\{\exp[t_1(X_1 + X_2) + t_2(X_1 - X_2)]\} = M(t_1 + t_2)M(t_1 - t_2)$. Express each member of $m(t_1, t_2) = m(t_1, 0)m(0, t_2)$ in terms of M; differentiate twice with respect to t_2; set $t_2 = 0$; and solve the resulting differential equation in M.

4.8 The Distributions of $\bar{X}$ and nS^2/σ^2

Let X_1, X_2, ..., X_n denote a random sample of size $n \geq 2$ from a distribution that is $n(\mu, \sigma^2)$. In this section we shall investigate the distributions of the mean and the variance of this random sample, that is, the distributions of the two statistics $\bar{X} = \sum_{1}^{n} X_i/n$ and $S^2 = \sum_{1}^{n} (X_i - \bar{X})^2/n$.

The problem of the distribution of $\bar{X}$, the mean of the sample, is solved by the use of Theorem 1 of Section 4.7. We have here, in the notation of the statement of that theorem, $\mu_1 = \mu_2 = \cdots = \mu_n = \mu$, $\sigma_1^2 = \sigma_2^2 = \cdots = \sigma_n^2 = \sigma^2$, and $k_1 = k_2 = \cdots = k_n = 1/n$. Accordingly, $Y = \bar{X}$ has a normal distribution with mean and variance given by

$$\sum_{1}^{n} \left(\frac{1}{n}\mu\right) = \mu, \qquad \sum_{1}^{n} \left[\left(\frac{1}{n}\right)^2 \sigma^2\right] = \frac{\sigma^2}{n},$$

respectively. That is, $\bar{X}$ is $n(\mu, \sigma^2/n)$.

Example 1. Let $\bar{X}$ be the mean of a random sample of size 25 from a distribution that is $n(75, 100)$. Thus $\bar{X}$ is $n(75, 4)$. Then, for instance,

$$\Pr(71 < \bar{X} < 79) = N\left(\frac{79 - 75}{2}\right) - N\left(\frac{71 - 75}{2}\right)$$
$$= N(2) - N(-2) = 0.954.$$

We now take up the problem of the distribution of S^2, the variance of a random sample $X_1, \ldots, X_n$ from a distribution that is $n(\mu, \sigma^2)$. To do this, let us first consider the joint distribution of $Y_1 = \bar{X}$, $Y_2 = X_2, \ldots, Y_n = X_n$. The corresponding transformation

$$x_1 = ny_1 - y_2 - \cdots - y_n$$
$$x_2 = y_2$$
$$\vdots \qquad \vdots$$
$$x_n = y_n$$

has Jacobian n. Since

$$\sum_{1}^{n} (x_i - \mu)^2 = \sum_{1}^{n} (x_i - \bar{x} + \bar{x} - \mu)^2$$
$$= \sum_{1}^{n} (x_i - \bar{x})^2 + n(\bar{x} - \mu)^2$$

because $2(\bar{x} - \mu) \sum_{1}^{n} (x_i - \bar{x}) = 0$, the joint p.d.f. of $X_1, X_2, \ldots, X_n$ can be written

$$\left(\frac{1}{\sqrt{2\pi}\sigma}\right)^n \exp\left[-\frac{\sum (x_i - \bar{x})^2}{2\sigma^2} - \frac{n(\bar{x} - \mu)^2}{2\sigma^2}\right],$$

where $\bar{x}$ represents $(x_1 + x_2 + \cdots + x_n)/n$ and $-\infty < x_i < \infty$, $i = 1, 2, \ldots, n$. Accordingly, with $y_1 = \bar{x}$, we find that the joint p.d.f. of $Y_1, Y_2, \ldots, Y_n$ is

$$n\left(\frac{1}{\sqrt{2\pi}\sigma}\right)^n \exp\left[-\frac{(ny_1 - y_2 - \cdots - y_n - y_1)^2}{2\sigma^2}\right.$$

$$\left. - \frac{\sum\limits_{2}^{n}(y_i - y_1)^2}{2\sigma^2} - \frac{n(y_1 - \mu)^2}{2\sigma^2}\right],$$

$-\infty < y_i < \infty$, $i = 1, 2, \ldots, n$. The quotient of this joint p.d.f. and the p.d.f.

$$\frac{\sqrt{n}}{\sqrt{2\pi}\sigma} \exp\left[-\frac{n(y_1 - \mu)^2}{2\sigma^2}\right]$$

of $Y_1 = \bar{X}$ is the conditional p.d.f. of $Y_2, Y_3, \ldots, Y_n$, given $Y_1 = y_1$,

$$\sqrt{n}\left(\frac{1}{\sqrt{2\pi}\sigma}\right)^{n-1} \exp\left(-\frac{q}{2\sigma^2}\right),$$

where $q = (ny_1 - y_2 - \cdots - y_n - y_1)^2 + \sum\limits_{2}^{n}(y_i - y_1)^2$. Since this is a joint conditional p.d.f., it must be, for all $\sigma > 0$, that

$$\int_{-\infty}^{\infty} \cdots \int_{-\infty}^{\infty} \sqrt{n}\left(\frac{1}{\sqrt{2\pi}\sigma}\right)^{n-1} \exp\left(-\frac{q}{2\sigma^2}\right) dy_2 \cdots dy_n = 1.$$

Now consider

$$nS^2 = \sum_{1}^{n}(X_i - \bar{X})^2$$

$$= (nY_1 - Y_2 - \cdots - Y_n - Y_1)^2 + \sum_{2}^{n}(Y_i - Y_1)^2 = Q.$$

The conditional moment-generating function of $nS^2/\sigma^2 = Q/\sigma^2$, given $Y_1 = y_1$, is

$$E(e^{tQ/\sigma^2}|y_1) = \int_{-\infty}^{\infty} \cdots \int_{-\infty}^{\infty} \sqrt{n}\left(\frac{1}{\sqrt{2\pi}\sigma}\right)^{n-1} \exp\left[-\frac{(1 - 2t)q}{2\sigma^2}\right] dy_2 \cdots dy_n$$

$$= \left(\frac{1}{1 - 2t}\right)^{(n-1)/2} \int_{-\infty}^{\infty} \cdots \int_{-\infty}^{\infty} \sqrt{n}\left[\frac{1 - 2t}{2\pi\sigma^2}\right]^{(n-1)/2}$$

$$\times \exp\left[-\frac{(1 - 2t)q}{2\sigma^2} dy_2 \cdots dy_n,\right]$$

where $0 < 1 - 2t$, or $t < \frac{1}{2}$. However, this latter integral is exactly the same as that of the conditional p.d.f. of $Y_2, Y_3, \ldots, Y_n$, given $Y_1 = y_1$, with σ^2 replaced by $\sigma^2/(1 - 2t) > 0$, and thus must equal 1. Hence the conditional moment-generating function of nS^2/σ^2, given $Y_1 = y_1$ or equivalently $\bar{X} = \bar{x}$, is

$$E(e^{tnS^2/\sigma^2}|\bar{x}) = (1 - 2t)^{-(n-1)/2}, \qquad t < \tfrac{1}{2}.$$

That is, the conditional distribution of nS^2/σ^2, given $\bar{X} = \bar{x}$, is $\chi^2(n - 1)$. Moreover, since it is clear that this conditional distribution does not depend upon $\bar{x}$, $\bar{X}$ and nS^2/σ^2 must be stochastically independent or, equivalently, $\bar{X}$ and S^2 are stochastically independent.

To summarize, we have established, in this section, three important properties of $\bar{X}$ and S^2 when the sample arises from a distribution which is $n(\mu, \sigma^2)$:

(a) $\bar{X}$ is $n(\mu, \sigma^2/n)$.
(b) nS^2/σ^2 is $\chi^2(n - 1)$.
(c) $\bar{X}$ and S^2 are stochastically independent.

Determination of the p.d.f. of S^2 is left as an exercise.

EXERCISES

4.83. Let $\bar{X}$ be the mean of a random sample of size 5 from a normal distribution with $\mu = 0$ and $\sigma^2 = 125$. Determine c so that $\Pr(\bar{X} < c) = 0.90$.

4.84. If $\bar{X}$ is the mean of a random sample of size n from a normal distribution with mean μ and variance 100, find n so that $\Pr(\mu - 5 < \bar{X} < \mu + 5) = 0.954$.

4.85. Let $X_1, X_2, \ldots, X_{25}$ and $Y_1, Y_2, \ldots, Y_{25}$ be two random samples from two independent normal distributions $n(0, 16)$ and $n(1, 9)$, respectively. Let $\bar{X}$ and $\bar{Y}$ denote the corresponding sample means. Compute $\Pr(\bar{X} > \bar{Y})$.

4.86. Find the mean and variance of $S^2 = \sum_1^n (X_i - \bar{X})^2/n$, where $X_1, X_2, \ldots, X_n$ is a random sample from $n(\mu, \sigma^2)$. *Hint.* Find the mean and variance of nS^2/σ^2.

4.87. Let S^2 be the variance of a random sample of size 6 from the normal distribution $n(\mu, 12)$. Find $\Pr(2.30 < S^2 < 22.2)$.

4.88. Find the p.d.f. of the sample variance S^2, provided that the distribution from which the sample arises is $n(\mu, \sigma^2)$.

4.89. Let $\bar{X}$ and S^2 be the mean and the variance of a random sample of size 25 from a distribution which is $n(3, 100)$. Evaluate $\Pr(0 < \bar{X} < 6, 55.2 < S^2 < 145.6)$.

4.9 Expectations of Functions of Random Variables

Let $X_1, X_2, \ldots, X_n$ denote random variables that have the joint p.d.f. $f(x_1, x_2, \ldots, x_n)$. Let the random variable Y be defined by $Y = u(X_1, X_2, \ldots, X_n)$. We found in Section 4.7 that we could compute expectations of functions of Y without first finding the p.d.f. of Y. Indeed, this fact was the basis of the moment-generating-function procedure for finding the p.d.f. of Y. We can take advantage of this fact in a number of other instances. Some illustrative examples will be given.

Example 1. Given that W is $n(0, 1)$, that V is $\chi^2(r)$ with $r \geq 2$, and let W and V be stochastically independent. The mean of the random variable $T = W\sqrt{r/V}$ exists and is zero because the graph of the p.d.f. of T (see Section 4.4) is symmetric about the vertical axis through $t = 0$. The variance of T, when it exists, could be computed by integrating the product of t^2 and the p.d.f. of T. But it seems much simpler to compute

$$\sigma_T^2 = E(T^2) = E\left(W^2 \frac{r}{V}\right) = E(W^2)E\left(\frac{r}{V}\right).$$

Now W^2 is $\chi^2(1)$, so $E(W^2) = 1$. Furthermore,

$$E\left(\frac{r}{V}\right) = \int_0^\infty \frac{r}{v} \frac{1}{2^{r/2}\Gamma(r/2)} v^{r/2-1} e^{-v/2} \, dv$$

exists if $r > 2$ and is given by

$$\frac{r\Gamma[(r-2)/2]}{2\Gamma(r/2)} = \frac{r\Gamma[(r-2)/2]}{2[(r-2)/2]\Gamma[(r-2)/2]} = \frac{r}{r-2}.$$

Thus $\sigma_T^2 = r/(r-2)$, $r > 2$.

Example 2. Let X_i denote a random variable with mean μ_i and variance σ_i^2, $i = 1, 2, \ldots, n$. Let $X_1, X_2, \ldots, X_n$ be mutually stochastically independent and let $k_1, k_2, \ldots, k_n$ denote real constants. We shall compute the mean and variance of the linear function $Y = k_1X_1 + k_2X_2 + \cdots + k_nX_n$. Because E is a linear operator, the mean of Y is given by

$$\begin{aligned}
\mu_Y &= E(k_1X_1 + k_2X_2 + \cdots + k_nX_n) \\
&= k_1E(X_1) + k_2E(X_2) + \cdots + k_nE(X_n) \\
&= k_1\mu_1 + k_2\mu_2 + \cdots + k_n\mu_n = \sum_1^n k_i\mu_i.
\end{aligned}$$

The variance of Y is given by

$$
\begin{aligned}
\sigma_Y^2 &= E\{[(k_1 X_1 + \cdots + k_n X_n) - (k_1 \mu_1 + \cdots + k_n \mu_n)]^2\} \\
&= E\{[k_1(X_1 - \mu_1) + \cdots + k_n(X_n - \mu_n)]^2\} \\
&= E\left\{ \sum_{i=1}^{n} k_i^2 (X_i - \mu_i)^2 + 2 \sum\sum_{i<j} k_i k_j (X_i - \mu_i)(X_j - \mu_j) \right\} \\
&= \sum_{i=1}^{n} k_i^2 E[(X_i - \mu_i)^2] + 2 \sum\sum_{i<j} k_i k_j E[(X_i - \mu_i)(X_j - \mu_j)].
\end{aligned}
$$

Consider $E[(X_i - \mu_i)(X_j - \mu_j)]$, $i < j$. Because X_i and X_j are stochastically independent, we have

$$
E[(X_i - \mu_i)(X_j - \mu_j)] = E(X_i - \mu_i)E(X_j - \mu_j) = 0.
$$

Finally, then,

$$
\sigma_Y^2 = \sum_{i=1}^{n} k_i^2 E[(X_i - \mu_i)^2] = \sum_{i=1}^{n} k_i^2 \sigma_i^2.
$$

We can obtain a more general result if, in Example 2, we remove the hypothesis of mutual stochastic independence of $X_1, X_2, \ldots, X_n$. We shall do this and we shall let ρ_{ij} denote the correlation coefficient of X_i and X_j. Thus for easy reference to Example 2, we write

$$
E[(X_i - \mu_i)(X_j - \mu_j)] = \rho_{ij} \sigma_i \sigma_j, \qquad i < j.
$$

If we refer to Example 2, we see that again $\mu_Y = \sum_1^n k_i \mu_i$. But now

$$
\sigma_Y^2 = \sum_1^n k_i^2 \sigma_i^2 + 2 \sum\sum_{i<j} k_i k_j \rho_{ij} \sigma_i \sigma_j.
$$

Thus we have the following theorem.

Theorem 4. *Let $X_1, \ldots, X_n$ denote random variables that have means $\mu_1, \ldots, \mu_n$ and variances $\sigma_1^2, \ldots, \sigma_n^2$. Let ρ_{ij}, $i \neq j$, denote the correlation coefficient of X_i and X_j and let $k_1, \ldots, k_n$ denote real constants. The mean and the variance of the linear function*

$$
Y = \sum_1^n k_i X_i
$$

are, respectively,

$$
\mu_Y = \sum_1^n k_i \mu_i
$$

and

$$
\sigma_Y^2 = \sum_1^n k_i^2 \sigma_i^2 + 2 \sum\sum_{i<j} k_i k_j \rho_{ij} \sigma_i \sigma_j.
$$

The following corollary of this theorem is quite useful.

Corollary. *Let* $X_1, \ldots, X_n$ *denote the items of a random sample of size n from a distribution that has mean* μ *and variance* σ^2. *The mean and the variance of* $Y = \sum_1^n k_i X_i$ *are, respectively,* $\mu_Y = \left(\sum_1^n k_i \right) \mu$ *and* $\sigma_Y^2 = \left(\sum_1^n k_i^2 \right) \sigma^2$.

Example 3. Let $\bar{X} = \sum_1^n X_i/n$ denote the mean of a random sample of size n from a distribution that has mean μ and variance σ^2. In accordance with the Corollary, we have $\mu_{\bar{X}} = \mu \sum_1^n (1/n) = \mu$ and $\sigma_{\bar{X}}^2 = \sigma^2 \sum_1^n (1/n)^2 = \sigma^2/n$. We have seen, in Section 4.8, that if our sample is from a distribution that is $n(\mu, \sigma^2)$, then $\bar{X}$ is $n(\mu, \sigma^2/n)$. It is interesting that $\mu_{\bar{X}} = \mu$ and $\sigma_{\bar{X}}^2 = \sigma^2/n$ whether the sample is or is not from a normal distribution.

EXERCISES

4.90. Let X_1, X_2, X_3, X_4 be four mutually stochastically independent random variables having the same p.d.f. $f(x) = 2x$, $0 < x < 1$, zero elsewhere. Find the mean and variance of the sum Y of these four random variables.

4.91. Let X_1 and X_2 be two stochastically independent random variables so that the variances of X_1 and X_2 are $\sigma_1^2 = k$ and $\sigma_2^2 = 2$, respectively. Given that the variance of $Y = 3X_2 - X_1$ is 25, find k.

4.92. If the stochastically independent variables X_1 and X_2 have means μ_1, μ_2 and variances σ_1^2, σ_2^2, respectively, show that the mean and variance of the product $Y = X_1 X_2$ are $\mu_1 \mu_2$ and $\sigma_1^2 \sigma_2^2 + \mu_1^2 \sigma_2^2 + \mu_2^2 \sigma_1^2$, respectively.

4.93. Find the mean and variance of the sum Y of the items of a random sample of size 5 from the distribution having p.d.f. $f(x) = 6x(1 - x)$, $0 < x < 1$, zero elsewhere.

4.94. Determine the mean and variance of the mean $\bar{X}$ of a random sample of size 9 from a distribution having p.d.f. $f(x) = 4x^3$, $0 < x < 1$, zero elsewhere.

4.95. Let X and Y be random variables with $\mu_1 = 1$, $\mu_2 = 4$, $\sigma_1^2 = 4$, $\sigma_2^2 = 6$, $\rho = \frac{1}{2}$. Find the mean and variance of $Z = 3X - 2Y$.

4.96. Let X and Y be stochastically independent random variables with means μ_1, μ_2 and variances σ_1^2, σ_2^2. Determine the correlation coefficient of X and $Z = X - Y$ in terms of μ_1, μ_2, σ_1^2, σ_2^2.

4.97. Let μ and σ^2 denote the mean and variance of the random variable X. Let $Y = c + bX$, where b and c are real constants. Show that the mean and the variance of Y are, respectively, $c + b\mu$ and $b^2 \sigma^2$.

4.98. Let X and Y be random variables with means μ_1, μ_2; variances σ_1^2, σ_2^2; and correlation coefficient ρ. Show that the correlation coefficient of $W = aX + b$, $a > 0$, and $Z = cY + d$, $c > 0$, is ρ.

4.99. A person rolls a die, tosses a coin, and draws a card from an ordinary deck. He receives \$3 for each point up on the die, \$10 for a head, \$0 for a tail, and \$1 for each spot on the card (jack $= 11$, queen $= 12$, king $= 13$). If we assume that the three random variables involved are mutually stochastically independent and uniformly distributed, compute the mean and variance of the amount to be received.

4.100. Let U and V be two stochastically independent chi-square variables with r_1 and r_2 degrees of freedom, respectively. Find the mean and variance of $F = (r_2 U)/(r_1 V)$. What restriction is needed on the parameters r_1 and r_2 in order to ensure the existence of both the mean and the variance of F?

4.101. Let $X_1, X_2, \ldots, X_n$ be a random sample of size n from a distribution with mean μ and variance σ^2. Show that $E(S^2) = (n - 1)\sigma^2/n$, where S^2 is the variance of the random sample. *Hint.* Write $S^2 = (1/n) \sum_1^n (X_i - \mu)^2 - (\bar{X} - \mu)^2$.

4.102. Let X_1 and X_2 be stochastically independent random variables with nonzero variances. Find the correlation coefficient of $Y = X_1 X_2$ and X_1 in terms of the means and variances of X_1 and X_2.

4.103. Let X_1 and X_2 have a joint distribution with parameters μ_1, μ_2, σ_1^2, σ_2^2, and ρ. Find the correlation coefficient of the linear functions $Y = a_1 X_1 + a_2 X_2$ and $Z = b_1 X_1 + b_2 X_2$ in terms of the real constants a_1, a_2, b_1, b_2, and the parameters of the distribution.

4.104. Let $X_1, X_2, \ldots, X_n$ be a random sample of size n from a distribution which has mean μ and variance σ^2. Use Chebyshev's inequality to show, for every $\epsilon > 0$, that $\lim_{n \to \infty} \Pr(|\bar{X} - \mu| < \epsilon) = 1$; this is another form of the law of large numbers.

4.105. Let X_1, X_2, and X_3 be random variables with equal variances but with correlation coefficients $\rho_{12} = 0.3$, $\rho_{13} = 0.5$, and $\rho_{23} = 0.2$. Find the correlation coefficient of the linear functions $Y = X_1 + X_2$ and $Z = X_2 + X_3$.

4.106. Find the variance of the sum of 10 random variables if each has variance 5 and if each pair has correlation coefficient 0.5.

4.107. Let $X_1, \ldots, X_n$ be random variables that have means $\mu_1, \ldots, \mu_n$ and variances $\sigma_1^2, \ldots, \sigma_n^2$. Let ρ_{ij}, $i \neq j$, denote the correlation coefficient of X_i and X_j. Let $a_1, \ldots, a_n$ and $b_1, \ldots, b_n$ be real constants. Show that the covariance of $Y = \sum_{i=1}^{n} a_i X_i$ and $Z = \sum_{j=1}^{n} b_j X_j$ is $\sum_{j=1}^{n} \sum_{i=1}^{n} a_i b_j \sigma_i \sigma_j \rho_{ij}$, where $\rho_{ii} = 1$, $i = 1, 2, \ldots, n$.

4.108. Let X_1 and X_2 have a bivariate normal distribution with parameters μ_1, μ_2, σ_1^2, σ_2^2, and ρ. Compute the means, the variances, and the correlation coefficient of $Y_1 = \exp(X_1)$ and $Y_2 = \exp(X_2)$. *Hint.* Various moments of Y_1 and Y_2 can be found by assigning appropriate values to t_1 and t_2 in $E[\exp(t_1 X_1 + t_2 X_2)]$.

4.109. Let X be $n(\mu, \sigma^2)$ and consider the transformation $X = \ln Y$ or, equivalently, $Y = e^X$.

(a) Find the mean and the variance of Y by first determining $E(e^X)$ and $E[(e^X)^2]$.

(b) Find the p.d.f. of Y. This is called the *lognormal distribution*.

4.110. Let X_1 and X_2 have a trinomial distribution with parameters n, p_1, p_2.

(a) What is the distribution of $Y = X_1 + X_2$?

(b) From the equality $\sigma_Y^2 = \sigma_1^2 + \sigma_2^2 + 2\rho\sigma_1\sigma_2$, once again determine the correlation coefficient ρ of X_1 and X_2.

4.111. Let $Y_1 = X_1 + X_2$ and $Y_2 = X_2 + X_3$, where X_1, X_2, and X_3 are three stochastically independent random variables. Find the joint moment-generating function and the correlation coefficient of Y_1 and Y_2 provided that:

(a) X_i has a Poisson distribution with mean μ_i, $i = 1, 2, 3$.

(b) X_i is $n(\mu_i, \sigma_i^2)$, $i = 1, 2, 3$.

Chapter *5*
Limiting Distributions

5.1 Limiting Distributions

In some of the preceding chapters it has been demonstrated by example that the distribution of a random variable (perhaps a statistic) often depends upon a positive integer n. For example, if the random variable X is $b(n, p)$, the distribution of X depends upon n. If $\bar{X}$ is the mean of a random sample of size n from a distribution that is $n(\mu, \sigma^2)$, then $\bar{X}$ is itself $n(\mu, \sigma^2/n)$ and the distribution of $\bar{X}$ depends upon n. If S^2 is the variance of this random sample from the normal distribution to which we have just referred, the random variable nS^2/σ^2 is $\chi^2(n-1)$, and so the distribution of this random variable depends upon n.

We know from experience that the determination of the p.d.f. of a random variable can, upon occasion, present rather formidable computational difficulties. For example, if $\bar{X}$ is the mean of a random sample $X_1, X_2, \ldots, X_n$ from a distribution that has the p.d.f.

$$f(x) = 1, \qquad 0 < x < 1,$$
$$= 0 \text{ elsewhere,}$$

then (Exercise 4.74) the moment-generating function of $\bar{X}$ is given by $[M(t/n)]^n$, where here

$$M(t) = \int_0^1 e^{tx} \, dx = \frac{e^t - 1}{t}, \qquad t \neq 0,$$
$$= 1, \qquad t = 0.$$

Hence

$$E(e^{t\bar{X}}) = \left(\frac{e^{t/n} - 1}{t/n}\right)^n, \qquad t \neq 0,$$

$$= 1, \qquad t = 0.$$

Since the moment-generating function of $\bar{X}$ depends upon n, the distribution of $\bar{X}$ depends upon n. It is true that various mathematical techniques can be used to determine the p.d.f. of $\bar{X}$ for a fixed, but arbitrarily fixed, positive integer n. But the p.d.f. is so complicated that few, if any, of us would be interested in using it to compute probabilities about $\bar{X}$. One of the purposes of this chapter is to provide ways of approximating, for large values of n, some of these complicated probability density functions.

Consider a distribution that depends upon the positive integer n. Clearly, the distribution function F of that distribution will also depend upon n. Throughout this chapter, we denote this fact by writing the distribution function as F_n and the corresponding p.d.f. as f_n. Moreover, to emphasize the fact that we are working with sequences of distribution functions, we place a subscript n on the random variables. For example, we shall write

$$F_n(\bar{x}) = \int_{-\infty}^{\bar{x}} \frac{1}{\sqrt{1/n}\sqrt{2\pi}} e^{-nw^2/2} \, dw$$

for the distribution function of the mean $\bar{X}_n$ of a random sample of size n from a normal distribution with mean zero and variance 1.

We now define a limiting distribution of a random variable whose distribution depends upon n.

Definition 1. Let the distribution function $F_n(y)$ of the random variable Y_n depend upon n, a positive integer. If $F(y)$ is a distribution function and if $\lim_{n \to \infty} F_n(y) = F(y)$ for every point y at which $F(y)$ is continuous, then the random variable Y_n is said to have a *limiting distribution* with distribution function $F(y)$.

The following examples are illustrative of random variables that have limiting distributions.

Example 1. Let Y_n denote the nth order statistic of a random sample $X_1, X_2, \ldots, X_n$ from a distribution having p.d.f.

$$f(x) = \frac{1}{\theta}, \qquad 0 < x < \theta, 0 < \theta < \infty,$$

$$= 0 \text{ elsewhere.}$$

The p.d.f. of Y_n is

$$g_n(y) = \frac{ny^{n-1}}{\theta^n}, \qquad 0 < y < \theta,$$

$$= 0 \text{ elsewhere,}$$

and the distribution function of Y_n is

$$F_n(y) = 0, \qquad y < 0,$$

$$= \int_0^y \frac{nz^{n-1}}{\theta^n} \, dz = \left(\frac{y}{\theta}\right)^n, \qquad 0 \le y < \theta,$$

$$= 1, \qquad \theta \le y < \infty.$$

Then

$$\lim_{n \to \infty} F_n(y) = 0, \qquad -\infty < y < \theta,$$

$$= 1, \qquad \theta \le y < \infty.$$

Now

$$F(y) = 0, \qquad -\infty < y < \theta,$$

$$= 1, \qquad \theta \le y < \infty,$$

is a distribution function. Moreover, $\lim_{n \to \infty} F_n(y) = F(y)$ at each point of continuity of $F(y)$. In accordance with the definition of a limiting distribution, the random variable Y_n has a limiting distribution with distribution function $F(y)$. Recall that a distribution of the discrete type which has a probability of 1 at a single point has been called a *degenerate distribution*. Thus in this example the limiting distribution of Y_n is degenerate. Sometimes this is the case, sometimes a limiting distribution is not degenerate, and sometimes there is no limiting distribution at all.

Example 2. Let X_n have the distribution function

$$F_n(\bar{x}) = \int_{-\infty}^{\bar{x}} \frac{1}{\sqrt{1/n}\sqrt{2\pi}} e^{-nw^2/2} \, dw.$$

If the change of variable $v = \sqrt{n}w$ is made, we have

$$F_n(\bar{x}) = \int_{-\infty}^{\sqrt{n}\bar{x}} \frac{1}{\sqrt{2\pi}} e^{-v^2/2} \, dv.$$

It is clear that

$$\lim_{n \to \infty} F_n(\bar{x}) = 0, \qquad \bar{x} < 0,$$

$$= \tfrac{1}{2}, \qquad \bar{x} = 0,$$

$$= 1, \qquad \bar{x} > 0.$$

Now the function

$$F(\bar{x}) = 0, \qquad \bar{x} < 0,$$
$$= 1, \qquad \bar{x} \geq 0,$$

is a distribution function and $\lim_{n \to \infty} F_n(\bar{x}) = F(\bar{x})$ at every point of continuity of $F(\bar{x})$. To be sure, $\lim_{n \to \infty} F_n(0) \neq F(0)$, but $F(\bar{x})$ is not continuous at $\bar{x} = 0$. Accordingly, the random variable $\bar{X}_n$ has a limiting distribution with distribution function $F(\bar{x})$. Again, this limiting distribution is degenerate and has all the probability at the one point $\bar{x} = 0$.

Example 3. The fact that limiting distributions, if they exist, cannot in general be determined by taking the limit of the p.d.f. will now be illustrated. Let X_n have the p.d.f.

$$f_n(x) = 1, \qquad x = 2 + \frac{1}{n}$$

$$= 0 \text{ elsewhere.}$$

Clearly, $\lim_{n \to \infty} f_n(x) = 0$ for all values of x. This may suggest that X_n has no limiting distribution. However, the distribution function of X_n is

$$F_n(x) = 0, \qquad x < 2 + \frac{1}{n},$$

$$= 1, \qquad x \geq 2 + \frac{1}{n},$$

and

$$\lim_{n \to \infty} F_n(x) = 0, \qquad x \leq 2,$$

$$= 1, \qquad x > 2.$$

Since

$$F(x) = 0, \qquad x < 2,$$
$$= 1, \qquad x \geq 2,$$

is a distribution function, and since $\lim_{n \to \infty} F_n(x) = F(x)$ at all points of continuity of $F(x)$, there is a limiting distribution of X_n with distribution function $F(x)$.

Example 4. Let Y_n denote the nth order statistic of a random sample from the uniform distribution of Example 1. Let $Z_n = n(\theta - Y_n)$. The p.d.f. of Z_n is

$$h_n(z) = \frac{(\theta - z/n)^{n-1}}{\theta^n}, \qquad 0 < z < n\theta,$$

$$= 0 \text{ elsewhere,}$$

and the distribution function of Z_n is

$$G_n(z) = 0, \quad z < 0,$$

$$= \int_0^z \frac{(\theta - w/n)^{n-1}}{\theta^n} \, dw = 1 - \left(1 - \frac{z}{n\theta}\right)^n, \quad 0 \le z < n\theta,$$

$$= 1, \quad n\theta \le z.$$

Hence

$$\lim_{n \to \infty} G_n(z) = 0, \quad z \le 0$$

$$= 1 - e^{-z/\theta}, \quad 0 < z < \infty.$$

Now

$$G(z) = 0, \quad z < 0,$$

$$= 1 - e^{-z/\theta}, \quad 0 \le z,$$

is a distribution function that is everywhere continuous and $\lim_{n \to \infty} G_n(z) = G(z)$ at all points. Thus Z_n has a limiting distribution with distribution function $G(z)$. This affords us an example of a limiting distribution that is not degenerate.

EXERCISES

5.1. Let $\bar{X}_n$ denote the mean of a random sample of size n from a distribution that is $n(\mu, \sigma^2)$. Find the limiting distribution of $\bar{X}_n$.

5.2. Let Y_1 denote the first order statistic of a random sample of size n from a distribution that has the p.d.f. $f(x) = e^{-(x - \theta)}$, $\theta < x < \infty$, zero elsewhere. Let $Z_n = n(Y_1 - \theta)$. Investigate the limiting distribution of Z_n.

5.3. Let Y_n denote the nth order statistic of a random sample from a distribution of the continuous type that has distribution function $F(x)$ and p.d.f. $f(x) = F'(x)$. Find the limiting distribution of $Z_n = n[1 - F(Y_n)]$.

5.4. Let Y_2 denote the second order statistic of a random sample of size n from a distribution of the continuous type that has distribution function $F(x)$ and p.d.f. $f(x) = F'(x)$. Find the limiting distribution of $W_n = nF(Y_2)$.

5.5. Let the p.d.f. of Y_n be $f_n(y) = 1$, $y = n$, zero elsewhere. Show that Y_n does not have a limiting distribution. (In this case, the probability has "escaped" to infinity.)

5.6. Let $X_1, X_2, \ldots, X_n$ be a random sample of size n from a distribution which is $n(\mu, \sigma^2)$, where $\mu > 0$. Show that the sum $Z_n = \sum_1^n X_i$ does not have a limiting distribution.

5.2 Stochastic Convergence

When the limiting distribution of a random variable is degenerate, the random variable is said to *converge stochastically* to the constant that has a probability of 1. Thus Examples 1 to 3 of Section 5.1 illustrate not only the notion of a limiting distribution but also the concept of stochastic convergence. In Example 1, the nth order statistic Y_n converges stochastically to θ; in Example 2, the statistic $\bar{X}_n$ converges stochastically to zero, the mean of the normal distribution from which the sample was taken; and in Example 3, the random variable X_n converges stochastically to 2. We shall show that in some instances the inequality of Chebyshev can be used to advantage in proving stochastic convergence. But first we shall prove the following theorem.

Theorem 1. *Let $F_n(y)$ denote the distribution function of a random variable Y_n whose distribution depends upon the positive integer n. Let c denote a constant which does not depend upon n. The random variable Y_n converges stochastically to the constant c if and only if, for every $\epsilon > 0$, the*

$$\lim_{n \to \infty} \Pr(|Y_n - c| < \epsilon) = 1.$$

Proof. First, assume that the $\lim_{n \to \infty} \Pr(|Y_n - c| < \epsilon) = 1$ for every $\epsilon > 0$. We are to prove that the random variable Y_n converges stochastically to the constant c. This means we must prove that

$$\lim_{n \to \infty} F_n(y) = 0, \qquad y < c,$$
$$= 1, \qquad y > c.$$

Note that we do not need to know anything about the $\lim_{n \to \infty} F_n(c)$. For if the limit of $F_n(y)$ is as indicated, then Y_n has a limiting distribution with distribution function

$$F(y) = 0, \qquad y < c,$$
$$= 1, \qquad y \geq c.$$

Now

$$\Pr(|Y_n - c| < \epsilon) = F_n[(c + \epsilon)-] - F_n(c - \epsilon),$$

where $F_n[(c + \epsilon)-]$ is the left-hand limit of $F_n(y)$ at $y = c + \epsilon$. Thus we have

$$1 = \lim_{n \to \infty} \Pr(|Y_n - c| < \epsilon) = \lim_{n \to \infty} F_n[(c + \epsilon)-] - \lim_{n \to \infty} F_n(c - \epsilon).$$

Because $0 \leq F_n(y) \leq 1$ for all values of y and for every positive integer n, it must be that

$$\lim_{n \to \infty} F_n(c - \epsilon) = 0, \qquad \lim_{n \to \infty} F_n[(c + \epsilon) -] = 1.$$

Since this is true for every $\epsilon > 0$, we have

$$\lim_{n \to \infty} F_n(y) = 0, \qquad y < c,$$
$$= 1, \qquad y > c,$$

as we were required to show.

To complete the proof of Theorem 1, we assume that

$$\lim_{n \to \infty} F_n(y) = 0, \qquad y < c,$$
$$= 1, \qquad y > c.$$

We are to prove that $\lim_{n \to \infty} \Pr \left(|Y_n - c| < \epsilon \right) = 1$ for every $\epsilon > 0$. Because

$$\Pr \left(|Y_n - c| < \epsilon \right) = F_n[(c + \epsilon) -] - F_n(c - \epsilon),$$

and because it is given that

$$\lim_{n \to \infty} F_n[(c + \epsilon) -] = 1,$$
$$\lim_{n \to \infty} F_n(c - \epsilon) = 0,$$

for every $\epsilon > 0$, we have the desired result. This completes the proof of the theorem.

We should like to point out a simple but useful fact. Clearly,

$$\Pr \left(|Y_n - c| < \epsilon \right) + \Pr \left(|Y_n - c| \geq \epsilon \right) = 1.$$

Thus the limit of $\Pr \left(|Y_n - c| < \epsilon \right)$ is equal to 1 when and only when

$$\lim_{n \to \infty} \Pr \left(|Y_n - c| \geq \epsilon \right) = 0.$$

That is, this last limit is also a necessary and sufficient condition for the stochastic convergence of the random variable Y_n to the constant c.

Example 1. Let $\bar{X}_n$ denote the mean of a random sample of size n from a distribution that has mean μ and positive variance σ^2. Then the mean and variance of $\bar{X}_n$ are μ and σ^2/n. Consider, for every fixed $\epsilon > 0$, the probability

$$\Pr \left(|\bar{X}_n - \mu| \geq \epsilon \right) = \Pr \left(|\bar{X}_n - \mu| \geq \frac{k\sigma}{\sqrt{n}} \right),$$

where $k = \epsilon\sqrt{n}/\sigma$. In accordance with the inequality of Chebyshev, this probability is less than or equal to $1/k^2 = \sigma^2/n\epsilon^2$. So, for every fixed $\epsilon > 0$, we have

$$\lim_{n \to \infty} \Pr\left(|\bar{X}_n - \mu| \geq \epsilon\right) \leq \lim_{n \to \infty} \frac{\sigma^2}{n\epsilon^2} = 0.$$

Hence $\bar{X}_n$ converges stochastically to μ if σ^2 is finite. In a more advanced course, the student will learn that μ finite is sufficient to ensure this stochastic convergence.

Remark. The condition $\lim_{n \to \infty} \Pr\left(|Y_n - c| < \epsilon\right) = 1$ is often used as the definition of convergence in probability and one says that Y_n converges to c *in probability*. Thus stochastic convergence and convergence in probability are equivalent. A stronger type of convergence is given by $\Pr\left(\lim_{n \to \infty} Y_n = c\right) = 1$; in this case we say that Y_n converges to c *with probability 1*. Although we do not consider this type of convergence, it is known that the mean $\bar{X}_n$ of a random sample converges with probability 1 to the mean μ of the distribution, provided that the latter exists. This is one form of the *strong law of large numbers*.

EXERCISES

5.7. Let the random variable Y_n have a distribution that is $b(n, p)$. (a) Prove that Y_n/n converges stochastically to p. This result is one form of the weak law of large numbers. (b) Prove that $1 - Y_n/n$ converges stochastically to $1 - p$.

5.8. Let S_n^2 denote the variance of a random sample of size n from a distribution that is $n(\mu, \sigma^2)$. Prove that $nS_n^2/(n - 1)$ converges stochastically to σ^2.

5.9. Let W_n denote a random variable with mean μ and variance b/n^p, where $p > 0$, μ, and b are constants (not functions of n). Prove that W_n converges stochastically to μ. *Hint.* Use Chebyshev's inequality.

5.10. Let Y_n denote the nth order statistic of a random sample of size n from a uniform distribution on the interval $(0, \theta)$, as in Example 1 of Section 5.1. Prove that $Z_n = \sqrt{Y_n}$ converges stochastically to $\sqrt{\theta}$.

5.3 Limiting Moment-Generating Functions

To find the limiting distribution function of a random variable Y_n by use of the definition of limiting distribution function obviously requires that we know $F_n(y)$ for each positive integer n. But, as indicated in the introductory remarks of Section 5.1, this is precisely the problem we should like to avoid. If it exists, the moment-generating function that corresponds to the distribution function $F_n(y)$ often provides a convenient method of determining the limiting distribution function. To emphasize that the distribution of a random variable Y_n depends upon the positive integer n, in this chapter we shall write the moment-generating function of Y_n in the form $M(t; n)$.

The following theorem, which is essentially Curtiss' modification of a theorem of Lévy and Cramér, explains how the moment-generating function may be used in problems of limiting distributions. A proof of the theorem requires a knowledge of that same facet of analysis that permitted us to assert that a moment-generating function, when it exists, uniquely determines a distribution. Accordingly, no proof of the theorem will be given.

Theorem 2. *Let the random variable Y_n have the distribution function $F_n(y)$ and the moment-generating function $M(t; n)$ that exists for $-h < t < h$ for all n. If there exists a distribution function $F(y)$, with corresponding moment-generating function $M(t)$, defined for $|t| \leq h_1 < h$, such that $\lim_{n \to \infty} M(t; n) = M(t)$, then Y_n has a limiting distribution with distribution function $F(y)$.*

In this and the subsequent section are several illustration of the use of Theorem 2. In some of these examples it is convenient to use a certain limit that is established in some courses in advanced calculus. We refer to a limit of the form

$$\lim_{n \to \infty} \left[1 + \frac{b}{n} + \frac{\psi(n)}{n} \right]^{cn},$$

where b and c do not depend upon n and where $\lim_{n \to \infty} \psi(n) = 0$. Then

$$\lim_{n \to \infty} \left[1 + \frac{b}{n} + \frac{\psi(n)}{n} \right]^{cn} = \lim_{n \to \infty} \left(1 + \frac{b}{n} \right)^{cn} = e^{bc}.$$

For example,

$$\lim_{n \to \infty} \left(1 - \frac{t^2}{n} + \frac{t^3}{n^{3/2}} \right)^{-n/2} = \lim_{n \to \infty} \left(1 - \frac{t^2}{n} + \frac{t^3/\sqrt{n}}{n} \right)^{-n/2}.$$

Here $b = -t^2$, $c = -\frac{1}{2}$, and $\psi(n) = t^3/\sqrt{n}$. Accordingly, for every fixed value of t, the limit is $e^{t^2/2}$.

Example 1. Let Y_n have a distribution that is $b(n, p)$. Suppose that the mean $\mu = np$ is the same for every n; that is, $p = \mu/n$, where μ is a constant.

We shall find the limiting distribution of the binomial distribution, when $p = \mu/n$, by finding the limit of $M(t; n)$. Now

$$M(t; n) = E(e^{tY_n}) = [(1 - p) + pe^t]^n = \left[1 + \frac{\mu(e^t - 1)}{n}\right]^n$$

for all real values of t. Hence we have

$$\lim_{n \to \infty} M(t; n) = e^{\mu(e^t - 1)}$$

for all real values of t. Since there exists a distribution, namely the Poisson distribution with mean μ, that has this moment-generating function $e^{\mu(e^t - 1)}$, then in accordance with the theorem and under the conditions stated, it is seen that Y_n has a limiting Poisson distribution with mean μ.

Whenever a random variable has a limiting distribution, we may, if we wish, use the limiting distribution as an approximation to the exact distribution function. The result of this example enables us to use the Poisson distribution as an approximation to the binomial distribution when n is large and p is small. This is clearly an advantage, for it is easy to provide tables for the one-parameter Poisson distribution. On the other hand, the binomial distribution has two parameters, and tables for this distribution are very ungainly. To illustrate the use of the approximation, let Y have a binomial distribution with $n = 50$ and $p = \frac{1}{25}$. Then

$$\Pr\,(Y \leq 1) = (\tfrac{24}{25})^{50} + 50(\tfrac{1}{25})(\tfrac{24}{25})^{49} = 0.400,$$

approximately. Since $\mu = np = 2$, the Poisson approximation to this probability is

$$e^{-2} + 2e^{-2} = 0.406.$$

Example 2. Let Z_n be $\chi^2(n)$. Then the moment-generating function of Z_n is $(1 - 2t)^{-n/2}$, $t < \frac{1}{2}$. The mean and the variance of Z_n are, respectively, n and $2n$. The limiting distribution of the random variable $Y_n = (Z_n - n)/\sqrt{2n}$ will be investigated. Now the moment-generating function of Y_n is

$$M(t; n) = E\left\{\exp\left[t\left(\frac{Z_n - n}{\sqrt{2n}}\right)\right]\right\}$$

$$= e^{-tn/\sqrt{2n}} E(e^{tZ_n/\sqrt{2n}})$$

$$= \exp\left[-\left(t\sqrt{\frac{2}{n}}\right)\left(\frac{n}{2}\right)\right]\left(1 - 2\frac{t}{\sqrt{2n}}\right)^{-n/2}, \qquad t < \frac{\sqrt{2n}}{2}.$$

This may be written in the form

$$M(t; n) = \left(e^{t\sqrt{2/n}} - t\sqrt{\frac{2}{n}}\,e^{t\sqrt{2/n}}\right)^{-n/2}, \qquad t < \sqrt{\frac{n}{2}}.$$

In accordance with Taylor's formula, there exists a number $\xi(n)$, between 0 and $t\sqrt{2/n}$, such that

$$e^{t\sqrt{2/n}} = 1 + t\sqrt{\frac{2}{n}} + \frac{1}{2}\left(t\sqrt{\frac{2}{n}}\right)^2 + \frac{e^{\xi(n)}}{6}\left(t\sqrt{\frac{2}{n}}\right)^3.$$

If this sum is substituted for $e^{t\sqrt{2/n}}$ in the last expression for $M(t; n)$, it is seen that

$$M(t; n) = \left(1 - \frac{t^2}{n} + \frac{\psi(n)}{n}\right)^{-n/2},$$

where

$$\psi(n) = \frac{\sqrt{2}t^3 e^{\xi(n)}}{3\sqrt{n}} - \frac{\sqrt{2}t^3}{\sqrt{n}} - \frac{2t^4 e^{\xi(n)}}{3n}.$$

Since $\xi(n) \to 0$ as $n \to \infty$, then $\lim \psi(n) = 0$ for every fixed value of t. In accordance with the limit proposition cited earlier in this section, we have

$$\lim_{n \to \infty} M(t; n) = e^{t^2/2}$$

for all real values of t. That is, the random variable $Y_n = (Z_n - n)/\sqrt{2n}$ has a limiting normal distribution with mean zero and variance 1.

EXERCISES

5.11. Let X_n have a gamma distribution with parameter $\alpha = n$ and β, where β is not a function of n. Let $Y_n = X_n/n$. Find the limiting distribution of Y_n.

5.12. Let Z_n be $\chi^2(n)$ and let $W_n = Z_n/n^2$. Find the limiting distribution of W_n.

5.13. Let X be $\chi^2(50)$. Approximate $\Pr(40 < X < 60)$.

5.14. Let $p = 0.95$ be the probability that a man, in a certain age group, lives at least 5 years.

(a) If we are to observe 60 such men and if we assume independence, find the probability that at least 56 of them live 5 or more years.

(b) Find an approximation to the result of part (a) by using the Poisson distribution. *Hint.* Redefine p to be 0.05 and $1 - p = 0.95$.

5.15. Let the random variable Z_n have a Poisson distribution with parameter $\mu = n$. Show that the limiting distribution of the random variable $Y_n = (Z_n - n)/\sqrt{n}$ is normal with mean zero and variance 1.

5.16. Let S_n^2 denote the variance of a random sample of size n from a distribution that is $n(\mu, \sigma^2)$. It has been proved that $nS_n^2/(n - 1)$ converges stochastically to σ^2. Prove that S_n^2 converges stochastically to σ^2.

5.17. Let X_n and Y_n have a bivariate normal distribution with parameters μ_1, μ_2, σ_1^2, σ_2^2 (free of n) but $\rho = 1 - 1/n$. Consider the conditional distribution of Y_n, given $X_n = x$. Investigate the limit of this conditional distribution as $n \to \infty$. What is the limiting distribution if $\rho = -1 + 1/n$? Reference to these facts was made in the Remark, Section 2.3.

5.18. Let $\bar{X}_n$ denote the mean of a random sample of size n from a Poisson distribution with parameter $\mu = 1$.

(a) Show that the moment-generating function of $Y_n = \sqrt{n}(\bar{X}_n - \mu)/\sigma = \sqrt{n}(\bar{X}_n - 1)$ is given by $\exp[-t\sqrt{n} + n(e^{t/\sqrt{n}} - 1)]$.

(b) Investigate the limiting distribution of Y_n as $n \to \infty$. *Hint.* Replace, by its MacLaurin's series, the expression $e^{t/\sqrt{n}}$, which is in the exponent of the moment-generating function of Y_n.

5.19. Let $\bar{X}_n$ denote the mean of a random sample of size n from a distribution that has p.d.f. $f(x) = e^{-x}$, $0 < x < \infty$, zero elsewhere.

(a) Show that the moment-generating function $M(t; n)$ of $Y_n = \sqrt{n}(\bar{X}_n - 1)$ is equal to $[e^{t/\sqrt{n}} - (t/\sqrt{n})e^{t/\sqrt{n}}]^{-n}$, $t < \sqrt{n}$.

(b) Find the limiting distribution of Y_n as $n \to \infty$.

This exercise and the immediately preceding one are special instances of an important theorem that will be proved in the next section.

5.4 The Central Limit Theorem

It was seen (Section 4.8) that, if $X_1, X_2, \ldots, X_n$ is a random sample from a normal distribution with mean μ and variance σ^2, the random variable

$$\frac{\sum_1^n X_i - n\mu}{\sigma\sqrt{n}} = \frac{\sqrt{n}(\bar{X}_n - \mu)}{\sigma}$$

is, for every positive integer n, normally distributed with zero mean and unit variance. In probability theory there is a very elegant theorem called the *central limit theorem*. A special case of this theorem asserts the remarkable and important fact that if $X_1, X_2, \ldots, X_n$ denote the items of a random sample of size n from any distribution having positive variance σ^2 (and hence finite mean μ), then the random variable $\sqrt{n}(\bar{X}_n - \mu)/\sigma$ has a limiting normal distribution with zero mean and unit variance. If this fact can be established, it will imply, whenever the conditions of the theorem are satisfied, that (for fixed n) the random variable $\sqrt{n}(\bar{X} - \mu)/\sigma$ has an approximate normal distribution with mean zero and variance 1. It will then be possible to use this approximate normal distribution to compute approximate probabilities concering $\bar{X}$.

The more general form of the theorem is stated, but it is proved only in the modified case. However, this is exactly the proof of the theorem that would be given if we could use the characteristic function in place of the moment-generating function.

Theorem 3. *Let $X_1, X_2, \ldots, X_n$ denote the items of a random sample from a distribution that has mean μ and positive variance σ^2. Then the random variable $Y_n = \left(\sum_1^n X_i - n\mu\right)\big/\sqrt{n}\,\sigma = \sqrt{n}(\bar{X}_n - \mu)/\sigma$ has a limiting distribution that is normal with mean zero and variance 1.*

Proof. In the modification of the proof, we assume the existence of the moment-generating function $M(t) = E(e^{tX})$, $-h < t < h$, of the distribution. However, this proof is essentially the same one that would be given for this theorem in a more advanced course by replacing the moment-generating function by the characteristic function $\varphi(t) = E(e^{itX})$.

The function

$$m(t) = E[e^{t(X-\mu)}] = e^{-\mu t}M(t)$$

also exists for $-h < t < h$. Since $m(t)$ is the moment-generating function for $X - \mu$, it must follow that $m(0) = 1$, $m'(0) = E(X - \mu) = 0$, and $m''(0) = E[(X - \mu)^2] = \sigma^2$. By Taylor's formula there exists a number ξ between 0 and t such that

$$m(t) = m(0) + m'(0)t + \frac{m''(\xi)t^2}{2}$$

$$= 1 + \frac{m''(\xi)t^2}{2}.$$

If $\sigma^2 t^2/2$ is added and subtracted, then

$$m(t) = 1 + \frac{\sigma^2 t^2}{2} + \frac{[m''(\xi) - \sigma^2]t^2}{2}.$$

Next consider $M(t; n)$, where

$$M(t; n) = E\left[\exp\left(t\,\frac{\sum X_i - n\mu}{\sigma\sqrt{n}}\right)\right]$$

$$= E\left[\exp\left(t\,\frac{X_1 - \mu}{\sigma\sqrt{n}}\right)\exp\left(t\,\frac{X_2 - \mu}{\sigma\sqrt{n}}\right)\cdots\exp\left(t\,\frac{X_n - \mu}{\sigma\sqrt{n}}\right)\right]$$

$$= E\left[\exp\left(t\,\frac{X_1 - \mu}{\sigma\sqrt{n}}\right)\right]\cdots E\left[\exp\left(t\,\frac{X_n - \mu}{\sigma\sqrt{n}}\right)\right]$$

$$= \left\{E\left[\exp\left(t\,\frac{X - \mu}{\sigma\sqrt{n}}\right)\right]\right\}^n$$

$$= \left[m\left(\frac{t}{\sigma\sqrt{n}}\right)\right]^n, \qquad -h < \frac{t}{\sigma\sqrt{n}} < h.$$

In $m(t)$, replace t by $t/\sigma\sqrt{n}$ to obtain

$$m\left(\frac{t}{\sigma\sqrt{n}}\right) = 1 + \frac{t^2}{2n} + \frac{[m''(\xi) - \sigma^2]t^2}{2n\sigma^2},$$

where now ξ is between 0 and $t/\sigma\sqrt{n}$ with $-h\sigma\sqrt{n} < t < h\sigma\sqrt{n}$. Accordingly,

$$M(t; n) = \left\{1 + \frac{t^2}{2n} + \frac{[m''(\xi) - \sigma^2]t^2}{2n\sigma^2}\right\}^n.$$

Since $m''(t)$ is continuous at $t = 0$ and since $\xi \to 0$ as $n \to \infty$, we have

$$\lim_{n \to \infty} [m''(\xi) - \sigma^2] = 0.$$

The limit proposition cited in Section 5.3 shows that

$$\lim_{n \to \infty} M(t; n) = e^{t^2/2}$$

for all real values of t. This proves that the random variable $Y_n = \sqrt{n}(\overline{X}_n - \mu)/\sigma$ has a limiting normal distribution with mean zero and variance 1.

We interpret this theorem as saying, with n a fixed positive integer, that the random variable $\sqrt{n}(\overline{X} - \mu)/\sigma$ has an approximate normal distribution with mean zero and variance 1; and in applications we use the approximate normal p.d.f. as though it were the exact p.d.f. of $\sqrt{n}(\overline{X} - \mu)/\sigma$.

Some illustrative examples, here and later, will help show the importance of this version of the central limit theorem.

Example 1. Let $\overline{X}$ denote the mean of a random sample of size 75 from the distribution that has the p.d.f.

$$f(x) = 1, \qquad 0 < x < 1,$$
$$= 0 \text{ elsewhere.}$$

It was stated in Section 5.1 that the exact p.d.f. of $\overline{X}$, say $g(\overline{x})$, is rather complicated. It can be shown that $g(\overline{x})$ has a graph at points of positive probability density that is composed of arcs of 75 different polynomials of degree 74. The computation of such a probability as $\Pr(0.45 < \overline{X} < 0.55)$ would be extremely laborious. The conditions of the theorem are satisfied, since $M(t)$ exists for all real values of t. Moreover, $\mu = \frac{1}{2}$ and $\sigma^2 = \frac{1}{12}$, so we have approximately

$$\Pr (0.45 < \bar{X} < 0.55) = \Pr \left[\frac{\sqrt{n}(0.45 - \mu)}{\sigma} < \frac{\sqrt{n}(\bar{X} - \mu)}{\sigma} < \frac{\sqrt{n}(0.55 - \mu)}{\sigma} \right]$$

$$= \Pr [-1.5 < 30(\bar{X} - 0.5) < 1.5]$$

$$= 0.866,$$

from Table III in Appendix B.

Example 2. Let $X_1, X_2, \ldots, X_n$ denote a random sample from a distribution that is $b(1, p)$. Here $\mu = p$, $\sigma^2 = p(1 - p)$, and $M(t)$ exists for all real values of t. If $Y_n = X_1 + \cdots + X_n$, it is known that Y_n is $b(n, p)$. Calculation of probabilities concerning Y_n, when we do not use the Poisson approximation, can be greatly simplified by making use of the fact that $(Y_n - np)/\sqrt{np(1 - p)} = \sqrt{n}(\bar{X}_n - p)/\sqrt{p(1 - p)} = \sqrt{n}(\bar{X}_n - \mu)/\sigma$ has a limiting distribution that is normal with mean zero and variance 1. Let $n = 100$ and $p = \frac{1}{2}$, and suppose that we wish to compute $\Pr (Y = 48, 49, 50, 51, 52)$. Since Y is a random variable of the discrete type, the events $Y = 48, 49, 50, 51, 52$ and $47.5 < Y < 52.5$ are equivalent. That is, $\Pr (Y = 48, 49, 50, 51, 52) = \Pr (47.5 < Y < 52.5)$. Since $np = 50$ and $np(1 - p) = 25$, the latter probability may be written

$$\Pr (47.5 < Y < 52.5) = \Pr \left(\frac{47.5 - 50}{5} < \frac{Y - 50}{5} < \frac{52.5 - 50}{5} \right)$$

$$= \Pr \left(-0.5 < \frac{Y - 50}{5} < 0.5 \right).$$

Since $(Y - 50)/5$ has an approximate normal distribution with mean zero and variance 1, Table III shows this probability to be approximately 0.382.

The convention of selecting the event $47.5 < Y < 52.5$, instead of, say, $47.8 < Y < 52.3$, as the event equivalent to the event $Y = 48, 49, 50, 51, 52$ seems to have originated in the following manner: The probability, $\Pr (Y = 48, 49, 50, 51, 52)$, can be interpreted as the sum of five rectangular areas where the rectangles have bases 1 but the heights are, respectively, $\Pr (Y = 48), \ldots, \Pr (Y = 52)$. If these rectangles are so located that the midpoints of their bases are, respectively, at the points $48, 49, \ldots, 52$ on a horizontal axis, then in approximating the sum of these areas by an area bounded by the horizontal axis, the graph of a normal p.d.f., and two ordinates, it seems reasonable to take the two ordinates at the points 47.5 and 52.5.

EXERCISES

5.20. Let $\bar{X}$ denote the mean of a random sample of size 100 from a distribution that is $\chi^2(50)$. Compute an approximate value of $\Pr (49 < \bar{X} < 51)$.

5.21. Let $\bar{X}$ denote the mean of a random sample of size 128 from a gamma distribution with $\alpha = 2$ and $\beta = 4$. Approximate $\Pr(7 < \bar{X} < 9)$.

5.22. Let Y be $b(72, \frac{1}{3})$. Approximate $\Pr(22 \le Y \le 28)$.

5.23. Compute an approximate probability that the mean of a random sample of size 15 from a distribution having p.d.f. $f(x) = 3x^2$, $0 < x < 1$, zero elsewhere, is between $\frac{3}{5}$ and $\frac{4}{5}$.

5.24. Let Y denote the sum of the items of a random sample of size 12 from a distribution having p.d.f. $f(x) = \frac{1}{6}$, $x = 1, 2, 3, 4, 5, 6$, zero elsewhere. Compute an approximate value of $\Pr(36 \le Y \le 48)$. *Hint.* Since the event of interest is $Y = 36, 37, \ldots, 48$, rewrite the probability as $\Pr(35.5 < Y < 48.5)$.

5.25. Let Y be $b(400, \frac{1}{5})$. Compute an approximate value of $\Pr(0.25 < Y/n)$.

5.26. If Y is $b(100, \frac{1}{2})$, approximate the value of $\Pr(Y = 50)$.

5.27. Let Y be $b(n, 0.55)$. Find the smallest value of n so that (approximately) $\Pr(Y/n > \frac{1}{2}) \ge 0.95$.

5.28. Let $f(x) = 1/x^2$, $1 < x < \infty$, zero elsewhere, be the p.d.f. of a random variable X. Consider a random sample of size 72 from the distribution having this p.d.f. Compute approximately the probability that more than 50 of the items of the random sample are less than 3.

5.29. Forty-eight measurements are recorded to several decimal places. Each of these 48 numbers is rounded off to the nearest integer. The sum of the original 48 numbers is approximated by the sum of these integers. If we assume that the errors made by rounding off are stochastically independent and have uniform distributions over the interval $(-\frac{1}{2}, \frac{1}{2})$, compute approximately the probability that the sum of the integers is within 2 units of the true sum.

´5.5 Some Theorems on Limiting Distributions

In this section we shall present some theorems that can often be used to simplify the study of certain limiting distributions.

Theorem 4. *Let $F_n(u)$ denote the distribution function of a random variable U_n whose distribution depends upon the positive integer n. Let U_n converge stochastically to the constant $c \ne 0$. The random variable U_n/c converges stochastically to 1.*

The proof of this theorem is very easy and is left as an exercise.

Theorem 5. *Let $F_n(u)$ denote the distribution function of a random variable U_n whose distribution depends upon the positive integer n. Further, let U_n converge stochastically to the positive constant c and let $\Pr(U_n < 0) = 0$ for every n. The random variable $\sqrt{U_n}$ converges stochastically to $\sqrt{c}$.*

Proof. We are given that the $\lim_{n \to \infty} \Pr(|U_n - c| \geq \epsilon) = 0$ for every $\epsilon > 0$. We are to prove that the $\lim_{n \to \infty} \Pr(|\sqrt{U_n} - \sqrt{c}| \geq \epsilon') = 0$ for every $\epsilon' > 0$. Now the probability

$$\Pr(|U_n - c| \geq \epsilon) = \Pr[|(\sqrt{U_n} - \sqrt{c})(\sqrt{U_n} + \sqrt{c})| \geq \epsilon]$$

$$= \Pr\left(|\sqrt{U_n} - \sqrt{c}| \geq \frac{\epsilon}{\sqrt{U_n} + \sqrt{c}}\right)$$

$$\geq \Pr\left(|\sqrt{U_n} - \sqrt{c}| \geq \frac{\epsilon}{\sqrt{c}}\right) \geq 0.$$

If we let $\epsilon' = \epsilon/\sqrt{c}$, and if we take the limit, as n becomes infinite, we have

$$0 = \lim_{n \to \infty} \Pr(|U_n - c| \geq \epsilon) \geq \lim_{n \to \infty} \Pr(|\sqrt{U_n} - \sqrt{c}| \geq \epsilon') = 0$$

for every $\epsilon' > 0$. This completes the proof.

The conclusions of Theorems 4 and 5 are very natural ones and they certainly appeal to our intuition. There are many other theorems of this flavor in probability theory. As exercises, it is to be shown that if the random variables U_n and V_n converge stochastically to the respective constants c and d, then $U_n V_n$ converges stochastically to the constant cd, and U_n/V_n converges stochastically to the constant c/d, provided that $d \neq 0$. However, we shall accept, without proof, the following theorem.

Theorem 6. *Let $F_n(u)$ denote the distribution function of a random variable U_n whose distribution depends upon the positive integer n. Let U_n have a limiting distribution with distribution function $F(u)$. Let $H_n(v)$ denote the distribution function of a random variable V_n whose distribution depends upon the positive integer n. Let V_n converge stochastically to 1. The limiting distribution of the random variable $W_n = U_n/V_n$ is the same as that of U_n; that is, W_n has a limiting distribution with distribution function $F(w)$.*

Example 1. Let Y_n denote a random variable that is $b(n, p)$, $0 < p < 1$. We know that

$$U_n = \frac{Y_n - np}{\sqrt{np(1 - p)}}$$

has a limiting distribution that is $n(0, 1)$. Moreover, it has been proved that Y_n/n and $1 - Y_n/n$ converge stochastically to p and $1 - p$, respectively; thus $(Y_n/n)(1 - Y_n/n)$ converges stochastically to $p(1 - p)$. Then, by Theorem 4, $(Y_n/n)(1 - Y_n/n)/[p(1 - p)]$ converges stochastically to 1, and Theorem 5 asserts that the following does also:

$$V_n = \left[\frac{(Y_n/n)(1 - Y_n/n)}{p(1 - p)}\right]^{1/2}.$$

Thus, in accordance with Theorem 6, the ratio $W_n = U_n/V_n$, namely

$$\frac{Y_n - np}{\sqrt{n(Y_n/n)(1 - Y_n/n)}},$$

has a limiting distribution that is $n(0, 1)$. This fact enables us to write (with n a fixed positive integer)

$$\Pr\left[-2 < \frac{Y - np}{\sqrt{n(Y/n)(1 - Y/n)}} < 2\right] = 0.954,$$

approximately.

Example 2. Let $\bar{X}_n$ and S_n^2 denote, respectively, the mean and the variance of a random sample of size n from a distribution that is $n(\mu, \sigma^2)$, $\sigma^2 > 0$. It has been proved that $\bar{X}_n$ converges stochastically to μ and that S_n^2 converges stochastically to σ^2. Theorem 5 asserts that S_n converges stochastically to σ and Theorem 4 tells us that S_n/σ converges stochastically to 1. In accordance with Theorem 6, the random variable $W_n = \sigma\bar{X}_n/S_n$ has the same limiting distribution as does $\bar{X}_n$. That is, $\sigma\bar{X}_n/S_n$ converges stochastically to μ.

EXERCISES

5.30. Prove Theorem 4. *Hint.* Note that $\Pr(|U_n/c - 1| < \epsilon) = \Pr(|U_n - c| < \epsilon|c|)$, for every $\epsilon > 0$. Then take $\epsilon' = \epsilon|c|$.

5.31. Let $\bar{X}_n$ denote the mean of a random sample of size n from a gamma distribution with parameters $\alpha = \mu > 0$ and $\beta = 1$. Show that the limiting distribution of $\sqrt{n}(\bar{X}_n - \mu)/\sqrt{\bar{X}_n}$ is $n(0, 1)$.

5.32. Let $T_n = (\bar{X}_n - \mu)/\sqrt{S_n^2/(n - 1)}$, where $\bar{X}_n$ and S_n^2 represent, respectively, the mean and the variance of a random sample of size n from a distribution that is $n(\mu, \sigma^2)$. Prove that the limiting distribution of T_n is $n(0, 1)$.

5.33. Let $X_1, \ldots, X_n$ and $Y_1, \ldots, Y_n$ be the items of two independent random samples, each of size n, from the distributions that have the respective means μ_1 and μ_2 and the common variance σ^2. Find the limiting distribution of

$$\frac{(\bar{X}_n - \bar{Y}_n) - (\mu_1 - \mu_2)}{\sigma \sqrt{2/n}},$$

where $\bar{X}_n$ and $\bar{Y}_n$ are the respective means of the samples. *Hint.* Let $\bar{Z}_n = \sum_1^n Z_i/n$, where $Z_i = X_i - Y_i$.

5.34. Let U_n and V_n converge stochastically to c and d, respectively. Prove the following.

(a) The sum $U_n + V_n$ converges stochastically to $c + d$. *Hint.* Show that $\Pr(|U_n + V_n - c - d| \geq \epsilon) \leq \Pr(|U_n - c| + |V_n - d| \geq \epsilon) \leq \Pr(|U_n - c| \geq \epsilon/2 \text{ or } |V_n - d| \geq \epsilon/2) \leq \Pr(|U_n - c| \geq \epsilon/2) + \Pr(|V_n - d| \geq \epsilon/2)$.

(b) The product $U_n V_n$ converges stochastically to cd.

(c) If $d \neq 0$, the ratio U_n/V_n converges stochastically to c/d.

5.35. Let U_n converge stochastically to c. If $h(u)$ is a continuous function at $u = c$, prove that $h(U_n)$ converges stochastically to $h(c)$. *Hint.* For each $\epsilon > 0$, there exists a $\delta > 0$ such that $\Pr[|h(U_n) - h(c)| < \epsilon] \geq \Pr[|U_n - c| < \delta]$. Why?

Chapter 6

Estimation

6.1 Point Estimation

The first five chapters of this book deal with certain concepts and problems of probability theory. Throughout we have carefully distinguished between a sample space $\mathscr{C}$ of outcomes and the space $\mathscr{A}$ of one or more random variables defined on $\mathscr{C}$. With this chapter we begin a study of some problems in statistics and here we are more interested in the number (or numbers) by which an outcome is represented than we are in the outcome itself. Accordingly, we shall adopt a frequently used convention. We shall refer to a random variable X as the outcome of a random experiment and we shall refer to the space of X as the sample space. Were it not so awkward, we would call X the numerical outcome. Once the experiment has been performed and it is found that $X = x$, we shall call x the experimental value of X for that performance of the experiment.

This convenient terminology can be used to advantage in more general situations. To illustrate this, let a random experiment be repeated n independent times and under identical conditions. Then the random variables $X_1, X_2, \ldots, X_n$ (each of which assigns a numerical value to an outcome) constitute (Section 4.1) the items of a random sample. If we are more concerned with the numerical representations of the outcomes than with the outcomes themselves, it seems natural to refer to $X_1, X_2, \ldots, X_n$ as the outcomes. And what more appropriate name can we give to the space of a random sample than the sample space? Once the experiment has been performed the indicated number of times and it is found that $X_1 = x_1, X_2 = x_2, \ldots, X_n = x_n$, we shall refer to $x_1, x_2, \ldots, x_n$ as the experimental values of $X_1, X_2, \ldots, X_n$ or as the sample data.

We shall use the terminology of the two preceding paragraphs, and in this section we shall give some examples of *statistical inference*. These examples will be built around the notion of a *point estimate* of an unknown parameter in a p.d.f.

Let a random variable X have a p.d.f. that is of known functional form but in which the p.d.f. depends upon an unknown parameter θ that may have any value in a set Ω. This will be denoted by writing the p.d.f. in the form $f(x; \theta)$, $\theta \in \Omega$. The set Ω will be called the *parameter space*. Thus we are confronted, not with one distribution of probability, but with a *family* of distributions. To each value of θ, $\theta \in \Omega$, there corresponds one member of the family. A family of probability density functions will be denoted by the symbol $\{f(x; \theta); \theta \in \Omega\}$. Any member of this family of probability density functions will be denoted by the symbol $f(x; \theta)$, $\theta \in \Omega$. We shall continue to use the special symbols that have been adopted for the normal, the chi-square, and the binomial distributions. We may, for instance, have the family $\{n(\theta, 1); \theta \in \Omega\}$, where Ω is the set $-\infty < \theta < \infty$. One member of this family of distributions is the distribution that is $n(0, 1)$. Any arbitrary member is $n(\theta, 1)$, $-\infty < \theta < \infty$.

Consider a family of probability density functions $\{f(x; \theta); \theta \in \Omega\}$. It may be that the experimenter needs to select precisely *one* member of the family as being the p.d.f. of his random variable. That is, he needs a *point* estimate of θ. Let $X_1, X_2, \ldots, X_n$ denote a random sample from a distribution that has a p.d.f. which is one member (but which member we do not know) of the family $\{f(x; \theta); \theta \in \Omega\}$ of probability density functions. That is, our sample arises from a distribution that has the p.d.f. $f(x; \theta)$; $\theta \in \Omega$. Our problem is that of defining a statistic $Y_1 = u_1(X_1, X_2, \ldots, X_n)$, so that if $x_1, x_2, \ldots, x_n$ are the observed experimental values of $X_1, X_2, \ldots, X_n$, then the number $y_1 = u_1(x_1, x_2, \ldots, x_n)$ will be a good point estimate of θ.

The following illustration should help motivate one principle that is often used in finding point estimates.

Example 1. Let $X_1, X_2, \ldots, X_n$ denote a random sample from the distribution with p.d.f.

$$f(x) = \theta^x(1 - \theta)^{1-x}, \qquad x = 0, 1,$$

$$= 0 \text{ elsewhere,}$$

where $0 \leq \theta \leq 1$. The probability that $X_1 = x_1$, $X_2 = x_2, \ldots, X_n = x_n$ is the joint p.d.f.

$$\theta^{x_1}(1 - \theta)^{1-x_1}\theta^{x_2}(1 - \theta)^{1-x_2}\cdots \theta^{x_n}(1 - \theta)^{1-x_n} = \theta^{\Sigma x_i}(1 - \theta)^{n-\Sigma x_i},$$

where x_i equals zero or 1, $i = 1, 2, \ldots, n$. This probability, which is the joint p.d.f. of $X_1, X_2, \ldots, X_n$, may be regarded as a function of θ and, when so regarded, is denoted by $L(\theta)$ and called the *likelihood function.*That is,

$$L(\theta) = \theta^{\Sigma x_i}(1 - \theta)^{n-\Sigma x_i}, \qquad 0 \leq \theta \leq 1.$$

We might ask what value of θ would maximize the probability $L(\theta)$ of obtaining this particular observed sample $x_1, x_2, \ldots, x_n$. Certainly, this maximizing value of θ would seemingly be a good estimate of θ because it would provide the largest probability of this particular sample. However, since the likelihood function $L(\theta)$ and its logarithm, $\ln L(\theta)$, are maximized for the same value θ, either $L(\theta)$ or $\ln L(\theta)$ can be used. Here

$$\ln L(\theta) = \left(\sum_1^n x_i\right) \ln \theta + \left(n - \sum_1^n x_i\right) \ln (1 - \theta);$$

so we have

$$\frac{d \ln L(\theta)}{d\theta} = \frac{\Sigma x_i}{\theta} - \frac{n - \Sigma x_i}{1 - \theta} = 0,$$

provided that θ is not equal to zero or 1. This is equivalent to the equation

$$(1 - \theta) \sum_1^n x_i = \theta\left(n - \sum_1^n x_i\right),$$

whose solution for θ is $\sum_1^n x_i/n$. That $\sum_1^n x_i/n$ actually maximizes $L(\theta)$ and $\ln L(\theta)$ can be easily checked, even in the cases in which all of $x_1, x_2, \ldots, x_n$ equal zero together or 1 together. That is, $\sum_1^n x_i/n$ is the value of θ that maximizes $L(\theta)$. The corresponding statistic,

$$\hat{\theta} = \frac{1}{n} \sum_{i=1}^n X_i,$$

is called the *maximum likelihood estimator* of θ. The observed value of $\hat{\theta}$, namely $\sum_1^n x_i/n$, is called the *maximum likelihood estimate* of θ. For a simple example, suppose that $n = 3$, and $x_1 = 1$, $x_2 = 0$, $x_3 = 1$, then $L(\theta) = \theta^2(1 - \theta)$ and the observed $\hat{\theta} = \frac{2}{3}$ is the maximum likelihood estimate of θ.

The principle of the *method of maximum likelihood* can now be formulated easily. Consider a random sample $X_1, X_2, \ldots, X_n$ from a distribution having p.d.f. $f(x; \theta)$, $\theta \in \Omega$. The joint p.d.f. of $X_1, X_2, \ldots, X_n$ is $f(x_1; \theta)f(x_2; \theta) \cdots f(x_n; \theta)$. This joint p.d.f. may be regarded as a function of θ. When so regarded, it is called the likelihood function L of the random sample, and we write

$$L(\theta; x_1, x_2, \ldots, x_n) = f(x_1; \theta)f(x_2; \theta) \cdots f(x_n; \theta), \qquad \theta \in \Omega.$$

Suppose that we can find a nontrivial function of $x_1, x_2, \ldots, x_n$, say $u(x_1, x_2, \ldots, x_n)$, such that, when θ is replaced by $u(x_1, x_2, \ldots, x_n)$, the likelihood function L is a maximum. That is, $L[u(x_1, x_2, \ldots, x_n);$ $x_1, x_2, \ldots, x_n]$ is at least as great as $L(\theta; x_1, x_2, \ldots, x_n)$ for every $\theta \in \Omega$. Then the statistic $u(X_1, X_2, \ldots, X_n)$ will be called a *maximum likelihood estimator* of θ and will be denoted by the symbol $\hat{\theta} = u(X_1, X_2, \ldots, X_n)$. We remark that in many instances there will be a unique maximum likelihood estimator $\hat{\theta}$ of a parameter θ, and often it may be obtained by the process of differentiation.

Example 2. Let $X_1, X_2, \ldots, X_n$ be a random sample from the normal distribution $n(\theta, 1)$, $-\infty < \theta < \infty$. Here

$$L(\theta; x_1, x_2, \ldots, x_n) = \left(\frac{1}{\sqrt{2\pi}}\right)^n \exp\left[-\sum_1^n (x_i - \theta)^2/2\right].$$

This function L can be maximized by setting the first derivative of L, with respect to θ, equal to zero and solving the resulting equation for θ. We note, however, that each of the functions L and $\ln L$ is a maximum for the same value θ. So it may be easier to solve

$$\frac{d \ln L(\theta; x_1, x_2, \ldots, x_n)}{d\theta} = 0.$$

For this example,

$$\frac{d \ln L(\theta; x_1, x_2, \ldots, x_n)}{d\theta} = \sum_1^n (x_i - \theta).$$

If this derivative is equated to zero, the solution for θ is $u(x_1, x_2, \ldots, x_n) = \sum_1^n x_i/n$. That $\sum_1^n x_i/n$ actually maximizes L is easily shown. Thus the statistic

$$\hat{\theta} = u(X_1, X_2, \ldots, X_n) = \frac{1}{n} \sum_1^n X_i = \bar{X}$$

is the unique maximum likelihood estimator of the mean θ.

It is interesting to note that in both Examples 1 and 2, it is true that $E(\hat{\theta}) = \theta$. That is, in each of these cases, the expected value of the estimator is equal to the corresponding parameter, which leads to the following definition.

Definition 1. Any statistic whose mathematical expectation is equal to a parameter θ is called an *unbiased* estimator of the parameter θ. Otherwise, the statistic is said to be *biased*.

Example 3. Let

$$f(x; \theta) = \frac{1}{\theta}, \qquad 0 < x \leq \theta, 0 < \theta < \infty,$$

$$= 0 \text{ elsewhere,}$$

and let $X_1, X_2, \ldots, X_n$ denote a random sample from this distribution. Note that we have taken $0 < x \leq \theta$ instead of $0 < x < \theta$ so as to avoid a discussion of supremum versus maximum. Here

$$L(\theta; x_1, x_2, \ldots, x_n) = \frac{1}{\theta^n}, \qquad 0 < x_i \leq \theta,$$

which is an ever-decreasing function of θ. The maximum of such functions cannot be found by differentiation but by selecting θ as small as possible. Now $\theta \geq$ each x_i; in particular, then, $\theta \geq \max(x_i)$. Thus L can be made no larger than

$$\frac{1}{[\max(x_i)]^n}$$

and the unique maximum likelihood estimator $\hat{\theta}$ of θ in this example is the nth order statistic $\max(X_i)$. It can be shown that $E[\max(X_i)] = n\theta/(n + 1)$. Thus, in this instance, the maximum likelihood estimator of the parameter θ is biased. That is, the property of unbiasedness is not in general a property of a maximum likelihood estimator.

While the maximum likelihood estimator $\hat{\theta}$ of θ in Example 3 is a biased estimator, results in Chapter 5 show that the nth order statistic $\hat{\theta} = \max(X_i) = Y_n$ converges stochastically to θ. Thus, in accordance with the following definition, we say that $\hat{\theta} = Y_n$ is a consistent estimator of θ.

Definition 2. Any statistic that converges stochastically to a parameter θ is called a *consistent* estimator of that parameter θ.

Consistency is a desirable property of an estimator; and, in all cases of practical interest, maximum likelihood estimators are consistent.

The preceding definitions and properties are easily generalized. Let $X, Y, \ldots, Z$ denote random variables that may or may not be stochastically independent and that may or may not be identically distributed. Let the joint p.d.f. $g(x, y, \ldots, z; \theta_1, \theta_2, \ldots, \theta_m)$, $(\theta_1, \theta_2, \ldots,$

$\theta_m) \in \Omega$, depend on m parameters. This joint p.d.f., when regarded as a function of $(\theta_1, \theta_2, \ldots, \theta_m) \in \Omega$, is called the likelihood function of the random variables. Those functions $u_1(x, y, \ldots, z), u_2(x, y, \ldots, z), \ldots, u_m(x, y, \ldots, z)$ that maximize this likelihood function with respect to $\theta_1, \theta_2, \ldots, \theta_m$, respectively, define the maximum likelihood estimators

$$\hat{\theta}_1 = u_1(X, Y, \ldots, Z), \qquad \hat{\theta}_2 = u_2(X, Y, \ldots, Z), \ldots,$$
$$\hat{\theta}_m = u_m(X, Y, \ldots, Z)$$

of the m parameters.

Example 4. Let $X_1, X_2, \ldots, X_n$ denote a random sample from a distribution that is $n(\theta_1, \theta_2)$, $-\infty < \theta_1 < \infty$, $0 < \theta_2 < \infty$. We shall find $\hat{\theta}_1$ and $\hat{\theta}_2$, the maximum likelihood estimators of θ_1 and θ_2. The logarithm of the likelihood function may be written in the form

$$\ln L(\theta_1, \theta_2; x_1, \ldots, x_n) = -\frac{\sum\limits_1^n (x_i - \theta_1)^2}{2\theta_2} - \frac{n \ln (2\pi\theta_2)}{2}.$$

We observe that we may maximize by differentiation. We have

$$\frac{\partial \ln L}{\partial \theta_1} = \frac{\sum\limits_1^n (x_i - \theta_1)}{\theta_2}, \qquad \frac{\partial \ln L}{\partial \theta_2} = \frac{\sum\limits_1^n (x_i - \theta_1)^2}{2\theta_2^2} - \frac{n}{2\theta_2}.$$

If we equate these partial derivatives to zero and solve simultaneously the two equations thus obtained, the solutions for θ_1 and θ_2 are found to be $\sum\limits_1^n x_i/n = \bar{x}$ and $\sum\limits_1^n (x_i - \bar{x})^2/n = s^2$, respectively. It can be verified that these solutions maximize L. Thus the maximum likelihood estimators of $\theta_1 = \mu$ and $\theta_2 = \sigma^2$ are, respectively, the mean and the variance of the sample, namely $\hat{\theta}_1 = \bar{X}$ and $\hat{\theta}_2 = S^2$. Whereas $\hat{\theta}_1$ is an unbiased estimator of θ_1, the estimator $\hat{\theta}_2 = S^2$ is biased because

$$E(\hat{\theta}_2) = \frac{\sigma^2}{n} E\left(\frac{n\hat{\theta}_2}{\sigma^2}\right) = \frac{\sigma^2}{n} E\left(\frac{nS^2}{\sigma^2}\right) = \frac{(n-1)\sigma^2}{n} = \frac{(n-1)\theta_2}{n}.$$

However, in Chapter 5 it has been shown that $\hat{\theta}_1 = \bar{X}$ and $\hat{\theta}_2 = S^2$ converge stochastically to θ_1 and θ_2, respectively, and thus they are consistent estimators of θ_1 and θ_2.

Sometimes it is impossible to find maximum likelihood estimators in a convenient closed form and numerical methods must be used to maximize the likelihood function. For illustration, suppose that $X_1, X_2, \ldots, X_n$ is a random sample from a gamma distribution with parameters $\alpha = \theta_1$ and $\beta = \theta_2$, where $\theta_1 > 0$, $\theta_2 > 0$. It is difficult to maximize

$$L(\theta_1, \theta_2; x_1, \ldots, x_n) = \left[\frac{1}{\Gamma(\theta_1)\theta_2^{\theta_1}}\right]^n (x_1 x_2 \cdots x_n)^{\theta_1 - 1} \exp\left(-\sum_1^n x_i/\theta_2\right)$$

with respect to θ_1 and θ_2, owing to the presence of the gamma function $\Gamma(\theta_1)$. However, to obtain easily point estimates of θ_1 and θ_2, let us simply equate the first two moments of the distribution to the corresponding moments of the sample. This seems like a reasonable way in which to find estimators, since the empirical distribution $F_n(x)$ converges stochastically to $F(x)$, and hence corresponding moments should be about equal. Here in this illustration we have

$$\theta_1 \theta_2 = \bar{X}, \qquad \theta_1 \theta_2^2 = S^2,$$

the solutions of which are

$$\tilde{\theta}_1 = \frac{\bar{X}^2}{S^2} \qquad \text{and} \qquad \tilde{\theta}_2 = \frac{S^2}{\bar{X}}.$$

We say that these latter two statistics, $\tilde{\theta}_1$ and $\tilde{\theta}_2$, are respective estimators of θ_1 and θ_2 found by the *method of moments*.

To generalize the discussion of the preceding paragraph, let $X_1, X_2, \ldots, X_n$ be a random sample of size n from a distribution with p.d.f. $f(x; \theta_1, \theta_2, \ldots, \theta_r)$, $(\theta_1, \ldots, \theta_r) \in \Omega$. The expectation $E(X^k)$ is frequently called the kth moment of the distribution, $k = 1, 2, 3, \ldots$. The sum $M_k = \sum_1^n X_i^k/n$ is the kth moment of the sample, $k = 1, 2, 3, \ldots$. The *method of moments* can be described as follows. Equate $E(X^k)$ to M_k, beginning with $k = 1$ and continuing until there are enough equations to provide unique solutions for $\theta_1, \theta_2, \ldots, \theta_r$, say $h_i(M_1, M_2, \ldots)$, $i = 1, 2, \ldots, r$, respectively. It should be noted that this could be done in an equivalent manner by equating $\mu = E(X)$ to $\bar{X}$ and $E[(X - \mu)^k]$ to $\sum_1^n (X_i - \bar{X})^k/n$, $k = 2, 3$, and so on until unique solutions for $\theta_1, \theta_2, \ldots, \theta_r$ are obtained. This alternative procedure was used in the preceding illustration. In most practical cases, the estimator $\tilde{\theta}_i = h_i(M_1, M_2, \ldots)$ of θ_i, found by the method of moments, is a consistent estimator of θ_i, $i = 1, 2, \ldots, r$.

EXERCISES

6.1. Let $X_1, X_2, \ldots, X_n$ represent a random sample from each of the distributions having the following probability density functions:

(a) $f(x; \theta) = \theta^x e^{-\theta}/x!$, $x = 0, 1, 2, \ldots, 0 \le \theta < \infty$, zero elsewhere, where $f(0; 0) = 1$.

(b) $f(x; \theta) = \theta x^{\theta - 1}$, $0 < x < 1, 0 < \theta < \infty$, zero elsewhere.

(c) $f(x; \theta) = (1/\theta)e^{-x/\theta}$, $0 < x < \infty$, $0 < \theta < \infty$, zero elsewhere.

(d) $f(x; \theta) = \frac{1}{2}e^{-|x - \theta|}$, $-\infty < x < \infty$, $-\infty < \theta < \infty$.

(e) $f(x; \theta) = e^{-(x - \theta)}$, $\theta \le x < \infty$, $-\infty < \theta < \infty$, zero elsewhere.

In each case find the maximum likelihood estimator $\hat{\theta}$ of θ.

6.2. Let $X_1, X_2, \ldots, X_n$ be a random sample from the distribution having p.d.f. $f(x; \theta_1, \theta_2) = (1/\theta_2)e^{-(x - \theta_1)/\theta_2}$, $\theta_1 \le x < \infty$, $-\infty < \theta_1 < \infty$, $0 < \theta_2 < \infty$, zero elsewhere. Find the maximum likelihood estimators of θ_1 and θ_2.

6.3. Let $Y_1 < Y_2 < \cdots < Y_n$ be the order statistics of a random sample from a distribution with p.d.f. $f(x; \theta) = 1, \theta - \frac{1}{2} \le x \le \theta + \frac{1}{2}, -\infty < \theta < \infty$, zero elsewhere. Show that every statistic $u(X_1, X_2, \ldots, X_n)$ such that

$$Y_n - \tfrac{1}{2} \le u(X_1, X_2, \ldots, X_n) \le Y_1 + \tfrac{1}{2}$$

is a maximum likelihood estimator of θ. In particular, $(4Y_1 + 2Y_n + 1)/6$, $(Y_1 + Y_n)/2$, and $(2Y_1 + 4Y_n - 1)/6$ are three such statistics. Thus uniqueness is not in general a property of a maximum likelihood estimator.

6.4. Let X_1, X_2, and X_3 have the multinomial distribution in which $n = 25, k = 4$, and the unknown probabilities are θ_1, θ_2, and θ_3, respectively. Here we can, for convenience, let $X_4 = 25 - X_1 - X_2 - X_3$ and $\theta_4 = 1 - \theta_1 - \theta_2 - \theta_3$. If the observed values of the random variables are $x_1 = 4, x_2 = 11$, and $x_3 = 7$, find the maximum likelihood estimates of θ_1, θ_2, and θ_3.

6.5. The *Pareto distribution* is frequently used as a model in study of incomes and has the distribution function

$$F(x; \theta_1, \theta_2) = 1 - (\theta_1/x)^{\theta_2}, \quad \theta_1 \le x, \text{zero elsewhere}, \quad \text{where } \theta_1 > 0 \text{ and } \theta_2 > 0.$$

If $X_1, X_2, \ldots, X_n$ is a random sample from this distribution, find the maximum likelihood estimators of θ_1 and θ_2.

6.6. Let Y_n be a statistic such that $\lim_{n \to \infty} E(Y_n) = \theta$ and $\lim_{n \to \infty} \sigma_{Y_n}^2 = 0$. Prove that Y_n is a consistent estimator of θ. *Hint.* $\Pr(|Y_n - \theta| \ge \epsilon) \le E[(Y_n - \theta)^2]/\epsilon^2$ and $E[(Y_n - \theta)^2] = [E(Y_n - \theta)]^2 + \sigma_{Y_n}^2$. Why?

6.7. For each of the distributions in Exercise 6.1, find an estimator of θ by the method of moments and show that it is consistent.

6.2 Measures of Quality of Estimators

Now it would seem that if $y = u(x_1, x_2, \ldots, x_n)$ is to qualify as a good point estimate of θ, there should be a great probability that the statistic $Y = u(X_1, X_2, \ldots, X_n)$ will be close to θ; that is, θ should be a sort of rallying point for the numbers $y = u(x_1, x_2, \ldots, x_n)$. This can be achieved in one way by selecting $Y = u(X_1, X_2, \ldots, X_n)$ in such a way that not only is Y an unbiased esimator of θ but also the variance of Y is as small as it can be made. We do this because the variance of Y is a measure of the intensity of the concentration of the probability for Y in the neighborhood of the point $\theta = E(Y)$. Accordingly, we define an unbiased minimum variance estimator of the parameter θ in the following manner.

Definition 3. For a given positive integer n, $Y = u(X_1, X_2, \ldots, X_n)$ will be called an *unbiased minimum variance* estimator of the parameter θ if Y is unbiased, that is $E(Y) = \theta$, and if the variance of Y is less than or equal to the variance of every other unbiased estimator of θ.

For illustration, let $X_1, X_2, \ldots, X_9$ denote a random sample from a distribution that is $n(\theta, 1)$, $-\infty < \theta < \infty$. Since the statistic $\bar{X} = (X_1 + X_2 + \cdots + X_9)/9$ is $n(\theta, \frac{1}{9})$, $\bar{X}$ is an unbiased estimator of θ. The statistic X_1 is $n(\theta, 1)$, so X_1 is also an unbiased estimator of θ. Although the variance $\frac{1}{9}$ of $\bar{X}$ is less than the variance 1 of X_1, we cannot say, with $n = 9$, that $\bar{X}$ is the unbiased minimum variance estimator of θ; that definition requires that the comparison be made with every unbiased estimator of θ. To be sure, it is quite impossible to tabulate all other unbiased estimators of this parameter θ, so other methods must be developed for making the comparisons of the variances. A beginning on this problem will be made in Chapter 10.

Let us now discuss the problem of point estimation of a parameter from a slightly different standpoint. Let $X_1, X_2, \ldots, X_n$ denote a random sample of size n from a distribution that has the p.d.f. $f(x; \theta)$, $\theta \in \Omega$. The distribution may be either of the continuous or the discrete type. Let $Y = u(X_1, X_2, \ldots, X_n)$ be a statistic on which we wish to base a point estimate of the parameter θ. Let $w(y)$ be that function of the observed value of the statistic Y which is the point estimate of θ. Thus the function w *decides* the value of our point estimate of θ and w is called a *decision function* or a *decision rule*. One value of the decision function, say $w(y)$, is called a *decision*. Thus a numerically determined point estimate of a parameter θ is a decision. Now a decision may be correct or it may be wrong. It would be useful to have a measure of the seriousness of the difference, if any, between the true value of θ and the point estimate $w(y)$. Accordingly, with each pair, $[\theta, w(y)]$, $\theta \in \Omega$, we associate a nonnegative number $\mathscr{L}[\theta, w(y)]$ that reflects this seriousness. We call the function $\mathscr{L}$ the *loss function*. The expected (mean) value of the loss function is called the *risk function*. If $g(y; \theta)$, $\theta \in \Omega$, is the p.d.f. of Y, the risk function $R(\theta, w)$ is given by

$$R(\theta, w) = E\{\mathscr{L}[\theta, w(Y)]\} = \int_{-\infty}^{\infty} \mathscr{L}[\theta, w(y)]g(y; \theta) \, dy$$

if Y is a random variable of the continuous type. It would be desirable to select a decision function that minimizes the risk $R(\theta, w)$ for all values of θ, $\theta \in \Omega$. But this is usually impossible because the decision function w that minimizes $R(\theta, w)$ for one value of θ may not minimize $R(\theta, w)$ for another value of θ. Accordingly, we need either to restrict our decision function to a certain class or to consider methods of ordering the risk functions. The following example, while very simple, dramatizes these difficulties.

Example 1. Let $X_1, X_2, \ldots, X_{25}$ be a random sample from a distribution that is $n(\theta, 1)$, $-\infty < \theta < \infty$. Let $Y = \bar{X}$, the mean of the random sample, and let $\mathscr{L}[\theta, w(y)] = [\theta - w(y)]^2$. We shall compare the two decision functions given by $w_1(y) = y$ and $w_2(y) = 0$ for $-\infty < y < \infty$. The corresponding risk functions are

$$R(\theta, w_1) = E[(\theta - Y)^2] = \tfrac{1}{25}$$

and

$$R(\theta, w_2) = E[(\theta - 0)^2] = \theta^2.$$

Obviously, if, in fact, $\theta = 0$, then $w_2(y) = 0$ is an excellent decision and we have $R(0, w_2) = 0$. However, if θ differs from zero by very much, it is equally clear that $w_2(y) = 0$ is a poor decision. For example, if, in fact, $\theta = 2$, $R(2, w_2) = 4 > R(2, w_1) = \tfrac{1}{25}$. In general, we see that $R(\theta, w_2) < R(\theta, w_1)$, provided that $-\tfrac{1}{5} < \theta < \tfrac{1}{5}$ and that otherwise $R(\theta, w_2) \geq R(\theta, w_1)$. That is, one of these decision functions is better than the other for some values of θ and the other decision function is better for other values of θ. If, however, we had restricted our consideration to decision functions w such that $E[w(Y)] = \theta$ for all values of θ, $\theta \in \Omega$, then the decision $w_2(y) = 0$ is not allowed. Under this restriction and with the given $\mathscr{L}[\theta, w(y)]$, the risk function is the variance of the unbiased estimator $w(Y)$, and we are confronted with the problem of finding the unbiased minimum variance estimator. In Chapter 10 we show that the solution is $w(y) = y = \bar{x}$.

Suppose, however, that we do not want to restrict ourselves to decision functions w such that $E[w(Y)] = \theta$ for all values of θ, $\theta \in \Omega$. Instead, let us say that the decision function that minimizes the maximum of the risk function is the best decision function. Because, in this example, $R(\theta, w_2) = \theta^2$

is unbounded, $w_2(y) = 0$ is not, in accordance, with this criterion, a good decision function. On the other hand, with $-\infty < \theta < \infty$, we have

$$\max_{\theta} R(\theta, w_1) = \max_{\theta} (\tfrac{1}{25}) = \tfrac{1}{25}.$$

Accordingly, $w_1(y) = y = \bar{x}$ seems to be a very good decision in accordance with this criterion because $\tfrac{1}{25}$ is small. As a matter of fact, it can be proved that w_1 is the best decision function, as measured by this *minimax criterion*, when the loss function is $\mathscr{L}[\theta, w(y)] = [\theta - w(y)]^2$.

In this example we illustrated the following:

(a) Without some restriction on the decision function, it is difficult to find a decision function that has a risk function which is uniformly less than the risk function of another decision function.

(b) A principle of selecting a best decision function, called the *minimax principle*. This principle may be stated as follows: If the decision function given by $w_0(y)$ is such that, for all $\theta \in \Omega$,

$$\max_{\theta} R[\theta, w_0(y)] \le \max_{\theta} R[\theta, w(y)]$$

for every other decision function $w(y)$, then $w_0(y)$ is called a *minimax decision function*.

With the restriction $E[w(Y)] = \theta$ and the loss function $\mathscr{L}[\theta, w(y)] = [\theta - w(y)]^2$, the decision function that minimizes the risk function yields an unbiased estimator with minimum variance. If, however, the restriction $E[w(Y)] = \theta$ is replaced by some other condition, the decision function $w(Y)$, if it exists, which minimizes $E\{[\theta - w(Y)]^2\}$ uniformly in θ is sometimes called the *minimum mean-square-error estimator*. Exercises 6.13, 6.14, and 6.15 provide examples of this type of estimator.

Another principle for selecting the decision function, which may be called a best decision function, will be stated in Section 6.6.

EXERCISES

6.8. Show that the mean $\bar{X}$ of a random sample of size n from a distribution having p.d.f. $f(x; \theta) = (1/\theta)e^{-(x/\theta)}$, $0 < x < \infty$, $0 < \theta < \infty$, zero elsewhere, is an unbiased estimator of θ and has variance θ^2/n.

6.9. Let $X_1, X_2, \ldots, X_n$ denote a random sample from a normal distribution with mean zero and variance θ, $0 < \theta < \infty$. Show that $\sum_1^n X_i^2/n$ is an unbiased estimator of θ and has variance $2\theta^2/n$.

6.10. Let $Y_1 < Y_2 < Y_3$ be the order statistics of a random sample of size 3 from the uniform distribution having p.d.f. $f(x; \theta) = 1/\theta$, $0 < x < \theta$, $0 < \theta < \infty$, zero elsewhere. Show that $4Y_1$, $2Y_2$, and $\frac{4}{3}Y_3$ are all unbiased estimators of θ. Find the variance of each of these unbiased estimators.

6.11. Let Y_1 and Y_2 be two stochastically independent unbiased estimators of θ. Say the variance of Y_1 is twice the variance of Y_2. Find the constants k_1 and k_2 so that $k_1 Y_1 + k_2 Y_2$ is an unbiased estimator with smallest possible variance for such a linear combination.

6.12. In Example 1 of this section, take $\mathscr{L}[\theta, w(y)] = |\theta - w(y)|$. Show that $R(\theta, w_1) = \frac{1}{3}\sqrt{2/\pi}$ and $R(\theta, w_2) = |\theta|$. Of these two decision functions w_1 and w_2, which yields the smaller maximum risk?

6.13. Let $X_1, X_2, \ldots, X_n$ denote a random sample from a Poisson distribution with parameter θ, $0 < \theta < \infty$. Let $Y = \sum_1^n X_i$ and let $\mathscr{L}[\theta, w(y)] = [\theta - w(y)]^2$. If we restrict our considerations to decision functions of the form $w(y) = b + y/n$, where b does not depend upon y, show that $R(\theta, w) = b^2 + \theta/n$. What decision function of this form yields a uniformly smaller risk than every other decision function of this form? With this solution, say w and $0 < \theta < \infty$, determine $\max_\theta R(\theta, w)$ if it exists.

6.14. Let $X_1, X_2, \ldots, X_n$ denote a random sample from a distribution that is $n(\mu, \theta)$, $0 < \theta < \infty$, where μ is unknown. Let $Y = \sum_1^n (X_i - \bar{X})^2/n = S^2$ and let $\mathscr{L}[\theta, w(y)] = [\theta - w(y)]^2$. If we consider decision functions of the form $w(y) = by$, where b does not depend upon y, show that $R(\theta, w) = (\theta^2/n^2)[(n^2 - 1)b^2 - 2n(n - 1)b + n^2]$. Show that $b = n/(n + 1)$ yields a minimum risk for decision functions of this form. Note that $nY/(n + 1)$ is not an unbiased estimator of θ. With $w(y) = ny/(n + 1)$ and $0 < \theta < \infty$, determine $\max_\theta R(\theta, w)$ if it exists.

6.15. Let $X_1, X_2, \ldots, X_n$ denote a random sample from a distribution that is $b(1, \theta)$, $0 \le \theta \le 1$. Let $Y = \sum_1^n X_i$ and let $\mathscr{L}[\theta, w(y)] = [\theta - w(y)]^2$. Consider decision functions of the form $w(y) = by$, where b does not depend upon y. Prove that $R(\theta, w) = b^2 n\theta(1 - \theta) + (bn - 1)^2\theta^2$. Show that

$$\max_\theta R(\theta, w) = \frac{b^4 n^2}{4[b^2 n - (bn - 1)^2]},$$

provided the value b is such that $b^2 n \ge 2(bn - 1)^2$. Prove that $b = 1/n$ does not minimize $\max_\theta R(\theta, w)$.

6.3 Confidence Intervals for Means

Suppose we are willing to accept as a fact that the (numerical) out-come X of a random experiment is a random variable that has a normal distribution with known variance σ^2 but unknown mean μ. That is, μ is some constant, but its value is unknown. To elicit some information about μ, we decide to repeat the random experiment n independent times, n being a fixed positive integer, and under identical conditions. Let the random variables $X_1, X_2, \ldots, X_n$ denote, respectively, the outcomes to be obtained on these n repetitions of the experiment. If our assumptions are fulfilled, we then have under consideration a random sample $X_1, X_2, \ldots, X_n$ from a distribution that is $n(\mu, \sigma^2)$, σ^2 known. Consider the maximum likelihood estimator of μ, namely $\hat{\mu} = \bar{X}$. Of course, $\bar{X}$ is $n(\mu, \sigma^2/n)$ and $(\bar{X} - \mu)/(\sigma/\sqrt{n})$ is $n(0, 1)$. Thus

$$\Pr\left(-2 < \frac{\bar{X} - \mu}{\sigma/\sqrt{n}} < 2\right) = 0.954.$$

However, the events

$$-2 < \frac{\bar{X} - \mu}{\sigma/\sqrt{n}} < 2,$$

$$\frac{-2\sigma}{\sqrt{n}} < \bar{X} - \mu < \frac{2\sigma}{\sqrt{n}},$$

and

$$\bar{X} - \frac{2\sigma}{\sqrt{n}} < \mu < \bar{X} + \frac{2\sigma}{\sqrt{n}}$$

are equivalent. Thus these events have the same probability. That is,

$$\Pr\left(\bar{X} - \frac{2\sigma}{\sqrt{n}} < \mu < \bar{X} + \frac{2\sigma}{\sqrt{n}}\right) = 0.954.$$

Since σ is a known number, each of the random variables $\bar{X} - 2\sigma/\sqrt{n}$ and $\bar{X} + 2\sigma/\sqrt{n}$ is a statistic. The interval $(\bar{X} - 2\sigma/\sqrt{n}, \bar{X} + 2\sigma/\sqrt{n})$ is a random interval. In this case, both end points of the interval are statistics. The immediately preceding probability statement can be read. Prior to the repeated independent performances of the random experiment, the probability is 0.954 that the random interval $(\bar{X} - 2\sigma/\sqrt{n}, \bar{X} + 2\sigma/\sqrt{n})$ includes the unknown fixed point (parameter) μ.

Up to this point, only probability has been involved; the determination of the p.d.f. of $\bar{X}$ and the determination of the random interval were problems of probability. Now the problem becomes statistical. Suppose the experiment yields $X_1 = x_1, X_2 = x_2, \ldots, X_n = x_n$. Then the sample value of $\bar{X}$ is $\bar{x} = (x_1 + x_2 + \cdots + x_n)/n$, a known number. Moreover, since σ is known, the interval $(\bar{x} - 2\sigma/\sqrt{n}, \bar{x} + 2\sigma/\sqrt{n})$ has known end points. Obviously, we cannot say that 0.954 is the probability that the particular interval $(\bar{x} - 2\sigma/\sqrt{n}, \bar{x} + 2\sigma/\sqrt{n})$ includes the parameter μ, for μ, although unknown, is some constant, and this particular interval either does or does not include μ. However, the fact that we had such a high degree of probability, prior to the performance of the experiment, that the random interval $(\bar{X} - 2\sigma/\sqrt{n}, \bar{X} + 2\sigma/\sqrt{n})$ includes the fixed point (parameter) μ leads us to have some reliance on the particular interval $(\bar{x} - 2\sigma/\sqrt{n}, \bar{x} + 2\sigma/\sqrt{n})$. This reliance is reflected by calling the known interval $(\bar{x} - 2\sigma/\sqrt{n}, \bar{x} + 2\sigma/\sqrt{n})$ a 95.4 per cent confidence interval for μ. The number 0.954 is called the *confidence coefficient*. The confidence coefficient is equal to the probability that the random interval includes the parameter. One may, of course, obtain an 80 per cent, a 90 per cent, or a 99 per cent confidence interval for μ by using 1.282, 1.645, or 2.576, respectively, instead of the constant 2.

A statistical inference of this sort is an example of *interval estimation* of a parameter. Note that the interval estimate of μ is found by taking a good (here maximum likelihood) estimate $\bar{x}$ of μ and adding and subtracting twice the standard deviation of $\bar{X}$, namely $2\sigma/\sqrt{n}$, which is small if n is large. If σ were not known, the end points of the random interval would not be statistics. Although the probability statement about the random interval remains valid, the sample data would not yield an interval with known end points.

Example 1. If in the preceding discussion $n = 40$, $\sigma^2 = 10$, and $\bar{x} = 7.164$, then $(7.164 - 1.282\sqrt{\frac{10}{40}}, 7.164 + 1.282\sqrt{\frac{10}{40}})$, or $(6.523, 7.805)$, is an 80 per cent confidence interval for μ. Thus we have an interval estimate of μ.

In the next example we shall show how the central limit theorem may be used to help us find an approximate confidence interval for μ when our sample arises from a distribution that is not normal.

Example 2. Let $\overline{X}$ denote the mean of a random sample of size 25 from a distribution that has a moment-generating function, variance $\sigma^2 = 100$, and mean μ. Since $\sigma/\sqrt{n} = 2$, then approximately

$$\Pr\left(-1.96 < \frac{\overline{X} - \mu}{2} < 1.96\right) = 0.95,$$

or

$$\Pr\left(\overline{X} - 3.92 < \mu < \overline{X} + 3.92\right) = 0.95.$$

Let the observed mean of the sample be $\overline{x} = 67.53$. Accordingly, the interval from $\overline{x} - 3.92 = 63.61$ to $\overline{x} + 3.92 = 71.45$ is an approximate 95 per cent confidence interval for the mean μ.

Let us now turn to the problem of finding a confidence interval for the mean μ of a normal distribution when we are not so fortunate as to know the variance σ^2. In Section 4.8 we found that nS^2/σ^2, where S^2 is the variance of a random sample of size n from a distribution that is $n(\mu, \sigma^2)$, is $\chi^2(n - 1)$. Thus we have $\sqrt{n}(\overline{X} - \mu)/\sigma$ to be $n(0, 1)$, nS^2/σ^2 to be $\chi^2(n - 1)$, and the two to be stochastically independent. In Section 4.4 the random variable T was defined in terms of two such random variables as these. In accordance with that section and the foregoing results, we know that

$$T = \frac{[\sqrt{n}(\overline{X} - \mu)]/\sigma}{\sqrt{nS^2/[\sigma^2(n - 1)]}} = \frac{\overline{X} - \mu}{S/\sqrt{n - 1}}$$

has a t distribution with $n - 1$ degrees of freedom, whatever the value of $\sigma^2 > 0$. For a given positive integer n and a probability of 0.95, say, we can find numbers $a < b$ from Table IV in Appendix B, such that

$$\Pr\left(a < \frac{\overline{X} - \mu}{S/\sqrt{n - 1}} < b\right) = 0.95.$$

Since the graph of the p.d.f. of the random variable T is symmetric about the vertical axis through the origin, we would doubtless take $a = -b, b > 0$. If the probability of this event is written (with $a = -b$) in the form

$$\Pr\left(\overline{X} - \frac{bS}{\sqrt{n - 1}} < \mu < \overline{X} + \frac{bS}{\sqrt{n - 1}}\right) = 0.95,$$

then the interval $[\overline{X} - (bS/\sqrt{n - 1}), \overline{X} + (bS/\sqrt{n - 1})]$ is a random interval having probability 0.95 of including the unknown fixed point (parameter) μ. If the experimental values of $X_1, X_2, \ldots, X_n$ are $x_1, x_2,$

$\ldots, x_n$ with $s^2 = \sum\limits_{1}^{n} (x_i - \bar{x})^2/n$, where $\bar{x} = \sum\limits_{1}^{n} x_i/n$, then the interval $[\bar{x} - (bs/\sqrt{n-1}),\ \bar{x} + (bs/\sqrt{n-1})]$ is a 95 per cent confidence interval for μ for every $\sigma^2 > 0$. Again this interval estimate of μ is found by adding and subtracting a quantity, here $bs/\sqrt{n-1}$, to the point estimate $\bar{x}$.

Example 3. If in the preceding discussion $n = 10, \bar{x} = 3.22$, and $s = 1.17$, then the interval $[3.22 - (2.262)(1.17)/\sqrt{9},\ 3.22 + (2.262)(1.17)/\sqrt{9}]$ or $(2.34, 4.10)$ is a 95 per cent confidence interval for μ.

Remark. If one wishes to find a confidence interval for μ and if the variance σ^2 of the nonnormal distribution is unknown (unlike Example 2 of this section), he may with large samples proceed as follows. If certain weak conditions are satisfied, then S^2, the variance of a random sample of size $n \geq 2$, converges stochastically to σ^2. Then in

$$\frac{\sqrt{n}(\bar{X} - \mu)/\sigma}{\sqrt{nS^2/(n-1)\sigma^2}} = \frac{\sqrt{n-1}(\bar{X} - \mu)}{S}$$

the numerator of the left-hand member has a limiting distribution that is $n(0, 1)$ and the denominator of that member converges stochastically to 1. Thus $\sqrt{n-1}(\bar{X} - \mu)/S$ has a limiting distribution that is $n(0, 1)$. This fact enables us to find approximate confidence intervals for μ when our conditions are satisfied. A similar procedure can be followed in the next section when seeking confidence intervals for the difference of the means of two independent nonnormal distributions.

We shall now consider the problem of determining a confidence interval for the unknown parameter p of a binomial distribution when the parameter n is known. Let Y be $b(n, p)$, where $0 < p < 1$ and n is known. Then p is the mean of Y/n. We shall use a result of Example 1, Section 5.5, to find an approximate 95.4 per cent confidence interval for the mean p. There we found that

$$\Pr\left[-2 < \frac{Y - np}{\sqrt{n(Y/n)(1 - Y/n)}} < 2\right] = 0.954,$$

approximately. Since

$$\frac{Y - np}{\sqrt{n(Y/n)(1 - Y/n)}} = \frac{(Y/n) - p}{\sqrt{(Y/n)(1 - Y/n)/n}},$$

the probability statement above can easily be written in the form

$$\Pr\left[\frac{Y}{n} - 2\sqrt{\frac{(Y/n)(1 - Y/n)}{n}} < p < \frac{Y}{n} + 2\sqrt{\frac{(Y/n)(1 - Y/n)}{n}}\right] = 0.954,$$

approximately. Thus, for large n, if the experimental value of Y is y, the interval

$$\left[\frac{y}{n} - 2\sqrt{\frac{(y/n)(1 - y/n)}{n}}, \frac{y}{n} + 2\sqrt{\frac{(y/n)(1 - y/n)}{n}}\right]$$

provides an approximate 95.4 per cent confidence interval for p.

A more complicated approximate 95.4 per cent confidence interval can be obtained from the fact that $Z = (Y - np)/\sqrt{np(1 - p)}$ has a limiting distribution that is $n(0, 1)$, and the fact that the event $-2 < Z < 2$ is equivalent to the event

(1)
$$\frac{Y + 2 - 2\sqrt{[Y(n - Y)/n] + 1}}{n + 4}$$

$$< p < \frac{Y + 2 + 2\sqrt{[Y(n - Y)/n] + 1}}{n + 4}.$$

The first of these facts was established in Example 2, Section 5.4; the proof of inequalities (1) is left as an exercise. Thus an experimental value y of Y may be used in inequalities (1) to determine an approximate 95.4 per cent confidence interval for p.

If one wishes a 95 per cent confidence interval for p that does not depend upon limiting distribution theory, he may use the following approach. (This approach is quite general and can be used in other instances.) Determine two *increasing* functions of p, say $c_1(p)$ and $c_2(p)$, such that for each value of p we have, at least approximately,

$$\Pr\left[c_1(p) < Y < c_2(p)\right] = 0.95.$$

The reason that this may be approximate is due to the fact that Y has a distribution of the discrete type and thus it is, in general, impossible to achieve the probability 0.95 exactly. With $c_1(p)$ and $c_2(p)$ increasing functions, they have single-valued inverses, say $d_1(y)$ and $d_2(y)$, respectively. Thus the events $c_1(p) < Y < c_2(p)$ and $d_2(Y) < p < d_1(Y)$ are equivalent and we have, at least approximately,

$$\Pr\left[d_2(Y) < p < d_1(Y)\right] = 0.95.$$

In the case of the binomial distribution, the functions $c_1(p)$, $c_2(p)$, $d_2(y)$, and $d_1(y)$ cannot be found explicitly, but a number of books provide tables of $d_2(y)$ and $d_1(y)$ for various values of n.

Example 4. If, in the preceding discussion, we take $n = 100$ and $y = 20$, the first approximate 95.4 per cent confidence interval is given by

$(0.2 - 2\sqrt{(0.2)(0.8)/100}, 0.2 + 2\sqrt{(0.2)(0.8)/100}$ or $(0.12, 0.28)$. The approximate 95.4 per cent confidence interval provided by inequalities (1) is

$$\left(\frac{22 - 2\sqrt{(1600/100) + 1}}{104}, \frac{22 + 2\sqrt{(1600/100) + 1}}{104}\right)$$

or $(0.13, 0.29)$. By referring to the appropriate tables found elsewhere, we find that an approximate 95 per cent confidence interval has the limits $d_2(20) = 0.13$ and $d_1(20) = 0.29$. Thus in this example we see that all three methods yield results that are in substantial agreement.

Remark. The fact that the variance of Y/n is a function of p caused us some difficulty in finding a confidence interval for p. Another way of handling the problem is to try to find a function $u(Y/n)$ of Y/n, whose variance is essentially free of p. Since Y/n converges stochastically to p, we can approximate $u(Y/n)$ by the first two terms of its Taylor's expansion about p, namely by

$$v\left(\frac{Y}{n}\right) = u(p) + \left(\frac{Y}{n} - p\right)u'(p).$$

Of course, $v(Y/n)$ is a linear function of Y/n and thus also has an approximate normal distribution; clearly, it has mean $u(p)$ and variance

$$[u'(p)]^2 \frac{p(1 - p)}{n}.$$

But it is the latter that we want to be essentially free of p; thus we set it equal to a constant, obtaining the differential equation

$$u'(p) = \frac{c}{\sqrt{p(1 - p)}}.$$

A solution of this is

$$u(p) = (2c) \arcsin \sqrt{p}.$$

If we take $c = \frac{1}{2}$, we have, since $u(Y/n)$ is approximately equal to $v(Y/n)$, that

$$u\left(\frac{Y}{n}\right) = \arcsin \sqrt{\frac{Y}{n}}.$$

This has an approximate normal distribution with mean $\arcsin \sqrt{p}$ and variance $1/4n$. Hence we could find an approximate 95.4 per cent confidence interval by using

$$\Pr\left(-2 < \frac{\arcsin \sqrt{Y/n} - \arcsin \sqrt{p}}{\sqrt{1/4n}} < 2\right) = 0.954$$

and solving the inequalities for p.

EXERCISES

6.16. Let the observed value of the mean $\bar{X}$ of a random sample of size 20 from a distribution that is $n(\mu, 80)$ be 81.2. Find a 95 per cent confidence interval for μ.

6.17. Let $\bar{X}$ be the mean of a random sample of size n from a distribution that is $n(\mu, 9)$. Find n such that $\Pr(\bar{X} - 1 < \mu < \bar{X} + 1) = 0.90$, approximately.

6.18. Let a random sample of size 17 from the normal distribution $n(\mu, \sigma^2)$ yield $\bar{x} = 4.7$ and $s^2 = 5.76$. Determine a 90 per cent confidence interval for μ.

6.19. Let $\bar{X}$ denote the mean of a random sample of size n from a distribution that has mean μ, variance $\sigma^2 = 10$, and a moment-generating function. Find n so that the probability is approximately 0.954 that the random interval $(\bar{X} - \frac{1}{2}, \bar{X} + \frac{1}{2})$ includes μ.

6.20. Let $X_1, X_2, \ldots, X_9$ be a random sample of size 9 from a distribution that is $n(\mu, \sigma^2)$.

(a) If σ is known, find the length of a 95 per cent confidence interval for μ if this interval is based on the random variable $\sqrt{9}(\bar{X} - \mu)/\sigma$.

(b) If σ is unknown, find the expected value of the length of a 95 per cent confidence interval for μ if this interval is based on the random variable $\sqrt{8}(\bar{X} - \mu)/S$.

(c) Compare these two answers. *Hint.* Write $E(S) = (\sigma/\sqrt{n})E[(nS^2/\sigma^2)^{1/2}]$.

6.21. Let $X_1, X_2, \ldots, X_n, X_{n+1}$ be a random sample of size $n + 1$, $n > 1$, from a distribution that is $n(\mu, \sigma^2)$. Let $\bar{X} = \sum_1^n X_i/n$ and $S^2 = \sum_1^n (X_i - \bar{X})^2/n$. Find the constant c so that the statistic $c(\bar{X} - X_{n+1})/S$ has a t distribution. If $n = 8$, determine k such that $\Pr(\bar{X} - kS < X_9 < \bar{X} + kS) = 0.80$. The observed interval $(\bar{x} - ks, \bar{x} + ks)$ is often called an 80 per cent *prediction interval* for X_9.

6.22. Let Y be $b(300, p)$. If the observed value of Y is $y = 75$, find an approximate 90 per cent confidence interval for p.

6.23. Let $\bar{X}$ be the mean of a random sample of size n from a distribution that is $n(\mu, \sigma^2)$, where the positive variance σ^2 is known. Use the fact that $N(2) - N(-2) = 0.954$ to find, for each μ, $c_1(\mu)$ and $c_2(\mu)$ such that $\Pr[c_1(\mu) < \bar{X} < c_2(\mu)] = 0.954$. Note that $c_1(\mu)$ and $c_2(\mu)$ are increasing functions of μ. Solve for the respective functions $d_1(\bar{x})$ and $d_2(\bar{x})$; thus we also have that $\Pr[d_2(\bar{X}) < \mu < d_1(\bar{X})] = 0.954$. Compare this with the answer obtained previously in the text.

6.24. In the notation of the discussion of the confidence interval for p, show that the event $-2 < Z < 2$ is equivalent to inequalities (1). *Hint.* First observe that $-2 < Z < 2$ is equivalent to $Z^2 < 4$, which can be written as an inequality involving a quadratic expression in p.

6.25. Let X denote the mean of a random sample of size 25 from a gamma-type distribution with $\alpha = 4$ and $\beta > 0$. Use the central limit theorem to find an approximate 0.954 confidence interval for μ, the mean of the gamma distribution. *Hint.* Base the confidence interval on the random variable $(\bar{X} - 4\beta)/(4\beta^2/25)^{1/2} = 5\bar{X}/2\beta - 10$.

6.4 Confidence Intervals for Differences of Means

The random variable T may also be used to obtain a confidence interval for the difference $\mu_1 - \mu_2$ between the means of two independent normal distributions, say $n(\mu_1, \sigma^2)$ and $n(\mu_2, \sigma^2)$, when the distributions have the same, but unknown, variance σ^2.

Remark. Let X have a normal distribution with unknown parameters μ_1 and σ^2. A modification can be made in conducting the experiment so that the variance of the distribution will remain the same but the mean of the distribution will be changed; say, increased. After the modification has been effected, let the random variable be denoted by Y, and let Y have a normal distribution with unknown parameters μ_2 and σ^2. Naturally, it is hoped that μ_2 is greater than μ_1, that is, that $\mu_1 - \mu_2 < 0$. Accordingly, one seeks a confidence interval for $\mu_1 - \mu_2$ in order to make a statistical inference.

A confidence interval for $\mu_1 - \mu_2$ may be obtained as follows: Let $X_1, X_2, \ldots, X_n$ and $Y_1, Y_2, \ldots, Y_m$ denote, respectively, independent random samples from the two independent distributions having, respectively, the probability density functions $n(\mu_1, \sigma^2)$ and $n(\mu_2, \sigma^2)$. Denote the means of the samples by $\bar{X}$ and $\bar{Y}$ and the variances of the samples by S_1^2 and S_2^2, respectively. It should be noted that these four statistics are mutually stochastically independent. The stochastic independence of $\bar{X}$ and S_1^2 (and, inferentially that of $\bar{Y}$ and S_2^2) was established in Section 4.8; the assumption that X and Y have independent distributions accounts for the stochastic independence of the others. Thus $\bar{X}$ and $\bar{Y}$ are normally and stochastically independently distributed with means μ_1 and μ_2 and variances σ^2/n and σ^2/m, respectively. In accordance with Section 4.7, their difference $\bar{X} - \bar{Y}$ is normally distributed with mean $\mu_1 - \mu_2$ and variance $\sigma^2/n + \sigma^2/m$. Then the random variable

$$\frac{(\bar{X} - \bar{Y}) - (\mu_1 - \mu_2)}{\sqrt{\sigma^2/n + \sigma^2/m}}$$

is normally distributed with zero mean and unit variance. This random variable may serve as the numerator of a T random variable. Further, nS_1^2/σ^2 and mS_2^2/σ^2 have stochastically independent chi-square distributions with $n - 1$ and $m - 1$ degrees of freedom, respectively, so that their sum $(nS_1^2 + mS_2^2)/\sigma^2$ has a chi-square distribution with $n + m - 2$ degrees of freedom, provided that $m + n - 2 > 0$. Because of the mutual stochastic independence of $\bar{X}$, $\bar{Y}$, S_1^2, and S_2^2, it is seen that

$$\sqrt{\frac{nS_1^2 + mS_2^2}{\sigma^2(n + m - 2)}}$$

may serve as the denominator of a T random variable. That is, the random variable

$$T = \frac{(\bar{X} - \bar{Y}) - (\mu_1 - \mu_2)}{\sqrt{\dfrac{nS_1^2 + mS_2^2}{n + m - 2}\left(\dfrac{1}{n} + \dfrac{1}{m}\right)}}$$

has a t distribution with $n + m - 2$ degrees of freedom. As in the previous section, we can (once n and m are specified positive integers with $n + m - 2 > 0$) find a positive number b from Table IV of Appendix B such that

$$\Pr(-b < T < b) = 0.95.$$

If we set

$$R = \sqrt{\frac{nS_1^2 + mS_2^2}{n + m - 2}\left(\frac{1}{n} + \frac{1}{m}\right)},$$

this probability may be written in the form

$$\Pr[(\bar{X} - \bar{Y}) - bR < \mu_1 - \mu_2 < (\bar{X} - \bar{Y}) + bR] = 0.95.$$

It follows that the random interval

$$\left[(\bar{X} - \bar{Y}) - b\sqrt{\frac{nS_1^2 + mS_2^2}{n + m - 2}\left(\frac{1}{n} + \frac{1}{m}\right)},\right.$$

$$\left.(\bar{X} - \bar{Y}) + b\sqrt{\frac{nS_1^2 + mS_2^2}{n + m - 2}\left(\frac{1}{n} + \frac{1}{m}\right)}\right]$$

has probability 0.95 of including the unknown fixed point $(\mu_1 - \mu_2)$. As usual, the experimental values of $\bar{X}$, $\bar{Y}$, S_1^2, and S_2^2, namely $\bar{x}$, $\bar{y}$, s_1^2, and s_2^2, will provide a 95 per cent confidence interval for $\mu_1 - \mu_2$ when the variances of the two independent normal distributions are unknown but equal. A consideration of the difficulty encountered when the unknown variances of the two independent normal distributions are not equal is assigned to one of the exercises.

Example 1. It may be verified that if in the preceding discussion $n = 10$, $m = 7$, $\bar{x} = 4.2$, $\bar{y} = 3.4$, $s_1^2 = 49$, $s_2^2 = 32$, then the interval $(-5.16,\ 6.76)$ is a 90 per cent confidence interval for $\mu_1 - \mu_2$.

Let Y_1 and Y_2 be two stochastically independent random variables with binomial distributions $b(n_1, p_1)$ and $b(n_2, p_2)$, respectively. Let us now turn to the problem of finding a confidence interval for the difference $p_1 - p_2$ of the means of Y_1/n_1 and Y_2/n_2 when n_1 and n_2 are known. Since the mean and the variance of $Y_1/n_1 - Y_2/n_2$ are, respectively, $p_1 - p_2$ and $p_1(1 - p_1)/n_1 + p_2(1 - p_2)/n_2$, then the random variable given by the ratio

$$\frac{(Y_1/n_1 - Y_2/n_2) - (p_1 - p_2)}{\sqrt{p_1(1 - p_1)/n_1 + p_2(1 - p_2)/n_2}}$$

has mean zero and variance 1 for all positive integers n_1 and n_2. Moreover, since both Y_1 and Y_2 have approximate normal distributions for large n_1 and n_2, one suspects that the ratio has an approximate normal distribution. This is actually the case, but it will not be proved here. Moreover, if $n_1/n_2 = c$, where c is a fixed positive constant, the result of Exercise 6.31 shows that the random variable

(1)
$$\frac{(Y_1/n_1)(1 - Y_1/n_1)/n_1 + (Y_2/n_2)(1 - Y_2/n_2)/n_2}{p_1(1 - p_1)/n_1 + p_2(1 - p_2)/n_2}$$

converges stochastically to 1 as $n_2 \to \infty$ (and thus $n_1 \to \infty$, since $n_1/n_2 = c$, $c > 0$). In accordance with Theorem 6, Section 5.5, the random variable

$$W = \frac{(Y_1/n_1 - Y_2/n_2) - (p_1 - p_2)}{U},$$

where

$$U = \sqrt{(Y_1/n_1)(1 - Y_1/n_1)/n_1 + (Y_2/n_2)(1 - Y_2/n_2)/n_2}\ ,$$

has a limiting distribution that is $n(0, 1)$. The event $-2 < W < 2$, the probability of which is approximately equal to 0.954, is equivalent to the event

$$\frac{Y_1}{n_1} - \frac{Y_2}{n_2} - 2U < p_1 - p_2 < \frac{Y_1}{n_1} - \frac{Y_2}{n_2} + 2U.$$

Accordingly, the experimental values y_1 and y_2 of Y_1 and Y_2, respectively, will provide an approximate 95.4 per cent confidence interval for $p_1 - p_2$.

Example 2. If, in the preceding discussion, we take $n_1 = 100$, $n_2 = 400$, $y_1 = 30$, $y_2 = 80$, then the experimental values of $Y_1/n_1 - Y_2/n_2$ and U are 0.1 and $\sqrt{(0.3)(0.7)/100 + (0.2)(0.8)/400} = 0.05$, respectively. Thus the interval (0, 0.2) is an approximate 95.4 per cent confidence interval for $p_1 - p_2$.

EXERCISES

6.26. Let two independent random samples, each of size 10, from two independent normal distributions $n(\mu_1, \sigma^2)$ and $n(\mu_2, \sigma^2)$ yield $\bar{x} = 4.8$, $s_1^2 = 8.64$, $\bar{y} = 5.6$, $s_2^2 = 7.88$. Find a 95 per cent confidence interval for $\mu_1 - \mu_2$.

6.27. Let two stochastically independent random variables Y_1 and Y_2, with binomial distributions that have parameters $n_1 = n_2 = 100$, p_1, and p_2, respectively, be observed to be equal to $y_1 = 50$ and $y_2 = 40$. Determine an approximate 90 per cent confidence interval for $p_1 - p_2$.

6.28. Discuss the problem of finding a confidence interval for the difference $\mu_1 - \mu_2$ between the two means of two independent normal distributions if the variances σ_1^2 and σ_2^2 are known but not necessarily equal.

6.29. Discuss Exercise 6.28 when it is assumed that the variances are unknown and unequal. This is a very difficult problem, and the discussion should point out exactly where the difficulty lies. If, however, the variances are unknown but their ratio σ_1^2/σ_2^2 is a known constant k, then a statistic that is a T random variable can again be used. Why?

6.30. Let $\bar{X}$ and $\bar{Y}$ be the means of two independent random samples, each of size n, from the respective distributions $n(\mu_1, \sigma^2)$ and $n(\mu_2, \sigma^2)$, where the common variance is known. Find n such that $\Pr(\bar{X} - \bar{Y} - \sigma/5 < \mu_1 - \mu_2 < \bar{X} - \bar{Y} + \sigma/5) = 0.90$.

6.31. Under the conditions given, show that the random variable defined by ratio (1) of the text converges stochastically to 1.

6.5 Confidence Intervals for Variances

Let the random variable X be $n(\mu, \sigma^2)$. We shall discuss the problem of finding a confidence interval for σ^2. Our discussion will consist of two parts: first, when μ is a known number, and second, when μ is unknown.

Let $X_1, X_2, \ldots, X_n$ denote a random sample of size n from a distribution that is $n(\mu, \sigma^2)$, where μ is a known number. The maximum likelihood estimator of σ^2 is $\sum_1^n (X_i - \mu)^2/n$, and the variable $Y = \sum_1^n (X_i - \mu)^2/\sigma^2$ is $\chi^2(n)$. Let us select a probability, say 0.95, and for the fixed positive integer n determine values of a and b, $a < b$, from Table II, such that

$$\Pr (a < Y < b) = 0.95.$$

Thus

$$\Pr \left[a < \frac{\sum_1^n (X_i - \mu)^2}{\sigma^2} < b \right] = 0.95,$$

or

$$\Pr \left[\frac{\sum_1^n (X_i - \mu)^2}{b} < \sigma^2 < \frac{\sum_1^n (X_i - \mu)^2}{a} \right] = 0.95.$$

Since μ, a, and b are known constants, each of $\sum_1^n (X_i - \mu)^2/b$ and $\sum_1^n (X_i - \mu)^2/a$ is a statistic. Moreover, the interval

$$\left[\frac{\sum_1^n (X_i - \mu)^2}{b}, \frac{\sum_1^n (X_i - \mu)^2}{a} \right]$$

is a random interval having probability of 0.95 that it includes the unknown fixed point (parameter) σ^2. Once the random experiment has been performed, and it is found that $X_1 = x_1, X_2 = x_2, \ldots, X_n = x_n$, then the particular interval

$$\left[\frac{\sum_1^n (x_i - \mu)^2}{b}, \frac{\sum_1^n (x_i - \mu)^2}{a} \right]$$

is a 95 per cent confidence interval for σ^2.

The reader will immediately observe that there are no unique numbers $a < b$ such that $\Pr (a < Y < b) = 0.95$. A common method of procedure is to find a and b such that $\Pr (Y < a) = 0.025$ and $\Pr (b < Y) = 0.025$. That procedure will be followed in this book.

Example 1. If in the preceding discussion $\mu = 0$, $n = 10$, and $\sum_1^{10} x_i^2 = 106.6$, then the interval $(106.6/20.5, 106.6/3.25)$, or $(5.2, 32.8)$, is a 95 per cent confidence interval for the variance σ^2, since $\Pr (Y < 3.25) = 0.025$ and

Pr $(20.5 < Y) = 0.025$, provided that Y has a chi-square distribution with 10 degrees of freedom.

We now turn to the case in which μ is not known. This case can be handled by making use of the facts that S^2 is the maximum likelihood estimator of σ^2 and nS^2/σ^2 is $\chi^2(n-1)$. For a fixed positive integer $n \geq 2$, we can find, from Table II, values of a and b, $a < b$, such that

$$\Pr\left(a < \frac{nS^2}{\sigma^2} < b\right) = 0.95.$$

Here, of course, we would find a and b by using a chi-square distribution with $n - 1$ degrees of freedom. In accordance with the convention previously adopted, we would select a and b so that

$$\Pr\left(\frac{nS^2}{\sigma^2} < a\right) = 0.025 \quad \text{and} \quad \Pr\left(\frac{nS^2}{\sigma^2} > b\right) = 0.025.$$

We then have

$$\Pr\left(\frac{nS^2}{b} < \sigma^2 < \frac{nS^2}{a}\right) = 0.95$$

so that $(nS^2/b, nS^2/a)$ is a random interval having probability 0.95 of including the fixed but unknown point (parameter) σ^2. After the random experiment has been performed and we find, say, $X_1 = x_1, X_2 = x_2, \ldots, X_n = x_n$, with $s^2 = \sum_1^n (x_i - \bar{x})^2/n$, we have, as a 95 per cent confidence interval for σ^2, the interval $(ns^2/b, ns^2/a)$.

Example 2. If, in the preceding discussion, we have $n = 9$, $s^2 = 7.63$, then the interval $[9(7.63)/17.5, 9(7.63)/2.18]$ or $(3.92, 31.50)$ is a 95 per cent confidence interval for the variance σ^2.

Next, let X and Y denote stochastically independent random variables that are $n(\mu_1, \sigma_1^2)$ and $n(\mu_2, \sigma_2^2)$, respectively. We shall determine a confidence interval for the ratio σ_2^2/σ_1^2 when μ_1 and μ_2 are unknown.

Remark. Consider a situation in which a random variable X has a normal distribution with variance σ_1^2. Although σ_1^2 is not known, it is found that the experimental values of X are quite widely dispersed, so that σ_1^2 must be fairly large. It is believed that a certain modification in conducting the experiment may reduce the variance. After the modification has been effected, let the random variable be denoted by Y, and let Y have a normal distribution with variance σ_2^2. Naturally, it is hoped that σ_2^2 is less than σ_1^2, that is, that $\sigma_2^2/\sigma_1^2 < 1$. In order to make a statistical inference, we find a confidence interval for the ratio σ_2^2/σ_1^2.

Consider a random sample $X_1, X_2, \ldots, X_n$ of size $n \geq 2$ from the distribution of X and a random sample $Y_1, Y_2, \ldots, Y_m$ of size $m \geq 2$ from the independent distribution of Y. Here n and m may or may not be equal. Let the means of the two samples be denoted by $\bar{X}$ and $\bar{Y}$, and the variances of the two samples by $S_1^2 = \sum_1^n (X_i - \bar{X})^2/n$ and $S_2^2 = \sum_1^m (Y_i - \bar{Y})^2/m$, respectively. The stochastically independent random variables nS_1^2/σ_1^2 and mS_2^2/σ_2^2 have chi-square distributions with $n - 1$ and $m - 1$ degrees of freedom, respectively. In Section 4.4 a random variable called F was defined, and through the change-of-variable technique the p.d.f. of F was obtained. If nS_1^2/σ_1^2 is divided by $n - 1$, the number of degrees of freedom, and if mS_2^2/σ_2^2 is divided by $m - 1$, then, by definition of an F random variable, we have that

$$F = \frac{nS_1^2/[\sigma_1^2(n - 1)]}{mS_2^2/[\sigma_2^2(m - 1)]}$$

has an F distribution with parameters $n - 1$ and $m - 1$. For numerically given values of n and m and with a preassigned probability, say 0.95, we can determine from Table V of Appendix B, in accordance with our convention, numbers $0 < a < b$ such that

$$\Pr\left[a < \frac{nS_1^2/[\sigma_1^2(n - 1)]}{mS_2^2/[\sigma_2^2(m - 1)]} < b\right] = 0.95.$$

If the probability of this event is written in the form

$$\Pr\left[a\,\frac{mS_2^2/(m - 1)}{nS_1^2/(n - 1)} < \frac{\sigma_2^2}{\sigma_1^2} < b\,\frac{mS_2^2/(m - 1)}{nS_1^2/(n - 1)}\right] = 0.95,$$

it is seen that the interval

$$\left[a\,\frac{mS_2^2/(m - 1)}{nS_1^2/(n - 1)}, \, b\,\frac{mS_2^2/(m - 1)}{nS_1^2/(n - 1)}\right]$$

is a random interval having probability 0.95 of including the fixed but unknown point σ_2^2/σ_1^2. If the experimental values of $X_1, X_2, \ldots, X_n$ and of $Y_1, Y_2, \ldots, Y_m$ are denoted by $x_1, x_2, \ldots, x_n$ and $y_1, y_2, \ldots, y_m$, respectively, and if $ns_1^2 = \sum_1^n (x_i - \bar{x})^2$, $ms_2^2 = \sum_1^m (y_i - \bar{y})^2$, then the interval with known end points, namely

$$\left[a \, \frac{ms_2^2/(m-1)}{ns_1^2/(n-1)}, \; b \, \frac{ms_2^2/(m-1)}{ns_1^2/(n-1)} \right],$$

is a 95 per cent confidence interval for the ratio σ_2^2/σ_1^2 of the two unknown variances.

Example 3. If in the preceding discussion $n = 10$, $m = 5$, $s_1^2 = 20.0$, $s_2^2 = 35.6$, then the interval

$$\left[\left(\frac{1}{4.72} \right) \frac{5(35.6)/4}{10(20.0)/9}, \; (8.90) \frac{5(35.6)/4}{10(20.0)/9} \right]$$

or (0.4, 17.8) is a 95 per cent confidence interval for σ_2^2/σ_1^2.

EXERCISES

6.32. If 8.6, 7.9, 8.3, 6.4, 8.4, 9.8, 7.2, 7.8, 7.5 are the observed values of a random sample of size 9 from a distribution that is $n(8, \sigma^2)$, construct a 90 per cent confidence interval for σ^2.

6.33. Let $X_1, X_2, \ldots, X_n$ be a random sample from the distribution $n(\mu, \sigma^2)$. Let $0 < a < b$. Show that the mathematical expectation of the length of the random interval $\left[\sum_1^n (X_i - \mu)^2/b, \sum_1^n (X_i - \mu)^2/a \right]$ is $(b - a) \times (n\sigma^2/ab)$.

6.34. A random sample of size 15 from the normal distribution $n(\mu, \sigma^2)$ yields $\bar{x} = 3.2$ and $s^2 = 4.24$. Determine a 90 per cent confidence interval for σ^2.

6.35. Let S^2 be the variance of a random sample of size n taken from a distribution that is $n(\mu, \sigma^2)$ where μ and σ^2 are unknown. Let $g(z)$ be the p.d.f. of $Z = nS^2/\sigma^2$, which is $\chi^2(n - 1)$. Let a and b be such that the observed interval $(ns^2/b, ns^2/a)$ is a 95 percent confidence interval for σ^2. If its length $ns^2(b - a)/ab$ is to be a minimum, show that a and b must satisfy the condition that $a^2 g(a) = b^2 g(b)$. *Hint.* If $G(z)$ is the distribution function of Z, then differentiate both $G(b) - G(a) = 0.95$ and $(b - a)/ab$ with respect to b, recalling that, from the first equation, a must be a function of b. Then equate the latter derivative to zero.

6.36. Let two independent random samples of sizes $n = 16$ and $m = 10$, taken from two independent normal distributions $n(\mu_1, \sigma_1^2)$ and $n(\mu_2, \sigma_2^2)$, respectively, yield $\bar{x} = 3.6$, $s_1^2 = 4.14$, $\bar{y} = 13.6$, $s_2^2 = 7.26$. Find a 90 per cent confidence interval for σ_2^2/σ_1^2 when μ_1 and μ_2 are unknown.

6.37. Discuss the problem of finding a confidence interval for the ratio σ_2^2/σ_1^2 of the two unknown variances of two independent normal distributions if the means μ_1 and μ_2 are known.

6.38. Let $X_1, X_2, \ldots, X_6$ be a random sample of size 6 from a gamma distribution with parameters $\alpha = 1$ and unknown $\beta > 0$. Discuss the construction of a 98 per cent confidence interval for β. *Hint.* What is the distribution of $2 \sum_1^6 X_i/\beta$?

6.39. Let S_1^2 and S_2^2 denote, respectively, the variances of random samples, of sizes n and m, from two independent distributions that are $n(\mu_1, \sigma^2)$ and $n(\mu_2, \sigma^2)$. Use the fact that $(nS_1^2 + mS_2^2)/\sigma^2$ is $\chi^2(n + m - 2)$ to find a confidence interval for the common unknown variance σ^2.

6.40. Let Y_4 be the nth order statistic of a random sample, $n = 4$, from a continuous-type uniform distribution on the interval $(0, \theta)$. Let $0 < c_1 < c_2 \le 1$ be selected so that $\Pr(c_1\theta < Y_4 < c_2\theta) = 0.95$. Verify that $c_1 = \sqrt[4]{0.05}$ and $c_2 = 1$ satisfy these conditions. What, then, is a 95 per cent confidence interval for θ?

6.6 Bayesian Estimates

In Sections 6.3, 6.4, and 6.5 we constructed two statistics, say U and V, $U < V$, such that we have a preassigned probability p that the random interval (U, V) contains a fixed but unknown point (parameter). We then adopted this *principle*: Use the experimental results to compute the values of U and V, say u and v; then call the interval (u, v) a $100p$ per cent confidence interval for the parameter. Adoption of this principle provided us with one method of interval estimation. This method of interval estimation is widely used in statistical literature and in the applications. But it is important for us to understand that other principles can be adopted. The student should constantly keep in mind that as long as he is working with probability, he is in the realm of mathematics; but once he begins to make inferences or to draw conclusions about a random experiment, which inferences are based upon experimental data, he is in the field of statistics.

We shall now describe another approach to the problem of interval estimation. This approach takes into account any prior knowledge of the experiment that the statistician has and it is one application of a principle of statistical inference that may be called *Bayesian statistics*. Consider a random variable X that has a distribution of probability that depends upon the symbol θ, where θ is an element of a well-defined set Ω. For example, if the symbol θ is the mean of a normal distribution, Ω may be the real line. We have previously looked upon θ as being some constant, although an unknown constant. Let us now introduce a random variable Θ that has a distribution of probability over the set Ω; and, just as we look upon x as a possible value of the random variable X, we now look upon θ as a possible value of the random variable Θ. Thus the distribution of X depends upon θ, a random determination of the random variable Θ. We shall denote the p.d.f. of Θ by $h(\theta)$ and we take $h(\theta) = 0$ when θ is not an element of Ω. Let $X_1, X_2, \ldots, X_n$ denote a random sample from this distribution of X, and let Y denote a statistic that is a function of $X_1, X_2, \ldots, X_n$. We can find the p.d.f. of Y for every given θ; that is, we can find the conditional p.d.f. of Y, given $\Theta = \theta$, which we denote by $g(y|\theta)$. Thus the joint p.d.f. of Y and Θ is given by

$$k(y, \theta) = h(\theta)g(y|\theta).$$

If Θ is a random variable of the continuous type, the marginal p.d.f. of Y is given by

$$k_1(y) = \int_{-\infty}^{\infty} h(\theta)g(y|\theta) \, d\theta.$$

If Θ is a random variable of the discrete type, integration would be replaced by summation. In either case the conditional p.d.f. of Θ, given $Y = y$, is

$$k(\theta|y) = \frac{k(y, \theta)}{k_1(y)} = \frac{h(\theta)g(y|\theta)}{k_1(y)}, \qquad k_1(y) > 0.$$

This relationship is one form of Bayes' formula (see Exercise 2.7, Section 2.1).

In Bayesian statistics, the p.d.f. $h(\theta)$ is called the *prior p.d.f.* of Θ, and the conditional p.d.f. $k(\theta|y)$ is called the *posterior p.d.f.* of Θ. This is because $h(\theta)$ is the p.d.f. of Θ prior to the observation of Y, whereas $k(\theta|y)$ is the p.d.f. of Θ after the observation of Y has been made. In many instances, $h(\theta)$ is not known; yet the choice of $h(\theta)$ affects the p.d.f. $k(\theta|y)$. In these instances the statistician takes into account all prior knowledge of the experiment and *assigns* the prior p.d.f. $h(\theta)$. This, of course, injects the problem of *personal* or *subjective probability* (see the Remark, Section 1.1).

Suppose that we want a point estimate of θ. From the Bayesian viewpoint, this really amounts to selecting a decision function w so that $w(y)$ is a predicted value of θ (an experimental value of the random variable Θ) when both the computed value y and the conditional p.d.f. $k(\theta|y)$ are known. Now, in general, how would we predict an experimental value of any random variable, say W, if we want our prediction to be "reasonably close" to the value to be observed? Many statisticians would predict the mean, $E(W)$, of the distribution of W; others would predict a median (perhaps unique) of the distribution of W; some would predict a mode (perhaps unique) of the distribution of W; and some would have other predictions. However, it seems desirable that the choice of the decision function should depend upon the loss function $\mathcal{L}[\theta, w(y)]$. One way in which this dependence upon the loss function can be reflected is to select the decision function w in such a way that the conditional expectation of the loss is a minimum. A *Bayes' solution* is a decision function w that minimizes

$$E\{\mathcal{L}[\Theta, w(y)]|Y = y\} = \int_{-\infty}^{\infty} \mathcal{L}[\theta, w(y)]k(\theta|y) \, d\theta,$$

if Θ is a random variable of the continuous type. The usual modification of the right-hand member of this equation is made for random variables of the discrete type. If, for example, the loss function is given by $\mathcal{L}[\theta, w(y)] = [\theta - w(y)]^2$, the Bayes' solution is given by $w(y) = E(\Theta|y)$, the mean of the conditional distribution of Θ, given $Y = y$. This follows from the fact (Exercise 1.91) that $E[(W - b)^2]$, if it exists, is a minimum when $b = E(W)$. If the loss function is given by $\mathcal{L}[\theta, w(y)] = |\theta - w(y)|$, then a median of the conditional distribution of Θ, given $Y = y$, is the Bayes' solution. This follows from the fact (Exercise 1.81) that $E(|W - b|)$, if it exists, is a minimum when b is equal to any median of the distribution of W.

The conditional expectation of the loss, given $Y = y$, defines a random variable that is a function of the statistic Y. The expected value of that function of Y, in the notation of this section, is given by

$$\int_{-\infty}^{\infty} \left\{ \int_{-\infty}^{\infty} \mathcal{L}[\theta, w(y)]k(\theta|y) \, d\theta \right\} k_1(y) \, dy$$
$$= \int_{-\infty}^{\infty} \left\{ \int_{-\infty}^{\infty} \mathcal{L}[\theta, w(y)]g(y|\theta) \, dy \right\} h(\theta) \, d\theta,$$

in the continuous case. The integral within the braces in the latter expression is, for every given $\theta \in \Omega$, the risk function $R(\theta, w)$; accordingly, the latter expression is the mean value of the risk, or the expected risk. Because a Bayes' solution minimizes

$$\int_{-\infty}^{\infty} \mathcal{L}[\theta, w(y)]k(\theta|y) \, d\theta$$

for every y for which $k_1(y) > 0$, it is evident that a Bayes' solution

$w(y)$ minimizes this mean value of the risk. We now give an illustrative example.

Example 1. Let $X_1, X_2, \ldots, X_n$ denote a random sample from a distribution that is $b(1, \theta)$, $0 < \theta < 1$. We seek a decision function w that is a Bayes' solution. If $Y = \sum_1^n X_i$, then Y is $b(n, \theta)$. That is, the conditional p.d.f. of Y, given $\Theta = \theta$, is

$$g(y|\theta) = \binom{n}{y} \theta^y (1 - \theta)^{n-y}, \qquad y = 0, 1, \ldots, n,$$

$$= 0 \text{ elsewhere.}$$

We take the prior p.d.f. of the random variable Θ to be

$$h(\theta) = \frac{\Gamma(\alpha + \beta)}{\Gamma(\alpha)\Gamma(\beta)} \theta^{\alpha-1}(1 - \theta)^{\beta-1}, \qquad 0 < \theta < 1,$$

$$= 0 \text{ elsewhere.}$$

where α and β are assigned positive constants. Thus the joint p.d.f. of Y and Θ is given by $g(y|\theta)h(\theta)$ and the marginal p.d.f. of Y is

$$k_1(y) = \int_0^1 h(\theta)g(y|\theta) \, d\theta$$

$$= \binom{n}{y} \frac{\Gamma(\alpha + \beta)}{\Gamma(\alpha)\Gamma(\beta)} \int_0^1 \theta^{y+\alpha-1}(1 - \theta)^{n-y+\beta-1} \, d\theta$$

$$= \binom{n}{y} \frac{\Gamma(\alpha + \beta)\Gamma(\alpha + y)\Gamma(n + \beta - y)}{\Gamma(\alpha)\Gamma(\beta)\Gamma(n + \alpha + \beta)}, \qquad y = 0, 1, 2, \ldots, n,$$

$$= 0 \text{ elsewhere.}$$

Finally, the conditional p.d.f. of Θ, given $Y = y$, is, at points of positive probability density,

$$k(\theta|y) = \frac{g(y|\theta)h(\theta)}{k_1(y)}$$

$$= \frac{\Gamma(n + \alpha + \beta)}{\Gamma(\alpha + y)\Gamma(n + \beta - y)} \theta^{\alpha+y-1}(1 - \theta)^{\beta+n-y-1}, \qquad 0 < \theta < 1,$$

and $y = 0, 1, \ldots, n$. We take the loss function to be $\mathscr{L}[\theta, w(y)] = [\theta - w(y)]^2$. Because Y is a random variable of the discrete type, whereas Θ is of the continuous type, we have for the expected risk,

$$\int_0^1 \left\{ \sum_{y=0}^n [\theta - w(y)]^2 \binom{n}{y} \theta^y (1 - \theta)^{n-y} \right\} h(\theta) \, d\theta$$

$$= \sum_{y=0}^n \left\{ \int_0^1 [\theta - w(y)]^2 k(\theta|y) \, d\theta \right\} k_1(y).$$

The Bayes' solution $w(y)$ is the mean of the conditional distribution of Θ, given $Y = y$. Thus

$$w(y) = \int_0^1 \theta k(\theta|y)\, d\theta$$

$$= \frac{\Gamma(n + \alpha + \beta)}{\Gamma(\alpha + y)\Gamma(n + \beta - y)} \int_0^1 \theta^{\alpha + y}(1 - \theta)^{\beta + n - y - 1}\, d\theta$$

$$= \frac{\alpha + y}{\alpha + \beta + n}.$$

This decision function $w(y)$ minimizes

$$\int_0^1 [\theta - w(y)]^2 k(\theta|y)\, d\theta$$

for $y = 0, 1, \ldots, n$ and, accordingly, it minimizes the expected risk. It is very instructive to note that this Bayes' solution can be written as

$$w(y) = \left(\frac{n}{\alpha + \beta + n}\right)\frac{y}{n} + \left(\frac{\alpha + \beta}{\alpha + \beta + n}\right)\frac{\alpha}{\alpha + \beta}$$

which is a weighted average of the maximum likelihood estimate y/n of θ and the mean $\alpha/(\alpha + \beta)$ of the prior p.d.f. of the parameter. Moreover, the respective weights are $n/(\alpha + \beta + n)$ and $(\alpha + \beta)/(\alpha + \beta + n)$. Thus we see that α and β should be selected so that not only is $\alpha/(\alpha + \beta)$ the desired prior mean, but the sum $\alpha + \beta$ indicates the worth of the prior opinion, relative to a sample of size n. That is, if we want our prior opinion to have as much weight as a sample size of 20, we would take $\alpha + \beta = 20$. So if our prior mean is $\frac{3}{4}$; we have that α and β are selected so that $\alpha = 15$ and $\beta = 5$.

In Example 1 it is extremely convenient to notice that it is not really necessary to determine $k_1(y)$ to find $k(\theta|y)$. If we divide $g(y|\theta)h(\theta)$ by $k_1(y)$ we must get the product of a factor, which depends upon y but does *not* depend upon θ, say $c(y)$, and

$$\theta^{y + \alpha - 1}(1 - \theta)^{n - y + \beta - 1}.$$

That is,

$$k(\theta|y) = c(y)\theta^{y + \alpha - 1}(1 - \theta)^{n - y + \beta - 1}, \qquad 0 < \theta < 1,$$

and $y = 0, 1, \ldots, n$. However, $c(y)$ must be that "constant" needed to make $k(\theta|y)$ a p.d.f., namely

$$c(y) = \frac{\Gamma(n + \alpha + \beta)}{\Gamma(y + \alpha)\Gamma(n - y + \beta)}.$$

Accordingly, Bayesian statisticians frequently write that $k(\theta|y)$ is proportional to $g(y|\theta)h(\theta)$; that is,

$$k(\theta|y) \propto g(y|\theta)h(\theta).$$

Then to actually form the p.d.f. $k(\theta|y)$, they simply find a "constant," which is some function of y, so that the expression integrates to 1. This is now illustrated.

Example 2. Suppose that $Y = \overline{X}$ is the mean of a random sample of size n that arises from the normal distribution $n(\theta, \sigma^2)$, where σ^2 is known. Then $g(y|\theta)$ is $n(\theta, \sigma^2/n)$. Further suppose that we are able to assign prior knowledge to θ through a prior p.d.f. $h(\theta)$ that is $n(\theta_0, \sigma_0^2)$. Then we have that

$$k(\theta|y) \propto \frac{1}{\sqrt{2\pi}\sigma/\sqrt{n}} \frac{1}{\sqrt{2\pi}\sigma_0} \exp\left[-\frac{(y - \theta)^2}{2(\sigma^2/n)} - \frac{(\theta - \theta_0)^2}{2\sigma_0^2}\right].$$

If we eliminate all constant factors (including factors involving y only), we have

$$k(\theta|y) \propto \exp\left[-\frac{(\sigma_0^2 + \sigma^2/n)\theta^2 - 2(y\sigma_0^2 + \theta_0\sigma^2/n)\theta}{2(\sigma^2/n)\sigma_0^2}\right].$$

This can be simplified, by completing the square, to read (after eliminating factors not involving θ)

$$k(\theta|y) \propto \exp\left[-\frac{\left(\theta - \dfrac{y\sigma_0^2 + \theta_0\sigma^2/n}{\sigma_0^2 + \sigma^2/n}\right)^2}{\dfrac{2(\sigma^2/n)\sigma_0^2}{(\sigma_0^2 + \sigma^2/n)}}\right].$$

That is, the posterior p.d.f. of the parameter is obviously normal with mean

$$\frac{y\sigma_0^2 + \theta_0\sigma^2/n}{\sigma_0^2 + \sigma^2/n} = \left(\frac{\sigma_0^2}{\sigma_0^2 + \sigma^2/n}\right)y + \left(\frac{\sigma^2/n}{\sigma_0^2 + \sigma^2/n}\right)\theta_0$$

and variance $(\sigma^2/n)\sigma_0^2/(\sigma_0^2 + \sigma^2/n)$. If the square-error loss function is used, this posterior mean is the Bayes' solution. Again, note that it is a weighted average of the maximum likelihood estimate $y = \overline{x}$ and the prior mean θ_0. Observe here and in Example 1 that the Bayes' solution gets closer to the maximum likelihood estimate as n increases. Thus the Bayesian procedures permit the decision maker to enter his or her prior opinions into the solution in a very formal way such that the influences of those prior notions will be less and less as n increases.

In Bayesian statistics all the information is contained in the posterior p.d.f. $k(\theta|y)$. In Examples 1 and 2 we found Bayesian point estimates using the square-error loss function. It should be noted that if $\mathscr{L}[w(y), \theta] = |w(y) - \theta|$, the absolute value of the error, then the Bayes' solution would be the median of the posterior distribution of the parameter, which is given by $k(\theta|y)$. Hence the Bayes' solution changes, *as it should*, with different loss functions.

If an interval estimate of θ is desired, we can now find two functions $u(y)$ and $v(y)$ so that the conditional probability

$$\Pr\left[u(y) < \Theta < v(y) \mid Y = y\right] = \int_{u(y)}^{v(y)} k(\theta|y)\, d\theta,$$

is large, say 0.95. The experimental values of $X_1, X_2, \ldots, X_n$, say $x_1, x_2, \ldots, x_n$, provide us with an experimental value of Y, say y. Then the interval $u(y)$ to $v(y)$ is an interval estimate of θ in the sense that the conditional probability of Θ belonging to that interval is equal to 0.95. For illustration, in Example 2 where the posterior p.d.f. of the parameter was normal, the interval, whose end points are found by taking the mean of that distribution and adding and subtracting 1.96 of its standard deviation,

$$\frac{y\sigma_0^2 + \theta_0\sigma^2/n}{\sigma_0^2 + \sigma^2/n} \pm 1.96\sqrt{\frac{(\sigma^2/n)\sigma_0^2}{\sigma_0^2 + \sigma^2/n}}$$

serves as an interval estimate for θ with posterior probability of 0.95.

Finally, it should be noted that in Bayesian statistics it is really better to begin with the sample items $X_1, X_2, \ldots, X_n$ rather than some statistic Y. We used the latter approach for convenience of notation. If $X_1, X_2, \ldots, X_n$ are used, then in our discussion, replace $g(y|\theta)$ by $f(x_1|\theta)f(x_2|\theta)\cdots f(x_n|\theta)$ and $k(\theta|y)$ by $k(\theta|x_1, x_2, \ldots, x_n)$. Thus we find that

$$k(\theta|x_1, x_2, \ldots, x_n) \propto h(\theta)f(x_1|\theta)f(x_2|\theta)\cdots f(x_n|\theta).$$

If the statistic Y is chosen correctly (namely, as a sufficient statistic, as explained in Chapter 10), we find that

$$k(\theta|x_1, x_2, \ldots, x_n) = k(\theta|y).$$

This is illustrated by Exercise 6.44. Of course, these Bayesian procedures can easily be extended to the case of several parameters, as demonstrated by Exercise 6.45.

EXERCISES

6.41. Let $X_1, X_2, \ldots, X_n$ denote a random sample from a distribution that is $n(\theta, \sigma^2)$, $-\infty < \theta < \infty$, where σ^2 is a given positive number. Let $Y = \bar{X}$, the mean of the random sample. Take the loss function to be $\mathscr{L}[\theta, w(y)] = |\theta - w(y)|$. If θ is an observed value of the random variable Θ that is $n(\mu, \tau^2)$, where $\tau^2 > 0$ and μ are known numbers, find the Bayes' solution $w(y)$ for a point estimate of θ.

6.42. Let $X_1, X_2, \ldots, X_n$ denote a random sample from a Poisson distribution with mean θ, $0 < \theta < \infty$. Let $Y = \sum_1^n X_i$ and take the loss function to $\mathscr{L}[\theta, w(y)] = [\theta - w(y)]^2$. Let θ be an observed value of the random variable Θ. If Θ has the p.d.f. $h(\theta) = \theta^{\alpha-1}e^{-\theta/\beta}/\Gamma(\alpha)\beta^{\alpha}$, $0 < \theta < \infty$, zero elsewhere, where $\alpha > 0$, $\beta > 0$ are known numbers, find the Bayes' solution $w(y)$ for a point estimate of θ.

6.43. Let Y_n be the nth order statistic of a random sample of size n from a distribution with p.d.f. $f(x|\theta) = 1/\theta$, $0 < x < \theta$, zero elsewhere. Take the loss function to be $\mathscr{L}[\theta, w(y_n)] = [\theta - w(y_n)]^2$. Let θ be an observed value of the random variable Θ, which has p.d.f. $h(\theta) = \beta\alpha^{\beta}/\theta^{\beta+1}$, $\alpha < \theta < \infty$, zero elsewhere, with $\alpha > 0$, $\beta > 0$. Find the Bayes' solution $w(y_n)$ for a point estimate of θ.

6.44. Let $X_1, X_2, \ldots, X_n$ be a random sample from a distribution that is $b(1, \theta)$. Let the prior p.d.f. of Θ be a beta one with parameters α and β. Show that the posterior p.d.f. $k(\theta|x_1, x_2, \ldots, x_n)$ is exactly the same as $k(\theta|y)$ given in Example 1. This demonstrates that we get exactly the same result whether we begin with the statistic Y or with the sample items. *Hint.* Note that $k(\theta|x_1, x_2, \ldots, x_n)$ is proportional to the product of the joint p.d.f. of $X_1, X_2, \ldots, X_n$ and the prior p.d.f. of θ.

6.45. Let Y_1 and Y_2 be statistics that have a trinomial distribution with parameters n, θ_1, and θ_2. Here θ_1 and θ_2 are observed values of the random variables Θ_1 and Θ_2, which have a Dirichlet distribution with known parameters α_1, α_2, and α_3 (see Example 1, Section 4.5). Show that the conditional distribution of Θ_1 and Θ_2 is Dirichlet and determine the conditional means $E(\Theta_1|y_1, y_2)$ and $E(\Theta_2|y_1, y_2)$.

6.46. Let X be $n(0, 1/\theta)$. Assume that the unknown θ is a value of a random variable Θ which has a gamma distribution with parameters $\alpha = r/2$ and $\beta = 2/r$, where r is a positive integer. Show that X has a marginal t distribution with r degrees of freedom. This procedure is called one of *compounding*, and it may be used by a Bayesian statistician as a way of first presenting the t distribution, as well as other distributions.

6.47. Let X have a Poisson distribution with parameter θ. Assume that the unknown θ is a value of a random variable Θ that has a gamma distribution with parameters $\alpha = r$ and $\beta = (1 - p)/p$, where r is a positive integer and $0 < p < 1$. Show, by the procedure of compounding, that X has a marginal distribution which is negative binomial, a distribution that was introduced earlier (Section 3.1) under very different assumptions.

6.48. In Example 1 let $n = 30$, $\alpha = 10$, and $\beta = 5$ so that $w(y) = (10 + y)/45$ is the Bayes' estimate of θ.

(a) If Y has the binomial distribution $b(30, \theta)$, compute the risk $E\{[\theta - w(Y)]^2\}$.

(b) Determine those values of θ for which the risk of part (a) is less than $\theta(1 - \theta)/30$, the risk associated with the maximum likelihood estimator Y/n of θ.

Chapter 7

Statistical Hypotheses

7.1 Some Examples and Definitions

The two principal areas of statistical inference are the areas of estimation of parameters and of tests of statistical hypotheses. The problem of estimation of parameters, both point and interval estimation, has been treated. In this chapter some aspects of statistical hypotheses and tests of statistical hypotheses will be considered. The subject will be introduced by way of example.

Example 1. Let it be known that the outcome X of a random experiment is $n(\theta, 100)$. For instance, X may denote a score on a test, which score we assume to be normally distributed with mean θ and variance 100. Let us say that past experience with this random experiment indicates that $\theta = 75$. Suppose, owing possibly to some research in the area pertaining to this experiment, some changes are made in the method of performing this random experiment. It is then suspected that no longer does $\theta = 75$ but that now $\theta > 75$. There is as yet no formal experimental evidence that $\theta > 75$; hence the statement $\theta > 75$ is a conjecture or a *statistical hypothesis*. In admitting that the statistical hypothesis $\theta > 75$ may be false, we allow, in effect, the possibility that $\theta \leq 75$. Thus there are actually two statistical hypotheses. First, that the unknown parameter $\theta \leq 75$; that is, there has been no increase in θ. Second, that the unknown parameter $\theta > 75$. Accordingly, the parameter space is $\Omega = \{\theta; -\infty < \theta < \infty\}$. We denote the first of these hypotheses by the symbols $H_0: \theta \leq 75$ and the second by the symbols $H_1: \theta > 75$. Since the values $\theta > 75$ are alternatives to those where $\theta \leq 75$, the hypothesis $H_1: \theta > 75$ is called the *alternative hypothesis*. Needless to say, H_0 could be called the alternative H_1; however, the conjecture, here $\theta > 75$, that is made by the research worker is usually taken to be the alternative hypothesis. In any case the problem is to decide which of these hypotheses is to be accepted. To reach a decision, the random experiment is to be repeated a number of independent times, say n, and the results observed. That is, we consider a random sample $X_1, X_2, \ldots, X_n$ from a distribution that is $n(\theta, 100)$, and we devise a rule that will tell us what decision to make once the experimental values, say $x_1, x_2, \ldots, x_n$, have been determined. Such a rule is called a *test* of the hypothesis $H_0: \theta \leq 75$ against the alternative hypothesis $H_1: \theta > 75$. There is no bound on the number of rules or tests that can be constructed. We shall consider three such tests. Our tests will be constructed around the following notion. We shall partition the sample space $\mathcal{A}$ into a subset C and its complement $C*$. If the experimental values of $X_1, X_2, \ldots, X_n$, say $x_1, x_2, \ldots, x_n$, are such that the point $(x_1, x_2, \ldots, x_n) \in C$, we shall reject the hypothesis H_0 (accept the hypothesis H_1). If we have $(x_1, x_2, \ldots, x_n) \in C*$, we shall accept the hypothesis H_0 (reject the hypothesis H_1).

Test 1. Let $n = 25$. The sample space $\mathcal{A}$ is the set

$$\{(x_1, x_2, \ldots, x_{25}); -\infty < x_i < \infty, i = 1, 2, \ldots, 25\}.$$

Let the subset C of the sample space be

$$C = \{(x_1, x_2, \ldots, x_{25}); x_1 + x_2 + \cdots + x_{25} > (25)(75)\}.$$

We shall reject the hypothesis H_0 if and only if our 25 experimental values are such that $(x_1, x_2, \ldots, x_{25}) \in C$. If $(x_1, x_2, \ldots, x_{25})$ is not an element of C, we shall accept the hypothesis H_0. This subset C of the sample space that leads to the rejection of the hypothesis $H_0: \theta \leq 75$ is called the *critical region* of Test 1. Now $\sum_1^{25} x_i > (25)(75)$ if and only if $\bar{x} > 75$, where $\bar{x} = \sum_1^{25} x_i/25$. Thus we can much more conveniently say that we shall reject the hypothesis $H_0: \theta \leq 75$ and accept the hypothesis $H_1: \theta > 75$ if and only if the experimentally determined value of the sample mean $\bar{x}$ is greater than 75. If $\bar{x} \leq 75$, we accept the hypothesis $H_0: \theta \leq 75$. Our test then amounts to this: We shall reject the hypothesis $H_0: \theta \leq 75$ if the mean of the sample exceeds the maximum value of the mean of the distribution when the hypothesis H_0 is true.

It would help us to evaluate a test of a statistical hypothesis if we knew the probability of rejecting that hypothesis (and hence of accepting the alternative hypothesis). In our Test 1, this means that we want to compute the probability

$$\Pr\left[(X_1, \ldots, X_{25}) \in C\right] = \Pr\left(\bar{X} > 75\right).$$

Obviously, this probability is a function of the parameter θ and we shall denote it by $K_1(\theta)$. The function $K_1(\theta) = \Pr(\bar{X} > 75)$ is called the *power*

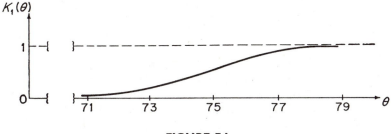

FIGURE 7.1

function of Test 1, and the value of the power function at a parameter point is called the *power* of Test 1 at that point. Because $\overline{X}$ is $n(\theta, 4)$, we have

$$K_1(\theta) = \Pr\left(\frac{\overline{X} - \theta}{2} > \frac{75 - \theta}{2}\right) = 1 - N\left(\frac{75 - \theta}{2}\right).$$

So, for illustration, we have, by Table III of Appendix B, the power at $\theta = 75$ to be $K_1(75) = 0.500$. Other powers are $K_1(73) = 0.159$, $K_1(77) = 0.841$, and $K_1(79) = 0.977$. The graph of $K_1(\theta)$ of Test 1 is depicted in Figure 7.1. Among other things, this means that, if $\theta = 75$, the probability of rejecting the hypothesis $H_0: \theta \leq 75$ is $\frac{1}{2}$. That is, if $\theta = 75$ so that H_0 is true, the probability of rejecting this true hypothesis H_0 is $\frac{1}{2}$. Many statisticians and research workers find it very undesirable to have such a high probability as $\frac{1}{2}$ assigned to this kind of mistake: namely the rejection of H_0 when H_0 is a true hypothesis. Thus Test 1 does not appear to be a very satisfactory test. Let us try to devise another test that does not have this objectionable feature. We shall do this by making it more difficult to reject the hypothesis H_0, with the hope that this will give a smaller probability of rejecting H_0 when that hypothesis is true.

Test 2. Let $n = 25$. We shall reject the hypothesis $H_0: \theta \leq 75$ and accept the hypothesis $H_1: \theta > 75$ if and only if $\bar{x} > 78$. Here the critical region is $C = \{(x_1, \ldots, x_{25}); x_1 + \cdots + x_{25} > (25)(78)\}$. The power function of Test 2 is, because $\overline{X}$ is $n(\theta, 4)$,

$$K_2(\theta) = \Pr\,(\overline{X} > 78) = 1 - N\left(\frac{78 - \theta}{2}\right).$$

Some values of the power function of Test 2 are $K_2(73) = 0.006$, $K_2(75) = 0.067$, $K_2(77) = 0.309$, and $K_2(79) = 0.691$. That is, if $\theta = 75$, the probability of rejecting $H_0: \theta \leq 75$ is 0.067; this is much more desirable than the corresponding probability $\frac{1}{2}$ that resulted from Test 1. However, if H_0 is false and, in fact, $\theta = 77$, the probability of rejecting $H_0: \theta \leq 75$ (and hence of accepting $H_1: \theta > 75$) is only 0.309. In certain instances, this low probability 0.309 of a correct decision (the acceptance of H_1 when H_1 is true) is objectionable. That is, Test 2 is not wholly satisfactory. Perhaps we can overcome the undesirable features of Tests 1 and 2 if we proceed as in Test 3.

Test 3. Let us first select a power function $K_3(\theta)$ that has the features of a small value at $\theta = 75$ and a large value at $\theta = 77$. For instance, take $K_3(75) = 0.159$ and $K_3(77) = 0.841$. To determine a test with such a power function, let us reject $H_0: \theta \leq 75$ if and only if the experimental value $\bar{x}$ of the mean of a random sample of size n is greater than some constant c. Thus the critical region is $C = \{(x_1, x_2, \ldots, x_n); x_1 + x_2 + \cdots + x_n > nc\}$. It should be noted that the sample size n and the constant c have not been determined as yet. However, since $\bar{X}$ is $n(\theta, 100/n)$, the power function is

$$K_3(\theta) = \Pr(\bar{X} > c) = 1 - N\left(\frac{c - \theta}{10/\sqrt{n}}\right).$$

The conditions $K_3(75) = 0.159$ and $K_3(77) = 0.841$ require that

$$1 - N\left(\frac{c - 75}{10/\sqrt{n}}\right) = 0.159, \qquad 1 - N\left(\frac{c - 77}{10/\sqrt{n}}\right) = 0.841.$$

Equivalently, from Table III of Appendix B, we have

$$\frac{c - 75}{10/\sqrt{n}} = 1, \qquad \frac{c - 77}{10/\sqrt{n}} = -1.$$

The solution to these two equations in n and c is $n = 100$, $c = 76$. With these values of n and c, other powers of Test 3 are $K_3(73) = 0.001$ and $K_3(79) = 0.999$. It is important to observe that although Test 3 has a more desirable power function than those of Tests 1 and 2, a certain "price" has been paid—a sample size of $n = 100$ is required in Test 3, whereas we had $n = 25$ in the earlier tests.

Remark. Throughout the text we frequently say that we accept the hypothesis H_0 if we do not reject H_0 in favor of H_1. If this decision is made, it certainly does not mean that H_0 is true or that we even believe that it is true. All it means is, based upon the data at hand, that we are not convinced that the hypothesis H_0 is wrong. Accordingly, the statement "We accept H_0" would possibly be better read as "We do not reject H_0." However, because it is in fairly common use, we use the statement "We accept H_0," but read it with this remark in mind.

We have now illustrated the following concepts:

(a) A statistical hypothesis.
(b) A test of a hypothesis against an alternative hypothesis and the associated concept of the critical region of the test.
(c) The power of a test.

These concepts will now be formally defined.

Definition 1. A *statistical hypothesis* is an assertion about the distribution of one or more random variables. If the statistical hypothesis completely specifies the distribution, it is called a *simple statistical hypothesis*; if it does not, it is called a *composite statistical hypothesis*.

If we refer to Example 1, we see that both $H_0: \theta \leq 75$ and $H_1: \theta > 75$ are composite statistical hypotheses, since neither of them completely specifies the distribution. If there, instead of $H_0: \theta \leq 75$, we had $H_0: \theta = 75$, then H_0 would have been a simple statistical hypothesis.

Definition 2. A *test* of a statistical hypothesis is a rule which, when the experimental sample values have been obtained, leads to a decision to accept or to reject the hypothesis under consideration.

Definition 3. Let C be that subset of the sample space which, in accordance with a prescribed test, leads to the rejection of the hypothesis under consideration. Then C is called the *critical region* of the test.

Definition 4. The *power function* of a test of a statistical hypothesis H_0 against an alternative hypothesis H_1 is that function, defined for all distributions under consideration, which yields the probability that the sample point falls in the critical region C of the test, that is, a function that yields the probability of rejecting the hypothesis under consideration. The value of the power function at a parameter point is called the *power* of the test at that point.

Definition 5. Let H_0 denote a hypothesis that is to be tested against an alternative hypothesis H_1 in accordance with a prescribed test. The *significance level* of the test (or the *size* of the critical region C) is the maximum value (actually supremum) of the power function of the test when H_0 is true.

If we refer again to Example 1, we see that the significance levels of Tests 1, 2, and 3 of that example are 0.500, 0.067, and 0.159, respectively. An additional example may help clarify these definitions.

Example 2. It is known that the random variable X has a p.d.f. of the form

$$f(x; \theta) = \frac{1}{\theta} e^{-x/\theta}, \qquad 0 < x < \infty,$$

$$= 0 \text{ elsewhere.}$$

It is desired to test the simple hypothesis $H_0: \theta = 2$ against the alternative simple hypothesis $H_1: \theta = 4$. Thus $\Omega = \{\theta; \theta = 2, 4\}$. A random sample X_1, X_2 of size $n = 2$ will be used. The test to be used is defined by taking the critical region to be $C = \{(x_1, x_2); 9.5 \leq x_1 + x_2 < \infty\}$. The power function of the test and the significance level of the test will be determined.

There are but two probability density functions under consideration, namely, $f(x; 2)$ specified by H_0 and $f(x; 4)$ specified by H_1. Thus the power function is defined at but two points $\theta = 2$ and $\theta = 4$. The power function of the test is given by $\Pr[(X_1, X_2) \in C]$. If H_0 is true, that is, $\theta = 2$, the joint p.d.f. of X_1 and X_2 is

$$f(x_1; 2)f(x_2; 2) = \tfrac{1}{4}e^{-(x_1 + x_2)/2}, \qquad 0 < x_1 < \infty, 0 < x_2 < \infty,$$
$$= 0 \text{ elsewhere,}$$

and

$$\Pr[(X_1, X_2) \in C] = 1 - \Pr[(X_1, X_2) \in C^*]$$
$$= 1 - \int_0^{9.5} \int_0^{9.5 - x_2} \tfrac{1}{4}e^{-(x_1 + x_2)/2} \, dx_1 \, dx_2$$
$$= 0.05, \text{ approximately.}$$

If H_1 is true, that is, $\theta = 4$, the joint p.d.f. of X_1 and X_2 is

$$f(x_1; 4)f(x_2; 4) = \tfrac{1}{16}e^{-(x_1 + x_2)/4}, \qquad 0 < x_1 < \infty, 0 < x_2 < \infty,$$
$$= 0 \text{ elsewhere,}$$

and

$$\Pr[(X_1, X_2) \in C] = 1 - \int_0^{9.5} \int_0^{9.5 - x_2} \tfrac{1}{16}e^{-(x_1 + x_2)/4} \, dx_1 \, dx_2$$
$$= 0.31, \text{ approximately.}$$

Thus the power of the test is given by 0.05 for $\theta = 2$ and by 0.31 for $\theta = 4$. That is, the probability of rejecting H_0 when H_0 is true is 0.05, and the probability of rejecting H_0 when H_0 is false is 0.31. Since the significance level of this test (or the size of the critical region) is the power of the test when H_0 is true, the significance level of this test is 0.05.

The fact that the power of this test, when $\theta = 4$, is only 0.31 immediately suggests that a search be made for another test which, with the same power when $\theta = 2$, would have a power greater than 0.31 when $\theta = 4$. However, Section 7.2 will make clear that such a search would be fruitless. That is, there is no test with a significance level of 0.05 and based on a random sample of size $n = 2$ that has a greater power at $\theta = 4$. The only manner in which the situation may be improved is to have recourse to a random sample of size n greater than 2.

Our computations of the powers of this test at the two points $\theta = 2$ and $\theta = 4$ were purposely done the hard way to focus attention on fundamental concepts. A procedure that is computationally simpler is the following. When the hypothesis H_0 is true, the random variable X is $\chi^2(2)$. Thus the random variable $X_1 + X_2 = Y$, say, is $\chi^2(4)$. Accordingly, the power of the test when H_0 is true is given by

$$\Pr (Y \geq 9.5) = 1 - \Pr (Y < 9.5) = 1 - 0.95 = 0.05,$$

from Table II of Appendix B. When the hypothesis H_1 is true, the random variable $X/2$ is $\chi^2(2)$; so the random variable $(X_1 + X_2)/2 = Z$, say, is $\chi^2(4)$. Accordingly, the power of the test when H_1 is true is given by

$$\Pr (X_1 + X_2 \geq 9.5) = \Pr (Z \geq 4.75)$$

$$= \int_{4.75}^{\infty} \tfrac{1}{4} z e^{-z/2} \, dz,$$

which is equal to 0.31, approximately.

Remark. The rejection of the hypothesis H_0 when that hypothesis is true is, of course, an incorrect decision or an error. This incorrect decision is often called a type I error; accordingly, the significance level of the test is the probability of committing an error of type I. The acceptance of H_0 when H_0 is false (H_1 is true) is called an error of type II. Thus the probability of a type II error is 1 minus the power of the test when H_1 is true. Frequently, it is disconcerting to the student to discover that there are so many names for the same thing. However, since all of them are used in the statistical literature, we feel obligated to point out that "significance level," "size of the critical region," "power of the test when H_0 is true," and "the probability of committing an error of type I" are all equivalent.

EXERCISES

7.1. Let X have a p.d.f. of the form $f(x; \theta) = \theta x^{\theta-1}$, $0 < x < 1$, zero elsewhere, where $\theta \in \{\theta; \theta = 1, 2\}$. To test the simple hypothesis $H_0: \theta = 1$ against the alternative simple hypothesis $H_1: \theta = 2$, use a random sample X_1, X_2 of size $n = 2$ and define the critical region to be $C = \{(x_1, x_2); 3/4 \leq x_1 x_2\}$. Find the power function of the test.

7.2. Let X have a binomial distribution with parameters $n = 10$ and $p \in \{p; p = \tfrac{1}{4}, \tfrac{1}{2}\}$. The simple hypothesis $H_0: p = \tfrac{1}{2}$ is rejected, and the alternative simple hypothesis $H_1: p = \tfrac{1}{4}$ is accepted, if the observed value of X_1, a random sample of size 1, is less than or equal to 3. Find the power function of the test.

7.3. Let X_1, X_2 be a random sample of size $n = 2$ from the distribution having p.d.f. $f(x; \theta) = (1/\theta)e^{-x/\theta}$, $0 < x < \infty$, zero elsewhere. We reject $H_0: \theta = 2$ and accept $H_1: \theta = 1$ if the observed values of X_1, X_2, say x_1, x_2, are such that

$$\frac{f(x_1; 2)f(x_2; 2)}{f(x_1; 1)f(x_2; 1)} \leq \frac{1}{2}.$$

Here $\Omega = \{\theta; \theta = 1, 2\}$. Find the significance level of the test and the power of the test when H_0 is false.

7.4. Sketch, as in Figure 9.1, the graphs of the power functions of Tests 1, 2, and 3 of Example 1 of this section.

7.5. Let us assume that the life of a tire in miles, say X, is normally distributed with mean θ and standard deviation 5000. Past experience indicates that $\theta = 30,000$. The manufacturer claims that the tires made by a new process have mean $\theta > 30,000$, and it is very possible that $\theta = 35,000$. Let us check his claim by testing $H_0: \theta \le 30,000$ against $H_1: \theta > 30,000$. We shall observe n independent values of X, say $x_1, \ldots, x_n$, and we shall reject H_0 (thus accept H_1) if and only if $\bar{x} \ge c$. Determine n and c so that the power function $K(\theta)$ of the test has the values $K(30,000) = 0.01$ and $K(35,000) = 0.98$.

7.6. Let X have a Poisson distribution with mean θ. Consider the simple hypothesis $H_0: \theta = \frac{1}{2}$ and the alternative composite hypothesis $H_1: \theta < \frac{1}{2}$. Thus $\Omega = \{\theta; 0 < \theta \le \frac{1}{2}\}$. Let $X_1, \ldots, X_{12}$ denote a random sample of size 12 from this distribution. We reject H_0 if and only if the observed value of $Y = X_1 + \cdots + X_{12} \le 2$. If $K(\theta)$ is the power function of the test, find the powers $K(\frac{1}{2})$, $K(\frac{1}{3})$, $K(\frac{1}{4})$, $K(\frac{1}{6})$, and $K(\frac{1}{12})$. Sketch the graph of $K(\theta)$. What is the significance level of the test?

7.2 Certain Best Tests

In this section we require that both the hypothesis H_0, which is to be tested, and the alternative hypothesis H_1 be simple hypotheses. Thus, in all instances, the parameter space is a set that consists of exactly two points. Under this restriction, we shall do three things:

(a) Define a best test for testing H_0 against H_1.

(b) Prove a theorem that provides a method of determining a best test.

(c) Give two examples.

Before we define a best test, one important observation should be made. Certainly, a test specifies a critical region; but it can also be said that a choice of a critical region defines a test. For instance, if one is given the critical region $C = \{(x_1, x_2, x_3); x_1^2 + x_2^2 + x_3^2 \ge 1\}$, the test is determined: Three random variables X_1, X_2, X_3 are to be considered; if the observed values are x_1, x_2, x_3, accept H_0 if $x_1^2 + x_2^2 + x_3^2 < 1$; otherwise, reject H_0. That is, the terms "test" and "critical region" can, in this sense, be used interchangeably. Thus, if we define a best critical region, we have defined a best test.

Let $f(x; \theta)$ denote the p.d.f. of a random variable X. Let X_1, X_2, $\ldots$, X_n denote a random sample from this distribution, and consider the two simple hypotheses $H_0: \theta = \theta'$ and $H_1: \theta = \theta''$. Thus $\Omega = \{\theta; \theta = \theta', \theta''\}$. We now define a best critical region (and hence a best test) for testing the simple hypothesis H_0 against the alternative simple hypothesis H_1. In this definition the symbols $\Pr[(X_1, X_2, \ldots, X_n) \in C; H_0]$ and $\Pr[(X_1, X_2, \ldots, X_n) \in C; H_1]$ mean $\Pr[(X_1, X_2, \ldots, X_n) \in C]$ when, respectively, H_0 and H_1 are true.

Definition 6. Let C denote a subset of the sample space. Then C is called a *best critical region* of size α for testing the simple hypothesis $H_0: \theta = \theta'$ against the alternative simple hypothesis $H_1: \theta = \theta''$ if, for every subset A of the sample space for which $\Pr[(X_1, \ldots, X_n) \in A; H_0] = \alpha$:

(a) $\Pr[(X_1, X_2, \ldots, X_n) \in C; H_0] = \alpha$.
(b) $\Pr[(X_1, X_2, \ldots, X_n) \in C; H_1] \geq \Pr[(X_1, X_2, \ldots, X_n) \in A; H_1]$.

This definition states, in effect, the following: First assume H_0 to be true. In general, there will be a multiplicity of subsets A of the sample space such that $\Pr[(X_1, X_2, \ldots, X_n) \in A] = \alpha$. Suppose that there is one of these subsets, say C, such that when H_1 is true, the power of the test associated with C is at least as great as the power of the test associated with each other A. Then C is defined as a best critical region of size α for testing H_0 against H_1.

In the following example we shall examine this definition in some detail and in a very simple case.

Example 1. Consider the one random variable X that has a binomial distribution with $n = 5$ and $p = \theta$. Let $f(x; \theta)$ denote the p.d.f. of X and let $H_0: \theta = \frac{1}{2}$ and $H_1: \theta = \frac{3}{4}$. The following tabulation gives, at points of positive probability density, the values of $f(x; \frac{1}{2})$, $f(x; \frac{3}{4})$, and the ratio $f(x; \frac{1}{2})/f(x; \frac{3}{4})$.

x	0	1	2	3	4	5
$f(x; \frac{1}{2})$	$\frac{1}{32}$	$\frac{5}{32}$	$\frac{10}{32}$	$\frac{10}{32}$	$\frac{5}{32}$	$\frac{1}{32}$
$f(x; \frac{3}{4})$	$\frac{1}{1024}$	$\frac{15}{1024}$	$\frac{90}{1024}$	$\frac{270}{1024}$	$\frac{405}{1024}$	$\frac{243}{1024}$
$\dfrac{f(x; \frac{1}{2})}{f(x; \frac{3}{4})}$	32	$\frac{32}{3}$	$\frac{32}{9}$	$\frac{32}{27}$	$\frac{32}{81}$	$\frac{32}{243}$

We shall use one random value of X to test the simple hypothesis $H_0: \theta = \frac{1}{2}$ against the alternative simple hypothesis $H_1: \theta = \frac{3}{4}$, and we shall first assign the significance level of the test to be $\alpha = \frac{1}{32}$. We seek a best critical region of size $\alpha = \frac{1}{32}$. If $A_1 = \{x; x = 0\}$ and $A_2 = \{x; x = 5\}$, then $\mathrm{Pr}\ (X \in A_1; H_0)$ $= \mathrm{Pr}\ (X \in A_2; H_0) = \frac{1}{32}$ and there is no other subset A_3 of the space $\{x; x = 0, 1, 2, 3, 4, 5\}$ such that $\mathrm{Pr}\ (X \in A_3; H_0) = \frac{1}{32}$. Then either A_1 or A_2 is the best critical region C of size $\alpha = \frac{1}{32}$ for testing H_0 against H_1. We note that $\mathrm{Pr}\ (X \in A_1; H_0) = \frac{1}{32}$ and that $\mathrm{Pr}\ (X \in A_1; H_1) = \frac{1}{1024}$. Thus, if the set A_1 is used as a critical region of size $\alpha = \frac{1}{32}$, we have the intolerable situation that the probability of rejecting H_0 when H_1 is true (H_0 is false) is much less than the probability of rejecting H_0 when H_0 is true.

On the other hand, if the set A_2 is used as a critical region, then $\mathrm{Pr}\ (X \in A_2; H_0) = \frac{1}{32}$ and $\mathrm{Pr}\ (X \in A_2; H_1) = \frac{243}{1024}$. That is, the probability of rejecting H_0 when H_1 is true is much greater than the probability of rejecting H_0 when H_0 is true. Certainly, this is a more desirable state of affairs, and actually A_2 is the best critical region of size $\alpha = \frac{1}{32}$. The latter statement follows from the fact that, when H_0 is true, there are but two subsets, A_1 and A_2, of the sample space, each of whose probability measure is $\frac{1}{32}$ and the fact that

$$\frac{243}{1024} = \mathrm{Pr}\ (X \in A_2; H_1) > \mathrm{Pr}\ (X \in A_1; H_1) = \frac{1}{1024}.$$

It should be noted, in this problem, that the best critical region $C = A_2$ of size $\alpha = \frac{1}{32}$ is found by including in C the point (or points) at which $f(x; \frac{1}{2})$ is *small* in comparison with $f(x; \frac{3}{4})$. This is seen to be true once it is observed that the ratio $f(x; \frac{1}{2})/f(x; \frac{3}{4})$ is a minimum at $x = 5$. Accordingly, the ratio $f(x; \frac{1}{2})/f(x; \frac{3}{4})$, which is given in the last line of the above tabulation, provides us with a precise tool by which to find a best critical region C for certain given values of α. To illustrate this, take $\alpha = \frac{6}{32}$. When H_0 is true, each of the subsets $\{x; x = 0, 1\}$, $\{x; x = 0, 4\}$, $\{x; x = 1, 5\}$, $\{x; x = 4, 5\}$ has probability measure $\frac{6}{32}$. By direct computation it is found that the best critical region of this size is $\{x; x = 4, 5\}$. This reflects the fact that the ratio $f(x; \frac{1}{2})/f(x; \frac{3}{4})$ has its two smallest values for $x = 4$ and $x = 5$. The power of this test, which has $\alpha = \frac{6}{32}$, is

$$\mathrm{Pr}\ (X = 4, 5; H_1) = \frac{405}{1024} + \frac{243}{1024} = \frac{648}{1024}.$$

The preceding example should make the following theorem, due to Neyman and Pearson, easier to understand. It is an important theorem because it provides a systematic method of determining a best critical region.

← Gossett

Neyman–Pearson Theorem. *Let* $X_1, X_2, \ldots, X_n$, *where n is a fixed positive integer, denote a random sample from a distribution that has p.d.f.* $f(x; \theta)$. *Then the joint p.d.f. of* $X_1, X_2, \ldots, X_n$ *is*

$$L(\theta; x_1, x_2, \ldots, x_n) = f(x_1; \theta)f(x_2; \theta) \cdots f(x_n; \theta).$$

Let θ' and θ'' be distinct fixed values of θ so that $\Omega = \{\theta; \theta = \theta', \theta''\}$, *and* *let k be a positive number. Let C be a subset of the sample space such that:*

(a) $\dfrac{L(\theta'; x_1, x_2, \ldots, x_n)}{L(\theta''; x_1, x_2, \ldots, x_n)} \le k,$ *for each point* $(x_1, x_2, \ldots, x_n) \in C.$

(b) $\dfrac{L(\theta'; x_1, x_2, \ldots, x_n)}{L(\theta''; x_1, x_2, \ldots, x_n)} \ge k,$ *for each point* $(x_1, x_2, \ldots, x_n) \in C^*.$

(c) $\alpha = \Pr\left[(X_1, X_2, \ldots, X_n) \in C; H_0\right].$

Then C is a best critical region of size α for testing the simple hypothesis H_0: $\theta = \theta'$ against the alternative simple hypothesis H_1: $\theta = \theta''$.

Proof. We shall give the proof when the random variables are of the continuous type. If C is the only critical region of size α, the theorem is proved. If there is another critical region of size α, denote it by A. For convenience, we shall let $\int \cdots \int_R L(\theta; x_1, \ldots, x_n)\, dx_1 \cdots dx_n$ be denoted by $\int_R L(\theta)$. In this notation we wish to show that

$$\int_C L(\theta'') - \int_A L(\theta'') \ge 0.$$

Since C is the union of the disjoint sets $C \cap A$ and $C \cap A^*$ and A is the union of the disjoint sets $A \cap C$ and $A \cap C^*$, we have

(1) $\displaystyle\int_C L(\theta'') - \int_A L(\theta'')$

$$= \int_{C \cap A} L(\theta'') + \int_{C \cap A^*} L(\theta'') - \int_{A \cap C} L(\theta'') - \int_{A \cap C^*} L(\theta'')$$

$$= \int_{C \cap A^*} L(\theta'') - \int_{A \cap C^*} L(\theta'').$$

However, by the hypothesis of the theorem, $L(\theta'') \ge (1/k)L(\theta')$ at each point of C, and hence at each point of $C \cap A^*$; thus

$$\int_{C \cap A^*} L(\theta'') \ge \frac{1}{k} \int_{C \cap A^*} L(\theta').$$

But $L(\theta'') \le (1/k)L(\theta')$ at each point of C^*, and hence at each point of $A \cap C^*$; accordingly,

$$\int_{A \cap C^*} L(\theta'') \le \frac{1}{k} \int_{A \cap C^*} L(\theta').$$

These inequalities imply that

$$\int_{C \cap A^*} L(\theta'') - \int_{A \cap C^*} L(\theta'') \ge \frac{1}{k} \int_{C \cap A^*} L(\theta') - \frac{1}{k} \int_{A \cap C^*} L(\theta');$$

and, from Equation (1), we obtain

(2) $\displaystyle \int_C L(\theta'') - \int_A L(\theta'') \geq \frac{1}{k}\left[\int_{C\cap A^*} L(\theta') - \int_{A\cap C^*} L(\theta')\right].$

However,

$\displaystyle \int_{C\cap A^*} L(\theta') - \int_{A\cap C^*} L(\theta')$

$\displaystyle \qquad = \int_{C\cap A^*} L(\theta') + \int_{C\cap A} L(\theta') - \int_{A\cap C} L(\theta') - \int_{A\cap C^*} L(\theta')$

$\displaystyle \qquad = \int_C L(\theta') - \int_A L(\theta')$

$\displaystyle \qquad = \alpha - \alpha = 0.$

If this result is substituted in inequality (2), we obtain the desired result,

$$\int_C L(\theta'') - \int_A L(\theta'') \geq 0.$$

If the random variables are of the discrete type, the proof is the same, with integration replaced by summation.

Remark. As stated in the theorem, conditions (a), (b), and (c) are sufficient ones for region C to be a best critical region of size α. However, they are also necessary. We discuss this briefly. Suppose there is a region A of size α that does not satisfy (a) and (b) and that is as powerful at $\theta = \theta''$ as C, which satisfies (a), (b), and (c). Then expression (1) would be zero, since the power at θ'' using A is equal to that using C. It can be proved that to have expression (1) equal zero A must be of the same form as C. As a matter of fact, in the continuous case, A and C would essentially be the same region; that is, they could differ only by a set having probability zero. However, in the discrete case, if $\Pr[L(\theta') = kL(\theta''); H_0]$ is positive, A and C could be different sets, but each would necessarily enjoy conditions (a), (b), and (c) to be a best critical region of size α.

One aspect of the theorem to be emphasized is that if we take C to be the set of all points $(x_1, x_2, \ldots, x_n)$ which satisfy

$$\frac{L(\theta'; x_1, x_2, \ldots, x_n)}{L(\theta''; x_1, x_2, \ldots, x_n)} \leq k, \qquad k > 0,$$

then, in accordance with the theorem, C will be a best critical region. This inequality can frequently be expressed in one of the forms (where c_1 and c_2 are constants)

$$u_1(x_1, x_2, \ldots, x_n; \theta', \theta'') \leq c_1,$$

or

$$u_2(x_1, x_2, \ldots, x_n; \theta', \theta'') \geq c_2.$$

Suppose that it is the first form, $u_1 \leq c_1$. Since θ' and θ'' are given constants, $u_1(X_1, X_2, \ldots, X_n; \theta', \theta'')$ is a statistic; and if the p.d.f. of this statistic can be found when H_0 is true, then the significance level of the test of H_0 against H_1 can be determined from this distribution. That is,

$$\alpha = \Pr\left[u_1(X_1, X_2, \ldots, X_n; \theta', \theta'') \leq c_1; H_0\right].$$

Moreover, the test may be based on this statistic; for, if the observed values of $X_1, X_2, \ldots, X_n$ are $x_1, x_2, \ldots, x_n$, we reject H_0 (accept H_1) if $u_1(x_1, x_2, \ldots, x_n) \leq c_1$.

A positive number k determines a best critical region C whose size is $\alpha = \Pr\left[(X_1, X_2, \ldots, X_n) \in C; H_0\right]$ for that particular k. It may be that this value of α is unsuitable for the purpose at hand; that is, it is too large or too small. However, if there is a statistic $u_1(X_1, X_2, \ldots, X_n)$, as in the preceding paragraph, whose p.d.f. can be determined when H_0 is true, we need not experiment with various values of k to obtain a desirable significance level. For if the distribution of the statistic is known, or can be found, we may determine c_1 such that $\Pr[u_1(X_1, X_2, \ldots, X_n) \leq c_1; H_0]$ is a desirable significance level.

An illustrative example follows.

Example 2. Let $X_1, X_2, \ldots, X_n$ denote a random sample from the distribution that has the p.d.f.

$$f(x; \theta) = \frac{1}{\sqrt{2\pi}} \exp\left(-\frac{(x - \theta)^2}{2}\right), \qquad -\infty < x < \infty.$$

It is desired to test the simple hypothesis $H_0: \theta = \theta' = 0$ against the alternative simple hypothesis $H_1: \theta = \theta'' = 1$. Now

$$\frac{L(\theta'; x_1, \ldots, x_n)}{L(\theta''; x_1, \ldots, x_n)} = \frac{(1/\sqrt{2\pi})^n \exp\left[-\left(\sum_1^n x_i^2\right)/2\right]}{(1/\sqrt{2\pi})^n \exp\left[-\left(\sum_1^n (x_i - 1)^2\right)/2\right]}$$

$$= \exp\left(-\sum_1^n x_i + \frac{n}{2}\right).$$

If $k > 0$, the set of all points $(x_1, x_2, \ldots, x_n)$ such that

$$\exp\left(-\sum_1^n x_i + \frac{n}{2}\right) \leq k$$

is a best critical region. This inequality holds if and only if

$$-\sum_1^n x_i + \frac{n}{2} \le \ln k$$

or, equivalently,

$$\sum_1^n x_i \ge \frac{n}{2} - \ln k = c.$$

In this case, a best critical region is the set $C = \{(x_1, x_2, \ldots, x_n); \sum_1^n x_i \ge c\}$, where c is a constant that can be determined so that the size of the critical region is a desired number α. The event $\sum_1^n X_i \ge c$ is equivalent to the event $\bar{X} \ge c/n = c_1$, say, so the test may be based upon the statistic $\bar{X}$. If H_0 is true, that is, $\theta = \theta' = 0$, then $\bar{X}$ has a distribution that is $n(0, 1/n)$. For a given positive integer n, the size of the sample, and a given significance level α, the number c_1 can be found from Table III in Appendix B, so that $\Pr(\bar{X} \ge c_1; H_0) = \alpha$. Hence, if the experimental values of $X_1, X_2, \ldots, X_n$ were, respectively, $x_1, x_2, \ldots, x_n$, we would compute $\bar{x} = \sum_1^n x_i/n$. If $\bar{x} \ge c_1$, the simple hypothesis $H_0: \theta = \theta' = 0$ would be rejected at the significance level α; if $\bar{x} < c_1$, the hypothesis H_0 would be accepted. The probability of rejecting H_0, when H_0 is true, is α; the probability of rejecting H_0, when H_0 is false, is the value of the power of the test at $\theta = \theta'' = 1$. That is,

$$\Pr(\bar{X} \ge c_1; H_1) = \int_{c_1}^{\infty} \frac{1}{\sqrt{2\pi}\sqrt{1/n}} \exp\left(-\frac{(\bar{x}-1)^2}{2(1/n)}\right) d\bar{x}.$$

For example, if $n = 25$ and if α is selected to be 0.05, then from Table III we find that $c_1 = 1.645/\sqrt{25} = 0.329$. Thus the power of this best test of H_0 against H_1 is 0.05, when H_0 is true, and is

$$\int_{0.329}^{\infty} \frac{1}{\sqrt{2\pi}\sqrt{\frac{1}{25}}} \exp\left[-\frac{(\bar{x}-1)^2}{2(\frac{1}{25})}\right] d\bar{x} = \int_{-3.355}^{\infty} \frac{1}{\sqrt{2\pi}} e^{-w^2/2}\, dw = 0.999+,$$

when H_1 is true.

There is another aspect of this theorem that warrants special mention. It has to do with the number of parameters that appear in the p.d.f. Our notation suggests that there is but one parameter. However, a careful review of the proof will reveal that nowhere was this needed or assumed. The p.d.f. may depend upon any finite number of parameters. What is essential is that the hypothesis H_0 and the alternative hypothesis H_1 be simple, namely that they completely specify the distributions. With this in mind, we see that the simple hypotheses H_0 and H_1 do not need to be hypotheses about the parameters of a distribution, nor, as a matter of fact, do the random variables X_1, X_2, ..., X_n need to be mutually stochastically independent. That is, if H_0 is the simple hypothesis that the joint p.d.f. is $g(x_1, x_2, \ldots, x_n)$, and if H_1 is the alternative simple hypothesis that the joint p.d.f. is $h(x_1, x_2, \ldots, x_n)$, then C is a best critical region of size α for testing H_0 against H_1 if, for $k > 0$:

(a)′ $\dfrac{g(x_1, x_2, \ldots, x_n)}{h(x_1, x_2, \ldots, x_n)} \le k$ for $(x_1, x_2, \ldots, x_n) \in C$.

(b)′ $\dfrac{g(x_1, x_2, \ldots, x_n)}{h(x_1, x_2, \ldots, x_n)} \ge k$ for $(x_1, x_2, \ldots, x_n) \in C^*$.

(c)′ $\alpha = \Pr\left[(X_1, X_2, \ldots, X_n) \in C; H_0\right]$.

An illustrative example follows.

Example 3. Let $X_1, \ldots, X_n$ denote a random sample from a distribution which has a p.d.f. $f(x)$ that is positive on and only on the nonnegative integers. It is desired to test the simple hypothesis

$$H_0: f(x) = \frac{e^{-1}}{x!}, \qquad x = 0, 1, 2, \ldots,$$

$$= 0 \text{ elsewhere,}$$

against the alternative simple hypothesis

$$H_1: f(x) = (\tfrac{1}{2})^{x+1}, \qquad x = 0, 1, 2, \ldots,$$

$$= 0 \text{ elsewhere.}$$

Here

$$\frac{g(x_1, \ldots, x_n)}{h(x_1, \ldots, x_n)} = \frac{e^{-n}/(x_1! \, x_2! \cdots x_n!)}{(\tfrac{1}{2})^n (\tfrac{1}{2})^{x_1 + x_2 + \cdots + x_n}}$$

$$= \frac{(2e^{-1})^n 2^{\Sigma x_i}}{\prod_1^n (x_i!)}.$$

If $k > 0$, the set of points $(x_1, x_2, \ldots, x_n)$ such that

$$\left(\sum_1^n x_i\right) \ln 2 - \ln\left[\prod_1^n (x_i!)\right] \le \ln k - n \ln (2e^{-1}) = c$$

is a best critical region C. Consider the case of $k = 1$ and $n = 1$. The preceding inequality may be written $2^{x_1}/x_1! \le e/2$. This inequality is satisfied by all points in the set $C = \{x_1; x_1 = 0, 3, 4, 5, \ldots\}$. Thus the power of the test when H_0 is true is

$$\Pr(X_1 \in C; H_0) = 1 - \Pr(X_1 = 1, 2; H_0) = 0.448,$$

approximately, in accordance with Table I of Appendix B. The power of the test when H_1 is true is given by

$$\Pr\,(X_1 \in C;\, H_1) = 1 - \Pr\,(X_1 = 1,\, 2;\, H_1)$$
$$= 1 - (\tfrac{1}{4} + \tfrac{1}{8}) = 0.625.$$

EXERCISES

7.7. In Example 2 of this section, let the simple hypotheses read $H_0 \colon \theta = \theta' = 0$ and $H_1 \colon \theta = \theta'' = -1$. Show that the best test of H_0 against H_1 may be carried out by use of the statistic $\bar{X}$, and that if $n = 25$ and $\alpha = 0.05$, the power of the test is $0.999+$ when H_1 is true.

7.8. Let the random variable X have the p.d.f. $f(x;\,\theta) = (1/\theta)e^{-x/\theta}$, $0 < x < \infty$, zero elsewhere. Consider the simple hypothesis $H_0 \colon \theta = \theta' = 2$ and the alternative hypothesis $H_1 \colon \theta = \theta'' = 4$. Let X_1, X_2 denote a random sample of size 2 from this distribution. Show that the best test of H_0 against H_1 may be carried out by use of the statistic $X_1 + X_2$ and that the assertion in Example 2 of Section 7.1 is correct.

7.9. Repeat Exercise 7.8 when $H_1 \colon \theta = \theta'' = 6$. Generalize this for every $\theta'' > 2$.

7.10. Let $X_1, X_2, \ldots, X_{10}$ be a random sample of size 10 from a normal distribution $n(0, \sigma^2)$. Find a best critical region of size $\alpha = 0.05$ for testing $H_0 \colon \sigma^2 = 1$ against $H_1 \colon \sigma^2 = 2$. Is this a best critical region of size 0.05 for testing $H_0 \colon \sigma^2 = 1$ against $H_1 \colon \sigma^2 = 4$? Against $H_1 \colon \sigma^2 = \sigma_1^2 > 1$?

7.11. If $X_1, X_2, \ldots, X_n$ is a random sample from a distribution having p.d.f. of the form $f(x;\,\theta) = \theta x^{\theta-1}$, $0 < x < 1$, zero elsewhere, show that a best critical region for testing $H_0 \colon \theta = 1$ against $H_1 \colon \theta = 2$ is $C = \left\{ (x_1, x_2, \ldots, x_n);\, c \le \prod\limits_{i=1}^{n} x_i \right\}$.

7.12. Let $X_1, X_2, \ldots, X_{10}$ be a random sample from a distribution that is $n(\theta_1, \theta_2)$. Find a best test of the simple hypothesis $H_0 \colon \theta_1 = \theta_1' = 0$, $\theta_2 = \theta_2' = 1$ against the alternative simple hypothesis $H_1 \colon \theta_1 = \theta_1'' = 1$, $\theta_2 = \theta_2'' = 4$.

7.13. Let $X_1, X_2, \ldots, X_n$ denote a random sample from a normal distribution $n(\theta, 100)$. Show that $C = \{ (x_1, x_2, \ldots, x_n);\, c \le \bar{x} = \sum\limits_{1}^{n} x_i/n \}$ is a best critical region for testing $H_0 \colon \theta = 75$ against $H_1 \colon \theta = 78$. Find n and c so that

$$\Pr\,[(X_1, X_2, \ldots, X_n) \in C;\, H_0] = \Pr\,(\bar{X} \ge c;\, H_0) = 0.05$$

and

$$\Pr\,[(X_1, X_2, \ldots, X_n) \in C;\, H_1] = \Pr\,(\bar{X} \ge c;\, H_1) = 0.90, \text{ approximately.}$$

7.14. Let $X_1, X_2, \ldots, X_n$ denote a random sample from a distribution having the p.d.f. $f(x; p) = p^x(1 - p)^{1-x}$, $x = 0, 1$, zero elsewhere. Show that $C = \{(x_1, \ldots, x_n); \sum_1^n x_i \le c\}$ is a best critical region for testing $H_0: p = \frac{1}{2}$ against $H_1: p = \frac{1}{3}$. Use the central limit theorem to find n and c so that approximately $\Pr\left(\sum_1^n X_i \le c; H_0\right) = 0.10$ and $\Pr\left(\sum_1^n X_i \le c; H_1\right) = 0.80$.

7.15. Let $X_1, X_2, \ldots, X_{10}$ denote a random sample of size 10 from a Poisson distribution with mean θ. Show that the critical region C defined by $\sum_1^{10} x_i \ge 3$ is a best critical region for testing $H_0: \theta = 0.1$ against $H_1: \theta = 0.5$. Determine, for this test, the significance level α and the power at $\theta = 0.5$.

7.3 Uniformly Most Powerful Tests

This section will take up the problem of a test of a simple hypothesis H_0 against an alternative composite hypothesis H_1. We begin with an example.

Example 1. Consider the p.d.f.

$$f(x; \theta) = \frac{1}{\theta} e^{-x/\theta}, \qquad 0 < x < \infty,$$

$$= 0 \text{ elsewhere,}$$

of Example 2, Section 7.1. It is desired to test the simple hypothesis $H_0: \theta = 2$ against the alternative composite hypothesis $H_1: \theta > 2$. Thus $\Omega = \{\theta; \theta \ge 2\}$. A random sample, X_1, X_2, of size $n = 2$ will be used, and the critical region is $C = \{(x_1, x_2); 9.5 \le x_1 + x_2 < \infty\}$. It was shown in the example cited that the significance level of the test is approximately 0.05 and that the power of the test when $\theta = 4$ is approximately 0.31. The power function $K(\theta)$ of the test for all $\theta \ge 2$ will now be obtained. We have

$$K(\theta) = 1 - \int_0^{9.5} \int_0^{9.5 - x_2} \frac{1}{\theta^2} \exp\left(-\frac{x_1 + x_2}{\theta}\right) dx_1 \, dx_2$$

$$= \left(\frac{\theta + 9.5}{\theta}\right) e^{-9.5/\theta}, \qquad 2 \le \theta.$$

For example, $K(2) = 0.05$, $K(4) = 0.31$, and $K(9.5) = 2/e$. It is known (Exercise 7.9) that $C = \{(x_1, x_2); 9.5 \le x_1 + x_2 < \infty\}$ is a best critical region of size 0.05 for testing the simple hypothesis $H_0: \theta = 2$ against each simple hypothesis in the composite hypothesis $H_1: \theta > 2$.

The preceding example affords an illustration of a test of a simple hypothesis H_0 that is a best test of H_0 against every simple hypothesis in the alternative composite hypothesis H_1. We now define a critical region, when it exists, which is a best critical region for testing a simple hypothesis H_0 against an alternative composite hypothesis H_1. It seems desirable that this critical region should be a best critical region for testing H_0 against each simple hypothesis in H_1. That is, the power function of the test that corresponds to this critical region should be at least as great as the power function of any other test with the same significance level for every simple hypothesis in H_1.

Definition 7. The critical region C is a *uniformly most powerful critical region* of size α for testing the simple hypothesis H_0 against an alternative composite hypothesis H_1 if the set C is a best critical region of size α for testing H_0 against each simple hypothesis in H_1. A test defined by this critical region C is called a *uniformly most powerful test*, with significance level α, for testing the simple hypothesis H_0 against the alternative composite hypothesis H_1.

As will be seen presently, uniformly most powerful tests do not always exist. However, when they do exist, the Neyman–Pearson theorem provides a technique for finding them. Some illustrative examples are given here.

Example 2. Let $X_1, X_2, \ldots, X_n$ denote a random sample from a distribution that is $n(0, \theta)$, where the variance θ is an unknown positive number. It will be shown that there exists a uniformly most powerful test with significance level α for testing the simple hypothesis H_0: $\theta = \theta'$, where θ' is a fixed positive number, against the alternative composite hypothesis H_1: $\theta > \theta'$. Thus $\Omega = \{\theta; \theta \geq \theta'\}$. The joint p.d.f. of $X_1, X_2, \ldots, X_n$ is

$$L(\theta; x_1, x_2, \ldots, x_n) = \left(\frac{1}{2\pi\theta}\right)^{n/2} \exp\left(-\frac{\sum_1^n x_i^2}{2\theta}\right).$$

Let θ'' represent a number greater than θ', and let k denote a positive number. Let C be the set of points where

$$\frac{L(\theta'; x_1, x_2, \ldots, x_n)}{L(\theta''; x_1, x_2, \ldots, x_n)} \leq k,$$

that is, the set of points where

$$\left(\frac{\theta''}{\theta'}\right)^{n/2} \exp\left[-\left(\frac{\theta'' - \theta'}{2\theta'\theta''}\right) \sum_1^n x_i^2\right] \leq k$$

or, equivalently,

$$\sum_1^n x_i^2 \geq \frac{2\theta'\theta''}{\theta'' - \theta'} \left[\frac{n}{2} \ln\left(\frac{\theta''}{\theta'}\right) - \ln k\right] = c.$$

The set $C = \{(x_1, x_2, \ldots, x_n); \sum_1^n x_i^2 \geq c\}$ is then a best critical region for testing the simple hypothesis H_0: $\theta = \theta'$ against the simple hypothesis $\theta = \theta''$. It remains to determine c so that this critical region has the desired size α. If H_0 is true, the random variable $\sum_1^n X_i^2/\theta'$ has a chi-square distribution with n degrees of freedom. Since $\alpha = \Pr\left(\sum_1^n X_i^2/\theta' \geq c/\theta'; H_0\right)$, c/θ' may be read from Table II in Appendix B and c determined. Then $C = \{(x_1, x_2, \ldots, x_n); \sum_1^n x_i^2 \geq c\}$ is a best critical region of size α for testing H_0: $\theta = \theta'$ against the hypothesis $\theta = \theta''$. Moreover, for each number θ'' greater than θ', the foregoing argument holds. That is, if θ''' is another number greater than θ', then $C = \{(x_1, \ldots, x_n); \sum_1^n x_i^2 \geq c\}$ is a best critical region of size α for testing H_0: $\theta = \theta'$ against the hypothesis $\theta = \theta'''$. Accordingly, $C = \{(x_1, \ldots, x_n); \sum_1^n x_i^2 \geq c\}$ is a uniformly most powerful critical region of size α for testing H_0: $\theta = \theta'$ against H_1: $\theta > \theta'$. If $x_1, x_2, \ldots, x_n$ denote the experimental values of $X_1, X_2, \ldots, X_n$, then H_0: $\theta = \theta'$ is rejected at the significance level α, and H_1: $\theta > \theta'$ is accepted, if $\sum_1^n x_i^2 \geq c$; otherwise, H_0: $\theta = \theta'$ is accepted.

If in the preceding discussion we take $n = 15$, $\alpha = 0.05$, and $\theta' = 3$, then here the two hypotheses will be H_0: $\theta = 3$ and H_1: $\theta > 3$. From Table II, $c/3 = 25$ and hence $c = 75$.

Example 3. Let $X_1, X_2, \ldots, X_n$ denote a random sample from a distribution that is $n(\theta, 1)$, where the mean θ is unknown. It will be shown that there is no uniformly most powerful test of the simple hypothesis H_0: $\theta = \theta'$, where θ' is a fixed number, against the alternative composite hypothesis H_1: $\theta \neq \theta'$. Thus $\Omega = \{\theta; -\infty < \theta < \infty\}$. Let θ'' be a number not equal to θ'. Let k be a positive number and consider

$$\frac{(1/2\pi)^{n/2} \exp\left[-\sum_1^n (x_i - \theta')^2/2\right]}{(1/2\pi)^{n/2} \exp\left[-\sum_1^n (x_i - \theta'')^2/2\right]} \leq k.$$

The preceding inequality may be written as

$$\exp\left\{-(\theta'' - \theta') \sum_1^n x_i + \frac{n}{2}[(\theta'')^2 - (\theta')^2]\right\} \leq k,$$

or

$$(\theta'' - \theta') \sum_1^n x_i \geq \frac{n}{2}[(\theta'')^2 - (\theta')^2] - \ln k.$$

This last inequality is equivalent to

$$\sum_1^n x_i \geq \frac{n}{2}(\theta'' + \theta') - \frac{\ln k}{\theta'' - \theta'},$$

provided $\theta'' > \theta'$, and it is equivalent to

$$\sum_1^n x_i \leq \frac{n}{2}(\theta'' + \theta') - \frac{\ln k}{\theta'' - \theta'}$$

if $\theta'' < \theta'$. The first of these two expressions defines a best critical region for testing H_0: $\theta = \theta'$ against the hypothesis $\theta = \theta''$ provided that $\theta'' > \theta'$, while the second expression defines a best critical region for testing H_0: $\theta = \theta'$ against the hypothesis $\theta = \theta''$ provided that $\theta'' < \theta'$. That is, a best critical region for testing the simple hypothesis against an alternative simple hypothesis, say $\theta = \theta' + 1$, will not serve as a best critical region for testing H_0: $\theta = \theta'$ against the alternative simple hypothesis $\theta = \theta' - 1$, say. By definition, then, there is no uniformly most powerful test in the case under consideration.

It should be noted that had the alternative composite hypothesis been either H_1: $\theta > \theta'$ or H_1: $\theta < \theta'$, a uniformly most powerful test would exist in each instance.

Example 4. In Exercise 7.15, the reader is asked to show that if a random sample of size $n = 10$ is taken from a Poisson distribution with mean θ, the critical region defined by $\sum_1^{10} x_i \geq 3$ is a best critical region for testing H_0: $\theta = 0.1$ against H_1: $\theta = 0.5$. This critical region is also a uniformly most powerful one for testing H_0: $\theta = 0.1$ against H_1: $\theta > 0.1$ because, with $\theta'' > 0.1$,

$$\frac{(0.1)^{\Sigma x_i} e^{-10(0.1)}/(x_1!\, x_2! \cdots x_n!)}{(\theta'')^{\Sigma x_i} e^{-10(\theta'')}/(x_1!\, x_2! \cdots x_n!)} \leq k$$

is equivalent to

$$\left(\frac{0.1}{\theta''}\right)^{\Sigma x_i} e^{-10(0.1 - \theta'')} \leq k.$$

The preceding inequality may be written as

$$\left(\sum_1^n x_i\right)(\ln 0.1 - \ln \theta'') \leq \ln k + 10(0.1 - \theta'')$$

or, since $\theta'' > 0.1$, equivalently as

$$\sum_1^n x_i \geq \frac{\ln k + 1 - 10\theta''}{\ln 0.1 - \ln \theta''}.$$

Of course, $\sum_1^{10} x_i \geq 3$ is of the latter form. The statistic $Y = \sum_1^{10} X_i$ has a Poisson distribution with mean 10θ. Thus, with $\theta = 0.1$ so that the mean of Y is 1, the significance level of the test is

$$\Pr(Y \geq 3) = 1 - \Pr(Y \leq 2) = 1 - 0.920 = 0.080.$$

If the uniformly most powerful critical region defined by $\sum_1^{10} x_i \geq 4$ is used, the significance level is

$$\alpha = \Pr(Y \geq 4) = 1 - \Pr(Y \leq 3) = 1 - 0.981 = 0.019.$$

If a significance level of about $\alpha = 0.05$, say, is desired, most statisticians would use one of these tests; that is, they would adjust the significance level to that of one of these convenient tests. However, a significance level of $\alpha = 0.05$ can be achieved exactly by rejecting H_0 if $\sum_1^{10} x_i \geq 4$ or if $\sum_1^{10} x_i = 3$ and if an auxiliary independent random experiment resulted in "success," where the probability of success is selected to be equal to

$$\frac{0.050 - 0.019}{0.080 - 0.019} = \frac{31}{61}.$$

This is due to the fact that, when $\theta = 0.1$ so that the mean of Y is 1,

$$\Pr(Y \geq 4) + \Pr(Y = 3 \text{ and success}) = 0.019 + \Pr(Y = 3)\Pr(\text{success})$$
$$= 0.019 + (0.061)\tfrac{31}{61} = 0.05.$$

The process of performing the auxiliary experiment to decide whether to reject or not when $Y = 3$ is sometimes referred to as a *randomized test*.

Remarks. Not many statisticians like randomized tests in practice, because the use of them means that two statisticians could make the same assumptions, observe the same data, apply the same test, and yet make different decisions. Hence they usually adjust their significance level so as not to randomize. As a matter of fact, many statisticians report what are commonly called *p*-values. For illustrations, if in Example 4 the observed Y is $y = 4$, the *p*-value is 0.019; and if it is $y = 3$, the *p*-value is 0.080. That is, the *p*-value is the observed "tail" probability of a statistic being at least as extreme as the particular observed value when H_0 is true. Hence, more generally, if $Y = u(X_1, X_2, \ldots, X_n)$ is the statistic to be used in a test of H_0 and if a uniformly most powerful critical region is of the form

$$u(x_1, x_2, \ldots, x_n) \leq c,$$

an observed value $u(x_1, x_2, \ldots, x_n) = d$ would mean that the

$$p\text{-value} = \Pr(Y \leq d; H_0).$$

That is, if $G(y)$ is the distribution function of $Y = u(X_1, X_2, \ldots, X_n)$ provided that H_0 is true, the *p*-value is equal to $G(d)$ in this case. However, $G(Y)$, in the continuous case, is uniformly distributed on the unit interval, so an observed value $G(d) \leq 0.05$ would be equivalent to selecting c so that

$$\Pr\left[u(X_1, X_2, \ldots, X_n) \leq c; H_0\right] = 0.05$$

and observing that $d \leq c$.

There is a final remark that should be made about uniformly most powerful tests. Of course, in Definition 7, the word *uniformly* is associated with θ; that is, C is a best critical region of size α for testing $H_0: \theta = \theta_0$ against all θ values given by the composite alternative H_1. However, suppose that the form of such a region is

$$u(x_1, x_2, \ldots, x_n) \leq c.$$

Then this form provides uniformly most powerful critical regions for all attainable α values by, of course, appropriately changing the value of c. That is, there is a certain uniformity property, also associated with α, that is not always noted in statistics texts.

EXERCISES

7.16. Let X have the p.d.f. $f(x; \theta) = \theta^x(1 - \theta)^{1-x}$, $x = 0$, 1, zero elsewhere. We test the simple hypothesis $H_0: \theta = \frac{1}{4}$ against the alternative composite hypothesis $H_1: \theta < \frac{1}{4}$ by taking a random sample of size 10 and rejecting $H_0: \theta = \frac{1}{4}$ if and only if the observed values $x_1, x_2, \ldots, x_{10}$ of the sample items are such that $\sum_1^{10} x_i \leq 1$. Find the power function $K(\theta)$, $0 < \theta \leq \frac{1}{4}$, of this test.

7.17. Let X have a p.d.f. of the form $f(x; \theta) = 1/\theta$, $0 < x < \theta$, zero elsewhere. Let $Y_1 < Y_2 < Y_3 < Y_4$ denote the order statistics of a random sample of size 4 from this distribution. Let the observed value of Y_4 be y_4. We reject $H_0: \theta = 1$ and accept $H_1: \theta \neq 1$ if either $y_4 \leq \frac{1}{2}$ or $y_4 \geq 1$. Find the power function $K(\theta)$, $0 < \theta$, of the test.

7.18. Consider a normal distribution of the form $n(\theta, 4)$. The simple hypothesis $H_0: \theta = 0$ is rejected, and the alternative composite hypothesis $H_1: \theta > 0$ is accepted if and only if the observed mean $\bar{x}$ of a random sample of size 25 is greater than or equal to $\frac{3}{5}$. Find the power function $K(\theta)$, $0 \leq \theta$, of this test.

7.19. Consider the two independent normal distributions $n(\mu_1, 400)$ and $n(\mu_2, 225)$. Let $\theta = \mu_1 - \mu_2$. Let $\bar{x}$ and $\bar{y}$ denote the observed means of two independent random samples, each of size n, from these two distributions. We reject $H_0: \theta = 0$ and accept $H_1: \theta > 0$ if and only if $\bar{x} - \bar{y} \geq c$. If $K(\theta)$ is the power function of this test, find n and c so that $K(0) = 0.05$ and $K(10) = 0.90$, approximately.

7.20. If, in Example 2 of this section, $H_0: \theta = \theta'$, where θ' is a fixed positive number, and H_1: $\theta < \theta'$, show that the set $\{(x_1, x_2, \ldots, x_n); \sum_1^n x_i^2 \leq c\}$ is a uniformly most powerful critical region for testing H_0 against H_1.

7.21. If, in Example 2 of this section, H_0: $\theta = \theta'$, where θ' is a fixed positive number, and H_1: $\theta \neq \theta'$, show that there is no uniformly most powerful test for testing H_0 against H_1.

7.22. Let $X_1, X_2, \ldots, X_{25}$ denote a random sample of size 25 from a normal distribution $n(\theta, 100)$. Find a uniformly most powerful critical region of size $\alpha = 0.10$ for testing H_0: $\theta = 75$ against H_1: $\theta > 75$.

7.23. Let $X_1, X_2, \ldots, X_n$ denote a random sample from a normal distribution $n(\theta, 16)$. Find the sample size n and a uniformly most powerful test of H_0: $\theta = 25$ against H_1: $\theta < 25$ with power function $K(\theta)$ so that approximately $K(25) = 0.10$ and $K(23) = 0.90$.

7.24. Consider a distribution having a p.d.f. of the form $f(x; \theta) = \theta^x(1 - \theta)^{1-x}$, $x = 0, 1$, zero elsewhere. Let H_0: $\theta = \frac{1}{20}$ and H_1: $\theta > \frac{1}{20}$. Use the central limit theorem to determine the sample size n of a random sample so that a uniformly most powerful test of H_0 against H_1 has a power function $K(\theta)$, with approximately $K(\frac{1}{20}) = 0.05$ and $K(\frac{1}{10}) = 0.90$.

7.25. Illustrative Example 1 of this section dealt with a random sample of size $n = 2$ from a gamma distribution with $\alpha = 1$, $\beta = \theta$. Thus the moment-generating function of the distribution is $(1 - \theta t)^{-1}$ $t < 1/\theta$, $\theta \geq 2$. Let $Z = X_1 + X_2$. Show that Z has a gamma distribution with $\alpha = 2$, $\beta = \theta$. Express the power function $K(\theta)$ of Example 1 in terms of a single integral. Generalize this for a random sample of size n.

7.26. Let X have the p.d.f. $f(x; \theta) = \theta^x(1 - \theta)^{1-x}$, $x = 0, 1$, zero elsewhere. We test H_0: $\theta = \frac{1}{2}$ against H_1: $\theta < \frac{1}{2}$ by taking a random sample $X_1, X_2, \ldots, X_5$ of size $n = 5$ and rejecting H_0 if $Y = \sum_1^5 X_i$ is observed to be less than or equal to a constant c.

(a) Show that this is a uniformly most powerful test.

(b) Find the significance level when $c = 1$.

(c) Find the significance level when $c = 0$.

(d) By using a *randomized test*, modify the tests given in part (b) and part (c) to find a test with significance level $\alpha = \frac{2}{32}$.

7.4 Likelihood Ratio Tests

The notion of using the magnitude of the ratio of two probability density functions as the basis of a best test or of a uniformly most powerful test can be modified, and made intuitively appealing, to provide a method of constructing a test of a composite hypothesis against an alternative composite hypothesis or of constructing a test of a simple hypothesis against an alternative composite hypothesis when a uniformly most powerful test does not exist. This method leads to tests called *likelihood ratio tests*. A likelihood ratio test, as just remarked, is not necessarily a uniformly most powerful test, but it has been proved in the literature that such a test often has desirable properties.

A certain terminology and notation will be introduced by means of an example.

Example 1. Let the random variable X be $n(\theta_1, \theta_2)$ and let the parameter space be $\Omega = \{(\theta_1, \theta_2); -\infty < \theta_1 < \infty, 0 < \theta_2 < \infty\}$. Let the composite hypothesis be $H_0: \theta_1 = 0, \theta_2 > 0$, and let the alternative composite hypothesis be $H_1: \theta_1 \neq 0, \theta_2 > 0$. The set $\omega = \{(\theta_1, \theta_2); \theta_1 = 0, 0 < \theta_2 < \infty\}$ is a subset of Ω and will be called the *subspace* specified by the hypothesis H_0. Then, for instance, the hypothesis H_0 may be described as $H_0: (\theta_1, \theta_2) \in \omega$. It is proposed that we test H_0 against all alternatives in H_1.

Let $X_1, X_2, \ldots, X_n$ denote a random sample of size $n > 1$ from the distribution of this example. The joint p.d.f. of $X_1, X_2, \ldots, X_n$ is, at each point in Ω,

$$L(\theta_1, \theta_2; x_1, \ldots, x_n) = \left(\frac{1}{2\pi\theta_2}\right)^{n/2} \exp\left[-\frac{\sum\limits_{1}^{n}(x_i - \theta_1)^2}{2\theta_2}\right] = L(\Omega).$$

At each point $(\theta_1, \theta_2) \in \omega$, the joint p.d.f. of $X_1, X_2, \ldots, X_n$ is

$$L(0, \theta_2; x_1, \ldots, x_n) = \left(\frac{1}{2\pi\theta_2}\right)^{n/2} \exp\left[-\frac{\sum\limits_{1}^{n}x_i^2}{2\theta_2}\right] = L(\omega).$$

The joint p.d.f., now denoted by $L(\omega)$, is not completely specified, since θ_2 may be any positive number; nor is the joint p.d.f., now denoted by $L(\Omega)$, completely specified, since θ_1 may be any real number and θ_2 any positive number. Thus the ratio of $L(\omega)$ to $L(\Omega)$ could not provide a basis for a test of H_0 against H_1. Suppose, however, that we modify this ratio in the following manner. We shall find the maximum of $L(\omega)$ in ω, that is, the maximum of $L(\omega)$ with respect to θ_2. And we shall find the maximum of $L(\Omega)$ in Ω; that is, the maximum of $L(\Omega)$ with respect to θ_1 and θ_2. The ratio of these maxima will be taken as the criterion for a test of H_0 against H_1. Let the maximum of $L(\omega)$ in ω be denoted by $L(\hat{\omega})$ and let the maximum of $L(\Omega)$ in Ω be denoted by $L(\hat{\Omega})$. Then the criterion for the test of H_0 against H_1 is the likelihood ratio

$$\lambda(x_1, x_2, \ldots, x_n) = \lambda = \frac{L(\hat{\omega})}{L(\hat{\Omega})}.$$

Since $L(\omega)$ and $L(\Omega)$ are probability density functions, $\lambda \geq 0$; and since ω is a subset of Ω, $\lambda \leq 1$.

In our example the maximum, $L(\hat{\omega})$, of $L(\omega)$ is obtained by first setting

$$\frac{d \ln L(\omega)}{d\theta_2} = -\frac{n}{2\theta_2} + \frac{\sum\limits_{1}^{n} x_i^2}{2\theta_2^2}$$

equal to zero and solving for θ_2. The solution for θ_2 is $\sum\limits_{1}^{n} x_i^2/n$, and this number maximizes $L(\omega)$. Thus the maximum is

$$L(\hat{\omega}) = \left(\frac{1}{2\pi \sum\limits_{1}^{n} x_i^2/n}\right)^{n/2} \exp\left(-\frac{\sum\limits_{1}^{n} x_i^2}{2 \sum\limits_{1}^{n} x_i^2/n}\right)$$

$$= \left(\frac{ne^{-1}}{2\pi \sum\limits_{1}^{n} x_i^2}\right)^{n/2}.$$

On the other hand, by using Example 4, Section 6.1, the maximum, $L(\hat{\Omega})$, of $L(\Omega)$ is obtained by replacing θ_1 and θ_2 by $\sum\limits_{1}^{n} x_i/n = \bar{x}$ and $\sum\limits_{1}^{n} (x_i - \bar{x})^2/n$, respectively. That is,

$$L(\hat{\Omega}) = \left[\frac{1}{2\pi \sum\limits_{1}^{n} (x_i - \bar{x})^2/n}\right]^{n/2} \exp\left[-\frac{\sum\limits_{1}^{n} (x_i - \bar{x})^2}{2 \sum\limits_{1}^{n} (x_i - \bar{x})^2/n}\right]$$

$$= \left[\frac{ne^{-1}}{2\pi \sum\limits_{1}^{n} (x_i - \bar{x})^2}\right]^{n/2}.$$

Thus here

$$\lambda = \left[\frac{\sum\limits_{1}^{n} (x_i - \bar{x})^2}{\sum\limits_{1}^{n} x_i^2}\right]^{n/2}.$$

Because $\sum\limits_{1}^{n} x_i^2 = \sum\limits_{1}^{n} (x_i - \bar{x})^2 + n\bar{x}^2$, λ may be written

$$\lambda = \frac{1}{\left\{1 + \left[n\bar{x}^2 / \sum\limits_{1}^{n} (x_i - \bar{x})^2\right]\right\}^{n/2}}.$$

Now the hypothesis H_0 is $\theta_1 = 0$, $\theta_2 > 0$. If the observed number $\bar{x}$ were zero, the experiment tends to confirm H_0. But if $\bar{x} = 0$ and $\sum\limits_{1}^{n} x_i^2 > 0$, then

$\lambda = 1$. On the other hand, if $\bar{x}$ and $n\bar{x}^2/\sum_{1}^{n}(x_i - \bar{x})^2$ deviate considerably from zero, the experiment tends to negate H_0. Now the greater the deviation of $n\bar{x}^2/\sum_{1}^{n}(x_i - \bar{x})^2$ from zero, the smaller λ becomes. That is, if λ is used as a test criterion, then an intuitively appealing critical region for testing H_0 is a set defined by $0 \leq \lambda \leq \lambda_0$, where λ_0 is a positive proper fraction. Thus we reject H_0 if $\lambda \leq \lambda_0$. A test that has the critical region $\lambda \leq \lambda_0$ is a *likelihood ratio test*. In this example $\lambda \leq \lambda_0$ when and only when

$$\frac{\sqrt{n}\,|\bar{x}|}{\sqrt{\sum_{1}^{n}(x_i - \bar{x})^2/(n-1)}} \geq \sqrt{(n-1)(\lambda_0^{-2/n} - 1)} = c.$$

If $H_0: \theta_1 = 0$ is true, the results in Section 6.3 show that the statistic

$$t(X_1, X_2, \ldots, X_n) = \frac{\sqrt{n}\,(\bar{X} - 0)}{\sqrt{\sum_{1}^{n}(X_i - \bar{X})^2/(n-1)}}$$

has a t distribution with $n-1$ degrees of freedom. Accordingly, in this example the likelihood ratio test of H_0 against H_1 may be based on a T statistic. For a given positive integer n, Table IV in Appendix B may be used (with $n-1$ degrees of freedom) to determine the number c such that $\alpha = \Pr\left[|t(X_1, X_2, \ldots, X_n)| \geq c; H_0\right]$ is the desired significance level of the test. If the experimental values of $X_1, X_2, \ldots, X_n$ are, respectively, $x_1, x_2, \ldots, x_n$, then we reject H_0 if and only if $|t(x_1, x_2, \ldots, x_n)| \geq c$. If, for instance, $n = 6$ and $\alpha = 0.05$, then from Table IV, $c = 2.571$.

The preceding example should make the following generalization easier to read: Let $X_1, X_2, \ldots, X_n$ denote n mutually stochastically independent random variables having, respectively, the probability density functions $f_i(x_i; \theta_1, \theta_2, \ldots, \theta_m)$, $i = 1, 2, \ldots, n$. The set that consists of all parameter points $(\theta_1, \theta_2, \ldots, \theta_m)$ is denoted by Ω, which we have called the parameter space. Let ω be a subset of the parameter space Ω. We wish to test the (simple or composite) hypothesis $H_0: (\theta_1, \theta_2, \ldots, \theta_m) \in \omega$ against all alternative hypotheses. Define the likelihood functions

$$L(\omega) = \prod_{i=1}^{n} f_i(x_i; \theta_1, \theta_2, \ldots, \theta_m), \qquad (\theta_1, \theta_2, \ldots, \theta_m) \in \omega,$$

and

$$L(\Omega) = \prod_{i=1}^{n} f_i(x_i; \theta_1, \theta_2, \ldots, \theta_m), \qquad (\theta_1, \theta_2, \ldots, \theta_m) \in \Omega.$$

Let $L(\hat{\omega})$ and $L(\hat{\Omega})$ be the maxima, which we assume to exist, of these two likelihood functions. The ratio of $L(\hat{\omega})$ to $L(\hat{\Omega})$ is called the *likelihood ratio* and is denoted by

$$\lambda(x_1, x_2, \ldots, x_n) = \lambda = \frac{L(\hat{\omega})}{L(\hat{\Omega})}.$$

Let λ_0 be a positive proper function. The *likelihood ratio test principle* states that the hypothesis $H_0: (\theta_1, \theta_2, \ldots, \theta_m) \in \omega$ is rejected if and only if

$$\lambda(x_1, x_2, \ldots, x_n) = \lambda \le \lambda_0.$$

The function λ defines a random variable $\lambda(X_1, X_2, \ldots, X_n)$, and the significance level of the test is given by

$$\alpha = \Pr\left[\lambda(X_1, X_2, \ldots, X_n) \le \lambda_0; H_0\right].$$

The likelihood ratio test principle is an intuitive one. However, the principle does lead to the same test, when testing a simple hypothesis H_0 against an alternative simple hypothesis H_1, as that given by the Neyman–Pearson theorem (Exercise 7.29). Thus it might be expected that a test based on this principle has some desirable properties.

An example of the preceding generalization will be given.

Example 2. Let the stochastically independent random variables X and Y have distributions that are $n(\theta_1, \theta_3)$ and $n(\theta_2, \theta_3)$, where the means θ_1 and θ_2 and common variance θ_3 are unknown. Then $\Omega = \{(\theta_1, \theta_2, \theta_3); -\infty < \theta_1 < \infty, -\infty < \theta_2 < \infty, 0 < \theta_3 < \infty\}$. Let $X_1, X_2, \ldots, X_n$ and $Y_1, Y_2, \ldots, Y_m$ denote independent random samples from these distributions. The hypothesis $H_0: \theta_1 = \theta_2$, unspecified, and θ_3 unspecified, is to be tested against all alternatives. Then $\omega = \{(\theta_1, \theta_2, \theta_3); -\infty < \theta_1 = \theta_2 < \infty, 0 < \theta_3 < \infty\}$. Here $X_1, X_2, \ldots, X_n, Y_1, Y_2, \ldots, Y_m$ are $n + m > 2$ mutually stochastically independent random variables having the likelihood functions

$$L(\omega) = \left(\frac{1}{2\pi\theta_3}\right)^{(n+m)/2} \exp\left[-\frac{\sum_1^n (x_i - \theta_1)^2 + \sum_1^m (y_i - \theta_1)^2}{2\theta_3}\right]$$

and

$$L(\Omega) = \left(\frac{1}{2\pi\theta_3}\right)^{(n+m)/2} \exp\left[-\frac{\sum_1^n (x_i - \theta_1)^2 + \sum_1^m (y_i - \theta_2)^2}{2\theta_3}\right].$$

If

$$\frac{\partial \ln L(\omega)}{\partial \theta_1} \quad \text{and} \quad \frac{\partial \ln L(\omega)}{\partial \theta_3}$$

are equated to zero, then (Exercise 7.30)

$$\sum_1^n (x_i - \theta_1) + \sum_1^m (y_i - \theta_1) = 0,$$

(1)

$$-(n + m) + \frac{1}{\theta_3}\left[\sum_1^n (x_i - \theta_1)^2 + \sum_1^m (y_i - \theta_1)^2\right] = 0.$$

The solutions for θ_1 and θ_3 are, respectively,

$$u = \frac{\sum_1^n x_i + \sum_1^m y_i}{n + m}$$

and

$$w = \frac{\sum_1^n (x_i - u)^2 + \sum_1^m (y_i - u)^2}{n + m},$$

and u and w maximize $L(\omega)$. The maximum is

$$L(\hat{\omega}) = \left(\frac{e^{-1}}{2\pi w}\right)^{(n+m)/2}$$

In like manner, if

$$\frac{\partial \ln L(\Omega)}{\partial \theta_1}, \qquad \frac{\partial \ln L(\Omega)}{\partial \theta_2}, \qquad \frac{\partial \ln L(\Omega)}{\partial \theta_3}$$

are equated to zero, then (Exercise 7.31)

$$\sum_1^n (x_i - \theta_1) = 0,$$

(2)

$$\sum_1^m (y_i - \theta_2) = 0,$$

$$-(n + m) + \frac{1}{\theta_3}\left[\sum_1^n (x_i - \theta_1)^2 + \sum_1^m (y_i - \theta_2)^2\right] = 0.$$

The solutions for θ_1, θ_2, and θ_3 are, respectively,

$$u_1 = \frac{\sum_1^n x_i}{n},$$

$$u_2 = \frac{\sum_1^m y_i}{m},$$

$$w' = \frac{\sum_1^n (x_i - u_1)^2 + \sum_1^m (y_i - u_2)^2}{n + m},$$

and u_1, u_2, and w' maximize $L(\Omega)$. The maximum is

$$L(\hat{\Omega}) = \left(\frac{e^{-1}}{2\pi w'}\right)^{(n+m)/2},$$

so that

$$\lambda(x_1, \ldots, x_n, y_1, \ldots, y_m) = \lambda = \frac{L(\hat{\omega})}{L(\hat{\Omega})} = \left(\frac{w'}{w}\right)^{(n+m)/2}.$$

The random variable defined by $\lambda^{2/(n+m)}$ is

$$\frac{\sum_1^n (X_i - \bar{X})^2 + \sum_1^m (Y_i - \bar{Y})^2}{\sum_1^n \{X_i - [(n\bar{X} + m\bar{Y})/(n + m)]\}^2 + \sum_1^m \{Y_i - [(n\bar{X} + m\bar{Y})/(n + m)]\}^2}.$$

Now

$$\sum_1^n \left(X_i - \frac{n\bar{X} + m\bar{Y}}{n + m}\right)^2 = \sum_1^n \left[(X_i - \bar{X}) + \left(\bar{X} - \frac{n\bar{X} + m\bar{Y}}{n + m}\right)\right]^2$$

$$= \sum_1^n (X_i - \bar{X})^2 + n\left(\bar{X} - \frac{n\bar{X} + m\bar{Y}}{n + m}\right)^2$$

and

$$\sum_1^m \left(Y_i - \frac{n\bar{X} + m\bar{Y}}{n + m}\right)^2 = \sum_1^m \left[(Y_i - \bar{Y}) + \left(\bar{Y} - \frac{n\bar{X} + m\bar{Y}}{n + m}\right)\right]^2$$

$$= \sum_1^m (Y_i - \bar{Y})^2 + m\left(\bar{Y} - \frac{n\bar{X} + m\bar{Y}}{n + m}\right)^2.$$

But

$$n\left(\bar{X} - \frac{n\bar{X} + m\bar{Y}}{n + m}\right)^2 = \frac{m^2 n}{(n + m)^2}(\bar{X} - \bar{Y})^2$$

and

$$m\left(\bar{Y} - \frac{n\bar{X} + m\bar{Y}}{n + m}\right)^2 = \frac{n^2 m}{(n + m)^2}(\bar{X} - \bar{Y})^2.$$

Hence the random variable defined by $\lambda^{2/(n+m)}$ may be written

$$\frac{\sum_1^n (X_i - \bar{X})^2 + \sum_1^m (Y_i - \bar{Y})^2}{\sum_1^n (X_i - \bar{X})^2 + \sum_1^m (Y_i - \bar{Y})^2 + [nm/(n + m)](\bar{X} - \bar{Y})^2}$$

$$= \frac{1}{1 + \dfrac{[nm/(n + m)](\bar{X} - \bar{Y})^2}{\sum_1^n (X_i - \bar{X})^2 + \sum_1^m (Y_i - \bar{Y})^2}}.$$

If the hypothesis $H_0: \theta_1 = \theta_2$ is true, the random variable

$$T = \frac{\sqrt{\dfrac{nm}{n+m}}\,(\bar{X} - \bar{Y})}{\sqrt{\dfrac{\sum\limits_1^n (X_i - \bar{X})^2 + \sum\limits_1^m (Y_i - \bar{Y})^2}{n+m-2}}}$$

has, in accordance with Section 6.4, a t distribution with $n + m - 2$ degrees of freedom. Thus the random variable defined by $\lambda^{2/(n+m)}$ is

$$\frac{n+m-2}{(n+m-2)+T^2}.$$

The test of H_0 against all alternatives may then be based on a t distribution with $n + m - 2$ degrees of freedom.

The likelihood ratio principle calls for the rejection of H_0 if and only if $\lambda \le \lambda_0 < 1$. Thus the significance level of the test is

$$\alpha = \Pr\left[(\lambda(X_1, \ldots, X_n, Y_1, \ldots, Y_m) \le \lambda_0; H_0\right].$$

However, $\lambda(X_1, \ldots, X_n, Y_1, \ldots, Y_m) \le \lambda_0$ is equivalent to $|T| \ge c$, and so

$$\alpha = \Pr\left(|T| \ge c; H_0\right).$$

For given values of n and m, the number c is determined from Table IV in the Appendix (with $n + m - 2$ degrees of freedom) in such a manner as to yield a desired α. Then H_0 is rejected at a significance level α if and only if $|t| \ge c$, where t is the experimental value of T. If, for instance, $n = 10$, $m = 6$, and $\alpha = 0.05$, then $c = 2.145$.

In each of the two examples of this section it was found that the likelihood ratio test could be based on a statistic which, when the hypothesis H_0 is true, has a t distribution. To help us compute the powers of these tests at parameter points other than those described by the hypothesis H_0, we turn to the following definition.

Definition 8. Let the random variable W be $n(\delta, 1)$; let the random variable V be $\chi^2(r)$, and W and V be stochastically independent. The quotient

$$T = \frac{W}{\sqrt{V/r}}$$

is said to have a *noncentral t distribution* with r degrees of freedom and noncentrality parameter δ. If $\delta = 0$, we say that T has a central t distribution.

In the light of this definition, let us reexamine the statistics of the examples of this section. In Example 1 we had

$$t(X_1, \ldots, X_n) = \frac{\sqrt{n}\,\bar{X}}{\sqrt{\sum_1^n (X_i - \bar{X})^2 / (n - 1)}}$$

$$= \frac{\sqrt{n}\,\bar{X}/\sigma}{\sqrt{\sum_1^n (X_i - \bar{X})^2 / [\sigma^2(n - 1)]}}.$$

Here $W_1 = \sqrt{n}\,\bar{X}/\sigma$ is $n(\sqrt{n}\theta_1/\sigma, 1)$, $V_1 = \sum_1^n (X_i - \bar{X})^2/\sigma^2$ is $\chi^2(n - 1)$, and W_1 and V_1 are stochastically independent. Thus, if $\theta_1 \neq 0$, we see, in accordance with the definition, that $t(X_1, \ldots, X_n)$ has a noncentral t distribution with $n - 1$ degrees of freedom and noncentrality parameter $\delta_1 = \sqrt{n}\theta_1/\sigma$. In Example 2 we had

$$T = \frac{W_2}{\sqrt{V_2/(n + m - 2)}},$$

where

$$W_2 = \sqrt{\frac{nm}{n + m}}\, (\bar{X} - \bar{Y})\Big/\sigma$$

and

$$V_2 = \left[\sum_1^n (X_i - \bar{X})^2 + \sum_1^m (Y_i - \bar{Y})^2\right]\Big/\sigma^2.$$

Here W_2 is $n[\sqrt{nm/(n + m)}(\theta_1 - \theta_2)/\sigma, 1]$, V_2 is $\chi^2(n + m - 2)$, and W_2 and V_2 are stochastically independent. Accordingly, if $\theta_1 \neq \theta_2$, T has a noncentral t distribution with $n + m - 2$ degrees of freedom and noncentrality parameter $\delta_2 = \sqrt{nm/(n + m)}(\theta_1 - \theta_2)/\sigma$. It is interesting to note that $\delta_1 = \sqrt{n}\theta_1/\sigma$ measures the deviation of θ_1 from $\theta_1 = 0$ in units of the standard deviation $\sigma/\sqrt{n}$ of $\bar{X}$. The noncentrality parameter $\delta_2 = \sqrt{nm/(n + m)}(\theta_1 - \theta_2)/\sigma$ is equal to the deviation of $\theta_1 - \theta_2$ from $\theta_1 - \theta_2 = 0$ in units of the standard deviation $\sigma\sqrt{(n + m)/nm}$ of $\bar{X} - \bar{Y}$.

There are various tables of the noncentral t distribution, but they are much too cumbersome to be included in this book. However, with the aid of such tables, we can determine the power functions of these tests as functions of the noncentrality parameters.

In Example 2, in testing the equality of the means of two independent normal distributions, it was assumed that the unknown variances of the distributions were equal. Let us now consider the problem of testing the equality of these two unknown variances.

Example 3. We are given the stochastically independent random samples $X_1, \ldots, X_n$ and $Y_1, \ldots, Y_m$ from the independent distributions, which are $n(\theta_1, \theta_3)$ and $n(\theta_2, \theta_4)$, respectively. We have

$$\Omega = \{(\theta_1, \theta_2, \theta_3, \theta_4); -\infty < \theta_1, \theta_2 < \infty, 0 < \theta_3, \theta_4 < \infty\}.$$

The hypothesis $H_0: \theta_3 = \theta_4$, unspecified, with θ_1 and θ_2 also unspecified, is to be tested against all alternatives. Then

$$\omega = \{(\theta_1, \theta_2, \theta_3, \theta_4); -\infty < \theta_1, \theta_2 < \infty, 0 < \theta_3 = \theta_4 < \infty\}.$$

It is easy to show (see Exercise 7.34) that the statistic defined by $\lambda = L(\hat{\omega})/L(\hat{\Omega})$ is a function of the statistic

$$F = \frac{\sum\limits_1^n (X_i - \bar{X})^2/(n-1)}{\sum\limits_1^m (Y_i - \bar{Y})^2/(m-1)}.$$

If $\theta_3 = \theta_4$, this statistic F has an F distribution with $n-1$ and $m-1$ degrees of freedom. The hypothesis that $(\theta_1, \theta_2, \theta_3, \theta_4) \in \omega$ is rejected if the computed $F \leq c_1$ or if the computed $F \geq c_2$. The constants c_1 and c_2 are usually selected so that, if $\theta_3 = \theta_4$,

$$\Pr(F \leq c_1) = \Pr(F \geq c_2) = \frac{\alpha_1}{2},$$

where α_1 is the desired significance level of this test.

EXERCISES

7.27. In Example 1 let $n = 10$, and let the experimental values of the random variables yield $\bar{x} = 0.6$ and $\sum\limits_1^{10} (x_i - \bar{x})^2 = 3.6$. If the test derived in that example is used, do we accept or reject $H_0: \theta_1 = 0$ at the 5 per cent significance level?

7.28. In Example 2 let $n = m = 8$, $\bar{x} = 75.2$, $\bar{y} = 78.6$, $\sum\limits_1^8 (x_i - \bar{x})^2 = 71.2$, $\sum\limits_1^8 (y_i - \bar{y})^2 = 54.8$. If we use the test derived in that example, do we accept or reject $H_0: \theta_1 = \theta_2$ at the 5 per cent significance level?

7.29. Show that the likelihood ratio principle leads to the same test, when testing a simple hypothesis H_0 against an alternative simple hypothesis H_1, as that given by the Neyman–Pearson theorem. Note that there are only two points in Ω.

7.30. Verify Equations (1) of Example 2 of this section.

7.31. Verify Equations (2) of Example 2 of this section.

7.32. Let $X_1, X_2, \ldots, X_n$ be a random sample from the normal distribution $n(\theta, 1)$. Show that the likelihood ratio principle for testing $H_0: \theta = \theta'$, where θ' is specified, against $H_1: \theta \neq \theta'$ leads to the inequality $|\bar{x} - \theta'| \geq c$. Is this a uniformly most powerful test of H_0 against H_1?

7.33. Let $X_1, X_2, \ldots, X_n$ be a random sample from the normal distribution $n(\theta_1, \theta_2)$. Show that the likelihood ratio principle for testing $H_0: \theta_2 = \theta_2'$ specified, and θ_1 unspecified, against $H_1: \theta_2 \neq \theta_2'$, θ_1 unspecified, leads to a test that rejects when $\sum_1^n (x_i - \bar{x})^2 \leq c_1$ or $\sum_1^n (x_i - \bar{x})^2 \geq c_2$, where $c_1 < c_2$ are selected appropriately.

7.34. Let $X_1, \ldots, X_n$ and $Y_1, \ldots, Y_m$ be random samples from the independent distributions $n(\theta_1, \theta_3)$ and $n(\theta_2, \theta_4)$, respectively.

(a) Show that the likelihood ratio for testing $H_0: \theta_1 = \theta_2, \theta_3 = \theta_4$ against all alternatives is given by

$$\frac{\left[\sum_1^n (x_i - \bar{x})^2/n\right]^{n/2} \left[\sum_1^m (y_i - \bar{y})^2/m\right]^{m/2}}{\left\{\left[\sum_1^n (x_i - u)^2 + \sum_1^m (y_i - u)^2\right]/(m+n)\right\}^{(n+m)/2}},$$

where $u = (n\bar{x} + m\bar{y})/(n + m)$.

(b) Show that the likelihood ratio test for testing $H_0: \theta_3 = \theta_4$, θ_1 and θ_2 unspecified, against $H_1: \theta_3 \neq \theta_4$, θ_1 and θ_2 unspecified, can be based on the random variable

$$F = \frac{\sum_1^n (X_i - X)^2/(n-1)}{\sum_1^m (Y_i - Y)^2/(m-1)}.$$

7.35. Let n independent trials of an experiment be such that $x_1, x_2, \ldots,$ x_k are the respective numbers of times that the experiment ends in the mutually exclusive and exhaustive events $A_1, A_2, \ldots, A_k$. If $p_i = P(A_i)$ is constant throughout the n trials, then the probability of that particular sequence of trials is $L = p_1^{x_1} p_2^{x_2} \cdots p_k^{x_k}$.

(a) Recalling that $p_1 + p_2 + \cdots + p_k = 1$, show that the likelihood ratio for testing $H_0: p_i = p_{i0} > 0$, $i = 1, 2, \ldots, k$, against all alternatives is given by

$$\lambda = \prod_{i=1}^k \left(\frac{(p_{i0})^{x_i}}{(x_i/n)^{x_i}}\right).$$

(b) Show that

$$-2 \ln \lambda = \sum_{i=1}^{k} \frac{x_i (x_i - np_{0i})^2}{(np_i')^2}$$

where p_i' is between p_{0i} and x_i/n. *Hint.* Expand $\ln p_{i0}$ in a Taylor's series with the remainder in the term involving $(p_{i0} - x_i/n)^2$.

(c) For large n, argue that $x_i/(np_i')^2$ is approximated by $1/(np_{i0})$ and hence

$$-2 \ln \lambda \approx \sum_{i=1}^{k} \frac{(x_i - np_{0i})^2}{np_{0i}}, \text{ when } H \text{ is true.}$$

Chapter *8*

Other Statistical Tests

8.1 Chi-Square Tests

In this section we introduce tests of statistical hypotheses called *chi-square tests*. A test of this sort was originally proposed by Karl Pearson in 1900, and it provided one of the earlier methods of statistical inference.

Let the random variable X_i be $n(\mu_i, \sigma_i^2)$, $i = 1, 2, \ldots, n$, and let $X_1, X_2, \ldots, X_n$ be mutually stochastically independent. Thus the joint p.d.f. of these variables is

$$\frac{1}{\sigma_1 \sigma_2 \cdots \sigma_n (2\pi)^{n/2}} \exp\left[-\frac{1}{2} \sum_1^n \left(\frac{x_i - \mu_i}{\sigma_i} \right)^2 \right], \qquad -\infty < x_i < \infty.$$

The random variable that is defined by the exponent (apart from the coefficient $-\frac{1}{2}$) is $\sum_1^n (X_i - \mu_i)^2 / \sigma_i^2$, and this random variable is $\chi^2(n)$. In Chapter 12 we shall generalize this joint normal distribution of probability to n random variables that are stochastically *dependent* and we shall call the distribution a *multivariate normal distribution*. It will then be shown that a certain exponent in the joint p.d.f. (apart from a coefficient of $-\frac{1}{2}$) defines a random variable that is $\chi^2(n)$. This fact is the mathematical basis of the chi-square tests.

Let us now discuss some random variables that have approximate chi-square distributions. Let X_1 be $b(n, p_1)$. Since the random variable $Y = (X_1 - np_1)/\sqrt{np_1(1 - p_1)}$ has, as $n \to \infty$, a limiting distribution that is $n(0, 1)$, we would strongly suspect that the limiting distribution of $Z = Y^2$ is $\chi^2(1)$. This is, in fact, the case, as will now be shown. If $G_n(y)$ represents the distribution function of Y, we know that

$$\lim_{n \to \infty} G_n(y) = N(y), \qquad -\infty < y < \infty,$$

where $N(y)$ is the distribution function of a distribution that is $n(0, 1)$. Let $H_n(z)$ represent, for each positive integer n, the distribution function of $Z = Y^2$. Thus, if $z \geq 0$,

$$H_n(z) = \Pr (Z \leq z) = \Pr (-\sqrt{z} \leq Y \leq \sqrt{z})$$
$$= G_n(\sqrt{z}) - G_n[(-\sqrt{z}) -].$$

Accordingly, since $N(y)$ is everywhere continuous,

$$\lim_{n \to \infty} H_n(z) = N(\sqrt{z}) - N(-\sqrt{z})$$

$$= 2 \int_0^{\sqrt{z}} \frac{1}{\sqrt{2\pi}} e^{-w^2/2} \, dw.$$

If we change the variable of integration in this last integral by writing $w^2 = v$, then

$$\lim_{n \to \infty} H_n(z) = \int_0^z \frac{1}{\Gamma(\frac{1}{2})2^{1/2}} v^{1/2 - 1} e^{-v/2} \, dv,$$

provided that $z \geq 0$. If $z < 0$, then $\lim_{n \to \infty} H_n(z) = 0$. Thus $\lim_{n \to \infty} H_n(z)$ is equal to the distribution function of a random variable that is $\chi^2(1)$. This is the desired result.

Let us now return to the random variable X_1 which is $b(n, p_1)$. Let $X_2 = n - X_1$ and let $p_2 = 1 - p_1$. If we denote Y^2 by Q_1 instead of Z, we see that Q_1 may be written as

$$Q_1 = \frac{(X_1 - np_1)^2}{np_1(1 - p_1)} = \frac{(X_1 - np_1)^2}{np_1} + \frac{(X_1 - np_1)^2}{n(1 - p_1)}$$

$$= \frac{(X_1 - np_1)^2}{np_1} + \frac{(X_2 - np_2)^2}{np_2}$$

because $(X_1 - np_1)^2 = (n - X_2 - n + np_2)^2 = (X_2 - np_2)^2$. Since Q_1 has a limiting chi-square distribution with 1 degree of freedom, we say, when n is a positive integer, that Q_1 has an approximate chi-square distribution with 1 degree of freedom. This result can be generalized as follows.

Let $X_1, X_2, \ldots, X_{k-1}$ have a multinomial distribution with parameters $n, p_1, \ldots, p_{k-1}$, as in Section 3.1. As a convenience, let $X_k =$

$n - (X_1 + \cdots + X_{k-1})$ and let $p_k = 1 - (p_1 + \cdots + p_{k-1})$. Define Q_{k-1} by

$$Q_{k-1} = \sum_{i=1}^{k} \frac{(X_i - np_i)^2}{np_i}.$$

It is proved in a more advanced course that, as $n \to \infty$, Q_{k-1} has a limiting distribution that is $\chi^2(k - 1)$. If we accept this fact, we can say that Q_{k-1} has an approximate chi-square distribution with $k - 1$ degrees of freedom when n is a positive integer. Some writers caution the user of this approximation to be certain that n is large enough that each np_i, $i = 1, 2, \ldots, k$, is at least equal to 5. In any case it is important to realize that Q_{k-1} does not have a chi-square distribution, only an approximate chi-square distribution.

The random variable Q_{k-1} may serve as the basis of the tests of certain statistical hypotheses which we now discuss. Let the sample space $\mathscr{A}$ of a random experiment be the union of a finite number k of mutually disjoint sets $A_1, A_2, \ldots, A_k$. Furthermore, let $P(A_i) = p_i$, $i = 1, 2, \ldots, k$, where $p_k = 1 - p_1 - \cdots - p_{k-1}$, so that p_i is the probability that the outcome of the random experiment is an element of the set A_i. The random experiment is to be repeated n independent times and X_i will represent the number of times the outcome is an element of the set A_i. That is, $X_1, X_2, \ldots, X_k = n - X_1 - \cdots - X_{k-1}$ are the frequencies with which the outcome is, respectively, an element of $A_1, A_2, \ldots, A_k$. Then the joint p.d.f. of $X_1, X_2, \ldots, X_{k-1}$ is the multinomial p.d.f. with the parameters $n, p_1, \ldots, p_{k-1}$. Consider the simple hypothesis (concerning this multinomial p.d.f.) $H_0: p_1 = p_{10}$, $p_2 = p_{20}, \ldots, p_{k-1} = p_{k-1,0}$ $(p_k = p_{k0} = 1 - p_{10} - \cdots - p_{k-1,0})$, where $p_{10}, \ldots, p_{k-1,0}$ are specified numbers. It is desired to test H_0 against all alternatives.

If the hypothesis H_0 is true, the random variable

$$Q_{k-1} = \sum_{1}^{k} \left[\frac{(X_i - np_{i0})^2}{np_{i0}} \right]$$

has an approximate chi-square distribution with $k - 1$ degrees of freedom. Since, when H_0 is true, np_{i0} is the expected value of X_i, one would feel intuitively that experimental values of Q_{k-1} should not be too large if H_0 is true. With this in mind, we may use Table II of the Appendix, with $k - 1$ degrees of freedom, and find c so that $\Pr(Q_{k-1} \geq c) = \alpha$, where α is the desired significance level of the test. If, then, the hypothesis H_0 is rejected when the observed value of Q_{k-1} is at least as great as c, the test of H_0 will have a significance level that is approximately equal to α.

Some illustrative examples follow.

Example 1. One of the first six positive integers is to be chosen by a random experiment (perhaps by the cast of a die). Let $A_i = \{x; x = i\}$, $i = 1, 2, \ldots, 6$. The hypothesis H_0: $P(A_i) = p_{i0} = \frac{1}{6}$, $i = 1, 2, \ldots, 6$, will be tested, at the approximate 5 per cent significance level, against all alternatives. To make the test, the random experiment will be repeated, under the same conditions, 60 independent times. In this example $k = 6$ and $np_{i0} = 60(\frac{1}{6}) = 10$, $i = 1, 2, \ldots, 6$. Let X_i denote the frequency with which the random experiment terminates with the outcome in A_i, $i = 1, 2, \ldots, 6$, and let $Q_5 = \sum_1^6 (X_i - 10)^2/10$. If H_0 is true, Table II, with $k - 1 = 6 - 1 = 5$ degrees of freedom, shows that we have $\Pr(Q_5 \geq 11.1) = 0.05$. Now suppose that the experimental frequencies of $A_1, A_2, \ldots, A_6$ are, respectively, 13, 19, 11, 8, 5, and 4. The observed value of Q_5 is

$$\frac{(13 - 10)^2}{10} + \frac{(19 - 10)^2}{10} + \frac{(11 - 10)^2}{10}$$

$$+ \frac{(8 - 10)^2}{10} + \frac{(5 - 10)^2}{10} + \frac{(4 - 10)^2}{10} = 15.6.$$

Since $15.6 > 11.1$, the hypothesis $P(A_i) = \frac{1}{6}$, $i = 1, 2, \ldots, 6$, is rejected at the (approximate) 5 per cent significance level.

Example 2. A point is to be selected from the unit interval $\{x; 0 < x < 1\}$ by a random process. Let $A_1 = \{x; 0 < x \leq \frac{1}{4}\}$, $A_2 = \{x; \frac{1}{4} < x \leq \frac{1}{2}\}$, $A_3 = \{x; \frac{1}{2} < x \leq \frac{3}{4}\}$, and $A_4 = \{x; \frac{3}{4} < x < 1\}$. Let the probabilities p_i, $i = 1, 2, 3, 4$, assigned to these sets under the hypothesis be determined by the p.d.f. $2x$, $0 < x < 1$, zero elsewhere. Then these probabilities are, respectively,

$$p_{10} = \int_0^{1/4} 2x\, dx = \tfrac{1}{16}, \qquad p_{20} = \tfrac{3}{16}, \qquad p_{30} = \tfrac{5}{16}, \qquad p_{40} = \tfrac{7}{16}.$$

Thus the hypothesis to be tested is that p_1, p_2, p_3, and $p_4 = 1 - p_1 - p_2 - p_3$ have the preceding values in a multinomial distribution with $k = 4$. This hypothesis is to be tested at an approximate 0.025 significance level by repeating the random experiment $n = 80$ independent times under the same conditions. Here the np_{i0}, $i = 1, 2, 3, 4$, are, respectively, 5, 15, 25, and 35. Suppose the observed frequencies of A_1, A_2, A_3, and A_4 to be 6, 18, 20, and 36, respectively. Then the observed value of $Q_3 = \sum_1^4 (X_i - np_{i0})^2/(np_{i0})$ is

$$\frac{(6 - 5)^2}{5} + \frac{(18 - 15)^2}{15} + \frac{(20 - 25)^2}{25} + \frac{(36 - 35)^2}{35} \doteq \frac{64}{35} = 1.83,$$

approximately. From Table II, with $4 - 1 = 3$ degrees of freedom, the value corresponding to a 0.025 significance level is $c = 9.35$. Since the observed value of Q_3 is less than 9.35, the hypothesis is accepted at the (approximate) 0.025 level of significance.

Thus far we have used the chi-square test when the hypothesis H_0 is a simple hypothesis. More often we encounter hypotheses H_0 in which the multinomial probabilities $p_1, p_2, \ldots, p_k$ are not completely specified by the hypothesis H_0. That is, under H_0, these probabilities are functions of unknown parameters. For illustration, suppose that a certain random variable Y can take on any real value. Let us partition the space $\{y; -\infty < y < \infty\}$ into k mutually disjoint sets $A_1, A_2, \ldots, A_k$ so that the events $A_1, A_2, \ldots, A_k$ are mutually exclusive and exhaustive. Let H_0 be the hypothesis that Y is $n(\mu, \sigma^2)$ with μ and σ^2 unspecified. Then each

$$p_i = \int_{A_i} \frac{1}{\sqrt{2\pi}\sigma} \exp\left[-(y - \mu)^2/2\sigma^2\right] dy, \qquad i = 1, 2, \ldots, k,$$

is a function of the unknown parameters μ and σ^2. Suppose that we take a random sample $Y_1, \ldots, Y_n$ of size n from this distribution. If we let X_i denote the frequency of A_i, $i = 1, 2, \ldots, k$, so that $X_1 + \cdots + X_k = n$, the random variable

$$Q_{k-1} = \sum_{i=1}^{k} \frac{(X_i - np_i)^2}{np_i}$$

cannot be computed once $X_1, \ldots, X_k$ have been observed, since each p_i, and hence Q_{k-1}, is a function of the unknown parameters μ and σ^2.

There is a way out of our trouble, however. We have noted that Q_{k-1} is a function of μ and σ^2. Accordingly, choose the values of μ and σ^2 that minimize Q_{k-1}. Obviously, these values depend upon the observed $X_1 = x_1, \ldots, X_k = x_k$ and are called *minimum chi-square estimates* of μ and σ^2. These point estimates of μ and σ^2 enable us to compute numerically the estimates of each p_i. Accordingly, if these values are used, Q_{k-1} can be computed once $Y_1, Y_2, \ldots, Y_n$, and hence $X_1, X_2, \ldots, X_k$, are observed. However, a very important aspect of the fact, which we accept without proof, is that now Q_{k-1} is approximately $\chi^2(k - 3)$. That is, the number of degrees of freedom of the limiting chi-square distribution of Q_{k-1} is reduced by one for each parameter estimated by the experimental data. This statement applies not only to the problem at hand but also to more general situations. Two examples will now be given. The first of these examples will deal with the test of the hypothesis that two multinomial distributions are the same.

Remark. In many instances, such as that involving the mean μ and the variance σ^2 of a normal distribution, minimum chi-square estimates are difficult to compute. Hence other estimates, such as the maximum likelihood estimates $\hat{\mu} = \bar{Y}$ and $\widehat{\sigma^2} = S^2$, are used to evaluate p_i and Q_{k-1}. In general, Q_{k-1} is not minimized by maximum likelihood estimates, and thus its computed value is somewhat greater than it would be if minimum chi-square estimates were used. Hence, when comparing it to a critical value listed in the chi-square table with $k - 3$ degrees of freedom, there is a greater chance of rejecting than there would be if the actual minimum of Q_{k-1} is used. Accordingly, the approximate significance level of such a test will be somewhat higher than that value found in the table. This modification should be kept in mind and, if at all possible, each p_i should be estimated using the frequencies $X_1, \ldots, X_k$ rather than using directly the items $Y_1, Y_2, \ldots, Y_n$ of the random sample.

Example 3. Let us consider two independent multinomial distributions with parameters $n_j, p_{1j}, p_{2j}, \ldots, p_{kj}$, $j = 1$, 2, respectively. Let $X_{ij}, i = 1, 2, \ldots, k$, $j = 1$, 2, represent the corresponding frequencies. If n_1 and n_2 are large, the random variable

$$\sum_{j=1}^{2} \sum_{i=1}^{k} \frac{(X_{ij} - n_j p_{ij})^2}{n_j p_{ij}}$$

is the sum of two stochastically independent random variables, each of which we treat as though it were $\chi^2(k - 1)$; that is, the random variable is approximately $\chi^2(2k - 2)$. Consider the hypothesis

$$H_0: p_{11} = p_{12}, p_{21} = p_{22}, \ldots, p_{k1} = p_{k2},$$

where each $p_{i1} = p_{i2}$, $i = 1, 2, \ldots, k$, is unspecified. Thus we need point estimates of these parameters. The maximum likelihood estimator of $p_{i1} = p_{i2}$, based upon the frequencies X_{ij}, is $(X_{i1} + X_{i2})/(n_1 + n_2)$, $i = 1, 2, \ldots, k$. Note that we need only $k - 1$ point estimates, because we have a point estimate of $p_{k1} = p_{k2}$ once we have point estimates of the first $k - 1$ probabilities. In accordance with the fact that has been stated, the random variable

$$\sum_{j=1}^{2} \sum_{i=1}^{k} \frac{\{X_{ij} - n_j[(X_{i1} + X_{i2})/(n_1 + n_2)]\}^2}{n_j[(X_{i1} + X_{i2})/(n_1 + n_2)]}$$

has an approximate χ^2 distribution with $2k - 2 - (k - 1) = k - 1$ degrees of freedom. Thus we are able to test the hypothesis that two multinomial distributions are the same; this hypothesis is rejected when the computed value of this random variable is at least as great as an appropriate number from Table II, with $k - 1$ degrees of freedom.

The second example deals with the subject of *contingency tables.*

Example 4. Let the result of a random experiment be classified by two attributes (such as the color of the hair and the color of the eyes). That is, one attribute of the outcome is one and only one of certain mutually exclusive and exhaustive events, say $A_1, A_2, \ldots, A_a$; and the other attribute of the outcome is also one and only one of certain mutually exclusive and exhaustive events, say $B_1, B_2, \ldots, B_b$. Let $p_{ij} = P(A_i \cap B_j)$, $i = 1, 2, \ldots, a$; $j = 1, 2, \ldots, b$. The random experiment is to be repeated n independent times and X_{ij} will denote the frequency of the event $A_i \cap B_j$. Since there are $k = ab$ such events as $A_i \cap B_j$, the random variable

$$Q_{ab-1} = \sum_{j=1}^{b} \sum_{i=1}^{a} \frac{(X_{ij} - np_{ij})^2}{np_{ij}}$$

has an approximate chi-square distribution with $ab - 1$ degrees of freedom, provided that n is large. Suppose that we wish to test the independence of the A attribute and the B attribute; that is, we wish to test the hypothesis $H_0: P(A_i \cap B_j) = P(A_i)P(B_j)$, $i = 1, 2, \ldots, a$; $j = 1, 2, \ldots, b$. Let us denote $P(A_i)$ by $p_{i\cdot}$ and $P(B_j)$ by $p_{\cdot j}$; thus

$$p_{i\cdot} = \sum_{j=1}^{b} p_{ij}, \qquad p_{\cdot j} = \sum_{i=1}^{a} p_{ij},$$

and

$$1 = \sum_{j=1}^{b} \sum_{i=1}^{a} p_{ij} = \sum_{j=1}^{b} p_{\cdot j} = \sum_{i=1}^{a} p_{i\cdot}.$$

Then the hypothesis can be formulated as $H_0: p_{ij} = p_{i\cdot} p_{\cdot j}$, $i = 1, 2, \ldots, a$; $j = 1, 2, \ldots, b$. To test H_0, we can use Q_{ab-1} with p_{ij} replaced by $p_{i\cdot} p_{\cdot j}$. But if $p_{i\cdot}$, $i = 1, 2, \ldots, a$, and $p_{\cdot j}$, $j = 1, 2, \ldots, b$, are unknown, as they frequently are in the applications, we cannot compute Q_{ab-1} once the frequencies are observed. In such a case we estimate these unknown parameters by

$$\hat{p}_{i\cdot} = \frac{X_{i\cdot}}{n}, \quad \text{where} \quad X_{i\cdot} = \sum_{j=1}^{b} X_{ij}, \qquad i = 1, 2, \ldots, a,$$

and

$$\hat{p}_{\cdot j} = \frac{X_{\cdot j}}{n}, \quad \text{where} \quad X_{\cdot j} = \sum_{i=1}^{a} X_{ij}, \qquad j = 1, 2, \ldots, b.$$

Since $\sum_i p_{i\cdot} = \sum_j p_{\cdot j} = 1$, we have estimated only $a - 1 + b - 1 = a + b - 2$ parameters. So if these estimates are used in Q_{ab-1}, with $p_{ij} = p_{i\cdot} p_{\cdot j}$, then, according to the rule that has been stated in this section, the random variable

$$\sum_{j=1}^{b} \sum_{i=1}^{a} \frac{[X_{ij} - n(X_{i.}/n)(X_{.j}/n)]^2}{n(X_{i.}/n)(X_{.j}/n)}$$

has an approximate chi-square distribution with $ab - 1 - (a + b - 2) = (a - 1)(b - 1)$ degrees of freedom provided that H_0 is true. The hypothesis H_0 is then rejected if the computed value of this statistic exceeds the constant c, where c is selected from Table II so that the test has the desired significance level α.

In each of the four examples of this section we have indicated that the statistic used to test the hypothesis H_0 has an approximate chi-square distribution, provided that n is sufficiently large and H_0 is true. To compute the power of any of these tests for values of the parameters not described by H_0, we need the distribution of the statistic when H_0 is not true. In each of these cases, the statistic has an approximate distribution called a *noncentral chi-square distribution*. The noncentral chi-square distribution will be discussed in Section 8.4.

EXERCISES

8.1. A number is to be selected from the interval $\{x; 0 < x < 2\}$ by a random process. Let $A_i = \{x; (i - 1)/2 < x \le i/2\}$, $i = 1, 2, 3$, and let $A_4 = \{x; \frac{3}{2} < x < 2\}$. A certain hypothesis assigns probabilities p_{i0} to these sets in accordance with $p_{i0} = \int_{A_i} (\frac{1}{2})(2 - x)\, dx$, $i = 1, 2, 3, 4$. This hypothesis (concerning the multinomial p.d.f. with $k = 4$) is to be tested, at the 5 per cent level of significance, by a chi-square test. If the observed frequencies of the sets A_i, $i = 1, 2, 3, 4$, are, respectively, 30, 30, 10, 10, would H_0 be accepted at the (approximate) 5 per cent level of significance?

8.2. Let the following sets be defined. $A_1 = \{x; -\infty < x \le 0\}$, $A_i = \{x; i - 2 < x \le i - 1\}$, $i = 2, \ldots, 7$, and $A_8 = \{x; 6 < x < \infty\}$. A certain hypothesis assigns probabilities p_{i0} to these sets A_i in accordance with

$$p_{i0} = \int_{A_i} \frac{1}{2\sqrt{2\pi}} \exp\left[-\frac{(x - 3)^2}{2(4)}\right] dx, \qquad i = 1, 2, \ldots, 7, 8.$$

This hypothesis (concerning the multinomial p.d.f. with $k = 8$) is to be tested, at the 5 per cent level of significance, by a chi-square test. If the observed frequencies of the sets A_i, $i = 1, 2, \ldots, 8$, are, respectively, 60, 96, 140, 210, 172, 160, 88, and 74, would H_0 be accepted at the (approximate) 5 per cent level of significance?

8.3. A die was cast $n = 120$ independent times and the following data resulted:

Spots up	1	2	3	4	5	6
Frequency	b	20	20	20	20	40–b

If we use a chi-square test, for what values of b would the hypothesis that the die is unbiased be rejected at the 0.025 significance level?

8.4. Consider the problem from genetics of crossing two types of peas. The Mendelian theory states that the probabilities of the classifications (a) round and yellow, (b) wrinkled and yellow, (c) round and green, and (d) wrinkled and green are $\frac{9}{16}$, $\frac{3}{16}$, $\frac{3}{16}$, and $\frac{1}{16}$, respectively. If, from 160 independent observations, the observed frequencies of these respective classifications are 86, 35, 26, and 13, are these data consistent with the Mendelian theory? That is, test, with $\alpha = 0.01$, the hypothesis that the respective probabilities are $\frac{9}{16}$, $\frac{3}{16}$, $\frac{3}{16}$, and $\frac{1}{16}$.

8.5. Two different teaching procedures were used on two different groups of students. Each group contained 100 students of about the same ability. At the end of the term, an evaluating team assigned a letter grade to each student. The results were tabulated as follows.

Group	A	B	C	D	F	Total
			Grade			
I	15	25	32	17	11	100
II	9	18	29	28	16	100

If we consider these data to be observations from two independent multinomial distributions with $k = 5$, test, at the 5 per cent significance level, the hypothesis that the two distributions are the same (and hence the two teaching procedures are equally effective).

8.6. Let the result of a random experiment be classified as one of the mutually exclusive and exhaustive ways A_1, A_2, A_3 and also as one of the mutually exclusive and exhaustive ways B_1, B_2, B_3, B_4. Two hundred independent trials of the experiment result in the following data:

	B_1	B_2	B_3	B_4
A_1	10	21	15	6
A_2	11	27	21	13
A_3	6	19	27	24

Test, at the 0.05 significance level, the hypothesis of independence of the A attribute and the B attribute, namely $H_0: P(A_i \cap B_j) = P(A_i)P(B_j)$, $i = 1, 2, 3$ and $j = 1, 2, 3, 4$, against the alternative of dependence.

8.7. A certain genetic model suggests that the probabilities of a particular trinomial distribution are, respectively, $p_1 = p^2$, $p_2 = 2p(1 - p)$, and $p_3 = (1 - p)^2$, where $0 < p < 1$. If X_1, X_2, X_3 represent the respective frequencies in n independent trials, explain how we could check on the adequacy of the genetic model.

8.2 The Distributions of Certain Quadratic Forms

A homogeneous polynomial of degree 2 in n variables is called a *quadratic* form in those variables. If both the variables and the coefficients are real, the form is called a *real quadratic* form. Only real quadratic forms will be considered in this book. To illustrate, the form $X_1^2 + X_1X_2 + X_2^2$ is a quadratic form in the two variables X_1 and X_2; the form $X_1^2 + X_2^2 + X_3^2 - 2X_1X_2$ is a quadratic form in the three variables X_1, X_2, and X_3; but the form $(X_1 - 1)^2 + (X_2 - 2)^2 = X_1^2 + X_2^2 - 2X_1 - 4X_2 + 5$ is not a quadratic form in X_1 and X_2, although it is a quadratic form in the variables $X_1 - 1$ and $X_2 - 2$.

Let $\bar{X}$ and S^2 denote, respectively, the mean and the variance of a random sample $X_1, X_2, \ldots, X_n$ from an arbitrary distribution. Thus

$$nS^2 = \sum_1^n (X_i - \bar{X})^2 = \sum_1^n \left(X_i - \frac{X_1 + X_2 + \cdots + X_n}{n} \right)^2$$

$$= \frac{n-1}{n} (X_1^2 + X_2^2 + \cdots + X_n^2)$$

$$- \frac{2}{n} (X_1X_2 + \cdots + X_1X_n + \cdots + X_{n-1}X_n)$$

is a quadratic form in the n variables $X_1, X_2, \ldots, X_n$. If the sample arises from a distribution that is $n(\mu, \sigma^2)$, we know that the random variable nS^2/σ^2 is $\chi^2(n - 1)$ regardless of the value of μ. This fact proved useful in our search for a confidence interval for σ^2 when μ is unknown.

It has been seen that tests of certain statistical hypotheses require a statistic that is a quadratic form. For instance, Example 2, Section 7.3, made use of the statistic $\sum_1^n X_i^2$, which is a quadratic form in the variables $X_1, X_2, \ldots, X_n$. Later in this chapter, tests of other statistical hypotheses will be investigated, and it will be seen that functions of statistics that are quadratic forms will be needed to carry out the tests in an expeditious manner. But first we shall make a study of the distribution of certain quadratic forms in normal and stochastically independent random variables.

The following theorem will be proved in Chapter 12.

Theorem 1. *Let* $Q = Q_1 + Q_2 + \cdots + Q_{k-1} + Q_k$, *where* $Q, Q_1, \ldots,$ Q_k *are* $k + 1$ *random variables that are real quadratic forms in* n *mutually stochastically independent random variables which are normally distributed with the means* $\mu_1, \mu_2, \ldots, \mu_n$ *and the same variance* σ^2. *Let* Q/σ^2, $Q_1/\sigma^2, \ldots, Q_{k-1}/\sigma^2$ *have chi-square distributions with degrees of freedom* $r, r_1, \ldots, r_{k-1}$, *respectively. Let* Q_k *be nonnegative. Then:*

(a) $Q_1, \ldots, Q_k$ *are mutually stochastically independent, and hence*

(b) Q_k/σ^2 *has a chi-square distribution with* $r - (r_1 + \cdots + r_{k-1}) = r_k$ *degrees of freedom.*

Three examples illustrative of the theorem will follow. Each of these examples will deal with a distribution problem that is based on the remarks made in the subsequent paragraph.

Let the random variable X have a distribution that is $n(\mu, \sigma^2)$. Let a and b denote positive integers greater than 1 and let $n = ab$. Consider a random sample of size $n = ab$ from this normal distribution. The items of the random sample will be denoted by the symbols

$$
\begin{array}{cccccc}
X_{11}, & X_{12}, & \ldots, & X_{1j}, & \cdots, & X_{1b} \\
X_{21}, & X_{22}, & \ldots, & X_{2j}, & \ldots, & X_{2b} \\
\vdots & & & & & \\
X_{i1}, & X_{i2}, & \ldots, & X_{ij}, & \ldots, & X_{ib} \\
\vdots & & & & & \\
X_{a1}, & X_{a2}, & \ldots, & X_{aj}, & \ldots, & X_{ab}.
\end{array}
$$

In this notation the first subscript indicates the row, and the second subscript indicates the column in which the item appears. Thus X_{ij} is in row i and column j, $i = 1, 2, \ldots, a$ and $j = 1, 2, \ldots, b$. By assumption these $n = ab$ random variables are mutually stochastically independent, and each has the same normal distribution with mean μ and variance σ^2. Thus, if we wish, we may consider each row as being a random sample of size b from the given distribution; and we may consider each column as being a random sample of size a from the given distribution. We now define $a + b + 1$ statistics. They are

$$X = \frac{X_{11} + \cdots + X_{1b} + \cdots + X_{a1} + \cdots + X_{ab}}{ab} = \frac{\sum\limits_{i=1}^{a} \sum\limits_{j=1}^{b} X_{ij}}{ab},$$

$$\bar{X}_{i\cdot} = \frac{X_{i1} + X_{i2} + \cdots + X_{ib}}{b} = \frac{\sum\limits_{j=1}^{b} X_{ij}}{b}, \qquad i = 1, 2, \ldots, a,$$

and

$$\bar{X}_{\cdot j} = \frac{X_{1j} + X_{2j} + \cdots + X_{aj}}{a} = \frac{\sum\limits_{i=1}^{a} X_{ij}}{a}, \qquad j = 1, 2, \ldots, b.$$

In some texts the statistic X is denoted by $\bar{X}_{\cdot\cdot}$, but we use $\bar{X}$ for simplicity. In any case, $\bar{X} = \bar{X}_{\cdot\cdot}$ is the mean of the random sample of size $n = ab$; the statistics $\bar{X}_{1\cdot}, \bar{X}_{2\cdot}, \ldots, \bar{X}_{a\cdot}$ are, respectively, the means of the rows; and the statistics $\bar{X}_{\cdot 1}, \bar{X}_{\cdot 2}, \ldots, \bar{X}_{\cdot b}$ are, respectively, the means of the columns. The examples illustrative of the theorem follow.

Example 1. Consider the variance S^2 of the random sample of size $n = ab$. We have the algebraic identity

$$abS^2 = \sum_{i=1}^{a} \sum_{j=1}^{b} (X_{ij} - \bar{X})^2$$

$$= \sum_{i=1}^{a} \sum_{j=1}^{b} [(X_{ij} - \bar{X}_{i\cdot}) + (\bar{X}_{i\cdot} - \bar{X})]^2$$

$$= \sum_{i=1}^{a} \sum_{j=1}^{b} (X_{ij} - \bar{X}_{i\cdot})^2 + \sum_{i=1}^{a} \sum_{j=1}^{b} (\bar{X}_{i\cdot} - \bar{X})^2$$

$$+ 2 \sum_{i=1}^{a} \sum_{j=1}^{b} (X_{ij} - \bar{X}_{i\cdot})(\bar{X}_{i\cdot} - \bar{X}).$$

The last term of the right-hand member of this identity may be written

$$2 \sum_{i=1}^{a} \left[(\bar{X}_{i\cdot} - \bar{X}) \sum_{j=1}^{b} (X_{ij} - \bar{X}_{i\cdot}) \right] = 2 \sum_{i=1}^{a} [(\bar{X}_{i\cdot} - \bar{X})(b\bar{X}_{i\cdot} - b\bar{X}_{i\cdot})] = 0,$$

and the term

$$\sum_{i=1}^{a} \sum_{j=1}^{b} (\bar{X}_{i\cdot} - \bar{X})^2$$

may be written

$$b \sum_{i=1}^{a} (\bar{X}_{i\cdot} - \bar{X})^2.$$

Thus

$$abS^2 = \sum_{i=1}^{a} \sum_{j=1}^{b} (X_{ij} - \bar{X}_{i\cdot})^2 + b \sum_{i=1}^{a} (\bar{X}_{i\cdot} - \bar{X})^2,$$

or, for brevity,

$$Q = Q_1 + Q_2.$$

Clearly, Q, Q_1, and Q_2 are quadratic forms in the $n = ab$ variables X_{ij}. We shall use the theorem with $k = 2$ to show that Q_1 and Q_2 are stochastically independent. Since S^2 is the variance of a random sample of size $n = ab$ from the given normal distribution, then abS^2/σ^2 has a chi-square distribution with $ab - 1$ degrees of freedom. Now

$$\frac{Q_1}{\sigma^2} = \sum_{i=1}^{a} \left[\frac{\sum_{j=1}^{b} (X_{ij} - \bar{X}_{i\cdot})^2}{\sigma^2} \right].$$

For each fixed value of i, $\sum_{j=1}^{b} (X_{ij} - \bar{X}_{i\cdot})^2/b$ is the variance of a random sample of size b from the given normal distribution, and, accordingly, $\sum_{j=1}^{b} (X_{ij} - \bar{X}_{i\cdot})^2/\sigma^2$ has a chi-square distribution with $b - 1$ degrees of freedom. Because the X_{ij} are mutually stochastically independent, Q_1/σ^2 is the sum of a mutually stochastically independent random variables, each having a chi-square distribution with $b - 1$ degrees of freedom. Hence Q_1/σ^2 has a chi-square distribution with $a(b - 1)$ degrees of freedom. Now $Q_2 = b \sum_{i=1}^{a} (\bar{X}_{i\cdot} - \bar{X})^2 \geq 0$. In accordance with the theorem, Q_1 and Q_2 are stochastically independent, and Q_2/σ^2 has a chi-square distribution with $ab - 1 - a(b - 1) = a - 1$ degrees of freedom.

Example 2. In abS^2 replace $X_{ij} - \bar{X}$ by $(X_{ij} - \bar{X}_{\cdot j}) + (\bar{X}_{\cdot j} - \bar{X})$ to obtain

$$abS^2 = \sum_{j=1}^{b} \sum_{i=1}^{a} [(X_{ij} - \bar{X}_{\cdot j}) + (\bar{X}_{\cdot j} - \bar{X})]^2,$$

or

$$abS^2 = \sum_{j=1}^{b} \sum_{i=1}^{a} (X_{ij} - \bar{X}_{\cdot j})^2 + a \sum_{j=1}^{b} (\bar{X}_{\cdot j} - \bar{X})^2,$$

or, for brevity,

$$Q = Q_3 + Q_4.$$

It is easy to show (Exercise 8.8) that Q_3/σ^2 has a chi-square distribution with $b(a - 1)$ degrees of freedom. Since $Q_4 = a \sum_{j=1}^{b} (\bar{X}_{\cdot j} - \bar{X})^2 \geq 0$, the theorem enables us to assert that Q_3 and Q_4 are stochastically independent and that Q_4/σ^2 has a chi-square distribution with $ab - 1 - b(a - 1) = b - 1$ degrees of freedom.

Example 3. In abS^2 replace $X_{ij} - X$ by $(\bar{X}_{i\cdot} - \bar{X}) + (\bar{X}_{\cdot j} - \bar{X}) + (X_{ij} - \bar{X}_{i\cdot} - \bar{X}_{\cdot j} + \bar{X})$ to obtain (Exercise 8.9)

$$abS^2 = b \sum_{i=1}^{a} (\bar{X}_{i\cdot} - \bar{X})^2 + a \sum_{j=1}^{b} (\bar{X}_{\cdot j} - \bar{X})^2$$

$$+ \sum_{j=1}^{b} \sum_{i=1}^{a} (X_{ij} - \bar{X}_{i\cdot} - \bar{X}_{\cdot j} + \bar{X})^2,$$

or, for brevity,

$$Q = Q_2 + Q_4 + Q_5,$$

where Q_2 and Q_4 are as defined in Examples 1 and 2. From Examples 1 and 2, Q/σ^2, Q_2/σ^2, and Q_4/σ^2 have chi-square distributions with $ab - 1$, $a - 1$, and $b - 1$ degrees of freedom, respectively. Since $Q_5 \geq 0$, the theorem asserts that Q_2, Q_4, and Q_5 are mutually stochastically independent and that Q_5/σ^2 has a chi-square distribution with $ab - 1 - (a - 1) - (b - 1) = (a - 1)(b - 1)$ degrees of freedom.

Once these quadratic form statistics have been shown to be stochastically independent, a multiplicity of F statistics can be defined. For instance,

$$\frac{Q_4/[\sigma^2(b - 1)]}{Q_3/[\sigma^2 b(a - 1)]} = \frac{Q_4/(b - 1)}{Q_3/[b(a - 1)]}$$

has an F distribution with $b - 1$ and $b(a - 1)$ degrees of freedom; and

$$\frac{Q_4/[\sigma^2(b - 1)]}{Q_5/[\sigma^2(a - 1)(b - 1)]} = \frac{Q_4}{Q_5/(a - 1)}$$

has an F distribution with $b - 1$ and $(a - 1)(b - 1)$ degrees of freedom. In the subsequent sections it will be seen that some likelihood ratio tests of certain statistical hypotheses can be based on these F statistics.

EXERCISES

8.8. In Example 2 verify that $Q = Q_3 + Q_4$ and that Q_3/σ^2 has a chi-square distribution with $b(a - 1)$ degrees of freedom.

8.9. In Example 3 verify that $Q = Q_2 + Q_4 + Q_5$.

8.10. Let $X_1, X_2, \ldots, X_n$ be a random sample from a normal distribution $n(\mu, \sigma^2)$. Show that

$$\sum_{i=1}^{n} (X_i - \bar{X})^2 = \sum_{i=2}^{n} (X_i - \bar{X}')^2 + \frac{n-1}{n} (X_1 - \bar{X}')^2,$$

where $\bar{X} = \sum_{i=1}^{n} X_i/n$ and $\bar{X}' = \sum_{i=2}^{n} X_i/(n - 1)$. *Hint.* Replace $X_i - \bar{X}$ by

$(X_i - X') - (X_1 - X')/n$. Show that $\sum\limits_{i=2}^{n} (X_i - X')^2/\sigma^2$ has a chi-square distribution with $n - 2$ degrees of freedom. Prove that the two terms in the right-hand member are stochastically independent. What then is the distribution of

$$\frac{[(n - 1)/n](X_1 - X')^2}{\sigma^2}?$$

8.11. Let X_{ijk}, $i = 1, \ldots, a$; $j = 1, \ldots, b$; $k = 1, \ldots, c$, be a random sample of size $n = abc$ from a normal distribution $n(\mu, \sigma^2)$. Let $X = \sum\limits_{k=1}^{c}\sum\limits_{j=1}^{b}\sum\limits_{i=1}^{a} X_{ijk}/n$ and $X_{i\cdot\cdot} = \sum\limits_{k=1}^{c}\sum\limits_{j=1}^{b} X_{ijk}/bc$. Show that

$$\sum_{i=1}^{a}\sum_{j=1}^{b}\sum_{k=1}^{c} (X_{ijk} - X)^2 = \sum_{i=1}^{a}\sum_{j=1}^{b}\sum_{k=1}^{c} (X_{ijk} - X_{i\cdot\cdot})^2 + bc \sum_{i=1}^{a} (X_{i\cdot\cdot} - X)^2.$$

Show that $\sum\limits_{i=1}^{a}\sum\limits_{j=1}^{b}\sum\limits_{k=1}^{c} (X_{ijk} - X_{i\cdot\cdot})^2/\sigma^2$ has a chi-square distribution with $a(bc - 1)$ degrees of freedom. Prove that the two terms in the right-hand member are stochastically independent. What, then, is the distribution of $bc \sum\limits_{i=1}^{a} (X_{i\cdot\cdot} - X)^2/\sigma^2$? Furthermore, let $X_{\cdot j\cdot} = \sum\limits_{k=1}^{c}\sum\limits_{i=1}^{a} X_{ijk}/ac$ and $X_{ij\cdot} = \sum\limits_{k=1}^{c} X_{ijk}/c$. Show that

$$\sum_{i=1}^{a}\sum_{j=1}^{b}\sum_{k=1}^{c} (X_{ijk} - X)^2$$

$$= \sum_{i=1}^{a}\sum_{j=1}^{b}\sum_{k=1}^{c} (X_{ijk} - X_{ij\cdot})^2 + bc \sum_{i=1}^{a} (X_{i\cdot\cdot} - X)^2 + ac \sum_{j=1}^{b} (X_{\cdot j\cdot} - X)^2$$

$$+ c \sum_{i=1}^{a}\sum_{j=1}^{b} (X_{ij\cdot} - X_{i\cdot\cdot} - X_{\cdot j\cdot} + X)^2.$$

Show that the four terms in the right-hand member, when divided by σ^2, are mutually stochastically independent chi-square variables with $ab(c - 1)$, $a - 1$, $b - 1$, and $(a - 1)(b - 1)$ degrees of freedom, respectively.

8.12. Let X_1, X_2, X_3, X_4 be a random sample of size $n = 4$ from the normal distribution $n(0, 1)$. Show that $\sum\limits_{i=1}^{4} (X_i - X)^2$ equals

$$\frac{(X_1 - X_2)^2}{2} + \frac{[X_3 - (X_1 + X_2)/2]^2}{3/2} + \frac{[X_4 - (X_1 + X_2 + X_3)/3]^2}{4/3}$$

and argue that these three terms are mutually stochastically independent, each with a chi-square distribution with 1 degree of freedom.

8.3 A Test of the Equality of Several Means

Consider b mutually stochastically independent random variables that have normal distributions with unknown means $\mu_1, \mu_2, \ldots, \mu_b$, respectively, and unknown but common variance σ^2. Let $X_{1j}, X_{2j}, \ldots,$ X_{aj} represent a random sample of size a from the normal distribution with mean μ_j and variance σ^2, $j = 1, 2, \ldots, b$. It is desired to test the composite hypothesis $H_0: \mu_1 = \mu_2 = \cdots = \mu_b = \mu$, μ unspecified, against all possible alternative hypotheses H_1. A likelihood ratio test will be used. Here the total parameter space is

$$\Omega = \{(\mu_1, \mu_2, \ldots, \mu_b, \sigma^2); -\infty < \mu_j < \infty, 0 < \sigma^2 < \infty\}$$

and

$$\omega = \{(\mu_1, \mu_2, \ldots, \mu_b, \sigma^2); -\infty < \mu_1 = \mu_2 = \cdots$$
$$= \mu_b = \mu < \infty, 0 < \sigma^2 < \infty\}.$$

The likelihood functions, denoted by $L(\omega)$ and $L(\Omega)$ are, respectively,

$$L(\omega) = \left(\frac{1}{2\pi\sigma^2}\right)^{ab/2} \exp\left[-\frac{1}{2\sigma^2} \sum_{j=1}^{b} \sum_{i=1}^{a} (x_{ij} - \mu)^2\right]$$

and

$$L(\Omega) = \left(\frac{1}{2\pi\sigma^2}\right)^{ab/2} \exp\left[-\frac{1}{2\sigma^2} \sum_{j=1}^{b} \sum_{i=1}^{a} (x_{ij} - \mu_j)^2\right].$$

Now

$$\frac{\partial \ln L(\omega)}{\partial \mu} = \frac{\sum\limits_{j=1}^{b} \sum\limits_{i=1}^{a} (x_{ij} - \mu)}{\sigma^2}$$

and

$$\frac{\partial \ln L(\omega)}{\partial (\sigma^2)} = -\frac{ab}{2\sigma^2} + \frac{1}{2\sigma^4} \sum_{j=1}^{b} \sum_{i=1}^{a} (x_{ij} - \mu)^2.$$

If we equate these partial derivatives to zero, the solutions for μ and σ^2 are, respectively, in ω,

(1)
$$\frac{\sum\limits_{j=1}^{b} \sum\limits_{i=1}^{a} x_{ij}}{ab} = \bar{x},$$

$$\frac{\sum\limits_{j=1}^{b} \sum\limits_{i=1}^{a} (x_{ij} - \bar{x})^2}{ab} = v,$$

and these numbers maximize $L(\omega)$. Furthermore,

$$\frac{\partial \ln L(\Omega)}{\partial \mu_j} = \frac{\sum\limits_{i=1}^{a} (x_{ij} - \mu_j)}{\sigma^2}, \qquad j = 1, 2, \ldots, b,$$

and

$$\frac{\partial \ln L(\Omega)}{\partial (\sigma^2)} = -\frac{ab}{2\sigma^2} + \frac{1}{2\sigma^4} \sum\limits_{j=1}^{b} \sum\limits_{i=1}^{a} (x_{ij} - \mu_j)^2.$$

If we equate these partial derivatives to zero, the solutions for μ_1, $\mu_2, \ldots, \mu_b$, and σ^2 are, respectively, in Ω,

(2)
$$\frac{\sum\limits_{i=1}^{a} x_{ij}}{a} = \bar{x}_{\cdot j}, \qquad j = 1, 2, \ldots, b,$$

$$\frac{\sum\limits_{j=1}^{b} \sum\limits_{i=1}^{a} (x_{ij} - \bar{x}_{\cdot j})^2}{ab} = w,$$

and these numbers maximize $L(\Omega)$. These maxima are, respectively,

$$L(\hat{\omega}) = \left[\frac{ab}{2\pi \sum\limits_{j=1}^{b} \sum\limits_{i=1}^{a} (x_{ij} - \bar{x})^2} \right]^{ab/2} \exp\left[-\frac{ab \sum\limits_{j=1}^{b} \sum\limits_{i=1}^{a} (x_{ij} - \bar{x})^2}{2 \sum\limits_{j=1}^{b} \sum\limits_{i=1}^{a} (x_{ij} - \bar{x})^2} \right]$$

$$= \left[\frac{ab}{2\pi \sum\limits_{j=1}^{b} \sum\limits_{i=1}^{a} (x_{ij} - \bar{x})^2} \right]^{ab/2} e^{-ab/2}$$

and

$$L(\hat{\Omega}) = \left[\frac{ab}{2\pi \sum\limits_{j=1}^{b} \sum\limits_{i=1}^{a} (x_{ij} - \bar{x}_{\cdot j})^2} \right]^{ab/2} e^{-ab/2}.$$

Finally,

$$\lambda = \frac{L(\hat{\omega})}{L(\hat{\Omega})} = \left[\frac{\sum\limits_{j=1}^{b} \sum\limits_{i=1}^{a} (x_{ij} - \bar{x}_{\cdot j})^2}{\sum\limits_{j=1}^{b} \sum\limits_{i=1}^{a} (x_{ij} - \bar{x})^2} \right]^{ab/2}.$$

In the notation of Section 8.2, the statistics defined by the functions $\bar{x}$ and v given by Equations (1) of this section are

$$X = \sum\limits_{j=1}^{b} \sum\limits_{i=1}^{a} \frac{X_{ij}}{ab} \qquad \text{and} \qquad S^2 = \sum\limits_{j=1}^{b} \sum\limits_{i=1}^{a} \frac{(X_{ij} - \bar{X})^2}{ab} = \frac{Q}{ab};$$

while the statistics defined by the functions $\bar{x}_{.1}, \bar{x}_{.2}, \ldots, \bar{x}_{.b}$ and w given by Equations (2) in this section are, respectively, $\bar{X}_{.j} = \sum\limits_{i=1}^{a} X_{ij}/a$, $j = 1, 2, \ldots, b$, and $Q_3/ab = \sum\limits_{j=1}^{b} \sum\limits_{i=1}^{a} (X_{ij} - \bar{X}_{.j})^2/ab$. Thus, in the notation of Section 8.2, $\lambda^{2/ab}$ defines the statistic Q_3/Q.

We reject the hypothesis H_0 if $\lambda \leq \lambda_0$. To find λ_0 so that we have a desired significance level α, we must assume that the hypothesis H_0 is true. If the hypothesis H_0 is true, the random variables X_{ij} constitute a random sample of size $n = ab$ from a distribution that is normal with mean μ and variance σ^2. This being the case, it was shown in Example 2, Section 8.2, that $Q = Q_3 + Q_4$, where $Q_4 = a \sum\limits_{j=1}^{b} (\bar{X}_{.j} - \bar{X})^2$; that Q_3 and Q_4 are stochastically independent; and that Q_3/σ^2 and Q_4/σ^2 have chi-square distributions with $b(a - 1)$ and $b - 1$ degrees of freedom, respectively. Thus the statistic defined by $\lambda^{2/ab}$ may be written

$$\frac{Q_3}{Q_3 + Q_4} = \frac{1}{1 + Q_4/Q_3}.$$

The significance level of the test of H_0 is

$$\alpha = \Pr\left[\frac{1}{1 + Q_4/Q_3} \leq \lambda_0^{2/ab}; H_0\right]$$

$$= \Pr\left[\frac{Q_4/(b - 1)}{Q_3/[b(a - 1)]} \geq c; H_0\right],$$

where

$$c = \frac{b(a - 1)}{b - 1} (\lambda_0^{-2/ab} - 1).$$

But

$$F = \frac{Q_4/[\sigma^2(b - 1)]}{Q_3/[\sigma^2 b(a - 1)]} = \frac{Q_4/(b - 1)}{Q_3/[b(a - 1)]}$$

has an F distribution with $b - 1$ and $b(a - 1)$ degrees of freedom. Hence the test of the composite hypothesis $H_0 : \mu_1 = \mu_2 = \cdots = \mu_b = \mu$, μ unspecified, against all possible alternatives may be based on an F statistic. The constant c is so selected as to yield the desired value of α.

Remark. It should be pointed out that a test of the equality of the b means $\mu_j, j = 1, 2, \ldots, b$, does not require that we take a random sample of size a from each of the b normal distributions. That is, the samples may be of different sizes, say $a_1, a_2, \ldots, a_b$. A consideration of this procedure is left to Exercise 8.13.

Suppose now that we wish to compute the power of the test of H_0 against H_1 when H_0 is false, that is, when we do not have $\mu_1 = \mu_2 = \cdots = \mu_b = \mu$. It will be seen in Section 8.4 that, when H_1 is true, no longer is Q_4/σ^2 a random variable that is $\chi^2(b-1)$. Thus we cannot use an F statistic to compute the power of the test when H_1 is true. This problem is discussed in Section 8.4.

An observation should be made in connection with maximizing a likelihood function with respect to certain parameters. Sometimes it is easier to avoid the use of the calculus. For example, $L(\Omega)$ of this section can be maximized with respect to μ_j, for every fixed positive σ^2, by minimizing

$$z = \sum_{j=1}^{b} \sum_{i=1}^{a} (x_{ij} - \mu_j)^2$$

with respect to μ_j, $j = 1, 2, \ldots, b$. Now z can be written as

$$z = \sum_{j=1}^{b} \sum_{i=1}^{a} [(x_{ij} - \bar{x}_{\cdot j}) + (\bar{x}_{\cdot j} - \mu_j)]^2$$

$$= \sum_{j=1}^{b} \sum_{i=1}^{a} (x_{ij} - \bar{x}_{\cdot j})^2 + a \sum_{j=1}^{b} (\bar{x}_{\cdot j} - \mu_j)^2.$$

Since each term in the right-hand member of the preceding equation is nonnegative, clearly z is a minimum, with respect to μ_j, if we take $\mu_j = \bar{x}_{\cdot j}$, $j = 1, 2, \ldots, b$.

EXERCISES

8.13. Let $X_{1j}, X_{2j}, \ldots, X_{a_j j}$ represent independent random samples of sizes a_j from normal distributions with means μ_j and variances σ^2, $j = 1, 2, \ldots, b$. Show that

$$\sum_{j=1}^{b} \sum_{i=1}^{a_j} (X_{ij} - \bar{X})^2 = \sum_{j=1}^{b} \sum_{i=1}^{a_j} (X_{ij} - \bar{X}_{\cdot j})^2 + \sum_{j=1}^{b} a_j (\bar{X}_{\cdot j} - \bar{X})^2,$$

or $Q' = Q_3' + Q_4'$. Here $\bar{X} = \sum_{j=1}^{b} \sum_{i=1}^{a_j} X_{ij} / \sum_{j=1}^{b} a_j$ and $\bar{X}_{\cdot j} = \sum_{i=1}^{a_j} X_{ij}/a_j$. If $\mu_1 = \mu_2 = \cdots = \mu_b$, show that Q'/σ^2 and Q_3'/σ^2 have chi-square distributions. Prove that Q_3' and Q_4' are stochastically independent, and hence Q_4'/σ^2 also has a chi-square distribution. If the likelihood ratio λ is used to test $H_0: \mu_1 = \mu_2 = \cdots = \mu_b = \mu$, μ unspecified and σ^2 unknown, against all possible alternatives, show that $\lambda \le \lambda_0$ is equivalent to the computed $F \ge c$, where

$$F = \frac{\left(\sum\limits_{j=1}^{b} a_j - b\right)Q_4'}{(b-1)Q_3'}.$$

What is the distribution of F when H_0 is true?

8.14. Using the notation of this section, assume that the means satisfy the condition that $\mu = \mu_1 + (b-1)d = \mu_2 - d = \mu_3 - d = \cdots = \mu_b - d$. That is, the last $b-1$ means are equal but differ from the first mean μ_1, provided that $d \ne 0$. Let a random sample of size a be taken from each of the b independent normal distributions with common unknown variance σ^2.

(a) Show that the maximum likelihood estimators of μ and d are $\hat\mu = X$ and

$$\hat{d} = \left[\sum_{j=2}^{b} X_{\cdot j}/(b-1) - X_{\cdot 1}\right]/b.$$

(b) Find Q_6 and $Q_7 = c\hat{d}^2$ so that when $d = 0$, Q_7/σ^2 is $\chi^2(1)$ and

$$\sum_{i=1}^{a}\sum_{j=1}^{b}(X_{ij} - X)^2 = Q_3 + Q_6 + Q_7.$$

(c) Argue that the three terms in the right-hand member of part (b), once divided by σ^2, are stochastically independent random variables with chi-square distributions, provided that $d = 0$.

(d) The ratio $Q_7/(Q_3 + Q_6)$ times what constant has an F distribution, provided that $d = 0$?

8.4 Noncentral χ^2 and Noncentral F

Let $X_1, X_2, \ldots, X_n$ denote mutually stochastically independent random variables that are $n(\mu_i, \sigma^2)$, $i = 1, 2, \ldots, n$, and let $Y = \sum\limits_{1}^{n} X_i^2/\sigma^2$. If each μ_i is zero, we know that Y is $\chi^2(n)$. We shall now investigate the distribution of Y when each μ_i is not zero. The moment-generating function of Y is given by

$$M(t) = E\left[\exp\left(t\sum_{i=1}^{n}\frac{X_i^2}{\sigma^2}\right)\right]$$

$$= \prod_{i=1}^{n} E\left[\exp\left(t\frac{X_i^2}{\sigma^2}\right)\right].$$

Consider

$$E\left[\exp\left(\frac{tX_i^2}{\sigma^2}\right)\right] = \int_{-\infty}^{\infty}\frac{1}{\sigma\sqrt{2\pi}}\exp\left[\frac{tx_i^2}{\sigma^2} - \frac{(x_i - \mu_i)^2}{2\sigma^2}\right]dx_i.$$

The integral exists if $t < \frac{1}{2}$. To evaluate the integral, note that

$$\frac{tx_i^2}{\sigma^2} - \frac{(x_i - \mu_i)^2}{2\sigma^2} = -\frac{x_i^2(1 - 2t)}{2\sigma^2} + \frac{2\mu_i x_i}{2\sigma^2} - \frac{\mu_i^2}{2\sigma^2}$$

$$= \frac{t\mu_i^2}{\sigma^2(1 - 2t)} - \frac{1 - 2t}{2\sigma^2}\left(x_i - \frac{\mu_i}{1 - 2t}\right)^2.$$

Accordingly, with $t < \frac{1}{2}$, we have

$$E\left[\exp\left(\frac{tX_i^2}{\sigma^2}\right)\right]$$

$$= \exp\left[\frac{t\mu_i^2}{\sigma^2(1 - 2t)}\right]\int_{-\infty}^{\infty} \frac{1}{\sigma\sqrt{2\pi}}\exp\left[-\frac{1 - 2t}{2\sigma^2}\left(x_i - \frac{\mu_i}{1 - 2t}\right)^2\right]dx_i.$$

If we multiply the integrand by $\sqrt{1 - 2t}$, $t < \frac{1}{2}$, we have the integral of a normal p.d.f. with mean $\mu_i/(1 - 2t)$ and variance $\sigma^2/(1 - 2t)$. Thus

$$E\left[\exp\left(\frac{tX_i^2}{\sigma^2}\right)\right] = \frac{1}{\sqrt{1 - 2t}}\exp\left[\frac{t\mu_i^2}{\sigma^2(1 - 2t)}\right],$$

and the moment-generating function of $Y = \sum_1^n X_i^2/\sigma^2$ is given by

$$M(t) = \frac{1}{(1 - 2t)^{n/2}}\exp\left[\frac{t\sum_1^n \mu_i^2}{\sigma^2(1 - 2t)}\right], \qquad t < \frac{1}{2}.$$

A random variable that has a moment-generating function of the functional form

$$M(t) = \frac{1}{(1 - 2t)^{r/2}}e^{t\theta/(1 - 2t)},$$

where $t < \frac{1}{2}$, $0 < \theta$, and r is a positive integer, is said to have a *noncentral chi-square distribution* with r degrees of freedom and noncentrality parameter θ. If one sets the noncentrality parameter $\theta = 0$, one has $M(t) = (1 - 2t)^{-r/2}$, which is the moment-generating function of a random variable that is $\chi^2(r)$. Such a random variable can appropriately be called a *central chi-square variable*. We shall use the symbol $\chi^2(r, \theta)$ to denote a noncentral chi-square distribution that has the parameters r and θ; and we shall say that a random variable is $\chi^2(r, \theta)$

to mean that the random variable has this kind of distribution. The symbol $\chi^2(r, 0)$ is equivalent to $\chi^2(r)$. Thus our random variable $Y = \sum_1^n X_i^2/\sigma^2$ of this section is $\chi^2\left(n, \sum_1^n \mu_i^2/\sigma^2\right)$. If each μ_i is equal to zero, then Y is $\chi^2(n, 0)$ or, more simply, Y is $\chi^2(n)$.

The noncentral chi-square variables in which we have interest are certain quadratic forms, in normally distributed variables, divided by a variance σ^2. In our example it is worth noting that the noncentrality parameter of $\sum_1^n X_i^2/\sigma^2$, which is $\sum_1^n \mu_i^2/\sigma^2$, may be computed by replacing each X_i in the quadratic form by its mean μ_i, $i = 1, 2, \ldots, n$. This is no fortuitous circumstance; any quadratic form $Q = Q(X_1, \ldots, X_n)$ in normally distributed variables, which is such that Q/σ^2 is $\chi^2(r, \theta)$, has $\theta = Q(\mu_1, \mu_2, \ldots, \mu_n)/\sigma^2$; and if Q/σ^2 is a chi-square variable (central or noncentral) for certain real values of $\mu_1, \mu_2, \ldots, \mu_n$, it is chi-square (central or noncentral) for *all* real values of these means.

It should be pointed out that Theorem 1, Section 8.2, is valid whether the random variables are central or noncentral chi-square variables.

We next discuss a noncentral F variable. If U and V are stochastically independent and are, respectively, $\chi^2(r_1)$ and $\chi^2(r_2)$, the random variable F has been defined by $F = r_2U/r_1V$. Now suppose, in particular, that U is $\chi^2(r_1, \theta)$, V is $\chi^2(r_2)$, and that U and V are stochastically independent. The random variable r_2U/r_1V is called a *noncentral F variable* with r_1 and r_2 degrees of freedom and with noncentrality parameter θ. Note that the noncentrality parameter of F is precisely the noncentrality parameter of the random variable U, which is $\chi^2(r_1, \theta)$.

Tables of noncentral chi-square and noncentral F are available in the literature. However, like those of noncentral t, they are too bulky to be put in this book.

EXERCISES

8.15. Let Y_i, $i = 1, 2, \ldots, n$, denote mutually stochastically independent random variables that are, respectively, $\chi^2(r_i, \theta_i)$, $i = 1, 2, \ldots, n$. Prove that $Z = \sum_1^n Y_i$ is $\chi^2\left(\sum_1^n r_i, \sum_1^n \theta_i\right)$.

8.16. Compute the mean and the variance of a random variable that is $\chi^2(r, \theta)$.

8.17. Compute the mean of a random variable that has a noncentral F

distribution with degrees of freedom r_1 and $r_2 > 2$ and noncentrality parameter θ.

8.18. Show that the square of a noncentral T random variable is a noncentral F random variable.

8.19. Let X_1 and X_2 be two stochastically independent random variables. Let X_1 and $Y = X_1 + X_2$ be $\chi^2(r_1, \theta_1)$ and $\chi^2(r, \theta)$, respectively. Here $r_1 < r$ and $\theta_1 \leq \theta$. Show that X_2 is $\chi^2(r - r_1, \theta - \theta_1)$.

8.5 The Analysis of Variance

The problem considered in Section 8.3 is an example of a method of statistical inference called the *analysis of variance*. This method derives its name from the fact that the quadratic form abS^2, which is a total sum of squares, is resolved into several component parts. In this section other problems in the analysis of variance will be investigated.

Let X_{ij}, $i = 1, 2, \ldots, a$ and $j = 1, 2, \ldots, b$, denote $n = ab$ random variables which are mutually stochastically independent and have normal distributions with common variance σ^2. The means of these normal distributions are $\mu_{ij} = \mu + \alpha_i + \beta_j$, where $\sum_1^a \alpha_i = 0$ and $\sum_1^b \beta_j = 0$. For example, take $a = 2$, $b = 3$, $\mu = 5$, $\alpha_1 = 1$, $\alpha_2 = -1$, $\beta_1 = 1$, $\beta_2 = 0$, and $\beta_3 = -1$. Then the $ab = $ six random variables have means

$$\mu_{11} = 7, \qquad \mu_{12} = 6, \qquad \mu_{13} = 5,$$
$$\mu_{21} = 5, \qquad \mu_{22} = 4, \qquad \mu_{23} = 3.$$

Had we taken $\beta_1 = \beta_2 = \beta_3 = 0$, the six random variables would have had means

$$\mu_{11} = 6, \qquad \mu_{12} = 6, \qquad \mu_{13} = 6,$$
$$\mu_{21} = 4, \qquad \mu_{22} = 4, \qquad \mu_{23} = 4.$$

Thus, if we wish to test the composite hypothesis that

$$\mu_{11} = \mu_{12} = \cdots = \mu_{1b},$$
$$\mu_{21} = \mu_{22} = \cdots = \mu_{2b},$$
$$\vdots$$
$$\mu_{a1} = \mu_{a2} = \cdots = \mu_{ab},$$

we could say that we are testing the composite hypothesis that $\beta_1 = \beta_2 = \cdots = \beta_b$ (and hence each $\beta_j = 0$, since their sum is zero). On the other hand, the composite hypothesis

$$\mu_{11} = \mu_{21} = \cdots = \mu_{a1},$$

$$\mu_{12} = \mu_{22} = \cdots = \mu_{a2},$$

$$\vdots$$

$$\mu_{1b} = \mu_{2b} = \cdots = \mu_{ab},$$

is the same as the composite hypothesis that $\alpha_1 = \alpha_2 = \cdots = \alpha_a = 0$.

Remarks. The model just described, and others similar to it, are widely used in statistical applications. Consider a situation in which it is desirable to investigate the effects of two factors that influence an outcome. Thus the variety of a grain and the type of fertilizer used influence the yield; or the teacher and the size of a class may influence the score on a standard test. Let X_{ij} denote the yield from the use of variety i of a grain and type j of fertilizer. A test of the hypothesis that $\beta_1 = \beta_2 = \cdots = \beta_b = 0$ would then be a test of the hypothesis that the mean yield of each variety of grain is the same regardless of the type of fertilizer used.

There is no loss of generality in assuming that $\sum_{1}^{a} \alpha_i = \sum_{1}^{b} \beta_j = 0$. To see this, let $\mu_{ij} = \mu' + \alpha_i' + \beta_j'$. Write $\bar{\alpha}' = \sum \alpha_i'/a$ and $\bar{\beta}' = \sum \beta_j'/b$. We have $\mu_{ij} = (\mu' + \bar{\alpha}' + \bar{\beta}') + (\alpha_i' - \bar{\alpha}') + (\beta_j' - \bar{\beta}') = \mu + \alpha_i + \beta_j$, where $\sum \alpha_i = \sum \beta_j = 0$.

To construct a test of the composite hypothesis $H_0: \beta_1 = \beta_2 = \cdots = \beta_b = 0$ against all alternative hypotheses, we could obtain the corresponding likelihood ratio. However, to gain more insight into such a test, let us reconsider the likelihood ratio test of Section 8.3, namely that of the equality of the means of b mutually independent distributions. There the important quadratic forms are Q, Q_3, and Q_4, which are related through the equation $Q = Q_4 + Q_3$. That is,

$$abS^2 = \sum_{j=1}^{b} \sum_{i=1}^{a} (\bar{X}_{.j} - \bar{X})^2 + \sum_{i=1}^{a} \sum_{j=1}^{b} (X_{ij} - \bar{X}_{.j})^2;$$

so we see that the total sum of squares, abS^2, is decomposed into a sum of squares, Q_4, *among* column means and a sum of squares, Q_3, *within* columns. The latter sum of squares, divided by $n = ab$, is the maximum likelihood estimator of σ^2, provided that the parameters are in Ω; and we denote it by $\widehat{\sigma_\Omega^2}$. Of course, S^2 is the maximum likelihood estimator of σ^2 under ω, here denoted by $\widehat{\sigma_\omega^2}$. So the likelihood ratio $\lambda = (\widehat{\sigma_\Omega^2}/\widehat{\sigma_\omega^2})^{ab/2}$ is a monotone function of the statistic

$$F = \frac{Q_4/(b-1)}{Q_3/[b(a-1)]}$$

upon which the test of the equality of means is based.

To help find a test for $H_0: \beta_1 = \beta_2 = \cdots = \beta_b = 0$, where $\mu_{ij} = \mu + \alpha_i + \beta_j$, return to the decomposition of Example 3, Section 8.2, namely $Q = Q_2 + Q_4 + Q_5$. That is,

$$abS^2 = \sum_{i=1}^{a} \sum_{j=1}^{b} (\bar{X}_{i\cdot} - \bar{X})^2 + \sum_{i=1}^{a} \sum_{j=1}^{b} (\bar{X}_{\cdot j} - \bar{X})^2$$

$$+ \sum_{i=1}^{a} \sum_{j=1}^{b} (X_{ij} - \bar{X}_{i\cdot} - \bar{X}_{\cdot j} + \bar{X})^2;$$

thus the total sum of squares, abS^2, is decomposed into that among *rows* (Q_2), that among *columns* (Q_4), and that *remaining* (Q_5). It is interesting to observe that $\widehat{\sigma_\Omega^2} = Q_5/ab$ is the maximum likelihood estimator of σ^2 under Ω and

$$\widehat{\sigma_\omega^2} = \frac{(Q_4 + Q_5)}{ab} = \sum_{i=1}^{a} \sum_{j=1}^{b} \frac{(X_{ij} - \bar{X}_{i\cdot})^2}{ab}$$

is that estimator under ω. A useful monotone function of the likelihood ratio $\lambda = (\widehat{\sigma_\Omega^2}/\widehat{\sigma_\omega^2})^{ab/2}$ is

$$F = \frac{Q_4/(b-1)}{Q_5/[(a-1)(b-1)]},$$

which has, under H_0, an F distribution with $b-1$ and $(a-1)(b-1)$ degrees of freedom. The hypothesis H_0 is rejected if $F \geq c$, where $\alpha = \Pr(F \geq c; H_0)$.

If we are to compute the power function of the test, we need the distribution of F when H_0 is not true. From Section 8.4 we know, when H_1 is true, that Q_4/σ^2 and Q_5/σ^2 are stochastically independent (central or noncentral) chi-square variables. We shall compute the non-centrality parameters of Q_4/σ^2 and Q_5/σ^2 when H_1 is true. We have $E(X_{ij}) = \mu + \alpha_i + \beta_j$, $E(\bar{X}_{i\cdot}) = \mu + \alpha_i$, $E(\bar{X}_{\cdot j}) = \mu + \beta_j$ and $E(\bar{X}) = \mu$. Accordingly, the noncentrality parameter of Q_4/σ^2 is

$$\frac{a \sum_{j=1}^{b} (\mu + \beta_j - \mu)^2}{\sigma^2} = \frac{a \sum_{j=1}^{b} \beta_j^2}{\sigma^2}$$

and that of Q_5/σ^2 is

$$\frac{\sum_{j=1}^{b} \sum_{i=1}^{a} (\mu + \alpha_i + \beta_j - \mu - \alpha_i - \mu - \beta_j + \mu)^2}{\sigma^2} = 0.$$

Thus, if the hypothesis H_0 is not true, F has a noncentral F distribution with $b - 1$ and $(a - 1)(b - 1)$ degrees of freedom and noncentrality parameter $a \sum_{j=1}^{b} \beta_j^2/\sigma^2$. The desired probabilities can then be found in tables of the noncentral F distribution.

A similar argument can be used to construct the F needed to test the equality of row means; that is, this F is essentially the ratio of the sum of squares among rows and Q_5. In particular, this F is defined by

$$F = \frac{Q_2/(a - 1)}{Q_5/[(a - 1)(b - 1)]}$$

and, under $H_0: \alpha_1 = \alpha_2 = \cdots = \alpha_a = 0$, has an F distribution with $a - 1$ and $(a - 1)(b - 1)$ degrees of freedom.

The analysis-of-variance problem that has just been discussed is usually referred to as a *two-way classification with one observation per cell*. Each combination of i and j determines a cell; thus there is a total of ab cells in this model. Let us now investigate another two-way classification problem, but in this case we take $c > 1$ stochastically independent observations per cell.

Let $X_{ijk}, i = 1, 2, \ldots, a, j = 1, 2, \ldots, b$, and $k = 1, 2, \ldots, c$, denote $n = abc$ random variables which are mutually stochastically independent and which have normal distributions with common, but unknown, variance σ^2. The mean of each $X_{ijk}, k = 1, 2, \ldots, c$, is $\mu_{ij} = \mu + \alpha_i + \beta_j + \gamma_{ij}$, where $\sum_{i=1}^{a} \alpha_i = 0, \sum_{j=1}^{b} \beta_j = 0, \sum_{i=1}^{a} \gamma_{ij} = 0$, and $\sum_{j=1}^{b} \gamma_{ij} = 0$. For example, take $a = 2, b = 3, \mu = 5, \alpha_1 = 1, \alpha_2 = -1, \beta_1 = 1, \beta_2 = 0, \beta_3 = -1, \gamma_{11} = 1, \gamma_{12} = 1, \gamma_{13} = -2, \gamma_{21} = -1, \gamma_{22} = -1$, and $\gamma_{23} = 2$. Then the means are

$$\mu_{11} = 8, \qquad \mu_{12} = 7, \qquad \mu_{13} = 3,$$
$$\mu_{21} = 4, \qquad \mu_{22} = 3, \qquad \mu_{23} = 5.$$

Note that, if each $\gamma_{ij} = 0$, then

$$\mu_{11} = 7, \qquad \mu_{12} = 6, \qquad \mu_{13} = 5,$$
$$\mu_{21} = 5, \qquad \mu_{22} = 4, \qquad \mu_{23} = 3.$$

That is, if $\gamma_{ij} = 0$, each of the means in the first row is 2 greater than the corresponding mean in the second row. In general, if each $\gamma_{ij} = 0$, the means of row i_1 differ from the corresponding means of row i_2 by a constant. This constant may be different for different choices of i_1 and i_2. A similar statement can be made about the means of columns j_1 and j_2. The parameter γ_{ij} is called the *interaction* associated with cell (i, j). That is, the interaction between the *i*th level of one classification and the *j*th level of the other classification is γ_{ij}. One interesting hypothesis to test is that each interaction is equal to zero. This will now be investigated.

From Exercise 8.11 of Section 8.2 we have that

$$\sum_{i=1}^{a} \sum_{j=1}^{b} \sum_{k=1}^{c} (X_{ijk} - \bar{X})^2 = bc \sum_{i=1}^{a} (\bar{X}_{i..} - \bar{X})^2 + ac \sum_{j=1}^{b} (\bar{X}_{.j.} - \bar{X})^2$$
$$+ c \sum_{i=1}^{a} \sum_{j=1}^{b} (\bar{X}_{ij.} - \bar{X}_{i..} - \bar{X}_{.j.} + \bar{X})^2$$
$$+ \sum_{i=1}^{a} \sum_{j=1}^{b} \sum_{k=1}^{c} (X_{ijk} - \bar{X}_{ij.})^2;$$

that is, the total sum of squares is decomposed into that due to *row* differences, that due to *column* differences, that due to *interaction*, and that *within cells*. The test of

$$H_0: \gamma_{ij} = 0, \qquad i = 1, 2, \ldots, a, j = 1, 2, \ldots, b,$$

against all possible alternatives is based upon an F with $(a-1)(b-1)$ and $ab(c-1)$ degrees of freedom,

$$F = \frac{\left[c \sum_{i=1}^{a} \sum_{j=1}^{b} (\bar{X}_{ij.} - \bar{X}_{i..} - \bar{X}_{.j.} + \bar{X})^2 \right] \Big/ [(a-1)(b-1)]}{[\sum \sum \sum (X_{ijk} - \bar{X}_{ij.})^2] / [ab(c-1)]}.$$

The reader should verify that the noncentrality parameter of this F distribution is equal to $c \sum_{j=1}^{b} \sum_{i=1}^{a} \gamma_{ij}^2 / \sigma^2$. Thus F is central when $H_0: \gamma_{ij} = 0$, $i = 1, 2, \ldots, a, j = 1, 2, \ldots, b$, is true.

EXERCISES

8.20. Show that

$$\sum_{j=1}^{b} \sum_{i=1}^{a} (X_{ij} - \bar{X}_{i.})^2 = \sum_{j=1}^{b} \sum_{i=1}^{a} (X_{ij} - \bar{X}_{i.} - \bar{X}_{.j} + \bar{X})^2 + a \sum_{j=1}^{b} (\bar{X}_{.j} - \bar{X})^2.$$

8.21. If at least one $\gamma_{ij} \neq 0$, show that the F, which is used to test that

each interaction is equal to zero, has noncentrality parameter equal to
$$c \sum_{j=1}^{b} \sum_{i=1}^{a} \gamma_{ij}^2/\sigma^2.$$

8.6 A Regression Problem

Consider a laboratory experiment the outcome of which depends upon the temperature; that is, the technician first sets a temperature dial at a fixed point c and subsequently observes the outcome of the experiment for that dial setting. From past experience, the technician knows that if he repeats the experiment with the temperature dial set at the same point c, he is not likely to observe precisely the same outcome. He then assumes that the outcome of his experiment is a random variable X whose distribution depends not only upon certain unknown parameters but also upon a nonrandom variable c which he can choose more or less at pleasure. Let $c_1, c_2, \ldots, c_n$ denote n arbitrarily selected values of c (but not all equal) and let X_i denote the outcome of the experiment when $c = c_i$, $i = 1, 2, \ldots, n$. We then have the n pairs $(X_1, c_1), \ldots, (X_n, c_n)$ in which the X_i are random variables but the c_i are known numbers and $i = 1, 2, \ldots, n$. Once the n experiments have been performed (the first with $c = c_1$, the second with $c = c_2$, and so on) and the outcome of each recorded, we have the n pairs of known numbers $(x_1, c_1), \ldots, (x_n, c_n)$. These numbers are to be used to make statistical inferences about the unknown parameters in the distribution of the random variable X. Certain problems of this sort are called *regression* problems and we shall study a particular one in some detail.

Let $c_1, c_2, \ldots, c_n$ be n given numbers, not all equal, and let $\bar{c} = \sum_{1}^{n} c_i/n$. Let $X_1, X_2, \ldots, X_n$ be n mutually stochastically independent random variables with joint p.d.f.

$L(\alpha, \beta, \sigma^2; x_1, x_2, \ldots, x_n)$

$$= \left(\frac{1}{2\pi\sigma^2}\right)^{n/2} \exp\left\{-\frac{1}{2\sigma^2} \sum_{1}^{n} [x_i - \alpha - \beta(c_i - \bar{c})]^2\right\}.$$

Thus each X_i has a normal distribution with the same variance σ^2, but the means of these distributions are $\alpha + \beta(c_i - \bar{c})$. Since the c_i are not all equal, in this regression problem the means of the normal distributions depend upon the choice of $c_1, c_2, \ldots, c_n$. We shall investigate ways of making statistical inferences about the parameters α, β, and σ^2.

It is easy to show (see Exercise 8.22) that the maximum likelihood estimators of α, β, and σ^2 are

$$\hat{\alpha} = \frac{\sum\limits_{1}^{n} X_i}{n} = \bar{X},$$

$$\hat{\beta} = \frac{\sum\limits_{1}^{n} (c_i - \bar{c})(X_i - \bar{X})}{\sum\limits_{1}^{n} (c_i - \bar{c})^2} = \frac{\sum\limits_{1}^{n} (c_i - \bar{c})X_i}{\sum\limits_{1}^{n} (c_i - \bar{c})^2},$$

and

$$\hat{\sigma}^2 = \frac{1}{n} \sum\limits_{1}^{n} [X_i - \hat{\alpha} - \hat{\beta}(c_i - \bar{c})]^2.$$

Since $\hat{\alpha}$ and $\hat{\beta}$ are linear functions of $X_1, X_2, \ldots, X_n$, each is normally distributed (Theorem 1, Section 4.7). It is easy to show (Exercise 8.23) that their respective means are α and β and their respective variances are σ^2/n and $\sigma^2/\sum\limits_{1}^{n} (c_i - \bar{c})^2$.

Consider next the algebraic identity (Exercise 8.24)

$$\sum\limits_{1}^{n} [X_i - \alpha - \beta(c_i - \bar{c})]^2 = \sum\limits_{1}^{n} \{(\hat{\alpha} - \alpha) + (\hat{\beta} - \beta)(c_i - \bar{c})$$
$$+ [X_i - \hat{\alpha} - \hat{\beta}(c_i - \bar{c})]\}^2$$
$$= n(\hat{\alpha} - \alpha)^2 + (\hat{\beta} - \beta)^2 \sum\limits_{1}^{n} (c_i - \bar{c})^2$$
$$+ \sum\limits_{1}^{n} [X_i - \hat{\alpha} - \hat{\beta}(c_i - \bar{c})]^2,$$

or

$$\sum\limits_{1}^{n} [X_i - \alpha - \beta(c_i - \bar{c})]^2 = n(\hat{\alpha} - \alpha)^2 + (\hat{\beta} - \beta)^2 \sum\limits_{1}^{n} (c_i - \bar{c})^2 + n\hat{\sigma}^2,$$

or, for brevity,

$$Q = Q_1 + Q_2 + Q_3.$$

Here Q, Q_1, Q_2, and Q_3 are real quadratic forms in the variables

$$X_i - \alpha - \beta(c_i - \bar{c}), \qquad i = 1, 2, \ldots, n.$$

In this equation, Q represents the sum of the squares of n mutually stochastically independent random variables that have normal distributions with means zero and variances σ^2. Thus Q/σ^2 has a chi-square distribution with n degrees of freedom. Each of the random variables $\sqrt{n}(\hat{\alpha} - \alpha)/\sigma$ and $\sqrt{\sum_1^n (c_i - \bar{c})^2}(\hat{\beta} - \beta)/\sigma$ has a normal distribution with zero mean and unit variance; thus each of Q_1/σ^2 and Q_2/σ^2 has a chi-square distribution with 1 degree of freedom. Since Q_3 is nonnegative, we have, in accordance with the theorem of Section 8.2, that Q_1, Q_2, and Q_3 are mutually stochastically independent, so that Q_3/σ^2 has a chi-square distribution with $n - 1 - 1 = n - 2$ degrees of freedom. Then each of the random variables

$$T_1 = \frac{[\sqrt{n}(\hat{\alpha} - \alpha)]/\sigma}{\sqrt{Q_3/[\sigma^2(n - 2)]}} = \frac{\hat{\alpha} - \alpha}{\sqrt{\hat{\sigma}^2/(n - 2)}}$$

and

$$T_2 = \frac{\left[\sqrt{\sum_1^n (c_i - \bar{c})^2}(\hat{\beta} - \beta)\right]/\sigma}{\sqrt{Q_3/[\sigma^2(n - 2)]}} = \frac{\hat{\beta} - \beta}{\sqrt{n\hat{\sigma}^2/\left[(n - 2)\sum_1^n (c_i - \bar{c})^2\right]}}$$

has a t distribution with $n - 2$ degrees of freedom. These facts enable us to obtain confidence intervals for α and β. The fact that $n\hat{\sigma}^2/\sigma^2$ has a chi-square distribution with $n - 2$ degrees of freedom provides a means of determining a confidence interval for σ^2. These are some of the statistical inferences about the parameters to which reference was made in the introductory remarks of this section.

Remark. The more discerning reader should quite properly question our constructions of T_1 and T_2 immediately above. We know that the *squares* of the linear forms are stochastically independent of $Q_3 = n\hat{\sigma}^2$, but we do not know, at this time, that the linear forms themselves enjoy this independence. This problem arises again in Section 8.7. In Exercise 12.15, a more general problem is proposed, of which the present case is a special instance.

EXERCISES

8.22. Verify that the maximum likelihood estimators of α, β, and σ^2 are the $\hat{\alpha}$, $\hat{\beta}$, and $\hat{\sigma}^2$ given in this section.

8.23. Show that $\hat{\alpha}$ and $\hat{\beta}$ have the respective means α and β and the respective variances σ^2/n and $\sigma^2/\sum_1^n (c_i - \bar{c})^2$.

8.24. Verify that $\sum_{1}^{n} [X_i - \alpha - \beta(c_i - \bar{c})]^2 = Q_1 + Q_2 + Q_3$, as stated in the text.

8.25. Let the mutually stochastically independent random variables $X_1, X_2, \ldots, X_n$ have, respectively, the probability density functions $n(\beta c_i, \gamma^2 c_i^2)$, $i = 1, 2, \ldots, n$, where the given numbers $c_1, c_2, \ldots, c_n$ are not all equal and no one is zero. Find the maximum likelihood estimators of β and γ^2.

8.26. Let the mutually stochastically independent random variables $X_1, \ldots, X_n$ have the joint p.d.f.

$$L(\alpha, \beta, \sigma^2; x_1, \ldots, x_n) = \left(\frac{1}{2\pi\sigma^2}\right)^{n/2} \exp\left\{-\frac{1}{2\sigma^2} \sum_{1}^{n} [x_i - \alpha - \beta(c_i - \bar{c})]^2\right\},$$

where the given numbers $c_1, c_2, \ldots, c_n$ are not all equal. Let $H_0: \beta = 0$ (α and σ^2 unspecified). It is desired to use a likelihood ratio test to test H_0 against all possible alternatives. Find λ and see whether the test can be based on a familiar statistic. *Hint.* In the notation of this section show that

$$\sum_{1}^{n} (X_i - \hat{\alpha})^2 = Q_3 + \hat{\beta}^2 \sum_{1}^{n} (c_i - \bar{c})^2.$$

8.27. Using the notation of Section 8.3, assume that the means μ_j satisfy a linear function of j, namely $\mu_j = c + d[j - (b + 1)/2]$. Let a random sample of size a be taken from each of the b independent normal distributions with common unknown variance σ^2.

(a) Show that the maximum likelihood estimators of c and d are, respectively, $\hat{c} = \bar{X}$ and

$$\hat{d} = \sum_{j=1}^{b} [j - (b + 1)/2](\bar{X}_{.j} - \bar{X}) \Big/ \sum_{j=1}^{b} [j - (b + 1)/2]^2.$$

(b) Show that

$$\sum_{i=1}^{a} \sum_{j=1}^{b} (X_{ij} - \bar{X})^2$$

$$= \sum_{i=1}^{a} \sum_{j=1}^{b} \left[X_{ij} - \bar{X} - \hat{d}\left(j - \frac{b+1}{2}\right)\right]^2 + \hat{d}^2 \sum_{j=1}^{b} a\left(j - \frac{b+1}{2}\right)^2.$$

(c) Argue that the two terms in the right-hand member of part (b), once divided by σ^2, are stochastically independent random variables with chi-square distributions provided that $d = 0$.

(d) What F statistic would be used to test the equality of the means, that is, $H_0: d = 0$?

8.7 A Test of Stochastic Independence

Let X and Y have a bivariate normal distribution with means μ_1 and μ_2, positive variances σ_1^2 and σ_2^2, and correlation coefficient ρ. We wish to test the hypothesis that X and Y are stochastically independent. Because two jointly normally distributed random variables are stochastically independent if and only if $\rho = 0$, we test the hypothesis $H_0 \colon \rho = 0$ against the hypothesis $H_1 \colon \rho \neq 0$. A likelihood ratio test will be used. Let $(X_1, Y_1), (X_2, Y_2), \ldots, (X_n, Y_n)$ denote a random sample of size $n > 2$ from the bivariate normal distribution; that is, the joint p.d.f. of these $2n$ random variables is given by

$$f(x_1, y_1) f(x_2, y_2) \cdots f(x_n, y_n).$$

Although it is fairly difficult to show, the statistic that is defined by the likelihood ratio λ is a function of the statistic

$$R = \frac{\displaystyle\sum_{i=1}^{n} (X_i - \bar{X})(Y_i - \bar{Y})}{\sqrt{\displaystyle\sum_{i=1}^{n} (X_i - \bar{X})^2 \sum_{i=1}^{n} (Y_i - \bar{Y})^2}}.$$

This statistic R is called the *correlation coefficient* of the random sample. The likelihood ratio principle, which calls for the rejection of H_0 if $\lambda \leq \lambda_0$, is equivalent to the computed value of $|R| \geq c$. That is, if the absolute value of the correlation coefficient of the sample is too large, we reject the hypothesis that the correlation coefficient of the distribution is equal to zero. To determine a value of c for a satisfactory significance level, it will be necessary to obtain the distribution of R, or a function of R, when H_0 is true. This will now be done.

Let $X_1 = x_1, X_2 = x_2, \ldots, X_n = x_n$, $n > 2$, where $x_1, x_2, \ldots, x_n$ and $\bar{x} = \sum_{1}^{n} x_i / n$ are fixed numbers such that $\sum_{1}^{n} (x_i - \bar{x})^2 > 0$. Consider the conditional p.d.f. of $Y_1, Y_2, \ldots, Y_n$, given that $X_1 = x_1, X_2 = x_2, \ldots, X_n = x_n$. Because $Y_1, Y_2, \ldots, Y_n$ are mutually stochastically independent and, with $\rho = 0$, are also mutually stochastically independent of $X_1, X_2, \ldots, X_n$, this conditional p.d.f. is given by

$$\left(\frac{1}{\sqrt{2\pi}\,\sigma_2}\right)^n \exp\left[-\frac{\sum_{1}^{n} (y_i - \mu_2)^2}{2\sigma_2^2}\right],$$

The conditional distribution of $\sum_{1}^{n} (Y_i - \bar{Y})^2 / \sigma_2^2$, given $X_1 = x_1, \ldots,$

$X_n = x_n$, is $\chi^2(n - 1)$. Moreover, the conditional distribution of the linear function W of $Y_1, Y_2, \ldots, Y_n$,

$$(1) \qquad W = \frac{\sum\limits_{1}^{n} [(x_i - \bar{x})(Y_i - \bar{Y})]}{\sqrt{\sum\limits_{1}^{n} (x_i - \bar{x})^2}}$$

$$= \frac{\sum\limits_{1}^{n} [(x_i - \bar{x})Y_i]}{\sqrt{\sum\limits_{1}^{n} (x_i - \bar{x})^2}},$$

is $n(0, \sigma_2^2)$ (see Exercise 8.30). Thus the conditional distribution of W^2/σ_2^2, given $X_1 = x_1, \ldots, X_n = x_n$, is $\chi^2(1)$. We have the algebraic identity (see Exercise 8.31)

$$(2) \quad \sum\limits_{1}^{n} (Y_i - \bar{Y})^2$$

$$= W^2 + \sum\limits_{1}^{n} \left\{ Y_i - \bar{Y} - \frac{\sum\limits_{1}^{n} [(x_i - \bar{x})(Y_i - \bar{Y})]}{\sum\limits_{1}^{n} (x_i - \bar{x})^2}(x_i - \bar{x}) \right\}^2.$$

The left-hand member of this equation and the first term of the right-hand member are, when divided by σ_2^2, respectively, conditionally $\chi^2(n - 1)$ and conditionally $\chi^2(1)$. In accordance with Theorem 1, the nonnegative quadratic form, say U, which is the second term of the right-hand member of Equation (2), is conditionally stochastically independent of W^2, and, when divided by σ_2^2, is conditionally $\chi^2(n - 2)$. Now W/σ_2 is $n(0, 1)$. Then (Remark, Section 8.6)

$$\frac{W/\sigma_2}{\sqrt{U/[\sigma_2^2(n - 2)]}} = \frac{W\sqrt{(n - 2)}}{\sqrt{U}}$$

has a conditional t distribution with $n - 2$ degrees of freedom. Let

$$R_c = \frac{\sum\limits_{1}^{n} (x_i - \bar{x})(Y_i - \bar{Y})}{\sqrt{\sum\limits_{1}^{n} (x_i - \bar{x})^2 \sum\limits_{1}^{n} (Y_i - \bar{Y})^2}}.$$

Then we have (Exercise 8.32)

$$(3) \qquad \frac{W\sqrt{n - 2}}{\sqrt{U}} = \frac{R_c\sqrt{n - 2}}{\sqrt{1 - R_c^2}};$$

this ratio has, given $X_1 = x_1, \ldots, X_n = x_n$, a conditional t distribution with $n - 2$ degrees of freedom. Note that the p.d.f., say $g(t)$, of this t distribution does not depend upon $x_1, x_2, \ldots, x_n$. Now the joint p.d.f. of $X_1, X_2, \ldots, X_n$ and $R\sqrt{n-2}/\sqrt{1-R^2}$, where

$$ R = \frac{\sum_1^n (X_i - \bar{X})(Y_i - \bar{Y})}{\sqrt{\sum_1^n (X_i - \bar{X})^2 \sum_1^n (Y_i - \bar{Y})^2}}, $$

is the product of $g(t)$ and the joint p.d.f. of $X_1, X_2, \ldots, X_n$. Integration on $x_1, x_2, \ldots, x_n$ yields the marginal p.d.f. of $R\sqrt{n-2}/\sqrt{1-R^2}$; because $g(t)$ does not depend upon $x_1, x_2, \ldots, x_n$ it is obvious that this marginal p.d.f. is $g(t)$, the conditional p.d.f. of $R_c\sqrt{n-2}/\sqrt{1-R_c^2}$. The change-of-variable technique can now be used to find the p.d.f. of R.

Remarks. Since R has, when $\rho = 0$, a conditional distribution that does not depend upon $x_1, x_2, \ldots, x_n$ (and hence that conditional distribution is, in fact, the marginal distribution of R), we have the remarkable fact that R is stochastically independent of $X_1, X_2, \ldots, X_n$. It follows that R is stochastically independent of *every function* of $X_1, X_2, \ldots, X_n$ alone, that is, a function that does not depend upon any Y_i. In like manner, R is stochastically independent of every function of $Y_1, Y_2, \ldots, Y_n$ alone. Moreover, a careful review of the argument reveals that nowhere did we use the fact that X has a normal marginal distribution. Thus, if X and Y are stochastically independent, and if Y has a normal distribution, then R has the same conditional distribution whatever be the distribution of X, subject to the condition $\sum_1^n (x_i - \bar{x})^2 > 0$. Moreover, if $\Pr\left[\sum_1^n (X_i - \bar{X})^2 > 0\right] = 1$, then R has the same marginal distribution whatever be the distribution of X.

If we write $T = R\sqrt{n-2}/\sqrt{1-R^2}$, where T has a t distribution with $n - 2 > 0$ degrees of freedom, it is easy to show, by the change-of-variable technique (Exercise 8.33), that the p.d.f. of R is given by

$$ (4) \quad g(r) = \frac{\Gamma[(n-1)/2]}{\Gamma(\frac{1}{2})\Gamma[(n-2)/2]} (1 - r^2)^{(n-4)/2}, \qquad -1 < r < 1, $$

$$ = 0 \text{ elsewhere.} $$

We have now solved the problem of the distribution of R, when $\rho = 0$ and $n > 2$, or, perhaps more conveniently, that of $R\sqrt{n-2}/\sqrt{1-R^2}$. The likelihood ratio test of the hypothesis $H_0: \rho = 0$ against all alternatives $H_1: \rho \neq 0$ may be based either on the statistic R or on the statistic $R\sqrt{n-2}/\sqrt{1-R^2} = T$. In either case the significance level of the test is

$$\alpha = \Pr\left(|R| \geq c_1; H_0\right) = \Pr\left(|T| \geq c_2; H_0\right),$$

where the constants c_1 and c_2 are chosen so as to give the desired value of α.

Remark. It is also possible to obtain an approximate test of size α by using the fact that

$$W = \frac{1}{2}\ln\left(\frac{1+R}{1-R}\right)$$

has an approximate normal distribution with mean $\frac{1}{2}\ln\left[(1+\rho)/(1-\rho)\right]$ and variance $1/(n-3)$. We accept this statement without proof. Thus a test of $H_0: \rho = 0$ can be based on the statistic

$$Z = \frac{\frac{1}{2}\ln\left[(1+R)/(1-R)\right] - \frac{1}{2}\ln\left[(1+\rho)/(1-\rho)\right]}{\sqrt{1/(n-3)}},$$

with $\rho = 0$ so that $\frac{1}{2}\ln\left[(1+\rho)/(1-\rho)\right] = 0$. However, using W, we can also test hypotheses like $H_0: \rho = \rho_0$ against $H_1: \rho \neq \rho_0$, where ρ_0 is not necessarily zero. In that case the hypothesized mean of W is

$$\frac{1}{2}\ln\left(\frac{1+\rho_0}{1-\rho_0}\right).$$

EXERCISES

8.28. Show that

$$R = \frac{\sum\limits_{1}^{n}(X_i - \bar{X})(Y_i - \bar{Y})}{\sqrt{\sum\limits_{1}^{n}(X_i - \bar{X})^2 \sum\limits_{1}^{n}(Y_i - \bar{Y})^2}} = \frac{\sum\limits_{1}^{n}X_iY_i - n\bar{X}\bar{Y}}{\sqrt{\left(\sum\limits_{1}^{n}X_i^2 - n\bar{X}^2\right)\left(\sum\limits_{1}^{n}Y_i^2 - n\bar{Y}^2\right)}}.$$

8.29. A random sample of size $n = 6$ from a bivariate normal distribution yields the value of the correlation coefficient to be 0.89. Would we accept or reject, at the 5 per cent significance level, the hypothesis that $\rho = 0$?

8.30. Verify that W of Equation (1) of this section is $n(0, \sigma_2^2)$.

8.31. Verify the algebraic identity (2) of this section.

8.32. Verify Equation (3) of this section.

8.33. Verify the p.d.f. (4) of this section.

Chapter 9
Nonparametric Methods

9.1 Confidence Intervals for Distribution Quantiles

We shall first define the concept of a quantile of a distribution of a random variable of the continuous type. Let X be a random variable of the continuous type with p.d.f. $f(x)$ and distribution function $F(x)$. Let p denote a positive proper fraction and assume that the equation $F(x) = p$ has a unique solution for x. This unique root is denoted by the symbol ξ_p and is called the *quantile* (of the distribution) *of order p*. Thus $\Pr(X \le \xi_p) = F(\xi_p) = p$. For example, the quantile of order $\frac{1}{2}$ is the median of the distribution and $\Pr(X \le \xi_{0.5}) = F(\xi_{0.5}) = \frac{1}{2}$.

In Chapter 6 we computed the probability that a certain random interval includes a special point. Frequently, this special point was a parameter of the distribution of probability under consideration. Thus we were led to the notion of an interval estimate of a parameter. If the parameter happens to be a quantile of the distribution, and if we work with certain functions of the order statistics, it will be seen that this method of statistical inference is applicable to all distributions of the continuous type. We call these methods *distribution-free* or *nonparametric* methods of inference.

To obtain a distribution-free confidence interval for ξ_p, the quantile of order p, of a distribution of the continuous type with distribution function $F(x)$, take a random sample $X_1, X_2, \ldots, X_n$ of size n from that distribution. Let $Y_1 < Y_2 < \cdots < Y_n$ be the order statistics of the sample. Take $Y_i < Y_j$ and consider the event $Y_i < \xi_p < Y_j$. For the ith order statistic Y_i to be less than ξ_p it must be true that at least i

304

of the X values are less than ξ_p. Moreover, for the jth order statistic to be greater than ξ_p, fewer than j of the X values are less than ξ_p. That is, if we say that we have a "success" when an individual X value is less than ξ_p, then, in the n independent trials, there must be at least i successes but fewer than j successes for the event $Y_i < \xi_p < Y_j$ to occur. But since the probability of success on each trial is Pr $(X < \xi_p)$ $= F(\xi_p) = p$, the probability of this event is

$$\Pr\left(Y_i < \xi_p < Y_j\right) = \sum_{w=i}^{j-1} \frac{n!}{w!\,(n-w)!}\, p^w (1-p)^{n-w},$$

the probability of having at least i, but less than j, successes. When particular values of n, i, and j are specified, this probability can be computed. By this procedure, suppose it has been found that $\gamma =$ Pr $(Y_i < \xi_p < Y_j)$. Then the probability is γ that the random interval (Y_i, Y_j) includes the quantile of order p. If the experimental values of Y_i and Y_j are, respectively, y_i and y_j, the interval (y_i, y_j) serves as a 100γ per cent confidence interval for ξ_p, the quantile of order p.

An illustrative example follows.

Example 1. Let $Y_1 < Y_2 < Y_3 < Y_4$ be the order statistics of a random sample of size 4 from a distribution of the continuous type. The probability that the random interval (Y_1, Y_4) includes the median $\xi_{0.5}$ of the distribution will be computed. We have

$$\Pr\left(Y_1 < \xi_{0.5} < Y_4\right) = \sum_{w=1}^{3} \frac{4!}{w!\,(4-w)!} \left(\frac{1}{2}\right)^4 = 0.875.$$

If Y_1 and Y_4 are observed to be $y_1 = 2.8$ and $y_4 = 4.2$, respectively, the interval $(2.8, 4.2)$ is an 87.5 per cent confidence interval for the median $\xi_{0.5}$ of the distribution.

For samples of fairly large size, we can approximate the binomial probabilities with those associated with normal distributions, as illustrated in the next example.

Example 2. Let the following numbers represent the order statistics of $n = 27$ observations obtained in a random sample from a certain distribution of the continuous type.

61, 69, 71, 74, 79, 80, 83, 84, 86, 87, 92, 93, 96, 100,

104, 105, 113, 121, 122, 129, 141, 143, 156, 164, 191, 217, 276.

Say that we are interested in estimating the 25th percentile $\xi_{0.25}$ (that is, the quantile of order 0.25) of the distribution. Since $(n + 1)p = 28(\frac{1}{4}) = 7$, the seventh order statistic, $y_7 = 83$, could serve as a point estimate of $\xi_{0.25}$. To get a confidence interval for $\xi_{0.25}$, consider two order statistics, one less than y_7 and the other greater, for illustration, y_4 and y_{10}. What is the confidence coefficient associated with the interval (y_4, y_{10})? Of course, before the sample is drawn, we know that

$$\gamma = \Pr\left(Y_4 < \xi_{0.25} < Y_{10}\right) = \sum_{w=4}^{9} \binom{27}{w}(0.25)^w(0.75)^{27-w}.$$

That is,

$$\gamma = \Pr\left(3.5 < W < 9.5\right),$$

where W is $b(27, \frac{1}{4})$ with mean $\frac{27}{4} = 6.75$ and variance $\frac{81}{16}$. Hence γ is approximately equal to

$$N\left(\frac{9.5 - 6.75}{\frac{9}{4}}\right) - N\left(\frac{3.5 - 6.75}{\frac{9}{4}}\right) = N\left(\frac{11}{9}\right) - N\left(-\frac{13}{9}\right) = 0.814.$$

Thus ($y_4 = 74$, $y_{10} = 87$) serves as an 81.4 per cent confidence interval for $\xi_{0.25}$. It should be noted that we could choose other intervals also, for illustration, ($y_3 = 71$, $y_{11} = 92$), and these would have different confidence coefficients. The persons involved in the study must select the desired confidence coefficient, and then the appropriate order statistics, Y_i and Y_j, are taken in such a way that i and j are fairly symmetrically located about $(n+1)p$.

EXERCISES

9.1. Let Y_n denote the nth order statistic of a random sample of size n from a distribution of the continuous type. Find the smallest value of n for which $\Pr\left(\xi_{0.9} < Y_n\right) \geq 0.75$.

9.2. Let $Y_1 < Y_2 < Y_3 < Y_4 < Y_5$ denote the order statistics of a random sample of size 5 from a distribution of the continuous type. Compute:
(a) $\Pr\left(Y_1 < \xi_{0.5} < Y_5\right)$.
(b) $\Pr\left(Y_1 < \xi_{0.25} < Y_3\right)$.
(c) $\Pr\left(Y_4 < \xi_{0.80} < Y_5\right)$.

9.3. Compute $\Pr\left(Y_3 < \xi_{0.5} < Y_7\right)$ if $Y_1 < \cdots < Y_9$ are the order statistics of a random sample of size 9 from a distribution of the continuous type.

9.4. Find the smallest value of n for which $\Pr\left(Y_1 < \xi_{0.5} < Y_n\right) \geq 0.99$, where $Y_1 < \cdots < Y_n$ are the order statistics of a random sample of size n from a distribution of the continuous type.

9.5. Let $Y_1 < Y_2$ denote the order statistics of a random sample of size 2 from a distribution which is $n(\mu, \sigma^2)$, where σ^2 is known.

(a) Show that $\Pr(Y_1 < \mu < Y_2) = \frac{1}{2}$ and compute the expected value of the random length $Y_2 - Y_1$.

(b) If $\bar{X}$ is the mean of this sample, find the constant c such that $\Pr(\bar{X} - c\sigma < \mu < \bar{X} + c\sigma) = \frac{1}{2}$, and compare the length of this random interval with the expected value of that of part (a). *Hint.* See Exercise 4.60, Section 4.6.

9.6. Let $Y_1 < Y_2 < \cdots < Y_{25}$ be the order statistics of a random sample of size $n = 25$ from a distribution of the continuous type. Compute approximately:

(a) $\Pr(Y_8 < \xi_{0.5} < Y_{18})$.
(b) $\Pr(Y_2 < \xi_{0.2} < Y_9)$.
(c) $\Pr(Y_{18} < \xi_{0.8} < Y_{23})$.

$\checkmark$ **9.7.** Let $Y_1 < Y_2 < \cdots < Y_{100}$ be the order statistics of a random sample of size $n = 100$ from a distribution of the continuous type. Find $i < j$ so that $\Pr(Y_i < \xi_{0.2} < Y_j)$ is about equal to 0.95.

9.2 Tolerance Limits for Distributions

We propose now to investigate a problem that has something of the same flavor as that treated in Section 9.1. Specifically, can we compute the probability that a certain random interval includes (or *covers*) a preassigned percentage of the probability for the distribution under consideration? And, by appropriate selection of the random interval, can we be led to an additional distribution-free method of statistical inference?

Let X be a random variable with distribution function $F(x)$ of the continuous type. The random variable $Z = F(X)$ is an important random variable, and its distribution is given in Example 1, Section 4.1. It is our purpose now to make an interpretation. Since $Z = F(X)$ has the p.d.f.

$$h(z) = 1, \qquad 0 < z < 1,$$
$$= 0 \text{ elsewhere,}$$

then, if $0 < p < 1$, we have

$$\Pr[F(X) \le p] = \int_0^p dz = p.$$

Now $F(x) = \Pr(X \le x)$. Since $\Pr(X = x) = 0$, then $F(x)$ is the fractional part of the probability for the distribution of X that is between $-\infty$ and x. If $F(x) \le p$, then no more than $100p$ per cent of the probability for the distribution of X is between $-\infty$ and x. But recall $\Pr[F(X) \le p] = p$. That is, the probability that the random variable $Z = F(X)$ is less than or equal to p is precisely the probability that the random interval $(-\infty, X)$ contains no more than $100p$ per cent of the probability for the distribution. For example, the probability that the random interval $(-\infty, X)$ contains no more than 70 per cent of the probability for the distribution is 0.70; and the probability that the random interval $(-\infty, X)$ contains more than 70 per cent of the probability for the distribution is $1 - 0.70 = 0.30$.

We now consider certain functions of the order statistics. Let $X_1, X_2, \ldots, X_n$ denote a random sample of size n from a distribution that has a positive and continuous p.d.f. $f(x)$ if and only if $a < x < b$; and let $F(x)$ denote the associated distribution function. Consider the random variables $F(X_1), F(X_2), \ldots, F(X_n)$. These random variables are mutually stochastically independent and each, in accordance with Example 1, Section 4.1, has a uniform distribution on the interval $(0, 1)$. Thus $F(X_1), F(X_2), \ldots, F(X_n)$ is a random sample of size n from a uniform distribution on the interval $(0, 1)$. Consider the order statistics of this random sample $F(X_1), F(X_2), \ldots, F(X_n)$. Let Z_1 be the smallest of these $F(X_i)$, Z_2 the next $F(X_i)$ in order of magnitude, $\ldots$, and Z_n the largest $F(X_i)$. If $Y_1, Y_2, \ldots, Y_n$ are the order statistics of the initial random sample $X_1, X_2, \ldots, X_n$, the fact that $F(x)$ is a nondecreasing (here, strictly increasing) function of x implies that $Z_1 = F(Y_1)$, $Z_2 = F(Y_2), \ldots, Z_n = F(Y_n)$. Thus the joint p.d.f. of $Z_1, Z_2, \ldots, Z_n$ is given by

$$h(z_1, z_2, \ldots, z_n) = n!, \qquad 0 < z_1 < z_2 < \cdots < z_n < 1,$$
$$= 0 \text{ elsewhere.}$$

This proves a special case of the following theorem.

Theorem 1. *Let $Y_1, Y_2, \ldots, Y_n$ denote the order statistics of a random sample of size n from a distribution of the continuous type that has p.d.f. $f(x)$ and distribution function $F(x)$. The joint p.d.f. of the random variables $Z_i = F(Y_i)$, $i = 1, 2, \ldots, n$, is*

$$h(z_1, z_2, \ldots, z_n) = n!, \qquad 0 < z_1 < z_2 < \cdots < z_n < 1,$$
$$= 0 \text{ elsewhere.}$$

Because the distribution function of $Z = F(X)$ is given by z, $0 < z < 1$, the marginal p.d.f. of $Z_k = F(Y_k)$ is the following beta p.d.f.:

$$(1) \quad h_k(z_k) = \frac{n!}{(k-1)!\,(n-k)!}\, z_k^{k-1}(1 - z_k)^{n-k}, \qquad 0 < z_k < 1,$$
$$= 0 \text{ elsewhere.}$$

Moreover, the joint p.d.f. of $Z_i = F(Y_i)$ and $Z_j = F(Y_j)$ is, with $i < j$, given by

$$(2) \quad h_{ij}(z_i, z_j) = \frac{n!}{(i-1)!\,(j-i-1)!\,(n-j)!}$$

$$\times z_i^{i-1}(z_j - z_i)^{j-i-1}(1 - z_j)^{n-j}, \qquad 0 < z_i < z_j < 1,$$

$$= 0 \text{ elsewhere.}$$

Consider the difference $Z_j - Z_i = F(Y_j) - F(Y_i)$, $i < j$. Now $F(y_j) = \Pr(X \le y_j)$ and $F(y_i) = \Pr(X \le y_i)$. Since $\Pr(X = y_i) = \Pr(X = y_j) = 0$, then the difference $F(y_j) - F(y_i)$ is that fractional part of the probability for the distribution of X that is between y_i and y_j. Let p denote a positive proper fraction. If $F(y_j) - F(y_i) \ge p$, then at least $100p$ per cent of the probability for the distribution of X is between y_i and y_j. Let it be given that $\gamma = \Pr[F(Y_j) - F(Y_i) \ge p]$. Then the random interval (Y_i, Y_j) has probability γ of containing at least $100p$ per cent of the probability for the distribution of X. If now y_i and y_j denote, respectively, experimental values of Y_i and Y_j, the interval (y_i, y_j) either does or does not contain at least $100p$ per cent of the probability for the distribution of X. However, we refer to the interval (y_i, y_j) as a 100γ per cent tolerance interval for $100p$ per cent of the probability for the distribution of X. In like vein, y_i and y_j are called 100γ per cent tolerance limits for $100p$ per cent of the probability for the distribution of X.

One way to compute the probability $\gamma = \Pr[F(Y_j) - F(Y_i) \ge p]$ is to use Equation (2), which gives the joint p.d.f. of $Z_i = F(Y_i)$ and $Z_j = F(Y_j)$. The required probability is then given by

$$\gamma = \Pr(Z_j - Z_i \ge p) = \int_0^{1-p} \int_{p+z_i}^{1} h_{ij}(z_i, z_j)\, dz_j\, dz_i.$$

Sometimes this is a rather tedious computation. For this reason and for the reason that *coverages* are important in distribution-free statistical inference, we choose to introduce at this time the concept of a coverage. Consider the random variables $W_1 = F(Y_1) = Z_1$, $W_2 = F(Y_2) - F(Y_1) = Z_2 - Z_1$, $W_3 = F(Y_3) - F(Y_2) = Z_3 - Z_2, \ldots, W_n = F(Y_n) - F(Y_{n-1}) = Z_n - Z_{n-1}$. The random variable W_1 is called a *coverage* of the random interval $\{x; -\infty < x < Y_1\}$ and the random variable W_i, $i = 2, 3, \ldots, n$, is called a *coverage* of the random interval $\{x; Y_{i-1} < x < Y_i\}$. We shall find the joint p.d.f. of the n coverages

$W_1, W_2, \ldots, W_n$. First we note that the inverse functions of the associated transformation are given by

$$z_1 = w_1,$$
$$z_2 = w_1 + w_2,$$
$$z_3 = w_1 + w_2 + w_3,$$
$$\vdots \qquad\qquad \vdots$$
$$z_n = w_1 + w_2 + w_3 + \cdots + w_n.$$

We also note that the Jacobian is equal to 1 and that the space of positive probability density is

$$\{(w_1, w_2, \ldots, w_n); 0 < w_i, i = 1, 2, \ldots, n, w_1 + \cdots + w_n < 1\}.$$

Since the joint p.d.f. of $Z_1, Z_2, \ldots, Z_n$ is $n!, 0 < z_1 < z_2 < \cdots < z_n < 1$, zero elsewhere, the joint p.d.f. of the n coverages is

$$k(w_1, \ldots, w_n) = n!, \qquad 0 < w_i, i = 1, \ldots, n, w_1 + \cdots + w_n < 1,$$
$$= 0 \text{ elsewhere.}$$

A reexamination of Example 1 of Section 4.5 reveals that this is a Dirichlet p.d.f. with $k = n$ and $\alpha_1 = \alpha_2 = \cdots = \alpha_{n+1} = 1$.

Because the p.d.f. $k(w_1, \ldots, w_n)$ is symmetric in $w_1, w_2, \ldots, w_n$, it is evident that the distribution of every sum of $r, r < n$, of these coverages $W_1, \ldots, W_n$ is exactly the same for each fixed value of r. For instance, if $i < j$ and $r = j - i$, the distribution of $Z_j - Z_i = F(Y_j) - F(Y_i) = W_{i+1} + W_{i+2} + \cdots + W_j$ is exactly the same as that of $Z_{j-i} = F(Y_{j-i}) = W_1 + W_2 + \cdots + W_{j-i}$. But we know that the p.d.f. of Z_{j-i} is the beta p.d.f. of the form

$$h_{j-i}(v) = \frac{\Gamma(n+1)}{\Gamma(j-i)\Gamma(n-j+i+1)} v^{j-i-1}(1-v)^{n-j+i}, \qquad 0 < v < 1,$$
$$= 0 \text{ elsewhere.}$$

Consequently, $F(Y_j) - F(Y_i)$ has this p.d.f. and

$$\Pr[F(Y_j) - F(Y_i) \geq p] = \int_p^1 h_{j-i}(v) \, dv.$$

Example 1. Let $Y_1 < Y_2 < \cdots < Y_6$ be the order statistics of a random sample of size 6 from a distribution of the continuous type. We want to use the observed interval (y_1, y_6) as a tolerance interval for 80 per cent of the distribution. Then

$$\gamma = \Pr[F(Y_6) - F(Y_1) \geq 0.8]$$
$$= 1 - \int_0^{0.8} 30v^4(1-v) \, dv,$$

because the integrand is the p.d.f. of $F(Y_6) - F(Y_1)$. Accordingly,

$$\gamma = 1 - 6(0.8)^5 + 5(0.8)^6 = 0.34,$$

approximately. That is, the observed values of Y_1 and Y_6 will define a 34 per cent tolerance interval for 80 per cent of the probability for the distribution.

Example 2. Each of the coverages W_i, $i = 1, 2, \ldots, n$, has the beta p.d.f.

$$k_1(w) = n(1 - w)^{n-1}, \qquad 0 < w < 1,$$
$$= 0 \text{ elsewhere,}$$

because $W_1 = Z_1 = F(Y_1)$ has this p.d.f. Accordingly, the mathematical expectation of each W_i is

$$\int_0^1 nw(1 - w)^{n-1}\, dw = \frac{1}{n + 1}.$$

Now the coverage W_i can be thought of as the area under the graph of the p.d.f. $f(x)$, above the x-axis, and between the lines $x = Y_{i-1}$ and $x = Y_i$. (We take $Y_0 = -\infty$.) Thus the expected value of each of these random areas W_i, $i = 1, 2, \ldots, n$, is $1/(n + 1)$. That is, the order statistics partition the probability for the distribution into $n + 1$ parts, and the expected value of each of these parts is $1/(n + 1)$. More generally, the expected value of $F(Y_j) - F(Y_i)$, $i < j$, is $(j - i)/(n + 1)$, since $F(Y_j) - F(Y_i)$ is the sum of $j - i$ of these coverages. This result provides a reason for calling Y_k, where $(n + 1)p = k$, the $(100p)$th *percentile of the sample*, since

$$E[F(Y_k)] = \frac{k}{n + 1} = \frac{(n + 1)p}{n + 1} = p.$$

EXERCISES

9.8. Let Y_1 and Y_n be, respectively, the first and nth order statistics of a random sample of size n from a distribution of the continuous type having distribution function $F(x)$. Find the smallest value of n such that $\Pr[F(Y_n) - F(Y_1) \geq 0.5]$ is at least 0.95.

9.9. Let Y_2 and Y_{n-1} denote the second and the $(n - 1)$st order statistics of a random sample of size n from a distribution of the continuous type having distribution function $F(x)$. Compute $\Pr[F(Y_{n-1}) - F(Y_2) \geq p]$, where $0 < p < 1$.

9.10. Let $Y_1 < Y_2 < \cdots < Y_{48}$ be the order statistics of a random sample of size 48 from a distribution of the continuous type. We want to use the observed interval (y_4, y_{45}) as a 100γ per cent tolerance interval for 75 per cent of the distribution.

(a) To what is γ equal?

(b) Approximate the integral in part (a) by noting that it can be written as a partial sum of a binomial p.d.f., which in turn can be approximated by probabilities associated with a normal distribution.

9.11. Let $Y_1 < Y_2 < \cdots < Y_n$ be the order statistics of a random sample of size n from a distribution of the continuous type having distribution function $F(x)$.

(a) What is the distribution of $U = 1 - F(Y_j)$?

(b) Determine the distribution of $V = F(Y_n) - F(Y_j) + F(Y_i) - F(Y_1)$, where $i < j$.

9.3 The Sign Test

Some of the chi-square tests of Section 8.1 are illustrative of the type of tests that we investigate in the remainder of this chapter. Recall, in that section, we tested the hypothesis that the distribution of a certain random variable X is a specified distribution. We did this in the following manner. The space of X was partitioned into k mutually disjoint sets $A_1, A_2, \ldots, A_k$. The probability p_{i0} that $X \in A_i$ was computed under the assumption that the specified distribution is the correct distribution, $i = 1, 2, \ldots, k$. The original hypothesis was then replaced by the hypothesis

$$H_0 \colon \Pr\,(X \in A_i) = p_{i0}, \qquad i = 1, 2, \ldots, k;$$

and a chi-square test, based upon a statistic that was denoted by Q_{k-1}, was used to test the hypothesis H_0 against all alternative hypotheses.

There is a certain subjective element in the use of this test, namely the choice of k and of $A_1, A_2, \ldots, A_k$. But it is important to note that the limiting distribution of Q_{k-1}, under H_0, is $\chi^2(k-1)$; that is, the distribution of Q_{k-1} is *free* of $p_{10}, p_{20}, \ldots, p_{k0}$ and, accordingly, of the specified distribution of X. Here, and elsewhere, "under H_0" means when H_0 is true. A test of a hypothesis H_0 based upon a statistic whose distribution, under H_0, does not depend upon the specified distribution or any parameters of that distribution is called a *distribution-free* or a *nonparametric test*.

Next, let $F(x)$ be the unknown distribution function of the random variable X. Let there be given two numbers ξ and p_0, where $0 < p_0 < 1$. We wish to test the hypothesis $H_0 \colon F(\xi) = p_0$, that is, the hypothesis that $\xi = \xi_{p_0}$, the quantile of order p_0 of the distribution of X. We could use the statistic Q_{k-1}, with $k = 2$, to test H_0 against all alternatives.

Suppose, however, that we are interested only in the alternative hypothesis, which is H_1: $F(\xi) > p_0$. One procedure is to base the test of H_0 against H_1 upon the random variable Y, which is the number of items less than or equal to ξ in a random sample of size n from the distribution. The statistic Y can be thought of as the number of "successes" throughout n independent trials. Then, if H_0 is true, Y is $b[n, p_0 = F(\xi)]$; whereas if H_0 is false, Y is $b[n, p = F(\xi)]$ whatever be the distribution function $F(x)$. We reject H_0 and accept H_1 if and only if the observed value $y \ge c$, where c is an integer selected such that Pr $(Y \ge c; H_0)$ is some reasonable significance level α. The power function of the test is given by

$$K(p) = \sum_{y=c}^{n} \binom{n}{y} p^y (1 - p)^{n-y}, \qquad p_0 \le p < 1,$$

where $p = F(\xi)$. In certain instances we may wish to approximate $K(p)$ by using an approximation to the binomial distribution.

Suppose that the alternative hypothesis to H_0: $F(\xi) = p_0$ is H_1: $F(\xi) < p_0$. Then the critical region is a set $\{y; y \le c_1\}$. Finally, if the alternative hypothesis is H_1: $F(\xi) \ne p_0$, the critical region is a set $\{y; y \le c_2 \text{ or } c_3 \le y\}$.

Frequently, $p_0 = \frac{1}{2}$ and, in that case, the hypothesis is that the given number ξ is a median of the distribution. In the following example, this value of p_0 is used.

Example 1. Let $X_1, X_2, \ldots, X_{10}$ be a random sample of size 10 from a distribution with distribution function $F(x)$. We wish to test the hypothesis H_0: $F(72) = \frac{1}{2}$ against the alternative hypothesis H_1: $F(72) > \frac{1}{2}$. Let Y be the number of sample items that are less than or equal to 72. Let the observed value of Y be y, and let the test be defined by the critical region $\{y; y \ge 8\}$. The power function of the test is given by

$$K(p) = \sum_{y=8}^{10} \binom{10}{y} p^y (1 - p)^{10-y}, \qquad \frac{1}{2} \le p < 1,$$

where $p = F(72)$. In particular, the significance level is

$$\alpha = K\left(\frac{1}{2}\right) = \left[\binom{10}{8} + \binom{10}{9} + \binom{10}{10}\right]\left(\frac{1}{2}\right)^{10} = \frac{7}{128}.$$

In many places in the literature the test that we have just described is called the *sign test*. The reason for this terminology is that the test is based upon a statistic Y that is equal to the number of nonpositive signs in the sequence $X_1 - \xi, X_2 - \xi, \ldots, X_n - \xi$. In the next section a distribution-free test, which considers both the sign and the magnitude of each deviation $X_i - \xi$, is studied.

EXERCISES

9.12. Suggest a chi-square test of the hypothesis which states that a distribution is one of the beta type, with parameters $\alpha = 2$ and $\beta = 2$. Further, suppose that the test is to be based upon a random sample of size 100. In the solution, give k, define $A_1, A_2, \ldots, A_k$, and compute each p_{i0}. If possible, compare your proposal with those of other students. Are any of them the same?

9.13. Let $X_1, X_2, \ldots, X_{48}$ be a random sample of size 48 from a distribution that has the distribution function $F(x)$. To test $H_0: F(41) = \frac{1}{4}$ against $H_1: F(41) < \frac{1}{4}$, use the statistic Y, which is the number of sample items less than or equal to 41. If the observed value of Y is $y \le 7$, reject H_0 and accept H_1. If $p = F(41)$, find the power function $K(p)$, $0 < p \le \frac{1}{4}$, of the test. Approximate $\alpha = K(\frac{1}{4})$.

9.14. Let $X_1, X_2, \ldots, X_{100}$ be a random sample of size 100 from a distribution that has distribution function $F(x)$. To test $H_0: F(90) - F(60) = \frac{4}{5}$ against $H_1: F(90) - F(60) > \frac{4}{5}$, use the statistic Y, which is the number of sample items less than or equal to 90 but greater than 60. If the observed value of Y, say y, is such that $y \ge c$, reject H_0. Find c so that $\alpha = 0.05$, approximately.

9.4 A Test of Wilcoxon

Suppose $X_1, X_2, \ldots, X_n$ is a random sample from a distribution with distribution function $F(x)$. We have considered a test of the hypothesis $F(\xi) = \frac{1}{2}$, ξ given, which is based upon the signs of the deviations $X_1 - \xi$, $X_2 - \xi, \ldots, X_n - \xi$. In this section a statistic is studied that takes into account not only these signs, but also the magnitudes of the deviations.

To find such a statistic that is distribution-free, we must make two additional assumptions:

(a) $F(x)$ is the distribution function of a continuous type of random variable X.

(b) The p.d.f. $f(x)$ of X has a graph that is symmetric about the vertical axis through $\xi_{0.5}$, the median (which we assume to be unique) of the distribution.

Thus

$$F(\xi_{0.5} - x) = 1 - F(\xi_{0.5} + x)$$

and

$$f(\xi_{0.5} - x) = f(\xi_{0.5} + x),$$

for all x. Moreover, the probability that any two items of a random sample are equal is zero, and in our discussion we shall assume that no two are equal.

The problem is to test the hypothesis that the median $\xi_{0.5}$ of the distribution is equal to a fixed number, say ξ. Thus we may, in all cases and without loss of generality, take $\xi = 0$. The reason for this is that if $\xi \neq 0$, then the fixed ξ can be subtracted from each sample item and the resulting variables can be used to test the hypothesis that their underlying distribution is symmetric about zero. Hence our conditions on $F(x)$ and $f(x)$ become $F(-x) = 1 - F(x)$ and $f(-x) = f(x)$, respectively.

To test the hypothesis H_0: $F(0) = \frac{1}{2}$, we proceed by first ranking $X_1, X_2, \ldots, X_n$ according to magnitude, disregarding their algebraic signs. Let R_i be the rank of $|X_i|$ among $|X_1|, |X_2|, \ldots, |X_n|$, $i = 1, 2, \ldots, n$. For example, if $n = 3$ and if we have $|X_2| < |X_3| < |X_1|$, then $R_1 = 3$, $R_2 = 1$, and $R_3 = 2$. Thus $R_1, R_2, \ldots, R_n$ is an arrangement of the first n positive integers $1, 2, \ldots, n$. Further, let $Z_i, i = 1, 2, \ldots, n$, be defined by

$$Z_i = -1, \qquad \text{if } X_i < 0,$$
$$= 1, \qquad \text{if } X_i > 0.$$

If we recall that $\Pr(X_i = 0) = 0$, we see that it does not change the probabilities whether we associate $Z_i = 1$ or $Z_i = -1$ with the outcome $X_i = 0$.

The statistic $W = \sum\limits_{i=1}^{n} Z_i R_i$ is the Wilcoxon statistic. Note that in computing this statistic we simply associate the sign of each X_i with the rank of its absolute value and sum the resulting n products.

If the alternative to the hypothesis H_0: $\xi_{0.5} = 0$ is H_1: $\xi_{0.5} > 0$, we reject H_0 if the observed value of W is an element of the set $\{w; w \geq c\}$. This is due to the fact that large positive values of W indicate that most of the large deviations from zero are positive. For alternatives $\xi_{0.5} < 0$ and $\xi_{0.5} \neq 0$ the critical regions are, respectively, the sets $\{w; w \leq c_1\}$ and $\{w; w \leq c_2 \text{ or } w \geq c_3\}$. To compute probabilities like $\Pr(W \geq c; H_0)$, we need to determine the distribution of W, under H_0.

To help us find the distribution of W, when H_0: $F(0) = \frac{1}{2}$ is true, we note the following facts:

(a) The assumption that $f(x) = f(-x)$ ensures that $\Pr(X_i < 0) = \Pr(X_i > 0) = \frac{1}{2}$, $i = 1, 2, \ldots, n$.

(b) Now $Z_i = -1$ if $X_i < 0$ and $Z_i = 1$ if $X_i > 0$, $i = 1, 2, \ldots, n$. Hence we have $\Pr(Z_i = -1) = \Pr(Z_i = 1) = \frac{1}{2}$, $i = 1, 2, \ldots, n$. Moreover, $Z_1, Z_2, \ldots, Z_n$ are mutually stochastically independent because $X_1, X_2, \ldots, X_n$ are mutually stochastically independent.

(c) The assumption that $f(x) = f(-x)$ also assures that the rank R_i of $|X_i|$ does not depend upon the sign Z_i of X_i. More generally, $R_1, R_2, \ldots, R_n$ are stochastically independent of $Z_1, Z_2, \ldots, Z_n$.

(d) A sum W is made up of the numbers $1, 2, \ldots, n$, each number with either a positive or a negative sign.

The preceding observations enable us to say that $W = \sum_1^n Z_i R_i$ has the same distribution as the random variable $V = \sum_1^n V_i$, where $V_1, V_2, \ldots, V_n$ are mutually stochastically independent and

$$\Pr(V_i = i) = \Pr(V_i = -i) = \frac{1}{2},$$

$i = 1, 2, \ldots, n$. That $V_1, V_2, \ldots, V_n$ are mutually stochastically independent follows from the fact that $Z_1, Z_2, \ldots, Z_n$ have that property; that is, the numbers $1, 2, \ldots, n$ always appear in a sum W and those numbers receive their algebraic signs by independent assignment. Thus each of $V_1, V_2, \ldots, V_n$ is like one and only one of $Z_1 R_1, Z_2 R_2, \ldots, Z_n R_n$.

Since W and V have the same distribution, the moment-generating function of W is that of V,

$$M(t) = E\left[\exp\left(t\sum_1^n V_i\right)\right] = \prod_{i=1}^n E(e^{tV_i})$$

$$= \prod_{i=1}^n \left(\frac{e^{-it} + e^{it}}{2}\right).$$

We can express $M(t)$ as the sum of terms of the form $(a_j/2^n)e^{b_j t}$. When $M(t)$ is written in this manner, we can determine by inspection the p.d.f. of the discrete-type random variable W. For example, the smallest value of W is found from the term $(1/2^n)e^{-t}e^{-2t}\cdots e^{-nt} = (1/2^n)e^{-n(n+1)t/2}$ and it is $-n(n+1)/2$. The probability of this value of W is the coefficient $1/2^n$. To make these statements more concrete, take $n = 3$. Then

$$M(t) = \left(\frac{e^{-t} + e^t}{2}\right)\left(\frac{e^{-2t} + e^{2t}}{2}\right)\left(\frac{e^{-3t} + e^{3t}}{2}\right)$$

$$= (\tfrac{1}{8})(e^{-6t} + e^{-4t} + e^{-2t} + 2 + e^{2t} + e^{4t} + e^{6t}).$$

Thus the p.d.f. of W, for $n = 3$, is given by

$$g(w) = \tfrac{1}{8}, \qquad w = -6, -4, -2, 2, 4, 6,$$
$$= \tfrac{2}{8}, \qquad w = 0,$$
$$= 0 \text{ elsewhere.}$$

The mean and the variance of W are more easily computed directly than by working with the moment-generating function $M(t)$. Because $V = \sum_1^n V_i$ and $W = \sum_1^n Z_i R_i$ have the same distribution, they have the same mean and the same variance. When the hypothesis H_0: $F(0) = \tfrac{1}{2}$ is true, it is easy to determine the values of these two characteristics of the distribution of W. Since $E(V_i) = 0$, $i = 1, 2, \ldots, n$, we have

$$\mu_W = E(W) = \sum_1^n E(V_i) = 0.$$

The variance of V_i is $(-i)^2(\tfrac{1}{2}) + (i)^2(\tfrac{1}{2}) = i^2$. Thus the variance of W is

$$\sigma_w^2 = \sum_1^n i^2 = \frac{n(n+1)(2n+1)}{6}.$$

For large values of n, the determination of the exact distribution of W becomes tedious. Accordingly, one looks for an approximating distribution. Although W is distributed as is the sum of n random variables that are mutually stochastically independent, our form of the central limit theorem cannot be applied because the n random variables do not have identical distributions. However, a more general theorem, due to Liapounov, states that if U_i has mean μ_i and variance σ_i^2, $i = 1, 2, \ldots, n$, if $U_1, U_2, \ldots, U_n$ are mutually stochastically independent, if $E(|U_i - \mu_i|^3)$ is finite for every i, and if

$$\lim_{n \to \infty} \frac{\sum_{i=1}^{n} E(|U_i - \mu_i|^3)}{\left(\sum_{i=1}^{n} \sigma_i^2\right)^{3/2}} = 0,$$

then

$$\frac{\sum_{i=1}^{n} U_i - \sum_{i=1}^{n} \mu_i}{\sqrt{\sum_{i=1}^{n} \sigma_i^2}}$$

has a limiting distribution that is $n(0, 1)$. For our variables $V_1, V_2, \ldots, V_n$ we have

$$E(|V_i - \mu_i|^3) = i^3(\tfrac{1}{2}) + i^3(\tfrac{1}{2}) = i^3;$$

and it is known that

$$\sum_{i=1}^{n} i^3 = \frac{n^2(n+1)^2}{4}.$$

Now

$$\lim_{n \to \infty} \frac{n^2(n+1)^2/4}{[n(n+1)(2n+1)/6]^{3/2}} = 0$$

because the numerator is of order n^4 and the denominator is of order $n^{9/2}$. Thus

$$\frac{W}{\sqrt{n(n+1)(2n+1)/6}}$$

is approximately $n(0, 1)$ when H_0 is true. This allows us to approximate probabilities like $\Pr(W \geq c; H_0)$ when the sample size n is large.

Example 1. Let $\xi_{0.5}$ be the median of a symmetric distribution that is of the continuous type. To test, with $\alpha = 0.01$, the hypothesis $H_0: \xi_{0.5} = 75$ against $H_1: \xi_{0.5} > 75$, we observed a random sample of size $n = 18$. Let it be given that the deviations of these 18 values from 75 are the following numbers:

$$1.5, \; -0.5, \; 1.6, \; 0.4, \; 2.3, \; -0.8, \; 3.2, \; 0.9, \; 2.9,$$

$$0.3, \; 1.8, \; -0.1, \; 1.2, \; 2.5, \; 0.6, \; -0.7, \; 1.9, \; 1.3.$$

The experimental value of the Wilcoxon statistic is equal to

$$w = 11 - 4 + 12 + 3 + 15 - 7 + 18 + 8 + 17 + 2 + 13 - 1$$

$$+ \, 9 + 16 + 5 - 6 + 14 + 10 = 135.$$

Since, with $n = 18$ so that $\sqrt{n(n+1)(2n+1)/6} = 45.92$, we have that

$$0.01 = \Pr\left(\frac{W}{45.92} \geq 2.326\right) = \Pr(W \geq 106.8).$$

Because $w = 135 > 106.8$, we reject H_0 at the approximate 0.01 significance level.

There are many modifications and generalizations of the Wilcoxon statistic. One generalization is the following: Let $c_1 \leq c_2 \leq \cdots \leq c_n$ be nonnegative numbers. Then, in the Wilcoxon statistic, replace the ranks $1, 2, \ldots, n$ by $c_1, c_2, \ldots, c_n$, respectively. For example, if $n = 3$ and if we have $|X_2| < |X_3| < |X_1|$, then $R_1 = 3$ is replaced by c_3, $R_2 = 1$ by c_1, and $R_3 = 2$ by c_2. In this example the generalized statistic is given by $Z_1 c_3 + Z_2 c_1 + Z_3 c_2$. Similar to the Wilcoxon statistic, this generalized statistic is distributed under H_0, as is the sum of n stochastically independent random variables, the ith of which takes each of the values $c_i \neq 0$ and $-c_i$ with probability $\frac{1}{2}$; if $c_i = 0$, that variable takes the value $c_i = 0$ with probability 1. Some special cases of this statistic are proposed in the Exercises.

EXERCISES

9.15. The observed values of a random sample of size 10 from a distribution that is symmetric about $\xi_{0.5}$ are 10.2, 14.1, 9.2, 11.3, 7.2, 9.8, 6.5, 11.8, 8.7, 10.8. Use Wilcoxon's statistic to test the hypothesis $H_0: \xi_{0.5} = 8$ against $H_1: \xi_{0.5} > 8$ if $\alpha = 0.05$. Even though n is small, use the normal approximation.

9.16. Find the distribution of W for $n = 4$ and $n = 5$. *Hint.* Multiply the moment-generating function of W, with $n = 3$, by $(e^{-4t} + e^{4t})/2$ to get that of W, with $n = 4$.

9.17. Let $X_1, X_2, \ldots, X_n$ be mutually stochastically independent. If the p.d.f. of X_i is uniform over the interval $(-2^{1-i}, 2^{1-i})$, $i = 1, 2, 3, \ldots$, show that Liapounov's condition is not satisfied. The sum $\sum_{i=1}^{n} X_i$ does not have an approximate normal distribution because the first random variables in the sum tend to dominate it.

9.18. If $n = 4$ and, in the notation of the text, $c_1 = 1$, $c_2 = 2$, $c_3 = c_4 = 3$, find the distribution of the generalization of the Wilcoxon statistic, say W_g. For a general n, find the mean and the variance of W_g if $c_i = i$, $i \leq n/2$, and $c_i = [n/2] + 1, i > n/2$, where $[z]$ is the greatest integer function. Does Liapounov's condition hold here?

9.19. A modification of Wilcoxon's statistic that is frequently used is achieved by replacing R_i by $R_i - 1$; that is, use the modification $W_m = \sum_{1}^{n} Z_i (R_i - 1)$. Show that $W_m / \sqrt{(n-1)n(2n-1)/6}$ has a limiting distribution that is $n(0, 1)$.

9.20. If, in the discussion of the generalization of the Wilcoxon statistic, we let $c_1 = c_2 = \cdots = c_n = 1$, show that we obtain a statistic equivalent to that used in the sign test.

9.21. If $c_1, c_2, \ldots, c_n$ are selected so that $i/(n+1) = \int_0^{c_i} \sqrt{2/\pi}\, e^{-x^2/2}\, dx$, $i = 1, 2, \ldots, n$, the generalized Wilcoxon W_g is an example of a *normal scores statistic*. If $n = 9$, compute the mean and the variance of this W_g.

9.22. If $c_i = 2^i$, $i = 1, 2, \ldots, n$, the corresponding W_g is called the *binary statistic*. Find the mean and the variance of this W_g. Is Liapounov's condition satisfied?

9.23. In the definition of Wilcoxon's statistic, let W_1 be the sum of the ranks of those items of the sample that are positive and let W_2 be the sum of the ranks of those items that are negative. Then $W = W_1 - W_2$.

(a) Show that $W = 2W_1 - n(n + 1)/2$ and $W = n(n + 1)/2 - 2W_2$.

(b) Compute the mean and the variance of each of W_1 and W_2.

9.5 The Equality of Two Distributions

In Sections 9.3 and 9.4 some tests of hypotheses about one distribution were investigated. In this section, as in the next section, various tests of the equality of two independent distributions are studied. By the equality of two distributions, we mean that the two distribution functions, say F and G, have $F(z) = G(z)$ for all values of z.

The first test that we discuss is a natural extension of the chi-square test. Let X and Y be stochastically independent variables with distribution functions $F(x)$ and $G(y)$, respectively. We wish to test the hypothesis that $F(z) = G(z)$, for all z. Let us partition the real line into k mutually disjoint sets $A_1, A_2, \ldots, A_k$. Define

$$p_{i1} = \Pr\,(X \in A_i), \qquad i = 1, 2, \ldots, k,$$

and

$$p_{i2} = \Pr\,(Y \in A_i), \qquad i = 1, 2, \ldots, k.$$

If $F(z) = G(z)$, for all z, then $p_{i1} = p_{i2}$, $i = 1, 2, \ldots, k$. Accordingly, the hypothesis that $F(z) = G(z)$, for all z, is replaced by the less restrictive hypothesis

$$H_0: p_{i1} = p_{i2}, \qquad i = 1, 2, \ldots, k.$$

But this is exactly the problem of testing the equality of two independent multinomial distributions that was considered in Example 3, Section 8.1, and the reader is referred to that example for the details.

Some statisticians prefer a procedure which eliminates some of the subjectivity of selecting the partitions. For a fixed positive integer k, proceed as follows. Consider a random sample of size m from the distribution of X and a random sample of size n from the independent distribution of Y. Let the experimental values be denoted by $x_1, x_2, \ldots, x_m$ and $y_1, y_2, \ldots, y_n$. Then combine the two samples into one sample of size $m + n$ and order the $m + n$ values (not their absolute values) in ascending order of magnitude. These ordered items are then partitioned into k parts in such a way that each part has the same number of items. (If the sample sizes are such that this is impossible, a partition with approximately the same number of items in each group suffices.) In effect, then, the partition $A_1, A_2, \ldots, A_k$ is determined by the experimental values themselves. This does not alter the fact that the statistic, discussed in Example 3, Section 8.1, has a limiting distribution that is $\chi^2(k - 1)$. Accordingly, the procedures used in that example may be used here.

Among the tests of this type there is one that is frequently used. It is essentially a test of the equality of the medians of two independent distributions. To simplify the discussion, we assume that $m + n$, the size of the combined sample, is an even number, say $m + n = 2h$, where h is a positive integer. We take $k = 2$ and the combined sample of size $m + n = 2h$, which has been ordered, is separated into two parts, a "lower half" and an "upper half," each containing $h = (m + n)/2$ of the experimental values of X and Y. The statistic, suggested by Example 3, Section 8.1, could be used because it has, when H_0 is true, a limiting distribution that is $\chi^2(1)$. However, it is more interesting to find the exact distribution of another statistic which enables us to test the hypothesis H_0 against the alternative $H_1: F(z) \geq G(z)$ or against the alternative $H_1: F(z) \leq G(z)$ as opposed to merely $F(z) \neq G(z)$. [Here, and in the sequel, alternatives $F(z) \geq G(z)$ and $F(z) \leq G(z)$ and $F(z) \neq G(z)$ mean that strict inequality holds on some set of positive probability measure.] This other statistic is V, which is the number of observed values of X that are in the lower half of the combined sample. If the observed value of V is quite large, one might suspect that the median of the distribution of X is smaller than that of the distribution of Y. Thus the critical region of this test of the hypothesis $H_0: F(z) = G(z)$, for all z, against $H_1: F(z) \geq G(z)$ is of the form $V \geq c$. Because our combined sample is of even size, there is no unique median of the sample. However, one can arbitrarily insert a number between the hth and $(h + 1)$st ordered items and call it the median of the sample. On this account, a test of the sort just described is called a *median test*. Incidentally, if the alternative hypothesis is $H_1: F(z) \leq G(z)$, the critical region is of the form $V \leq c$.

The distribution of V is quite easy to find if the distribution funtions $F(x)$ and $G(y)$ are of the continuous type and if $F(z) = G(z)$, for all z. We shall now show that V has a hypergeometric p.d.f. Let $m + n = 2h$, h a positive integer. To compute $\Pr(V = v)$, we need the probability that exactly v of $X_1, X_2, \ldots, X_m$ are in the lower half of the ordered combined sample. Under our assumptions, the probability is zero that any two of the $2h$ random variables are equal. The smallest

h of the $m + n = 2h$ items can be selected in any one of $\binom{2h}{h}$ ways.

Each of these ways has the same probability. Of these $\binom{2h}{h}$ ways, we need to count the number of those in which exactly v of the m values of X (and hence $h - v$ of the n values of Y) appear in the lower h items. But this is $\binom{m}{v}\binom{n}{h-v}$. Thus the p.d.f. of V is the hypergeometric p.d.f.

$$k(v) = \text{Pr}\,(V = v) = \frac{\binom{m}{v}\binom{n}{h-v}}{\binom{m+n}{h}}, \qquad v = 0, 1, 2, \ldots, m,$$

$$= 0 \text{ elsewhere,}$$

where $m + n = 2h$.

The reader may be momentarily puzzled by the meaning of $\binom{n}{h-v}$ for $v = 0, 1, 2, \ldots, m$. For example, let $m = 17$, $n = 3$, so that $h = 10$. Then we have $\binom{3}{10-v}$, $v = 0, 1, \ldots, 17$. However, we take $\binom{n}{h-v}$ to be zero if $h - v$ is negative or if $h - v > n$.

If $m + n$ is an odd number, say $m + n = 2h + 1$, it is left to the reader to show that the p.d.f. $k(v)$ gives the probability that exactly v of the m values of X are among the lower h of the combined $2h + 1$ values; that is, exactly v of the m values of X are less than the median of the combined sample.

If the distribution functions $F(x)$ and $G(y)$ are of the continuous type, there is another rather simple test of the hypothesis that $F(z) = G(z)$, for all z. This test is based upon the notion of *runs* of values of X and of values of Y. We shall now explain what we mean by runs. Let us again combine the sample of m values of X and the sample of n values of Y into one collection of $m + n$ ordered items arranged in ascending order of magnitude. With $m = 7$ and $n = 8$ we might find that the 15 ordered items were in the arrangement

$$\underline{x}\ \underline{yyy}\ \underline{xx}\ \underline{y}\ \underline{x}\ \underline{yy}\ \underline{xxx}\ \underline{yy}.$$

Note that in this ordering we have underscored the groups of successive values of the random variable X and those of the random variable Y. If we read from left to right, we would say that we have a *run* of one value of X, followed by a *run* of three values of Y, followed by a *run* of two values of X, and so on. In our example, there is a total of eight runs. Three are runs of length 1; three are runs of length 2; and two are runs of length 3. Note that the total number of runs is always one more than the number of unlike adjacent symbols.

Of what can runs be suggestive? Suppose that with $m = 7$ and $n = 8$ we have the following ordering:

$$\underline{xxxxx} \; \underline{y} \; \underline{xx} \; \underline{yyyyyyy}.$$

To us, this strongly suggests that $F(z) > G(z)$. For if, in fact, $F(z) = G(z)$ for all z, we would anticipate a greater number of runs. And if the first run of five values of X were interchanged with the last run of seven values of Y, this would suggest that $F(z) < G(z)$. But runs can be suggestive of other things. For example, with $m = 7$ and $n = 8$, consider the runs.

$$\underline{yyyy} \; \underline{xxxxxxx} \; \underline{yyyy}.$$

This suggests to us that the medians of the distributions of X and Y may very well be about the same, but that the "spread" (measured possibly by the standard deviation) of the distribution of X is considerably less than that of the distribution of Y.

Let the random variable R equal the number of runs in the combined sample, once the combined sample has been ordered. Because our random variables X and Y are of the continuous type, we may assume that no two of these sample items are equal. We wish to find the p.d.f. of R. To find this distribution, when $F(z) = G(z)$, we shall suppose that all arrangements of the m values of X and the n values of Y have equal probabilities. We shall show that

$$\Pr(R = 2k + 1) = \left\{ \binom{m-1}{k}\binom{n-1}{k-1} + \binom{m-1}{k-1}\binom{n-1}{k} \right\} \Big/ \binom{m+n}{m}$$

(1)

$$\Pr(R = 2k) = 2\binom{m-1}{k-1}\binom{n-1}{k-1} \Big/ \binom{m+n}{m}$$

when $2k$ and $2k + 1$ are elements of the space of R.

To prove formulas (1), note that we can select the m positions for the m values of X from the $m + n$ positions in any one of $\binom{m+n}{m}$ ways. Since each of these choices yields one arrangement, the probability of each arrangement is equal to $1 \Big/ \binom{m+n}{m}$. The problem is now to determine how many of these arrangements yield $R = r$, where r is an integer in the space of R. First, let $r = 2k + 1$, where k is a positive integer. This means that there must be $k + 1$ runs of the ordered values of X and k runs of the ordered values of Y or vice versa. Consider first the number of ways of obtaining $k + 1$ runs of the m values of X. We can form $k + 1$ of these runs by inserting k "dividers" into the $m - 1$ spaces between the values of X, with no more than one divider per space. This can be done in any one of $\binom{m-1}{k}$ ways. Similarly, we can construct k runs of the n values of Y by inserting $k - 1$ dividers into the $n - 1$ spaces between the values of Y, with no more than one divider per space. This can be done in any one of $\binom{n-1}{k-1}$ ways. The joint operation can be performed in any one of $\binom{m-1}{k}\binom{n-1}{k-1}$ ways. These two sets of runs can be placed together to form $r = 2k + 1$ runs. But we could also have k runs of the values of X and $k + 1$ runs of the values of Y. An argument similar to the preceding shows that this can be effected in any one of $\binom{m-1}{k-1}\binom{n-1}{k}$ ways. Thus

$$\Pr(R = 2k + 1) = \frac{\binom{m-1}{k}\binom{n-1}{k-1} + \binom{m-1}{k-1}\binom{n-1}{k}}{\binom{m+n}{m}},$$

which is the first of formulas (1).

If $r = 2k$, where k is a positive integer, we see that the ordered values of X and the ordered values of Y must each be separated into k runs. These operations can be performed in any one of $\binom{m-1}{k-1}$ and $\binom{n-1}{k-1}$ ways, respectively. These two sets of runs can be placed together to form $r = 2k$ runs. But we may begin with either a run of values of X or a run of values of Y. Accordingly, the probability of $2k$ runs is

$$\Pr(R = 2k) = \frac{2\binom{m-1}{k-1}\binom{n-1}{k-1}}{\binom{m+n}{m}},$$

which is the second of formulas (1).

If the critical region of this *run test* of the hypothesis H_0: $F(z) = G(z)$ for all z is of the form $R \le c$, it is easy to compute $\alpha = \Pr(R \le c; H_0)$, provided that m and n are small. Although it is not easy to show, the distribution of R can be approximated, with large sample sizes m and n, by a normal distribution with mean

$$\mu = E(R) = 2\frac{mn}{m+n} + 1$$

and variance

$$\sigma^2 = \frac{(\mu - 1)(\mu - 2)}{m + n - 1}.$$

The run test may also be used to test for *randomness*. That is, it can be used as a check to see if it is reasonable to treat $X_1, X_2, \ldots, X_s$ as a random sample of size s from some continuous distribution. To facilitate the discussion, take s to be even. We are given the s values of X to be $x_1, x_2, \ldots, x_s$, which are not ordered by magnitude but by the order in which they were observed. However, there are $s/2$ of these values, each of which is smaller than the remaining $s/2$ values. Thus we have a "lower half" and an "upper half" of these values. In the sequence $x_1, x_2, \ldots, x_s$, replace each value X that is in the lower half by the letter L and each value in the upper half by the letter U. Then, for example, with $s = 10$, a sequence such as

$$L\ L\ L\ L\ U\ L\ U\ U\ U\ U$$

may suggest a trend toward increasing values of X; that is, these values of X may not reasonably be looked upon as being the items of a random sample. If trend is the only alternative to randomness, we can make a test based upon R and reject the hypothesis of randomness if $R \le c$. To make this test, we would use the p.d.f. of R with $m = n = s/2$. On the other hand if, with $s = 10$, we find a sequence such as

$$L\ H\ L\ H\ L\ H\ L\ H\ L\ H,$$

our suspicions are aroused that there may be a nonrandom effect which is cyclic even though $R = 10$. Accordingly, to test for a trend or a cyclic effect, we could use a critical region of the form $R \le c_1$ or $R \ge c_2$.

If the sample size s is odd, the number of sample items in the "upper half" and the number in the "lower half" will differ by one. Then, for example, we could use the p.d.f. of R with $m = (s - 1)/2$ and $n = (s + 1)/2$, or vice versa.

EXERCISES

9.24. Let 3.1, 5.6, 4.7, 3.8, 4.2, 3.0, 5.1, 3.9, 4.8 and 5.3, 4.0, 4.9, 6.2, 3.7, 5.0, 6.5, 4.5, 5.5, 5.9, 4.4, 5.8 be observed samples of sizes $m = 9$ and $n = 12$ from two independent distributions. With $k = 3$, use a chi-square test to test, with $\alpha = 0.05$ approximately, the equality of the two distributions.

9.25. In the median test, with $m = 9$ and $n = 7$, find the p.d.f. of the random variable V, the number of values of X in the lower half of the combined sample. In particular, what are the values of the probabilities $\Pr(V = 0)$ and $\Pr(V = 9)$?

9.26. In the notation of the text, use the median test and the data given in Exercise 9.24 to test, with $\alpha = 0.05$, approximately, the hypothesis of the equality of the two independent distributions against the alternative hypothesis that $F(z) \geq G(z)$. If the exact probabilities are too difficult to determine for $m = 9$ and $n = 12$, approximate these probabilities.

9.27. Using the notation of this section, let U be the number of observed values of X in the smallest d items of the combined sample of $m + n$ items. Argue that

$$\Pr(U = u) = \binom{m}{u}\binom{n}{d - u} \bigg/ \binom{m + n}{d}, \qquad u = 0, 1, \ldots, m.$$

The statistic U could be used to test the equality of the $(100p)$th percentiles, where $(m + n)p = d$, of the distributions of X and Y.

9.28. In the discussion of the run test, let the random variables R_1 and R_2 be, respectively, the number of runs of the values of X and the number of runs of the values of Y. Then $R = R_1 + R_2$. Let the pair (r_1, r_2) of integers be in the space of (R_1, R_2); then $|r_1 - r_2| \leq 1$. Show that the joint p.d.f. of R_1 and R_2 is $2\binom{m - 1}{r_1 - 1}\binom{n - 1}{r_2 - 1} \big/ \binom{m + n}{m}$ if $r_1 = r_2$; that this joint p.d.f. is $\binom{m - 1}{r_1 - 1}\binom{n - 1}{r_2 - 1} \big/ \binom{m + n}{m}$ if $|r_1 - r_2| = 1$; and is zero elsewhere. Show that the marginal p.d.f. of R_1 is $\binom{m - 1}{r_1 - 1}\binom{n + 1}{r_1} \big/ \binom{m + n}{m}$ $r_1 = 1, \ldots, m$, and is zero elsewhere. Find $E(R_1)$. In a similar manner, find $E(R_2)$. Compute $E(R) = E(R_1) + E(R_2)$.

9.6 The Mann–Whitney–Wilcoxon Test

We return to the problem of testing the equality of two independent distributions of the continuous type. Let X and Y be stochastically independent random variables of the continuous type. Let $F(x)$ and

$G(y)$ denote, respectively, the distribution functions of X and Y and let $X_1, X_2, \ldots, X_m$ and $Y_1, Y_2, \ldots, Y_n$ denote independent samples from these distributions. We shall discuss the Mann–Whitney–Wilcoxon test of the hypothesis $H_0: F(z) = G(z)$ for all values of z.

Let us define

$$Z_{ij} = 1, \qquad X_i < Y_j,$$
$$= 0, \qquad X_i > Y_j,$$

and consider the statistic

$$U = \sum_{j=1}^{n} \sum_{i=1}^{m} Z_{ij}.$$

We note that

$$\sum_{i=1}^{m} Z_{ij}$$

counts the number of values of X that are less than $Y_j, j = 1, 2, \ldots, n$. Thus U is the sum of these n counts. For example, with $m = 4$ and $n = 3$, consider the observations

$$x_2 < y_3 < x_1 < x_4 < y_1 < x_3 < y_2.$$

There are three values of x that are less than y_1; there are four values of x that are less than y_2; and there is one value of x that is less than y_3. Thus the experimental value of U is $u = 3 + 4 + 1 = 8$.

Clearly, the smallest value which U can take is zero, and the largest value is mn. Thus the space of U is $\{u; u = 0, 1, 2, \ldots, mn\}$. If U is large, the values of Y tend to be larger than the values of X, and this suggests that $F(z) \geq G(z)$ for all z. On the other hand, a small value of U suggests that $F(z) \leq G(z)$ for all z. Thus, if we test the hypothesis $H_0: F(z) = G(z)$ for all z against the alternative hypothesis $H_1: F(z) \geq G(z)$ for all z, the critical region is of the form $U \geq c_1$. If the alternative hypothesis is $H_1: F(z) \leq G(z)$ for all z, the critical region is of the form $U \leq c_2$. To determine the size of a critical region, we need the distribution of U when H_0 is true.

If u belongs to the space of U, let us denote $\Pr(U = u)$ by the symbol $h(u; m, n)$. This notation focuses attention on the sample sizes m and n. To determine the probability $h(u; m, n)$, we first note that we have $m + n$ positions to be filled by m values of X and n values of Y. We can fill m positions with the values of X in any one of $\binom{m+n}{m}$ ways. Once this has been done, the remaining n positions can be filled with the values of Y. When H_0 is true, each of these arrangements has the same probability, $1 \Big/ \binom{m+n}{m}$. The final right-hand position of an arrangement may be either a value of X or a value of Y. This position can be filled in any one of $m + n$ ways, m of which are favorable to X and n of which are favorable to Y. Accordingly, the probability that an arrangement ends with a value of X is $m/(m + n)$ and the probability that an arrangement terminates with a value of Y is $n/(m + n)$.

Now U can equal u in two mutually exclusive and exhaustive ways: (1) The final right-hand position (the largest of the $m + n$ values) in the arrangement may be a value of X and the remaining $(m - 1)$ values of X and the n values of Y can be arranged so as to have $U = u$. The probability that $U = u$, given an arrangement that terminates with a value of X, is given by $h(u; m - 1, n)$. Or (2) the largest value in the arrangement can be a value of Y. This value of Y is greater than m values of X. If we are to have $U = u$, the sum of $n - 1$ counts of the m values of X with respect to the remaining $n - 1$ values of Y must be $u - m$. Thus the probability that $U = u$, given an arrangement that terminates in a value of Y, is given by $h(u - m; m, n - 1)$. Accordingly, the probability that $U = u$ is

$$h(u; m, n) = \left(\frac{m}{m + n} \right) h(u; m - 1, n) + \left(\frac{n}{m + n} \right) h(u - m; m, n - 1).$$

We impose the following reasonable restrictions upon the function $h(u; m, n)$:

$$h(u; 0, n) = 1, \qquad u = 0,$$
$$= 0, \qquad u > 0, n \geq 1,$$

and

$$h(u; m, 0) = 1, \qquad u = 0,$$
$$= 0, \qquad u > 0, m \geq 1,$$

and

$$h(u; m, n) = 0, \qquad u < 0, m \geq 0, n \geq 0.$$

Then it is easy, for small values m and n, to compute these probabilities. For example, if $m = n = 1$, we have

$$h(0; 1, 1) = \tfrac{1}{2}h(0; 0, 1) + \tfrac{1}{2}h(-1; 1, 0) = \tfrac{1}{2} \cdot 1 + \tfrac{1}{2} \cdot 0 = \tfrac{1}{2},$$
$$h(1; 1, 1) = \tfrac{1}{2}h(1; 0, 1) + \tfrac{1}{2}h(0; 1, 0) = \tfrac{1}{2} \cdot 0 + \tfrac{1}{2} \cdot 1 = \tfrac{1}{2};$$

and if $m = 1$, $n = 2$, we have

$$h(0; 1, 2) = \tfrac{1}{3}h(0; 0, 2) + \tfrac{2}{3}h(-1; 1, 1) = \tfrac{1}{3} \cdot 1 + \tfrac{2}{3} \cdot 0 = \tfrac{1}{3},$$

$$h(1; 1, 2) = \tfrac{1}{3}h(1; 0, 2) + \tfrac{2}{3}h(0; 1, 1) = \tfrac{1}{3} \cdot 0 + \tfrac{2}{3} \cdot \tfrac{1}{2} = \tfrac{1}{3},$$

$$h(2; 1, 2) = \tfrac{1}{3}h(2; 0, 2) + \tfrac{2}{3}h(1; 1, 1) = \tfrac{1}{3} \cdot 0 + \tfrac{2}{3} \cdot \tfrac{1}{2} = \tfrac{1}{3}.$$

In Exercise 9.29 the reader is to determine the distribution of U when $m = 2$, $n = 1$; $m = 2$, $n = 2$; $m = 1$, $n = 3$; and $m = 3$, $n = 1$.

For large values of m and n, it is desirable to use an approximate distribution of U. Consider the mean and the variance of U when the hypothesis H_0: $F(z) = G(z)$, for all values of z, is true. Since $U = \sum_{j=1}^{n} \sum_{i=1}^{m} Z_{ij}$, then

$$E(U) = \sum_{i=1}^{m} \sum_{j=1}^{n} E(Z_{ij}).$$

But

$$E(Z_{ij}) = (1) \Pr(X_i < Y_j) + (0) \Pr(X_i > Y_j) = \tfrac{1}{2}$$

because, when H_0 is true, $\Pr(X_i < Y_j) = \Pr(X_i > Y_j) = \tfrac{1}{2}$. Thus

$$E(U) = \sum_{i=1}^{m} \sum_{j=1}^{n} \left(\frac{1}{2}\right) = \frac{mn}{2}.$$

To compute the variance of U, we first find

$$E(U^2) = \sum_{k=1}^{n} \sum_{h=1}^{m} \sum_{j=1}^{n} \sum_{i=1}^{m} E(Z_{ij}Z_{hk})$$

$$= \sum_{j=1}^{n} \sum_{i=1}^{m} E(Z_{ij}^2) + \sum_{\substack{k=1 \\ k \neq j}}^{n} \sum_{j=1}^{n} \sum_{i=1}^{m} E(Z_{ij}Z_{ik})$$

$$+ \sum_{j=1}^{n} \sum_{\substack{h=1 \\ h \neq i}}^{m} \sum_{i=1}^{m} E(Z_{ij}Z_{hj}) + \sum_{\substack{k=1 \\ k \neq j}}^{n} \sum_{j=1}^{n} \sum_{\substack{h=1 \\ h \neq i}}^{m} \sum_{i=1}^{m} E(Z_{ij}Z_{hk}).$$

Note that there are mn terms in the first of these sums, $mn(n-1)$ in the second, $mn(m-1)$ in the third, and $mn(m-1)(n-1)$ in the fourth. When H_0 is true, we know that X_i, X_h, Y_j, and Y_k, $i \neq h$, $j \neq k$, are mutually stochastically independent and have the same distribution of the continuous type. Thus $\Pr(X_i < Y_j) = \tfrac{1}{2}$. Moreover, $\Pr(X_i < Y_j, X_i < Y_k) = \tfrac{1}{3}$ because this is the probability that a designated one of three items is less than each of the other two.

Similarly, $\Pr(X_i < Y_j, X_h < Y_j) = \frac{1}{3}$. Finally, $\Pr(X_i < Y_j, X_h < Y_k)$
$= \Pr(X_i < Y_j)\Pr(X_h < Y_k) = \frac{1}{4}$. Hence we have

$$E(Z_{ij}^2) = (1)^2 \Pr(X_i < Y_j) = \frac{1}{2},$$

$$E(Z_{ij}Z_{ik}) = (1)(1)\Pr(X_i < Y_j, X_i < Y_k) = \frac{1}{3}, \qquad j \neq k,$$

$$E(Z_{ij}Z_{hj}) = (1)(1)\Pr(X_i < Y_j, X_h < Y_j) = \frac{1}{3}, \qquad i \neq h,$$

and

$$E(Z_{ij}Z_{hk}) = (1)(1)\Pr(X_i < Y_j, X_h < Y_k) = \frac{1}{4}, \qquad i \neq h, j \neq k.$$

Thus

$$E(U^2) = \frac{mn}{2} + \frac{mn(n-1)}{3} + \frac{mn(m-1)}{3} + \frac{mn(m-1)(n-1)}{4}$$

and

$$\sigma_U^2 = mn\left[\frac{1}{2} + \frac{n-1}{3} + \frac{m-1}{3} + \frac{(m-1)(n-1)}{4} - \frac{mn}{4}\right]$$

$$= \frac{mn(m+n+1)}{12}.$$

Although it is fairly difficult to prove, it is true, when $F(z) = G(z)$ for all z, that

$$\frac{U - \dfrac{mn}{2}}{\sqrt{\dfrac{mn(m+n+1)}{12}}}$$

has, if each of m and n is large, an approximate distribution that is $n(0, 1)$. This fact enables us to compute, approximately, various significance levels.

Prior to the introduction of the statistic U in the statistical literature, it had been suggested that a test of $H_0: F(z) = G(z)$, for all z, be based upon the following statistic, say T (not Student's t). Let T be the sum of the ranks of $Y_1, Y_2, \ldots, Y_n$ among the $m + n$ items $X_1, \ldots, X_m, Y_1, \ldots, Y_n$, once this combined sample has been ordered. In Exercise 9.31 the reader is asked to show that

$$U = T - \frac{n(n+1)}{2}.$$

This formula provides another method of computing U and it shows that a test of H_0 based on U is equivalent to a test based on T. A generalization of T is considered in Section 9.8.

Example 1. With the assumptions and the notation of this section, let $m = 10$ and $n = 9$. Let the observed values of X be as given in the first row and the observed values of Y as in the second row of the following display:

$$4.3, \ 5.9, \ 4.9, \ 3.1, \ 5.3, \ 6.4, \ 6.2, \ 3.8, \ 7.5, \ 5.8,$$

$$5.5, \ 7.9, \ 6.8, \ 9.0, \ 5.6, \ 6.3, \ 8.5, \ 4.6, \ 7.1.$$

Since, in the combined sample, the ranks of the values of y are 4, 7, 8, 12, 14, 15, 17, 18, 19, we have the experimental value of T to be equal to $t = 114$. Thus $u = 114 - 45 = 69$. If $F(z) = G(z)$ for all z, then, approximately,

$$0.05 = \Pr\left(\frac{U - 45}{12.247} \geq 1.645\right) = \Pr\left(U \geq 65.146\right).$$

Accordingly, at the 0.05 significance level, we reject the hypothesis $H_0: F(z) = G(z)$, for all z, and accept the alternative hypothesis $H_1: F(z) \geq G(z)$, for all z.

EXERCISES

9.29. Compute the distribution of U in each of the following cases: (a) $m = 2$, $n = 1$; (b) $m = 2$, $n = 2$; (c) $m = 1$, $n = 3$; (d) $m = 3$, $n = 1$.

9.30. Suppose the hypothesis $H_0: F(z) = G(z)$, for all z, is not true. Let $p = \Pr(X_i < Y_j)$. Show that U/mn is an unbiased estimator of p and that it converges stochastically to p as $m \to \infty$ and $n \to \infty$.

9.31. Show that $U = T - [n(n + 1)]/2$. *Hint.* Let $Y_{(1)} < Y_{(2)} < \cdots < Y_{(n)}$ be the order statistics of the random sample $Y_1, Y_2, \ldots, Y_n$. If R_i is the rank of $Y_{(i)}$ in the combined ordered sample, note that $Y_{(i)}$ is greater than $R_i - i$ values of X.

9.32. In Example 1 of this section assume that the values came from two independent normal distributions with means μ_1 and μ_2, respectively, and with common variance σ^2. Calculate the Student's t which is used to test the hypothesis $H_0: \mu_1 = \mu_2$. If the alternative hypothesis is $H_1: \mu_1 < \mu_2$, do we accept or reject H_0 at the 0.05 significance level?

9.7 Distributions Under Alternative Hypotheses

In this section we shall discuss certain problems that are related to a nonparametric test when the hypothesis H_0 is not true. Let X and Y be stochastically independent random variables of the continuous type with distribution functions $F(x)$ and $G(y)$, respectively, and probability density functions $f(x)$ and $g(y)$. Let $X_1, X_2, \ldots, X_m$ and $Y_1, Y_2, \ldots, Y_n$ denote independent random samples from these distributions. Consider the hypothesis $H_0: F(z) = G(z)$ for all values of

z. It has been seen that the test of this hypothesis may be based upon the statistic U, which, when the hypothesis H_0 is true, has a distribution that does not depend upon $F(z) = G(z)$. Or this test can be based upon the statistic $T = U + n(n + 1)/2$, where T is the sum of the ranks of $Y_1, Y_2, \ldots, Y_n$ in the combined sample. To elicit some information about the distribution of T when the alternative hypothesis is true, let us consider the joint distribution of the ranks of these values of Y.

Let $Y_{(1)} < Y_{(2)} < \cdots < Y_{(n)}$ be the order statistics of the sample $Y_1, Y_2, \ldots, Y_n$. Order the combined sample, and let R_i be the rank of $Y_{(i)}, i = 1, 2, \ldots, n$. Thus there are $i - 1$ values of Y and $R_i - i$ values of X that are less than $Y_{(i)}$. Moreover, there are $R_i - R_{i-1} - 1$ values of X between $Y_{(i-1)}$ and $Y_{(i)}$. If it is given that $Y_{(1)} = y_1 < Y_{(2)} = y_2 < \cdots < Y_{(n)} = y_n$, then the conditional probability

(1) $\Pr (R_1 = r_1, R_2 = r_2, \ldots, R_n = r_n | y_1 < y_2 < \cdots < y_n),$

where $r_1 < r_2 < \cdots < r_n \leq m + n$ are positive integers, can be computed by using the multinomial p.d.f. in the following manner. Define the following sets: $A_1 = \{x; -\infty < x < y_1\}$, $A_i = \{x; y_{i-1} < x < y_i\}$, $i = 2, \ldots, n$, $A_{n+1} = \{x; y_n < x < \infty\}$. The conditional probabilities of these sets are, respectively, $p_1 = F(y_1)$, $p_2 = F(y_2) - F(y_1), \ldots,$ $p_n = F(y_n) - F(y_{n-1})$, $p_{n+1} = 1 - F(y_n)$. Then the conditional probability of display (1) is given by

$$\frac{m! \, p_1^{r_1-1} p_2^{r_2-r_1-1} \cdots p_n^{r_n-r_{n-1}-1} p_{n+1}^{m+n-r_n}}{(r_1 - 1)! \, (r_2 - r_1 - 1)! \cdots (r_n - r_{n-1} - 1)! \, (m + n - r_n)!}.$$

To find the unconditional probability $\Pr (R_1 = r_1, R_2 = r_2, \ldots, R_n = r_n)$, which we denote simply by $\Pr (r_1, \ldots, r_n)$, we multiply the conditional probability by the joint p.d.f. of $Y_{(1)} < Y_{(2)} < \cdots < Y_{(n)}$, namely $n! \, g(y_1) g(y_2) \cdots g(y_n)$, and then integrate on $y_1, y_2, \ldots, y_n$. That is,

$$\Pr (r_1, r_2, \ldots, r_n) = \int_{-\infty}^{\infty} \cdots \int_{-\infty}^{y_3} \int_{-\infty}^{y_2} \Pr (r_1, \ldots, r_n | y_1 < \cdots < y_n) n!$$
$$\times g(y_1) \cdots g(y_n) \, dy_1 \cdots dy_n,$$

where $\Pr (r_1, \ldots, r_n | y_1 < \cdots < y_n)$ denotes the conditional probability in display (1).

Now that we have the joint distribution of $R_1, R_2, \ldots, R_n$, we can find, theoretically, the distributions of functions of $R_1, R_2, \ldots, R_n$ and, in particular, the distribution of $T = \sum_1^n R_i$. From the latter we can find that of $U = T - n(n + 1)/2$. To point out the extremely tedious computational problems of distribution theory that we encounter, we give an example. In this example we use the assumptions of this section.

Example 1. Suppose that an hypothesis H_0 is not true but that in fact $f(x) = 1, 0 < x < 1$, zero elsewhere, and $g(y) = 2y, 0 < y < 1$, zero elsewhere. Let $m = 3$ and $n = 2$. Note that the space of U is the set $\{u; u = 0, 1, \ldots, 6\}$. Consider $\Pr(U = 5)$. This event $U = 5$ occurs when and only when $R_1 = 3, R_2 = 5$, since in this section $R_1 < R_2$ are the ranks of $Y_{(1)} < Y_{(2)}$ in the combined sample and $U = R_1 + R_2 - 3$. Because $F(x) = x, 0 < x \le 1$, we have

$$\Pr(U = 5) = \Pr(R_1 = 3, R_2 = 5)$$

$$= \int_0^1 \int_0^{y_2} \frac{3! \, y_1^2(y_2 - y_1)}{2! \, 1!} 2! \, (2y_1)(2y_2) \, dy_1 \, dy_2$$

$$= 24 \int_0^1 \left(\frac{y_2^6}{4} - \frac{y_2^6}{5}\right) dy_2 = \tfrac{6}{35}.$$

Consider next $\Pr(U = 4)$. The event $U = 4$ occurs if $R_1 = 2, R_2 = 5$ or if $R_1 = 3, R_2 = 4$. Thus

$$\Pr(U = 4) = \Pr(R_1 = 2, R_2 = 5) + \Pr(R_1 = 3, R_2 = 4);$$

the computation of each of these probabilities is similar to that of $\Pr(R_1 = 3, R_2 = 5)$. This procedure may be continued until we have computed $\Pr(U = u)$ for each $u \in \{u; u = 0, 1, \ldots, 6\}$.

In the preceding example the probability density functions and the sample sizes m and n were selected so as to provide relatively simple integrations. The reader can discover for himself how tedious, and even difficult, the computations become if the sample sizes are large or if the probability density functions are not of a simple functional form.

EXERCISES

9.33. Let the probability density functions of X and Y be those given in Example 1 of this section. Further let the sample sizes be $m = 5$ and $n = 3$. If $R_1 < R_2 < R_3$ are the ranks of $Y_{(1)} < Y_{(2)} < Y_{(3)}$ in the combined sample, compute $\Pr(R_1 = 2, R_2 = 6, R_3 = 8)$.

9.34. Let $X_1, X_2, \ldots, X_m$ be a random sample of size m from a distribution of the continuous type with distribution function $F(x)$ and p.d.f. $F'(x) = f(x)$. Let $Y_1, Y_2, \ldots, Y_n$ be a random sample from a distribution with distribution function $G(y) = [F(y)]^\theta$, $0 < \theta$. If $\theta \neq 1$, this distribution is called a *Lehmann alternative*. With $\theta = 2$, show that

$$\Pr(r_1, r_2, \ldots, r_n) = \frac{2^n r_1(r_2 + 1)(r_3 + 2)\cdots(r_n + n - 1)}{\dbinom{m + n}{m}(m + n + 1)(m + n + 2)\cdots(m + 2n)}.$$

9.35. To generalize the results of Exercise 9.34, let $G(y) = h[F(y)]$, where $h(z)$ is a differentiable function such that $h(0) = 0$, $h(1) = 1$, and $h'(z) > 0$, $0 < z < 1$. Show that

$$\Pr(r_1, r_2, \ldots, r_n) = \frac{E[h'(V_{r_1})h'(V_{r_2})\cdots h'(V_{r_n})]}{\dbinom{m + n}{m}},$$

where $V_1 < V_2 < \cdots < V_{m+n}$ are the order statistics of a random sample of size $m + n$ from the uniform distribution over the interval $(0, 1)$.

9.8 Linear Rank Statistics

In this section we consider a type of distribution-free statistic that is, among other things, a generalization of the Mann–Whitney–Wilcoxon statistic. Let $V_1, V_2, \ldots, V_N$ be a random sample of size N from a distribution of the continuous type. Let R_i be the rank of V_i among $V_1, V_2, \ldots, V_N$, $i = 1, 2, \ldots, N$; and let $c(i)$ be a scoring function defined on the first N positive integers—that is, let $c(1), c(2), \ldots, c(N)$ be some appropriately selected constants. If $a_1, a_2, \ldots, a_N$ are constants, then a statistic of the form

$$L = \sum_{i=1}^{N} a_i c(R_i)$$

is called a *linear rank statistic*.

To see that this type of statistic is actually a generalization of both the Mann–Whitney–Wilcoxon statistic and also that statistic associated with the median test, let $N = m + n$ and

$$V_1 = X_1, \ldots, V_m = X_m, V_{m+1} = Y_1, \ldots, V_N = Y_n.$$

These two special statistics result from the following respective assignments for $c(i)$ and $a_1, a_2, \ldots, a_N$:

(a) Take $c(i) = i$, $a_1 = \cdots = a_m = 0$ and $a_{m+1} = \cdots = a_N = 1$, so that

$$L = \sum_{i=1}^{N} a_i c(R_i) = \sum_{i=m+1}^{m+n} R_i,$$

which is the sum of the ranks of $Y_1, Y_2, \ldots, Y_n$ among the $m + n$ items (a statistic denoted by T in Section 9.6).

(b) Take $c(i) = 1$, provided that $i \leq (m + n)/2$, zero otherwise. If $a_1 = \cdots = a_m = 1$ and $a_{m+1} = \cdots = a_N = 0$, then

$$L = \sum_{i=1}^{N} a_i c(R_i) = \sum_{i=1}^{m} c(R_i),$$

which is equal to the number of the m values of X that are in the lower half of the combined sample of $m + n$ items (a statistic used in the median test of Section 9.5).

To determine the mean and the variance of L, we make some observations about the joint and marginal distributions of the ranks $R_1, R_2, \ldots, R_N$. Clearly, from the results of Section 4.6 on the distribution of order statistics of a random sample, we observe that each permutation of the ranks has the same probability,

$$\Pr (R_1 = r_1, R_2 = r_2, \ldots, R_N = r_N) = \frac{1}{N!},$$

where $r_1, r_2, \ldots, r_N$ is any permutation of the first N positive integers. This implies that the marginal p.d.f. of R_i is

$$g_i(r_i) = \frac{1}{N}, \qquad r_i = 1, 2, \ldots, N,$$

zero elsewhere, because the number of permutations in which $R_i = r_i$ is $(N - 1)!$ so that

$$\sum_{\text{all } (r_1, \ldots, r_{i-1}, r_{i+1}, \ldots, r_N)} \sum \cdots \sum \frac{1}{N!} = \frac{(N - 1)!}{N!} = \frac{1}{N}.$$

In a similar manner, the joint marginal p.d.f. of R_i and R_j, $i \neq j$, is

$$g_{ij}(r_i, r_j) = \frac{1}{N(N - 1)}, \qquad r_i \neq r_j,$$

zero elsewhere. That is, the $(n - 2)$-fold summation

$$\sum \cdots \sum \frac{1}{N!} = \frac{(N - 2)!}{N!} = \frac{1}{N(N - 1)},$$

where the summation is over all permutations in which $R_i = r_i$ and $R_j = r_j$.

Among other things these properties of the distribution of R_1, R_2, ..., R_N imply that

$$E[c(R_i)] = \sum_{r_i=1}^{N} c(r_i)\left(\frac{1}{N}\right) = \frac{c(1) + \cdots + c(N)}{N}.$$

If, for convenience, we let $c(k) = c_k$, then

$$E[c(R_i)] = \sum_{k=1}^{N} \left(\frac{c_k}{N}\right) = \bar{c},$$

say, for all $i = 1, 2, \ldots, N$. In addition, we have that

$$\sigma^2_{c(R_i)} = \sum_{r_i=1}^{N} [c(r_i) - \bar{c}]^2\left(\frac{1}{N}\right) = \sum_{k=1}^{N} \frac{(c_k - \bar{c})^2}{N},$$

for all $i = 1, 2, \ldots, N$.

A simple expression for the covariance of $c(R_i)$ and $c(R_j)$, $i \neq j$, is a little more difficult to determine. That covariance is

$$E\{[c(R_i) - \bar{c}][c(R_j) - \bar{c}]\} = \sum_{k \neq h}\sum \frac{(c_k - \bar{c})(c_h - \bar{c})}{N(N - 1)}.$$

However, since

$$0 = \left[\sum_{k=1}^{N}(c_k - \bar{c})\right]^2 = \sum_{k=1}^{N}(c_k - \bar{c})^2 + \sum_{k \neq h}\sum(c_k - \bar{c})(c_h - \bar{c}),$$

the covariance can be written simply as

$$E\{[c(R_i) - \bar{c}][c(R_j) - \bar{c}]\} = -\sum_{k=1}^{N}\frac{(c_k - \bar{c})^2}{N(N - 1)}.$$

With these results, we first observe that the mean of L is

$$\mu_L = E\left[\sum_{i=1}^{N} a_i c(R_i)\right] = \sum_{i=1}^{N} a_i E[c(R_i)] = \sum_{i=1}^{N} a_i \bar{c} = N\bar{a}\bar{c},$$

where $\bar{a} = (\sum a_i)/N$. Second, note that the variance of L is

$$\sigma^2_L = \sum_{i=1}^{N} a_i{}^2 \sigma^2_{c(R_i)} + \sum_{i \neq j}\sum a_i a_j E\{[c(R_i) - \bar{c}][c(R_j) - \bar{c}]\}$$

$$= \sum_{i=1}^{N} a_i^2 \sum_{k=1}^{N}\frac{(c_k - \bar{c})^2}{N} + \sum_{i \neq j}\sum a_i a_j\left[-\sum_{k=1}^{N}\frac{(c_k - \bar{c})^2}{N(N - 1)}\right]$$

$$= \left[\sum_{k=1}^{N}\frac{(c_k - \bar{c})^2}{N(N - 1)}\right]\left[(N - 1)\sum_{i=1}^{N} a_i^2 - \sum_{i \neq j}\sum a_i a_j\right].$$

However, we can determine a substitute for the second factor by observing that

$$N \sum_{i=1}^{N} (a_i - \bar{a})^2 = N \sum_{i=1}^{N} a_i^2 - N^2 \bar{a}^2$$

$$= N \sum_{i=1}^{N} a_i^2 - \left(\sum_{i=1}^{N} a_i \right)^2$$

$$= N \sum_{i=1}^{N} a_i^2 - \left[\sum_{i=1}^{N} a_i^2 + \sum\sum_{i \neq j} a_i a_j \right]$$

$$= (N - 1) \sum_{i=1}^{N} a_i^2 - \sum\sum_{i \neq j} a_i a_j.$$

So, making this substitution in σ_L^2, we finally have that

$$\sigma_L^2 = \left[\sum_{k=1}^{N} \frac{(c_k - \bar{c})^2}{N(N-1)} \right] \left[N \sum_{i=1}^{N} (a_i - \bar{a})^2 \right]$$

$$= \frac{1}{N-1} \sum_{i=1}^{N} (a_i - \bar{a})^2 \sum_{k=1}^{N} (c_k - \bar{c})^2.$$

In the special case in which $N = m + n$ and

$$L = \sum_{i=m+1}^{N} c(R_i),$$

the reader is asked to show that (Exercise 9.36)

$$\mu_L = n\bar{c}, \qquad \sigma_L^2 = \frac{mn}{N(N-1)} \sum_{k=1}^{N} (c_k - \bar{c})^2.$$

A further simplification when $c_k = c(k) = k$ yields

$$\mu_L = \frac{n(m+n+1)}{2}, \qquad \sigma_L^2 = \frac{mn(m+n+1)}{12};$$

these latter are, respectively, the mean and the variance of the statistic T as defined in Section 9.6.

As in the case of the Mann–Whitney–Wilcoxon statistic, the determination of the exact distribution of a linear rank statistic L can be very difficult. However, for many selections of the constants $a_1, a_2, \ldots, a_N$ and the scores $c(1), c(2), \ldots, c(N)$, the ratio $(L - \mu_L)/\sigma_L$ has, for large N, an approximate distribution that is $n(0, 1)$. This approximation is better if the scores $c(k) = c_k$ are like an ideal sample from a normal distribution, in particular, symmetric and without extreme values. For example, use of normal scores defined by

$$\frac{k}{N+1} = \int_{-\infty}^{c_k} \frac{1}{\sqrt{2\pi}} \exp\left(-\frac{w^2}{2} \right) dw$$

makes the approximation better. However, even with the use of ranks, $c(k) = k$, the approximation is reasonably good, provided that N is large enough, say around 30 or greater.

In addition to being a generalization of statistics such as those of Mann, Whitney, and Wilcoxon, we give two additional applications of linear rank statistics in the following illustrations.

Example 1. Let $X_1, X_2, \ldots, X_n$ denote n random variables. However, suppose that we question whether they are items of a random sample due either to possible lack of mutual stochastic independence or to the fact that $X_1, X_2, \ldots, X_n$ might not have the same distributions. In particular, say we suspect a trend toward larger and larger values in the sequence $X_1, X_2, \ldots, X_n$. If $R_i = \text{rank}(X_i)$, a statistic that could be used to test the alternative (trend) hypothesis is $L = \sum_{i=1}^{n} iR_i$. Under the assumption (H_0) that the n random variables are actually items of a random sample from a distribution of the continuous type, the reader is asked to show that (Exercise 9.37)

$$\mu_L = \frac{n(n+1)^2}{4}, \qquad \sigma_L^2 = \frac{n^2(n+1)^2(n-1)}{144}.$$

The critical region of the test is of the form $L \geq d$, and the constant d can be determined either by using the normal approximation or referring to a tabulated distribution of L so that $\Pr(L \geq d; H_0)$ is approximately equal to a desired significance level α.

Example 2. Let $(X_1, Y_1), (X_2, Y_2), \ldots, (X_n, Y_n)$ be a random sample from a bivariate distribution of the continuous type. Let R_i be the rank of X_i among $X_1, X_2, \ldots, X_n$ and Q_i be the rank of Y_i among $Y_1, Y_2, \ldots, Y_n$. If X and Y have a large positive correlation coefficient, we would anticipate that R_i and Q_i would tend to be large or small together. In particular, the correlation coefficient of $(R_1, Q_1), (R_2, Q_2), \ldots, (R_n, Q_n)$, namely the Spearman rank correlation coefficient,

$$\frac{\sum_{i=1}^{n} (R_i - \bar{R})(Q_i - \bar{Q})}{\sqrt{\sum_{i=1}^{n} (R_i - \bar{R})^2 \sum_{i=1}^{n} (Q_i - \bar{Q})^2}},$$

would tend to be large. Since $R_1, R_2, \ldots, R_n$ and $Q_1, Q_2, \ldots, Q_n$ are permutations of $1, 2, \ldots, n$, this correlation coefficient can be shown (Exercise 9.38) to equal

$$\frac{\sum_{i=1}^{n} R_i Q_i - n(n+1)^2/4}{n(n^2-1)/12},$$

which in turn equals

$$1 - \frac{6 \sum_{i=1}^{n} (R_i - Q_i)^2}{n(n^2 - 1)}.$$

From the first of these two additional expressions for Spearman's statistic, it is clear that $\sum_{i=1}^{n} R_i Q_i$ is an equivalent statistic for the purpose of testing the stochastic independence of X and Y, say H_0. However, note that if H_0 is true, then the distribution of $\sum_{i=1}^{n} Q_i R_i$, which is not a linear rank statistic, and $L = \sum_{i=1}^{n} i R_i$ are the same. The reason for this is that the ranks R_1, R_2, ..., R_n and the ranks $Q_1, Q_2, \ldots, Q_n$ are stochastically independent because of the stochastic independence of X and Y. Hence, under H_0, pairing R_1, $R_2, \ldots, R_n$ at random with $1, 2, \ldots, n$ is distributionally equivalent to pairing those ranks with $Q_1, Q_2, \ldots, Q_n$, which is simply a permutation of $1, 2, \ldots, n$. The mean and the variance of L is given in Example 1.

EXERCISES

9.36. Use the notation of this section.

(a) Show that the mean and the variance of $L = \sum_{i=m+1}^{N} c(R_i)$ are equal to the expressions in the text.

(b) In the special case in which $L = \sum_{i=m+1}^{N} R_i$, show that μ_L and σ_L^2 are those of T considered in Section 9.6. *Hint.* Recall that

$$\sum_{k=1}^{N} k^2 = \frac{N(N + 1)(2N + 1)}{6}.$$

9.37. If $X_1, X_2, \ldots, X_n$ is a random sample from a distribution of the continuous type and if $R_i = \text{rank } (X_i)$, show that the mean and the variance of $L = \sum i R_i$ are $n(n + 1)^2/4$ and $n^2(n + 1)^2(n - 1)/144$, respectively.

9.38. Verify that the two additional expressions, given in Example 2, for the Spearman rank correlation coefficient are equivalent to the first one. *Hint.* $\sum R_i^2 = n(n + 1)(2n + 1)/6$ and $\sum (R_i - Q_i)^2/2 = \sum (R_i^2 + Q_i^2)/2 - \sum R_i Q_i$.

9.39. Let $X_1, X_2, \ldots, X_6$ be a random sample of size $n = 6$ from a distribution of the continuous type. Let $R_i = \text{rank } (X_i)$ and take $a_1 = a_6 = 9$, $a_2 = a_5 = 4$, $a_3 = a_4 = 1$. Find the mean and the variance of $L = \sum_{i=1}^{6} a_i R_i$, a statistic that could be used to detect a parabolic trend in $X_1, X_2, \ldots, X_6$.

9.40. In the notation of this section show that the covariance of the two linear rank statistics, $L_1 = \sum_{i=1}^{N} a_i c(R_i)$ and $L_2 = \sum_{i=1}^{N} b_i d(R_i)$, is equal to

$$\sum_{i=1}^{N} (a_i - \bar{a})(b_i - \bar{b}) \sum_{k=1}^{N} (c_k - \bar{c})(d_k - \bar{d})/(N-1),$$

where, for convenience, $d_k = d(k)$.

Chapter 10

Sufficient Statistics

10.1 A Sufficient Statistic for a Parameter

In Section 6.2 we let $X_1, X_2, \ldots, X_n$ denote a random sample of size n from a distribution that has p.d.f. $f(x; \theta)$, $\theta \in \Omega$. In each of several examples and exercises there, we tried to determine a decision function w of a statistic $Y = u(X_1, X_2, \ldots, X_n)$ or, for simplicity, a function w of $X_1, X_2, \ldots, X_n$ such that the expected value of a loss function $\mathscr{L}(\theta, w)$ is a minimum. That is, we said that the "best" decision function $w(X_1, X_2, \ldots, X_n)$ for a given loss function $\mathscr{L}(\theta, w)$ is one that minimizes the risk $R(\theta, w)$, which for a distribution of the continuous type is given by

$$R(\theta, w) = \int_{-\infty}^{\infty} \cdots \int_{-\infty}^{\infty} \mathscr{L}[\theta, w(x_1, \ldots, x_n)] f(x_1; \theta) \cdots f(x_n; \theta) \, dx_1 \cdots dx_n.$$

In particular, if $E[w(X_1, \ldots, X_n)] = \theta$ and if $\mathscr{L}(\theta, w) = (\theta - w)^2$, the best decision function (statistic) is an unbiased minimum variance estimator of θ. For convenience of exposition in this chapter, we continue to call each unbiased minimum variance estimator of θ a *best estimator* of that parameter. However, the reader must recognize that "best" defined in this way is wholly arbitrary and could be changed by modifying the loss function or relaxing the unbiased assumption.

The purpose of establishing this definition of a best estimator of θ is to help us motivate, in a somewhat natural way, the study of an important class of statistics called *sufficient statistics*. For illustration, note that in Section 6.2 the mean $\bar{X}$ of a random sample of $X_1, X_2, \ldots,$

X_n of size $n = 9$ from a distribution that is $n(\theta, 1)$ is unbiased and has variance less than that of the unbiased estimator X_1. However, to claim that it is a best estimator requires that a comparison be made with the variance of each other unbiased estimator of θ. Certainly, it is impossible to do this by tabulation, and hence we must have some mathematical means that essentially does this. Sufficient statistics provide a beginning to the solution of this problem.

To understand clearly the definition of a sufficient statistic for a parameter θ, we start with an illustration.

Example 1. Let $X_1, X_2, \ldots, X_n$ denote a random sample from the distribution that has p.d.f.

$$f(x; \theta) = \theta^x(1 - \theta)^{1-x}, \qquad x = 0, 1; 0 < \theta < 1;$$
$$= 0 \text{ elsewhere.}$$

The statistic $Y_1 = X_1 + X_2 + \cdots + X_n$ has the p.d.f.

$$g_1(y_1; \theta) = \binom{n}{y_1}\theta^{y_1}(1 - \theta)^{n-y_1}, \qquad y_1 = 0, 1, \ldots, n,$$
$$= 0 \text{ elsewhere.}$$

What is the conditional probability

$$\Pr (X_1 = x_1, X_2 = x_2, \ldots, X_n = x_n | Y_1 = y_1) = P(A|B),$$

say, where $y_1 = 0, 1, 2, \ldots, n$? Unless the sum of the integers $x_1, x_2, \ldots, x_n$ (each of which equals zero or 1) is equal to y_1, this conditional probability obviously equals zero because $A \cap B = \varnothing$. But in the case $y_1 = \sum x_i$, we have that $A \subset B$ so that $A \cap B = A$ and $P(A|B) = P(A)/P(B)$; thus the conditional probability equals

$$\frac{\theta^{x_1}(1 - \theta)^{1-x_1}\theta^{x_2}(1 - \theta)^{1-x_2}\cdots\theta^{x_n}(1 - \theta)^{1-x_n}}{\binom{n}{y_1}\theta^{y_1}(1 - \theta)^{n-y_1}} = \frac{\theta^{\sum x_i}(1 - \theta)^{n-\sum x_i}}{\binom{n}{\sum x_i}\theta^{\sum x_i}(1 - \theta)^{n-\sum x_i}}$$

$$= \frac{1}{\binom{n}{\sum x_i}}.$$

Since $y_1 = x_1 + x_2 + \cdots + x_n$ equals the number of 1's in the n independent trials, this conditional probability is the probability of selecting a particular arrangement of y_1 1's and $(n - y_1)$ zeros. Note that this conditional probability does *not* depend upon the value of the parameter θ.

In general, let $g_1(y_1; \theta)$ be the p.d.f. of the statistic $Y_1 = u_1(X_1, X_2, \ldots, X_n)$, where $X_1, X_2, \ldots, X_n$ is a random sample arising from a distribution of the discrete type having p.d.f. $f(x; \theta)$, $\theta \in \Omega$. The conditional probability of $X_1 = x_1$, $X_2 = x_2, \ldots, X_n = x_n$, given $Y_1 = y_1$, equals

$$\frac{f(x_1; \theta)f(x_2; \theta) \cdots f(x_n; \theta)}{g_1[u_1(x_1, x_2, \ldots, x_n); \theta]},$$

provided that $x_1, x_2, \ldots, x_n$ are such that the fixed $y_1 = u_1(x_1, x_2, \ldots, x_n)$, and equals zero otherwise. We say that $Y_1 = u_1(X_1, X_2, \ldots, X_n)$ is a *sufficient statistic* for θ if and only if this ratio does not depend upon θ. While, with distributions of the continuous type, we cannot use the same argument, we do, in this case, accept the fact that if this ratio does not depend upon θ, then the conditional distribution of $X_1, X_2, \ldots, X_n$, given $Y_1 = y_1$, does not depend upon θ. Thus, in both cases, we use the same definition of a sufficient statistic for θ.

Definition 1. Let $X_1, X_2, \ldots, X_n$ denote a random sample of size n from a distribution that has p.d.f. $f(x; \theta)$, $\theta \in \Omega$. Let $Y_1 = u_1(X_1, X_2, \ldots, X_n)$ be a statistic whose p.d.f. is $g_1(y_1; \theta)$. Then Y_1 is a *sufficient statistic* for θ if and only if

$$\frac{f(x_1; \theta)f(x_2; \theta) \cdots f(x_n; \theta)}{g_1[u_1(x_1, x_2, \ldots, x_n); \theta]} = H(x_1, x_2, \ldots, x_n),$$

where $H(x_1, x_2, \ldots, x_n)$ does not depend upon $\theta \in \Omega$ for every fixed value of $y_1 = u_1(x_1, x_2, \ldots, x_n)$.

Remark. Why we use the terminology "sufficient statistic" can be explained as follows: If a statistic Y_1 satisfies the preceding definition, then the conditional joint p.d.f. of $X_1, X_2, \ldots, X_n$, given $Y_1 = y_1$, and hence of each other statistic, say $Y_2 = u_2(X_1, X_2, \ldots, X_n)$, does not depend upon the parameter θ. As a consequence, once given $Y_1 = y_1$, it is impossible to use Y_2 to make a statistical inference about θ; for example, we could not find a confidence interval for θ based on Y_2. In a sense, Y_1 exhausts all the information about θ that is contained in the sample. It is in this sense that we call Y_1 a sufficient statistic for θ. In some instances it is preferable to call Y_1 a sufficient statistic for the family $\{f(x; \theta); \theta \in \Omega\}$ of probability density functions.

We now give an example that is illustrative of the definition.

Example 2. Let $Y_1 < Y_2 < \cdots < Y_n$ denote the order statistics of a random sample $X_1, X_2, \ldots, X_n$ from the distribution that has p.d.f.

$$f(x; \theta) = e^{-(x-\theta)}, \qquad \theta < x < \infty, \ -\infty < \theta < \infty,$$
$$= 0 \text{ elsewhere.}$$

The p.d.f. of the statistic Y_1 is

$$g_1(y_1; \theta) = ne^{-n(y_1 - \theta)}, \qquad \theta < y_1 < \infty,$$

$$= 0 \text{ elsewhere.}$$

Thus we have that

$$\frac{e^{-(x_1 - \theta)}e^{-(x_2 - \theta)} \cdots e^{-(x_n - \theta)}}{g_1(\min x_i; \theta)} = \frac{e^{-x_1 - x_2 - \cdots - x_n}}{ne^{-n(\min x_i)}},$$

which is free of θ for each fixed $y_1 = \min (x_i)$, since $y_1 \leq x_i$, $i = 1, 2, \ldots, n$. That is, neither the formula nor the domain of the resulting ratio depends upon θ, and hence the first order statistic Y_1 is a sufficient statistic for θ.

If we are to show, by means of the definition, that a certain statistic Y_1 is or is not a sufficient statistic for a parameter θ, we must first of all know the p.d.f. of Y_1, say $g_1(y_1; \theta)$. In some instances it may be quite tedious to find this p.d.f. Fortunately, this problem can be avoided if we will but prove the following *factorization theorem* of Neyman.

Theorem 1. *Let* $X_1, X_2, \ldots, X_n$ *denote a random sample from a distribution that has p.d.f.* $f(x; \theta)$, $\theta \in \Omega$. *The statistic* $Y_1 = u_1(X_1, X_2, \ldots, X_n)$ *is a sufficient statistic for* θ *if and only if we can find two nonnegative functions,* k_1 *and* k_2, *such that*

$$f(x_1; \theta)f(x_2; \theta) \cdots f(x_n; \theta) = k_1[u_1(x_1, x_2, \ldots, x_n); \theta]k_2(x_1, x_2, \ldots, x_n),$$

where, for every fixed value of $y_1 = u_1(x_1, x_2, \ldots, x_n)$, $k_2(x_1, x_2, \ldots, x_n)$ *does not depend upon* θ.

Proof. We shall prove the theorem when the random variables are of the continuous type. Assume the factorization as stated in the theorem. In our proof we shall make the one-to-one transformation $y_1 = u_1(x_1, \ldots, x_n)$, $y_2 = u_2(x_1, \ldots, x_n), \ldots, y_n = u_n(x_1, \ldots, x_n)$ having the inverse functions $x_1 = w_1(y_1, \ldots, y_n)$, $x_2 = w_2(y_1, \ldots, y_n), \ldots, x_n = w_n(y_1, \ldots, y_n)$ and Jacobian J. The joint p.d.f. of the statistics $Y_1, Y_2, \ldots, Y_n$ is then given by

$$g(y_1, y_2, \ldots, y_n; \theta) = k_1(y_1; \theta)k_2(w_1, w_2, \ldots, w_n)|J|,$$

where $w_i = w_i(y_1, y_2, \ldots, y_n)$, $i = 1, 2, \ldots, n$. The p.d.f. of Y_1, say $g_1(y_1; \theta)$, is given by

$$g_1(y_1; \theta) = \int_{-\infty}^{\infty} \cdots \int_{-\infty}^{\infty} g(y_1, y_2, \ldots, y_n; \theta) \, dy_2 \cdots dy_n$$

$$= k_1(y_1; \theta) \int_{-\infty}^{\infty} \cdots \int_{-\infty}^{\infty} |J| k_2(w_1, w_2, \ldots, w_n) \, dy_2 \cdots dy_n.$$

Now the function k_2, for every fixed value of $y_1 = u_1(x_1, \ldots, x_n)$, does not depend upon θ. Nor is θ involved in either the Jacobian J or the limits of integration. Hence the $(n-1)$-fold integral in the right-hand member of the preceding equation is a function of y_1 alone, say $m(y_1)$. Thus

$$g_1(y_1; \theta) = k_1(y_1; \theta)m(y_1).$$

If $m(y_1) = 0$, then $g_1(y_1; \theta) = 0$. If $m(y_1) > 0$, we can write

$$k_1[u_1(x_1, \ldots, x_n); \theta] = \frac{g_1[u_1(x_1, \ldots, x_n); \theta]}{m[u_1(x_1, \ldots, x_n)]},$$

and the assumed factorization becomes

$$f(x_1; \theta) \cdots f(x_n; \theta) = g_1[u_1(x_1, \ldots, x_n); \theta]\frac{k_2(x_1, \ldots, x_n)}{m[u_1(x_1, \ldots, x_n)]}.$$

Since neither the function k_2 nor the function m depends upon θ, then in accordance with the definition, Y_1 is a sufficient statistic for the parameter θ.

Conversely, if Y_1 is a sufficient statistic for θ, the factorization can be realized by taking the function k_1 to be the p.d.f. of Y_1, namely the function g_1. This completes the proof of the theorem.

Example 3. Let $X_1, X_2, \ldots, X_n$ denote a random sample from a distribution that is $n(\theta, \sigma^2)$, $-\infty < \theta < \infty$, where the variance σ^2 is known. If $\bar{x} = \sum_1^n x_i/n$, then

$$\sum_{i=1}^n (x_i - \theta)^2 = \sum_{i=1}^n [(x_i - \bar{x}) + (\bar{x} - \theta)]^2 = \sum_{i=1}^n (x_i - \bar{x})^2 + n(\bar{x} - \theta)^2$$

because

$$2\sum_{i=1}^n (x_i - \bar{x})(\bar{x} - \theta) = 2(\bar{x} - \theta)\sum_{i=1}^n (x_i - \bar{x}) = 0.$$

Thus the joint p.d.f. of $X_1, X_2, \ldots, X_n$ may be written

$$\left(\frac{1}{\sigma\sqrt{2\pi}}\right)^n \exp\left[-\sum_{i=1}^n (x_i - \theta)^2/2\sigma^2\right]$$

$$= \{\exp[-n(\bar{x} - \theta)^2/2\sigma^2]\}\left\{\frac{\exp\left[-\sum_{i=1}^n (x_i - \bar{x})^2/2\sigma^2\right]}{(\sigma\sqrt{2\pi})^n}\right\}.$$

Since the first factor of the right-hand member of this equation depends upon $x_1, x_2, \ldots, x_n$ only through $\bar{x}$, and since the second factor does not depend upon θ, the factorization theorem implies that the mean $\bar{X}$ of the sample is, for any particular value of σ^2, a sufficient statistic for θ, the mean of the normal distribution.

We could have used the definition in the preceding example because we know that $\bar{X}$ is $n(\theta, \sigma^2/n)$. Let us now consider an example in which the use of the definition is inappropriate.

Example 4. Let $X_1, X_2, \ldots, X_n$ denote a random sample from a distribution with p.d.f.

$$f(x; \theta) = \theta x^{\theta-1}, \qquad 0 < x < 1,$$
$$= 0 \text{ elsewhere,}$$

where $0 < \theta$. We shall use the factorization theorem to prove that the product $u_1(X_1, X_2, \ldots, X_n) = X_1 X_2 \cdots X_n$ is a sufficient statistic for θ. The joint p.d.f. of $X_1, X_2, \ldots, X_n$ is

$$\theta^n (x_1 x_2 \cdots x_n)^{\theta-1} = [\theta^n (x_1 x_2 \cdots x_n)^{\theta}] \left(\frac{1}{x_1 x_2 \cdots x_n} \right),$$

where $0 < x_i < 1$, $i = 1, 2, \ldots, n$. In the factorization theorem let

$$k_1[u_1(x_1, x_2, \ldots, x_n); \theta] = \theta^n (x_1 x_2 \cdots x_n)^{\theta}$$

and

$$k_2(x_1, x_2, \ldots, x_n) = \frac{1}{x_1 x_2 \cdots x_n}.$$

Since $k_2(x_1, x_2, \ldots, x_n)$ does not depend upon θ, the product $X_1 X_2 \cdots X_n$ is a sufficient statistic for θ.

There is a tendency for some readers to apply incorrectly the factorization theorem in those instances in which the domain of positive probability density depends upon the parameter θ. This is due to the fact that they do not give proper consideration to the domain of the function $k_2(x_1, x_2, \ldots, x_n)$ for every fixed value of $y_1 = u_1(x_1, x_2, \ldots, x_n)$. This will be illustrated in the next example.

Example 5. In Example 2, with $f(x; \theta) = e^{-(x-\theta)}$, $\theta < x < \infty$, $-\infty < \theta < \infty$, it was found that the first order statistic Y_1 is a sufficient statistic for θ. To illustrate our point, take $n = 3$ so that the joint p.d.f. of X_1, X_2, X_3 is given by

$$e^{-(x_1-\theta)} e^{-(x_2-\theta)} e^{-(x_3-\theta)}, \qquad \theta < x_i < \infty,$$

$i = 1, 2, 3$. We can factor this in several ways. One way, with $n = 3$, is given in Example 2. Another way would be to write the joint p.d.f. as the product

$$e^{-(\max x_i) + 3\theta} e^{-x_1 - x_2 - x_3 + \max x_i}.$$

Certainly, there is no θ in the *formula* of the second factor, and it might be assumed that $Y_3 = \max X_i$ is itself a sufficient statistic for θ. But what is the *domain* of the second factor for every fixed value of $y_3 = \max x_i$? If $\max x_i = x_1$, the domain is $\theta < x_2 < x_1$, $\theta < x_3 < x_1$; if $\max x_i = x_2$, the domain is $\theta < x_1 < x_2$, $\theta < x_3 < x_2$; and if $\max x_i = x_3$, the domain is $\theta < x_1 < x_3$, $\theta < x_2 < x_3$. That is, for each fixed $y_3 = \max x_i$, the domain of the second factor depends upon θ. Thus the factorization theorem is not satisfied.

If the reader has some difficulty using the factorization theorem when the domain of positive probability density depends upon θ, we recommend use of the definition even though it may be somewhat longer.

Before taking the next step in our search for a best statistic for a parameter θ, let us consider an important property possessed by a sufficient statistic $Y_1 = u_1(X_1, X_2, \ldots, X_n)$ for θ. The conditional p.d.f. of another statistic, say $Y_2 = u_2(X_1, X_2, \ldots, X_n)$, given $Y_1 = y_1$, does not depend upon θ. On intuitive grounds, we might surmise that the conditional p.d.f. of Y_2, given some linear function $aY_1 + b$, $a \neq 0$, of Y_1, does not depend upon θ. That is, it seems as though the random variable $aY_1 + b$ is also a sufficient statistic for θ. This conjecture is correct. In fact, every function $Z = u(Y_1)$, or $Z = u[u_1(X_1, X_2, \ldots, X_n)] = v(X_1, X_2, \ldots, X_n)$, not involving θ, with a single-valued inverse $Y_1 = w(Z)$, is also a sufficient statistic for θ. To prove this, we write, in accordance with the factorization theorem,

$$f(x_1; \theta) \cdots f(x_n; \theta) = k_1[u_1(x_1, x_2, \ldots, x_n); \theta]k_2(x_1, x_2, \ldots, x_n).$$

However, $y_1 = w(z)$ or, equivalently, $u_1(x_1, x_2, \ldots, x_n) = w[v(x_1, x_2, \ldots, x_n)]$, which is not a function of θ. Hence

$$f(x_1; \theta) \cdots f(x_n; \theta) = k_1\{w[v(x_1, \ldots, x_n)]; \theta\}k_2(x_1, x_2, \ldots, x_n).$$

Since the first factor of the right-hand member of this equation is a function of $z = v(x_1, \ldots, x_n)$ and θ, while the second factor does not depend upon θ, the factorization theorem implies that $Z = u(Y_1)$ is also a sufficient statistic for θ.

The relationship of a sufficient statistic for θ to the maximum likelihood estimator of θ is contained in the following theorem.

Theorem 2. *Let $X_1, X_2, \ldots, X_n$ denote a random sample from a distribution that has p.d.f. $f(x; \theta)$, $\theta \in \Omega$. If a sufficient statistic $Y_1 = u_1(X_1, X_2, \ldots, X_n)$ for θ exists and if a maximum likelihood estimator $\hat{\theta}$ of θ also exists uniquely, then $\hat{\theta}$ is a function of $Y_1 = u_1(X_1, X_2, \ldots, X_n)$.*

Proof. Let $g_1(y_1; \theta)$ be the p.d.f. of Y_1. Then by the definition of sufficiency, the likelihood function

$$L(\theta; x_1, x_2, \ldots, x_n) = f(x_1; \theta)f(x_2; \theta)\cdots f(x_n; \theta)$$

$$= g_1[u_1(x_1, \ldots, x_n); \theta]H(x_1, \ldots, x_n),$$

where $H(x_1, \ldots, x_n)$ does not depend upon θ. Thus L and g_1, as functions of θ, are maximized simultaneously. Since there is one and only one value of θ that maximizes L and hence $g_1[u_1(x_1, \ldots, x_n); \theta]$, that value of θ must be a function of $u_1(x_1, x_2, \ldots, x_n)$. Thus the maximum likelihood estimator $\hat{\theta}$ is a function of the sufficient statistic $Y_1 = u_1(X_1, X_2, \ldots, X_n)$.

EXERCISES

10.1. Let $X_1, X_2, \ldots, X_n$ be a random sample from the normal distribution $n(0, \theta)$, $0 < \theta < \infty$. Show that $\sum_1^n X_i^2$ is a sufficient statistic for θ.

10.2. Prove that the sum of the items of a random sample of size n from a Poisson distribution having parameter θ, $0 < \theta < \infty$, is a sufficient statistic for θ.

10.3. Show that the nth order statistic of a random sample of size n from the uniform distribution having p.d.f. $f(x; \theta) = 1/\theta$, $0 < x < \theta$, $0 < \theta < \infty$, zero elsewhere, is a sufficient statistic for θ. Generalize this result by considering the p.d.f. $f(x; \theta) = Q(\theta)M(x)$, $0 < x < \theta$, $0 < \theta < \infty$, zero elsewhere. Here, of course,

$$\int_0^\theta M(x)\,dx = \frac{1}{Q(\theta)}.$$

10.4. Let $X_1, X_2, \ldots, X_n$ be a random sample of size n from a geometric distribution that has p.d.f. $f(x; \theta) = (1 - \theta)^x \theta$, $x = 0, 1, 2, \ldots$, $0 < \theta < 1$, zero elsewhere. Show that $\sum_1^n X_i$ is a sufficient statistic for θ.

10.5. Show that the sum of the items of a random sample of size n from a gamma distribution that has p.d.f. $f(x; \theta) = (1/\theta)e^{-x/\theta}$, $0 < x < \infty$, $0 < \theta < \infty$, zero elsewhere, is a sufficient statistic for θ.

10.6. In each of the Exercises 10.1, 10.2, 10.4, and 10.5, show that the maximum likelihood estimator of θ is a function of the sufficient statistic for θ.

10.7. Let $X_1, X_2, \ldots, X_n$ be a random sample of size n from a beta distribution with parameters $\alpha = \theta > 0$ and $\beta = 2$. Show that the product $X_1 X_2 \cdots X_n$ is a sufficient statistic for θ.

10.8. Let $X_1, X_2, \ldots, X_n$ be a random sample of size n from a distribution with p.d.f. $f(x; \theta) = 1/\pi[1 + (x - \theta)^2]$, $-\infty < x < \infty$, $-\infty < \theta < \infty$. Can the joint p.d.f. of $X_1, X_2, \ldots, X_n$ be written in the form given in Theorem 1? Does the parameter θ have a sufficient statistic?

10.2 The Rao–Blackwell Theorem

We shall prove the Rao–Blackwell theorem.

Theorem 3. *Let X and Y denote random variables such that Y has mean μ and positive variance σ_Y^2. Let $E(Y|x) = \varphi(x)$. Then $E[\varphi(X)] = \mu$ and $\sigma_{\varphi(X)}^2 \leq \sigma_Y^2$.*

Proof. We shall give the proof when the random variables are of the continuous type. Let $f(x, y), f_1(x), f_2(y)$, and $h(y|x)$ denote, respectively, the joint p.d.f. of X and Y, the two marginal probability density functions, and the conditional p.d.f. of Y, given $X = x$. Then

$$E(Y|x) = \int_{-\infty}^{\infty} yh(y|x)\, dy = \frac{\int_{-\infty}^{\infty} yf(x, y)\, dy}{f_1(x)} = \varphi(x),$$

so that

$$\int_{-\infty}^{\infty} yf(x, y)\, dy = \varphi(x)f_1(x).$$

We have

$$E[\varphi(X)] = \int_{-\infty}^{\infty} \varphi(x)f_1(x)\, dx = \int_{-\infty}^{\infty} \left[\int_{-\infty}^{\infty} yf(x, y)\, dy\right] dx$$

$$= \int_{-\infty}^{\infty} y\left[\int_{-\infty}^{\infty} f(x, y)\, dx\right] dy$$

$$= \int_{-\infty}^{\infty} yf_2(y)\, dy = \mu,$$

and the first part of the theorem is established. Consider next

$$\sigma_Y^2 = E[(Y - \mu)^2] = E\{[(Y - \varphi(X)) + (\varphi(X) - \mu)]^2\}$$

$$= E\{[Y - \varphi(X)]^2\} + E\{[\varphi(X) - \mu]^2\} + 2E\{[Y - \varphi(X)][\varphi(X) - \mu]\}.$$

We shall show that the last term of the right-hand member of the immediately preceding equation is zero. We have

$$E\{[Y - \varphi(X)][\varphi(X) - \mu]\} = \int_{-\infty}^{\infty}\int_{-\infty}^{\infty} [y - \varphi(x)][\varphi(x) - \mu]f(x, y)\, dy\, dx.$$

In this integral we shall write $f(x, y)$ in the form $h(y|x)f_1(x)$, and we shall integrate first on y to obtain

$$\int_{-\infty}^{\infty} [\varphi(x) - \mu]\left\{\int_{-\infty}^{\infty} [y - \varphi(x)]h(y|x)\, dy\right\}f_1(x)\, dx.$$

But $\varphi(x)$ is the mean of the conditional p.d.f. $h(y|x)$. Hence

$$\int_{-\infty}^{\infty} [y - \varphi(x)]h(y|x)\,dy = 0,$$

and, accordingly,

$$E\{[Y - \varphi(X)][\varphi(X) - \mu]\} = 0.$$

Moreover,

$$\sigma_{\varphi(X)}^2 = E\{[\varphi(X) - \mu]^2\}$$

and

$$E\{[Y - \varphi(X)]^2\} \geq 0.$$

Accordingly,

$$\sigma_Y^2 \geq \sigma_{\varphi(X)}^2,$$

and the theorem is proved when X and Y are random variables of the continuous type. The proof in the discrete case is identical to the proof given here with the exception that summation replaces integration.

It is interesting to note, in connection with the proof of the theorem, that unless the probability measure of the set $\{(x, y); y - \varphi(x) = 0\}$ is equal to 1 then $E\{[Y - \varphi(X)]^2\} > 0$, and we have the strict inequality $\sigma_Y^2 > \sigma_{\varphi(X)}^2$.

We shall give an illustrative example.

Example 1. Let X and Y have a bivariate normal distribution with means μ_1 and μ_2, with positive variances σ_1^2 and σ_2^2, and with correlation coefficient ρ. Here $E(Y) = \mu = \mu_2$ and $\sigma_Y^2 = \sigma_2^2$. Now $E(Y|x)$ is linear in x and it is given by

$$E(Y|x) = \varphi(x) = \mu_2 + \rho\frac{\sigma_2}{\sigma_1}(x - \mu_1).$$

Thus $\varphi(X) = \mu_2 + \rho(\sigma_2/\sigma_1)(X - \mu_1)$ and $E[\varphi(X)] = \mu_2$, as stated in the theorem. Moreover,

$$\begin{aligned}
\sigma_{\varphi(X)}^2 &= E\{[\varphi(X) - \mu_2]^2\} \\
&= E\left\{\left[\rho\frac{\sigma_2}{\sigma_1}(X - \mu_1)\right]^2\right\} \\
&= \rho^2\sigma_2^2.
\end{aligned}$$

With $-1 < \rho < 1$, we have the strict inequality $\sigma_2^2 > \rho^2 \sigma_2^2$. It should be observed that $\varphi(X)$ is not a statistic if at least one of the five parameters is unknown.

We shall use the Rao–Blackwell theorem to help us in our search for a best estimator of a parameter. Let $X_1, X_2, \ldots, X_n$ denote a random sample from a distribution that has p.d.f. $f(x; \theta)$, $\theta \in \Omega$, where it is known that $Y_1 = u_1(X_1, X_2, \ldots, X_n)$ is a sufficient statistic for the parameter θ. Let $Y_2 = u_2(X_1, X_2, \ldots, X_n)$ be another statistic (but not a function of Y_1 alone), which is an unbiased estimator of θ; that is, $E(Y_2) = \theta$. Consider $E(Y_2|y_1)$. This expectation is a function of y_1, say $\varphi(y_1)$. Since Y_1 is a sufficient statistic for θ, the conditional p.d.f. of Y_2, given $Y_1 = y_1$, does not depend upon θ, so $E(Y_2|y_1) = \varphi(y_1)$ is a function of y_1 alone. That is, here $\varphi(Y_1)$ is a statistic. In accordance with the Rao–Blackwell theorem, $\varphi(Y_1)$ is an unbiased estimator of θ; and because Y_2 is not a function of Y_1 alone, the variance of $\varphi(Y_1)$ is strictly less than the variance of Y_2. We shall summarize this discussion in the following theorem.

Theorem 4. *Let $X_1, X_2, \ldots, X_n$, n a fixed positive integer, denote a random sample from a distribution (continuous or discrete) that has p.d.f. $f(x; \theta)$, $\theta \in \Omega$. Let $Y_1 = u_1(X_1, X_2, \ldots, X_n)$ be a sufficient statistic for θ, and let $Y_2 = u_2(X_1, X_2, \ldots, X_n)$, not a function of Y_1 alone, be an unbiased estimator of θ. Then $E(Y_2|y_1) = \varphi(y_1)$ defines a statistic $\varphi(Y_1)$. This statistic $\varphi(Y_1)$ is a function of the sufficient statistic for θ; it is an unbiased estimator of θ; and its variance is less than that of Y_2.*

This theorem tells us that in our search for a best estimator of a parameter, we may, if a sufficient statistic for the parameter exists, restrict that search to functions of the sufficient statistic. For if we begin with an unbiased estimator Y_2 that is not a function of the sufficient statistic Y_1 alone, then we can always improve on this by computing $E(Y_2|y_1) = \varphi(y_1)$ so that $\varphi(Y_1)$ is an unbiased estimator with smaller variance than that of Y_2.

After Theorem 4 many students believe that it is necessary to find first some unbiased estimator Y_2 in their search for $\varphi(Y_1)$, an unbiased estimator of θ based upon the sufficient statistic Y_1. This is not the case at all, and Theorem 4 simply convinces us that we can restrict our search for a best estimator to functions of Y_1. It frequently happens that $E(Y_1) = a\theta + b$, where $a \neq 0$ and b are constants, and thus $(Y_1 - b)/a$ is a function of Y_1 that is an unbiased estimator of θ.

That is, we can usually find an unbiased estimator based on Y_1 without first finding an estimator Y_2. In the next two sections we discover that, in most instances, if there is one function $\varphi(Y_1)$ that is unbiased, $\varphi(Y_1)$ is the only unbiased estimator based on the sufficient statistic Y_1.

Remark. Since the unbiased estimator $\varphi(Y_1)$, where $\varphi(y_1) = E(Y_2|y_1)$, has variance smaller than that of the unbiased estimator Y_2 of θ, students sometimes reason as follows. Let the function $\Upsilon(y_3) = E[\varphi(Y_1)|Y_3 = y_3]$, where Y_3 is another statistic, which is not sufficient for θ. By the Rao-Blackwell theorem, we have that $E[\Upsilon(Y_3)] = \theta$ and $\Upsilon(Y_3)$ has a smaller variance than does $\varphi(Y_1)$. Accordingly, $\Upsilon(Y_3)$ must be better than $\varphi(Y_1)$ as an unbiased estimator of θ. But this is *not* true because Y_3 is not sufficient; thus θ is present in the conditional distribution of Y_1, given $Y_3 = y_3$, and the conditional mean $\Upsilon(y_3)$. So although indeed $E[\Upsilon(Y_3)] = \theta$, $\Upsilon(Y_3)$ is not even a statistic because it involves the unknown parameter θ and hence cannot be used as an estimator.

EXERCISES

10.9. Let $Y_1 < Y_2 < Y_3 < Y_4 < Y_5$ be the order statistics of a random sample of size 5 from the uniform distribution having p.d.f. $f(x; \theta) = 1/\theta$, $0 < x < \theta$, $0 < \theta < \infty$, zero elsewhere. Show that $2Y_3$ is an unbiased estimator of θ. Determine the joint p.d.f. of Y_3 and the sufficient statistic Y_5 for θ. Find the conditional expectation $E(2Y_3|y_5) = \varphi(y_5)$. Compare the variances of $2Y_3$ and $\varphi(Y_5)$. *Hint.* All of the integrals needed in this exercise can be evaluated by making a change of variable such as $z = y/\theta$ and using the results associated with the beta p.d.f.; see Example 5, Section 4.3.

10.10. If X_1, X_2 is a random sample of size 2 from a distribution having p.d.f. $f(x; \theta) = (1/\theta)e^{-x/\theta}$, $0 < x < \infty$, $0 < \theta < \infty$, zero elsewhere, find the joint p.d.f. of the sufficient statistic $Y_1 = X_1 + X_2$ for θ and $Y_2 = X_2$. Show that Y_2 is an unbiased estimator of θ with variance θ^2. Find $E(Y_2|y_1) = \varphi(y_1)$ and the variance of $\varphi(Y_1)$.

10.11. Let the random variables X and Y have the joint p.d.f. $f(x, y) = (2/\theta^2)e^{-(x+y)/\theta}$, $0 < x < y < \infty$, zero elsewhere.
(a) Show that the mean and the variance of Y are, respectively, $3\theta/2$ and $5\theta^2/4$.
(b) Show that $E(Y|x) = x + \theta$. In accordance with the Rao-Blackwell theorem, the expected value of $X + \theta$ is that of Y, namely, $3\theta/2$, and the variance of $X + \theta$ is less than that of Y. Show that the variance of $X + \theta$ is in fact $\theta^2/4$.

10.12. In each of Exercises 10.1, 10.2, and 10.5, compute the expected value of the given sufficient statistic and, in each case, determine an unbiased estimator of θ that is a function of that sufficient statistic alone.

10.3 Completeness and Uniqueness

Let $X_1, X_2, \ldots, X_n$ be a random sample from the distribution that has p.d.f.

$$f(x; \theta) = \frac{\theta^x e^{-\theta}}{x!}, \qquad x = 0, 1, 2, \ldots; \quad 0 < \theta$$

$$= 0 \text{ elsewhere.}$$

From Exercise 10.2 of Section 10.1 we know that $Y_1 = \sum_{i=1}^{n} X_i$ is a sufficient statistic for θ and its p.d.f. is

$$g_1(y_1; \theta) = \frac{(n\theta)^{y_1} e^{-n\theta}}{y_1!}, \qquad y_1 = 0, 1, 2, \ldots$$

$$= 0 \text{ elsewhere.}$$

Let us consider the family $\{g_1(y_1; \theta); 0 < \theta\}$ of probability density functions. Suppose that the function $u(Y_1)$ of Y_1 is such that $E[u(Y_1)] = 0$ for every $\theta > 0$. We shall show that this requires $u(y_1)$ to be zero at every point $y_1 = 0, 1, 2, \ldots$. That is,

$$E[u(Y_1)] = 0, \qquad 0 < \theta$$

implies that

$$0 = u(0) = u(1) = u(2) = u(3) = \cdots.$$

We have for all $\theta > 0$ that

$$0 = E[u(Y_1)] = \sum_{y_1=0}^{\infty} u(y_1) \frac{(n\theta)^{y_1} e^{-n\theta}}{y_1!}$$

$$= e^{-n\theta} \left[u(0) + u(1) \frac{n\theta}{1!} + u(2) \frac{(n\theta)^2}{2!} + \cdots \right].$$

Since $e^{-n\theta}$ does not equal zero, we have that

$$0 = u(0) + [nu(1)]\theta + \left[\frac{n^2 u(2)}{2} \right] \theta^2 + \cdots.$$

However, if such an infinite series converges to zero for all $\theta > 0$, then each of the coefficients must equal zero. That is,

$$u(0) = 0, \quad nu(1) = 0, \quad \frac{n^2 u(2)}{2} = 0, \ldots$$

and thus $0 = u(0) = u(1) = u(2) = \cdots$, as we wanted to show. Of course, the condition $E[u(Y_1)] = 0$ for all $\theta > 0$ does not place any restriction on $u(y_1)$ when y_1 is not a nonnegative integer. So we see that, in this illustration, $E[u(Y_1)] = 0$ for all $\theta > 0$ requires that $u(y_1)$ equals zero except on a set of points that has probability zero for each p.d.f. $g_1(y_1; \theta)$, $0 < \theta$. From the following definition we observe that the family $\{g_1(y_1; \theta); 0 < \theta\}$ is complete.

Definition 2. Let the random variable Z of either the continuous type or the discrete type have a p.d.f. that is one member of the family $\{h(z; \theta); \theta \in \Omega\}$. If the condition $E[u(Z)] = 0$, for every $\theta \in \Omega$, requires that $u(z)$ be zero except on a set of points that has probability zero for each p.d.f. $h(z; \theta)$, $\theta \in \Omega$, then the family $\{h(z; \theta); \theta \in \Omega\}$ is called a *complete family* of probability density functions.

Remark. In Section 1.9 it was noted that the existence of $E[u(X)]$ implies that the integral (or sum) converge absolutely. This absolute convergence was tacitly assumed in our definition of completeness and it is needed to prove that certain families of probability density functions are complete.

In order to show that certain families of probability density functions of the continuous type are complete, we must appeal to the same type of theorem in analysis that we used when we claimed that the moment-generating function uniquely determines a distribution. This is illustrated in the next example.

Example 1. Let Z have a p.d.f. that is a member of the family $\{h(z; \theta), 0 < \theta < \infty\}$, where

$$h(z; \theta) = \frac{1}{\theta} e^{-z/\theta}, \qquad 0 < z < \infty,$$

$$= 0 \text{ elsewhere.}$$

Let us say that $E[u(Z)] = 0$ for every $\theta > 0$. That is,

$$\frac{1}{\theta} \int_0^\infty u(z) e^{-z/\theta} \, dz = 0, \qquad \text{for } \theta > 0.$$

Readers acquainted with the theory of transforms will recognize the integral in the left-hand member as being essentially the Laplace transform of $u(z)$. In that theory we learn that the only function $u(z)$ transforming to a function of θ which is identically equal to zero is $u(z) = 0$, except (in our terminology) on a set of points that has probability zero for each $h(z; \theta)$, $0 < \theta$. That is, the family $\{h(z; \theta); 0 < \theta < \infty\}$ is complete.

Let the parameter θ in the p.d.f. $f(x; \theta)$, $\theta \in \Omega$, have a sufficient statistic $Y_1 = u_1(X_1, X_2, \ldots, X_n)$, where $X_1, X_2, \ldots, X_n$ is a random sample from this distribution. Let the p.d.f. of Y_1 be $g_1(y_1; \theta)$, $\theta \in \Omega$. It has been seen that, if there is any unbiased estimator Y_2 (not a function of Y_1 alone) of θ, then there is at least one function of Y_1 that is an unbiased estimator of θ, and our search for a best estimator of θ may be restricted to functions of Y_1. Suppose it has been verified that a certain function $\varphi(Y_1)$, not a function of θ, is such that $E[\varphi(Y_1)] = \theta$ for all values of θ, $\theta \in \Omega$. Let $\psi(Y_1)$ be another function of the sufficient statistic Y_1 alone so that we have also $E[\psi(Y_1)] = \theta$ for all values of θ, $\theta \in \Omega$. Hence

$$E[\varphi(Y_1) - \psi(Y_1)] = 0, \qquad \theta \in \Omega.$$

If the family $\{g_1(y_1; \theta); \theta \in \Omega\}$ is complete, the function $\varphi(y_1) - \psi(y_1) = 0$, except on a set of points that has probability zero. That is, for every other unbiased estimator $\psi(Y_1)$ of θ, we have

$$\varphi(y_1) = \psi(y_1) \qquad \textit{for all } \theta$$

except possibly at certain special points. Thus, in this sense [namely $\varphi(y_1) = \psi(y_1)$, except on a set of points with probability zero], $\varphi(Y_1)$ is the unique function of Y_1, which is an unbiased estimator of θ. In accordance with the Rao–Blackwell theorem, $\varphi(Y_1)$ has a smaller variance than every other unbiased estimator of θ. That is, the statistic $\varphi(Y_1)$ is the best estimator of θ. This fact is stated in the following theorem of Lehmann and Scheffé.

Theorem 5. *Let $X_1, X_2, \ldots, X_n$, n a fixed positive integer, denote a random sample from a distribution that has p.d.f. $f(x; \theta)$, $\theta \in \Omega$, let $Y_1 = u_1(X_1, X_2, \ldots, X_n)$ be a sufficient statistic for θ, and let the family $\{g_1(y_1; \theta); \theta \in \Omega\}$ of probability density functions be complete. If there is a function of Y_1 that is an unbiased estimator of θ, then this function of Y_1 is the unique best estimator of θ. Here "unique" is used in the sense described in the preceding paragraph.*

The statement that Y_1 is a sufficient statistic for a parameter θ, $\theta \in \Omega$, and that the family $\{g_1(y_1; \theta); \theta \in \Omega\}$ of probability density functions is complete is lengthy and somewhat awkward. We shall adopt the less descriptive, but more convenient, terminology that Y_1 is a *complete sufficient statistic* for θ. In the next section we shall study a fairly large class of probability density functions for which a complete sufficient statistic Y_1 for θ can be determined by inspection.

EXERCISES

10.13. If $az^2 + bz + c = 0$ for more than two values of z, then $a = b = c = 0$. Use this result to show that the family $\{b(2, \theta); 0 < \theta < 1\}$ is complete.

10.14. Show that each of the following families $\{f(x; \theta); 0 < \theta < \infty\}$ is not complete by finding at least one nonzero function $u(x)$ such that $E[u(X)] = 0$, for all $\theta > 0$.

(a) $f(x; \theta) = \dfrac{1}{2\theta}, \qquad -\theta < x < \theta,$

$\qquad = 0$ elsewhere.

(b) $n(0, \theta)$.

10.15. Let $X_1, X_2, \ldots, X_n$ represent a random sample from the discrete distribution having the probability density function

$$f(x; \theta) = \theta^x(1 - \theta)^{1-x}, \qquad x = 0, 1, 0 < \theta < 1,$$

$$= 0 \text{ elsewhere.}$$

Show that $Y_1 = \sum_1^n X_i$ is a complete sufficient statistic for θ. Find the unique function of Y_1 that is the best estimator of θ. *Hint.* Display $E[u(Y_1)] = 0$, show that the constant term $u(0)$ is equal to zero, divide both members of the equation by $\theta \neq 0$, and repeat the argument.

10.16. Consider the family of probability density functions $\{h(z; \theta); \theta \in \Omega\}$, where $h(z; \theta) = 1/\theta, 0 < z < \theta$, zero elsewhere.

(a) Show that the family is complete provided that $\Omega = \{\theta; 0 < \theta < \infty\}$. *Hint.* For convenience, assume that $u(z)$ is continuous and note that the derivative of $E[u(Z)]$ with respect to θ is equal to zero also.

(b) Show that this family is not complete if $\Omega = \{\theta; 1 < \theta < \infty\}$. *Hint.* Concentrate on the interval $0 < z < 1$ and find a nonzero function $u(z)$ on that interval such that $E[u(Z)] = 0$ for all $\theta > 1$.

10.17. Show that the first order statistic Y_1 of a random sample of size n from the distribution having p.d.f. $f(x; \theta) = e^{-(x-\theta)}, \theta < x < \infty, -\infty < \theta < \infty$, zero elsewhere, is a complete sufficient statistic for θ. Find the unique function of this statistic which is the best estimator of θ.

10.18. Let a random sample of size n be taken from a distribution of the discrete type with p.d.f. $f(x; \theta) = 1/\theta, x = 1, 2, \ldots, \theta$, zero elsewhere, where θ is an unknown positive integer.

(a) Show that the largest item, say Y, of the sample is a complete sufficient statistic for θ.

(b) Prove that

$$[Y^{n+1} - (Y - 1)^{n+1}]/[Y^n - (Y - 1)^n]$$

is the unique best estimator of θ.

10.4 The Exponential Class of Probability Density Functions

Consider a family $\{f(x; \theta); \theta \in \Omega\}$ of probability density functions, where Ω is the interval set $\Omega = \{\theta; \gamma < \theta < \delta\}$, where γ and δ are known constants, and where

(1) $f(x; \theta) = \exp[p(\theta)K(x) + S(x) + q(\theta)]$, $a < x < b$,

$= 0$ elsewhere.

A p.d.f. of the form (1) is said to be a member of the *exponential class* of probability density functions of the continuous type. If, in addition,

(a) neither a nor b depends upon θ, $\gamma < \theta < \delta$,
(b) $p(\theta)$ is a nontrivial continuous function of θ, $\gamma < \theta < \delta$.
(c) each of $K'(x) \not\equiv 0$ and $S(x)$ is a continuous function of x, $a < x < b$,

we say that we have a *regular case* of the exponential class. A p.d.f.

$f(x; \theta) = \exp[p(\theta)K(x) + S(x) + q(\theta)]$, $x = a_1, a_2, a_3, \ldots$,
$= 0$ elsewhere

is said to represent a regular case of the exponential class of probability density functions of the discrete type if

(a) The set $\{x; x = a_1, a_2, \ldots\}$ does not depend upon θ.
(b) $p(\theta)$ is a nontrivial continuous function of θ, $\gamma < \theta < \delta$.
(c) $K(x)$ is a nontrivial function of x on the set $\{x; x = a_1, a_2, \ldots\}$.

For example, each member of the family $\{f(x; \theta); 0 < \theta < \infty\}$, where $f(x; \theta)$ is $n(0, \theta)$, represents a regular case of the exponential class of the continuous type because

$$f(x; \theta) = \frac{1}{\sqrt{2\pi\theta}} e^{-x^2/2\theta}$$

$$= \exp\left(-\frac{1}{2\theta} x^2 - \ln \sqrt{2\pi\theta}\right), \qquad -\infty < x < \infty.$$

Let $X_1, X_2, \ldots, X_n$ denote a random sample from a distribution that has a p.d.f. which represents a regular case of the exponential class of the continuous type. The joint p.d.f. of $X_1, X_2, \ldots, X_n$ is

$$\exp\left[p(\theta) \sum_1^n K(x_i) + \sum_1^n S(x_i) + nq(\theta)\right]$$

for $a < x_i < b$, $i = 1, 2, \ldots, n$, $\gamma < \theta < \delta$, and is zero elsewhere. At points of positive probability density, this joint p.d.f. may be written as the product of the two nonnegative functions

$$\exp \left[p(\theta) \sum_1^n K(x_i) + nq(\theta) \right] \exp \left[\sum_1^n S(x_i) \right].$$

In accordance with the factorization theorem (Theorem 1, Section 10.1) $Y_1 = \sum_1^n K(X_i)$ is a sufficient statistic for the parameter θ. To prove that $Y_1 = \sum_1^n K(X_i)$ is a sufficient statistic for θ in the discrete case, we take the joint p.d.f. of $X_1, X_2, \ldots, X_n$ to be positive on a discrete set of points, say, when $x_i \in \{x; x = a_1, a_2, \ldots\}$, $i = 1, 2, \ldots, n$. We then use the factorization theorem. It is left as an exercise to show that in either the continuous or the discrete case the p.d.f. of Y_1 is of the form

$$g_1(y_1; \theta) = R(y_1) \exp [p(\theta)y_1 + nq(\theta)]$$

at points of positive probability density. The points of positive probability density and the function $R(y_1)$ do not depend upon θ.

At this time we use a theorem in analysis to assert that the family $\{g_1(y_1; \theta); \gamma < \theta < \delta\}$ of probability density functions is complete. This is the theorem we used when we asserted that a moment-generating function (when it exists) uniquely determines a distribution. In the present context it can be stated as follows.

Theorem 6. *Let $f(x; \theta)$, $\gamma < \theta < \delta$, be a p.d.f. which represents a regular case of the exponential class. Then if $X_1, X_2, \ldots, X_n$ (where n is a fixed positive integer) is a random sample from a distribution with p.d.f. $f(x; \theta)$, the statistic $Y_1 = \sum_1^n K(X_i)$ is a sufficient statistic for θ and the family $\{g_1(y_1; \theta); \gamma < \theta < \delta\}$ of probability density functions of Y_1 is complete. That is, Y_1 is a complete sufficient statistic for θ.*

This theorem has useful implications. In a regular case of form (1), we can see by inspection that the sufficient statistic is $Y_1 = \sum_1^n K(X_i)$. If we can see how to form a function of Y_1, say $\varphi(Y_1)$, so that $E[\varphi(Y_1)] = \theta$, then the statistic $\varphi(Y_1)$ is unique and is the best estimator of θ.

Example 1. Let $X_1, X_2, \ldots, X_n$ denote a random sample from a normal distribution that has p.d.f.

$$f(x; \theta) = \frac{1}{\sigma\sqrt{2\pi}} \exp\left[-\frac{(x - \theta)^2}{2\sigma^2}\right], \qquad -\infty < x < \infty, \ -\infty < \theta < \infty,$$

or

$$f(x; \theta) = \exp\left(\frac{\theta}{\sigma^2} x - \frac{x^2}{2\sigma^2} - \ln\sqrt{2\pi\sigma^2} - \frac{\theta^2}{2\sigma^2}\right).$$

Here σ^2 is any fixed positive number. This is a regular case of the exponential class with

$$p(\theta) = \frac{\theta}{\sigma^2}, \qquad K(x) = x,$$

$$S(x) = -\frac{x^2}{2\sigma^2} - \ln\sqrt{2\pi\sigma^2}, \qquad q(\theta) = -\frac{\theta^2}{2\sigma^2}.$$

Accordingly, $Y_1 = X_1 + X_2 + \cdots + X_n = n\bar{X}$ is a complete sufficient statistic for the mean θ of a normal distribution for every fixed value of the variance σ^2. Since $E(Y_1) = n\theta$, then $\varphi(Y_1) = Y_1/n = \bar{X}$ is the only function of Y_1 that is an unbiased estimator of θ; and being a function of the sufficient statistic Y_1, it has a minimum variance. That is, $\bar{X}$ is the unique best estimator of θ. Incidentally, since Y_1 is a single-valued function of $\bar{X}$, $\bar{X}$ itself is also a complete sufficient statistic for θ.

Example 2. Consider a Poisson distribution with parameter $\theta, 0 < \theta < \infty$. The p.d.f. of this distribution is

$$f(x; \theta) = \frac{\theta^x e^{-\theta}}{x!} = \exp\left[(\ln\theta)x - \ln(x!) - \theta\right], \qquad x = 0, 1, 2, \ldots,$$

$$= 0 \text{ elsewhere.}$$

In accordance with Theorem 6, $Y_1 = \sum_{1}^{n} X_i$ is a complete sufficient statistic for θ. Since $E(Y_1) = n\theta$, the statistic $\varphi(Y_1) = Y_1/n = \bar{X}$, which is also a complete sufficient statistic for θ, is the unique best estimator of θ.

EXERCISES

10.19. Write the p.d.f.

$$f(x; \theta) = \frac{1}{6\theta^4} x^3 e^{-x/\theta}, \qquad 0 < x < \infty, \ 0 < \theta < \infty,$$

zero elsewhere, in the exponential form. If $X_1, X_2, \ldots, X_n$ is a random sample from this distribution, find a complete sufficient statistic Y_1 for θ

and the unique function $\varphi(Y_1)$ of this statistic that is the best estimator of θ. Is $\varphi(Y_1)$ itself a complete sufficient statistic?

10.20. Let $X_1, X_2, \ldots, X_n$ denote a random sample of size $n > 2$ from a distribution with p.d.f. $f(x; \theta) = \theta e^{-\theta x}$, $0 < x < \infty$, zero elsewhere, and $\theta > 0$. Then $Y = \sum_1^n X_i$ is a sufficient statistic for θ. Prove that $(n-1)/Y$ is the best estimator of θ.

10.21. Let $X_1, X_2, \ldots, X_n$ denote a random sample of size n from a distribution with p.d.f. $f(x; \theta) = \theta x^{\theta-1}$, $0 < x < 1$, zero elsewhere, and $\theta > 0$.

(a) Show that the *geometric mean* $(X_1 X_2 \cdots X_n)^{1/n}$ of the sample is a complete sufficient statistic for θ.

(b) Find the maximum likelihood estimator of θ, and observe that it is a function of this geometric mean.

10.22. Let $\bar{X}$ denote the mean of the random sample $X_1, X_2, \ldots, X_n$ from a gamma-type distribution with parameters $\alpha > 0$ and $\beta = \theta > 0$. Compute $E[X_1|\bar{x}]$. *Hint.* Can you find directly a function $\psi(\bar{X})$ of $\bar{X}$ such that $E[\psi(\bar{X})] = \theta$? Is $E(X_1|\bar{x}) = \psi(\bar{x})$? Why?

10.23. Let X be a random variable with a p.d.f. of a regular case of the exponential class. Show that $E[K(X)] = -q'(\theta)/p'(\theta)$, provided these derivatives exist, by differentiating both members of the equality

$$\int_a^b \exp\left[p(\theta)K(x) + S(x) + q(\theta)\right] dx = 1$$

with respect to θ. By a second differentiation, find the variance of $K(X)$.

10.24. Given that $f(x; \theta) = \exp\left[\theta K(x) + S(x) + q(\theta)\right]$, $a < x < b$, $\gamma < \theta < \delta$, represents a regular case of the exponential class. Show that the moment-generating function $M(t)$ of $Y = K(X)$ is $M(t) = \exp[q(\theta) - q(\theta + t)]$, $\gamma < \theta + t < \delta$.

10.25. Given, in the preceding exercise, that $E(Y) = E[K(X)] = \theta$. Prove that Y is $n(\theta, 1)$. *Hint.* Consider $M'(0) = \theta$ and solve the resulting differential equation.

10.26. If $X_1, X_2, \ldots, X_n$ is a random sample from a distribution that has a p.d.f. which is a regular case of the exponential class, show that the p.d.f. of $Y_1 = \sum_1^n K(X_i)$ is of the form $g_1(y_1; \theta) = R(y_1) \exp[p(\theta)y_1 + nq(\theta)]$. *Hint.* Let $Y_2 = X_2, \ldots, Y_n = X_n$ be $n-1$ auxiliary random variables. Find the joint p.d.f. of $Y_1, Y_2, \ldots, Y_n$ and then the marginal p.d.f. of Y_1.

10.27. Let Y denote the median and let $\bar{X}$ denote the mean of a random sample of size $n = 2k + 1$ from a distribution that is $n(\mu, \sigma^2)$. Compute $E(Y|\bar{X} = \bar{x})$. *Hint.* See Exercise 10.22.

10.5 Functions of a Parameter

Up to this point we have sought an unbiased and minimum variance estimator of a parameter θ. Not always, however, are we interested in θ but rather in a function of θ. This will be illustrated in the following examples.

Example 1. Let $X_1, X_2, \ldots, X_n$ denote the items of a random sample of size $n > 1$ from a distribution that is $b(1, \theta)$, $0 < \theta < 1$. We know that if $Y = \sum_1^n X_i$, then Y/n is the unique best estimator of θ. Now the variance of Y/n is $\theta(1 - \theta)/n$. Suppose that an unbiased and minimum variance estimator of this variance is sought. Because Y is a sufficient statistic for θ, it is known that we can restrict our search to functions of Y. Consider the statistic $(Y/n)(1 - Y/n)/n$. This statistic is suggested by the fact that Y/n is the best estimator of θ. The expectation of this statistic is given by

$$\frac{1}{n} E\left[\frac{Y}{n}\left(1 - \frac{Y}{n}\right)\right] = \frac{1}{n^2} E(Y) - \frac{1}{n^3} E(Y^2).$$

Now $E(Y) = n\theta$ and $E(Y^2) = n\theta(1 - \theta) + n^2\theta^2$. Hence

$$\frac{1}{n} E\left[\frac{Y}{n}\left(1 - \frac{Y}{n}\right)\right] = \frac{n - 1}{n} \frac{\theta(1 - \theta)}{n}.$$

If we multiply both members of this equation by $n/(n - 1)$, we find that the statistic $(Y/n)(1 - Y/n)/(n - 1)$ is the unique best estimator of the variance of Y/n.

A somewhat different, but very important problem in point estimation is considered in the next example. In the example the distribution of a random variable X is described by a p.d.f. $f(x; \theta)$ that depends upon $\theta \in \Omega$. The problem is to estimate the fractional part of the probability for this distribution which is at or to the left of a fixed point c. Thus we seek an unbiased, minimum variance estimator of $F(c; \theta)$, where $F(x; \theta)$ is the distribution function of X.

Example 2. Let $X_1, X_2, \ldots, X_n$ be a random sample of size $n > 1$ from a distribution that is $n(\theta, 1)$. Suppose that we wish to find a best estimator of the function of θ defined by

$$\Pr(X \leq c) = \int_{-\infty}^{c} \frac{1}{\sqrt{2\pi}} e^{-(x - \theta)^2/2}\, dx = N(c - \theta),$$

where c is a fixed constant. There are many unbiased estimators of $N(c - \theta)$. We first exhibit one of these, say $u(X_1)$, a function of X_1 alone. We shall then compute the conditional expectation, $E[u(X_1)|\bar{X} = \bar{x}] = \varphi(\bar{x})$, of this un-biased statistic, given the sufficient statistic $\bar{X}$, the mean of the sample. In accordance with the theorems of Rao–Blackwell and Lehmann–Scheffé, $\varphi(\bar{X})$ is the unique best estimator of $N(c - \theta)$.

Consider the function $u(x_1)$, where

$$u(x_1) = 1, \qquad x_1 \le c,$$
$$= 0, \qquad x_1 > c.$$

The expected value of the random variable $u(X_1)$ is given by

$$E[u(X_1)] = \int_{-\infty}^{\infty} u(x_1) \frac{1}{\sqrt{2\pi}} \exp\left[-\frac{(x_1 - \theta)^2}{2}\right] dx_1$$

$$= \int_{-\infty}^{c} (1) \frac{1}{\sqrt{2\pi}} \exp\left[-\frac{(x_1 - \theta)^2}{2}\right] dx_1,$$

because $u(x_1) = 0$, $x_1 > c$. But the latter integral has the value $N(c - \theta)$. That is, $u(X_1)$ is an unbiased estimator of $N(c - \theta)$.

We shall next discuss the joint distribution of X_1 and $\bar{X}$ and the conditional distribution of X_1, given $\bar{X} = \bar{x}$. This conditional distribution will enable us to compute $E[u(X_1)|\bar{X} = \bar{x}] = \varphi(\bar{x})$. In accordance with Exercise 4.81, Section 4.7, the joint distribution of X_1 and $\bar{X}$ is bivariate normal with means θ and θ, variances $\sigma_1^2 = 1$ and $\sigma_2^2 = 1/n$, and correlation coefficient $\rho = 1/\sqrt{n}$. Thus the conditional p.d.f. of X_1, given $\bar{X} = \bar{x}$, is normal with linear conditional mean

$$\theta + \frac{\rho\sigma_1}{\sigma_2} (\bar{x} - \theta) = \bar{x}$$

and with variance

$$\sigma_1^2(1 - \rho^2) = \frac{n - 1}{n}.$$

The conditional expectation of $u(X_1)$, given $\bar{X} = \bar{x}$, is then

$$\varphi(\bar{x}) = \int_{-\infty}^{\infty} u(x_1) \sqrt{\frac{n}{n-1}} \frac{1}{\sqrt{2\pi}} \exp\left[-\frac{n(x_1 - \bar{x})^2}{2(n-1)}\right] dx_1$$

$$= \int_{-\infty}^{c} \sqrt{\frac{n}{n-1}} \frac{1}{\sqrt{2\pi}} \exp\left[-\frac{n(x_1 - \bar{x})^2}{2(n-1)}\right] dx_1.$$

The change of variable $z = \sqrt{n}(x_1 - \bar{x})/\sqrt{n-1}$ enables us to write, with $c' = \sqrt{n}(c - \bar{x})/\sqrt{n-1}$, this conditional expectation as

$$\varphi(\bar{x}) = \int_{-\infty}^{c'} \frac{1}{\sqrt{2\pi}} e^{-z^2/2} dz = N(c') = N\left[\frac{\sqrt{n}(c - \bar{x})}{\sqrt{n-1}}\right].$$

Thus the unique, unbiased, and minimum variance estimator of $N(c - \theta)$ is, for every fixed constant c, given by $\varphi(\bar{X}) = N[\sqrt{n}(c - \bar{X})/\sqrt{n-1}]$.

Remark. We should like to draw the attention of the reader to a rather important fact. This has to do with the adoption of a *principle*, such as the principle of unbiasedness and minimum variance. A principle is not a theorem; and seldom does a principle yield satisfactory results in all cases. So far, this principle has provided quite satisfactory results. To see that this is not always the case, let X have a Poisson distribution with parameter θ, $0 < \theta < \infty$. We may look upon X as a random sample of size 1 from this distribution. Thus X is a complete sufficient statistic for θ. We seek the best estimator of $e^{-2\theta}$, best in the sense of being unbiased and having minimum variance. Consider $Y = (-1)^X$. We have

$$E(Y) = E[(-1)^X] = \sum_{x=0}^{\infty} \frac{(-\theta)^x e^{-\theta}}{x!} = e^{-2\theta}.$$

Accordingly, $(-1)^X$ is the (unique) best estimator of $e^{-2\theta}$, in the sense described. Here our principle leaves much to be desired. We are endeavoring to elicit some information about the number $e^{-2\theta}$, where $0 < e^{-2\theta} < 1$. Yet our point estimate is either -1 or $+1$, each of which is a very poor estimate of a number between zero and 1. We do not wish to leave the reader with the impression that an unbiased, minimum variance estimator is *bad*. That is not the case at all. We merely wish to point out that if one tries hard enough, he can find instances where such a statistic is *not good*. Incidentally, the maximum likelihood estimator of $e^{-2\theta}$ is, in the case where the sample size equals 1, e^{-2X}, which is probably a much better estimator in practice than is the "best" estimator $(-1)^X$.

EXERCISES

10.28. Let $X_1, X_2, \ldots, X_n$ denote a random sample from a distribution that is $n(\theta, 1)$, $-\infty < \theta < \infty$. Find the best estimator of θ^2. *Hint.* First determine $E(\bar{X}^2)$.

10.29. Let $X_1, X_2, \ldots, X_n$ denote a random sample from a distribution that is $n(0, \theta)$. Then $Y = \sum X_i^2$ is a sufficient statistic for θ. Find the best estimator of θ^2.

10.30. In the notation of Example 2 of this section, is there a best estimator of $\Pr(-c \le X \le c)$? Here $c > 0$.

10.31. Let $X_1, X_2, \ldots, X_n$ be a random sample from a Poisson distribution with parameter $\theta > 0$. Find the best estimator of $\Pr(X \le 1) = (1 + \theta)e^{-\theta}$. *Hint.* Let $u(x_1) = 1$, $x_1 \le 1$, zero elsewhere, and find $E[u(X_1)|Y = y]$, where $Y = \sum_{1}^{n} X_i$. Make use of Example 2, Section 4.2.

10.32. Let $X_1, X_2, \ldots, X_n$ denote a random sample from a Poisson distribution with parameter $\theta > 0$. From the Remark of this section, we know that $E[(-1)^{X_1}] = e^{-2\theta}$.

(a) Show that $E[(-1)^{X_1}|Y_1 = y_1] = (1 - 2/n)^{y_1}$, where $Y_1 = X_1 + X_2 + \cdots + X_n$. *Hint.* First show that the conditional p.d.f. of $X_1, X_2, \ldots, X_{n-1}$, given $Y_1 = y_1$, is multinomial, and hence that of X_1 given $Y_1 = y_1$ is $b(y_1, 1/n)$.

(b) Show that the maximum likelihood estimator of $e^{-2\theta}$ is $e^{-2\bar{X}}$.

(c) Since $y_1 = n\bar{x}$, show that $(1 - 2/n)^{y_1}$ is approximately equal to $e^{-2\bar{x}}$ when n is large.

10.6 The Case of Several Parameters

In many of the interesting problems we encounter, the p.d.f. may not depend upon a single parameter θ, but perhaps upon two (or more) parameters, say θ_1 and θ_2, where $(\theta_1, \theta_2) \in \Omega$, a two-dimensional parameter space. We now define joint sufficient statistics for the parameters. For the moment we shall restrict ourselves to the case of two parameters.

Definition 3. Let $X_1, X_2, \ldots, X_n$ denote a random sample from a distribution that has p.d.f. $f(x; \theta_1, \theta_2)$, where $(\theta_1, \theta_2) \in \Omega$. Let $Y_1 = u_1(X_1, X_2, \ldots, X_n)$ and $Y_2 = u_2(X_1, X_2, \ldots, X_n)$ be two statistics whose joint p.d.f. is $g_{12}(y_1, y_2; \theta_1, \theta_2)$. The statistics Y_1 and Y_2 are called *joint sufficient statistics* for θ_1 and θ_2 if and only if

$$\frac{f(x_1; \theta_1, \theta_2)f(x_2; \theta_1, \theta_2) \cdots f(x_n; \theta_1, \theta_2)}{g_{12}[u_1(x_1, \ldots, x_n), u_2(x_1, \ldots, x_n); \theta_1, \theta_2]} = H(x_1, x_2, \ldots, x_n),$$

where, for every fixed $y_1 = u_1(x_1, \ldots, x_n)$ and $y_2 = u_2(x_1, \ldots, x_n)$, $H(x_1, x_2, \ldots, x_n)$ does not depend upon θ_1 or θ_2.

As may be anticipated, the factorization theorem can be extended. In our notation it can be stated in the following manner. The statistics $Y_1 = u_1(X_1, X_2, \ldots, X_n)$ and $Y_2 = u_2(X_1, X_2, \ldots, X_n)$ are joint sufficient statistics for the parameters θ_1 and θ_2 if and only if we can find two nonnegative functions k_1 and k_2 such that

$$f(x_1; \theta_1, \theta_2)f(x_2; \theta_1, \theta_2) \cdots f(x_n; \theta_1, \theta_2)$$
$$= k_1[u_1(x_1, x_2, \ldots, x_n), u_2(u_1, x_2, \ldots, x_n); \theta_1, \theta_2]k_2(x_1, x_2, \ldots, x_n),$$

where, for all fixed values of the functions $y_1 = u_1(x_1, x_2, \ldots, x_n)$ and $y_2 = u_2(x_1, x_2, \ldots, x_n)$, the function $k_2(x_1, x_2, \ldots, x_n)$ does not depend upon both or either of θ_1 and θ_2.

Example 1. Let $X_1, X_2, \ldots, X_n$ be a random sample from a distribution having p.d.f.

$$f(x; \theta_1, \theta_2) = \frac{1}{2\theta_2}, \qquad \theta_1 - \theta_2 < x < \theta_1 + \theta_2,$$

$$= 0 \text{ elsewhere,}$$

where $-\infty < \theta_1 < \infty, 0 < \theta_2 < \infty$. Let $Y_1 < Y_2 < \cdots < Y_n$ be the order statistics. The joint p.d.f. of Y_1 and Y_n is given by

$$g_{1n}(y_1, y_n; \theta_1, \theta_2) = \frac{n(n-1)}{(2\theta_2)^n} (y_n - y_1)^{n-2}, \qquad \theta_1 - \theta_2 < y_1 < y_n < \theta_1 + \theta_2,$$

and equals zero elsewhere. Accordingly, the joint p.d.f. of $X_1, X_2, \ldots, X_n$ can be written, for points of positive probability density,

$$\left(\frac{1}{2\theta_2}\right)^n = \frac{n(n-1)[\max(x_i) - \min(x_i)]^{n-2}}{(2\theta_2)^n}$$

$$\times \left(\frac{1}{n(n-1)[\max(x_i) - \min(x_i)]^{n-2}}\right).$$

Since the last factor does not depend upon the parameters, either the definition or the factorization theorem assures us that Y_1 and Y_n are joint sufficient statistics for θ_1 and θ_2.

The extension of the notion of joint sufficient statistics for more than two parameters is a natural one. Suppose that a certain p.d.f. depends upon m parameters. Let a random sample of size n be taken from the distribution that has this p.d.f. and define m statistics. These m statistics are called joint sufficient statistics for the m parameters if and only if the ratio of the joint p.d.f. of the items of the random sample and the joint p.d.f. of these m statistics does not depend upon the m parameters, whatever the fixed values of the m statistics. Again the factorization theorem is readily extended.

There is an extension of the Rao–Blackwell theorem that can be adapted to joint sufficient statistics for several parameters, but that extension will not be included in this book. However, the concept of a complete family of probability density functions is generalized as follows: Let

$$\{f(v_1, v_2, \ldots, v_k; \theta_1, \theta_2, \ldots, \theta_m); (\theta_1, \theta_2, \ldots, \theta_m) \in \Omega\}$$

denote a family of probability density functions of k random variables $V_1, V_2, \ldots, V_k$ that depends upon m parameters $(\theta_1, \theta_2, \ldots, \theta_m) \in \Omega$.

Let $u(v_1, v_2, \ldots, v_k)$ be a function of $v_1, v_2, \ldots, v_k$ (but not a function of any or all of the parameters). If

$$E[u(V_1, V_2, \ldots, V_k)] = 0$$

for all $(\theta_1, \theta_2, \ldots, \theta_m) \in \Omega$ implies that $u(v_1, v_2, \ldots, v_k) = 0$ at all points $(v_1, v_2, \ldots, v_k)$, except on a set of points that has probability zero for all members of the family of probability density functions, we shall say that the family of probability density functions is a complete family.

The remainder of our treatment of the case of several parameters will be restricted to probability density functions that represent what we shall call regular cases of the exponential class. Let $X_1, X_2, \ldots, X_n$, $n > m$, denote a random sample from a distribution that depends on m parameters and has a p.d.f. of the form

(1) $f(x; \theta_1, \theta_2, \ldots, \theta_m)$

$$= \exp \left[\sum_{j=1}^{m} p_j(\theta_1, \theta_2, \ldots, \theta_m) K_j(x) + S(x) + q(\theta_1, \theta_2, \ldots, \theta_m) \right]$$

for $a < x < b$, and equals zero elsewhere.

A p.d.f. of the form (1) is said to be a member of the *exponential class* of probability density functions of the continuous type. If, in addition,

(a) neither a nor b depends upon any or all of the parameters $\theta_1, \theta_2, \ldots, \theta_m$,

(b) the $p_j(\theta_1, \theta_2, \ldots, \theta_m)$, $j = 1, 2, \ldots, m$, are nontrivial, functionally independent, continuous functions of θ_j, $\gamma_j < \theta_j < \delta_j$, $j = 1, 2, \ldots, m$,

(c) the $K_j'(x)$, $j = 1, 2, \ldots, m$, are continuous for $a < x < b$ and no one is a linear homogeneous function of the others,

(d) $S(x)$ is a continuous function of x, $a < x < b$,

we say that we have a *regular case* of the exponential class.

The joint p.d.f. of $X_1, X_2, \ldots, X_n$ is given, at points of positive probability density, by

$$\exp \left[\sum_{j=1}^{m} p_j(\theta_1, \ldots, \theta_m) \sum_{i=1}^{n} K_j(x_i) + \sum_{i=1}^{n} S(x_i) + nq(\theta_1, \ldots, \theta_m) \right]$$

$$= \exp \left[\sum_{j=1}^{m} p_j(\theta_1, \ldots, \theta_m) \sum_{i=1}^{n} K_j(x_i) + nq(\theta_1, \ldots, \theta_m) \right] \exp \left[\sum_{i=1}^{n} S(x_i) \right].$$

In accordance with the factorization theorem, the statistics

$$Y_1 = \sum_{i=1}^{n} K_1(X_i), \quad Y_2 = \sum_{i=1}^{n} K_2(X_i), \ldots, Y_m = \sum_{i=1}^{n} K_m(X_i)$$

are joint sufficient statistics for the m parameters $\theta_1, \theta_2, \ldots, \theta_m$. It is left as an exercise to prove that the joint p.d.f. of $Y_1, \ldots, Y_m$ is of the form

$$(2) \qquad R(y_1, \ldots, y_m) \exp\left[\sum_{j=1}^{m} p_j(\theta_1, \ldots, \theta_m)y_j + nq(\theta_1, \ldots, \theta_m)\right]$$

at points of positive probability density. These points of positive probability density and the function $R(y_1, \ldots, y_m)$ do not depend upon any or all of the parameters $\theta_1, \theta_2, \ldots, \theta_m$. Moreover, in accordance with a theorem in analysis, it can be asserted that, in a regular case of the exponential class, the family of probability density functions of these joint sufficient statistics $Y_1, Y_2, \ldots, Y_m$ is complete when $n > m$. In accordance with a convention previously adopted, we shall refer to $Y_1, Y_2, \ldots, Y_m$ as *joint complete sufficient statistics for the parameters* $\theta_1, \theta_2, \ldots, \theta_m$.

Example 2. Let $X_1, X_2, \ldots, X_n$ denote a random sample from a distribution that is $n(\theta_1, \theta_2)$, $-\infty < \theta_1 < \infty$, $0 < \theta_2 < \infty$. Thus the p.d.f. $f(x; \theta_1, \theta_2)$ of the distribution may be written as

$$f(x; \theta_1, \theta_2) = \exp\left(\frac{-1}{2\theta_2} x^2 + \frac{\theta_1}{\theta_2} x - \frac{\theta_1^2}{2\theta_2} - \ln\sqrt{2\pi\theta_2}\right).$$

Therefore, we can take $K_1(x) = x^2$ and $K_2(x) = x$. Consequently, the statistics

$$Y_1 = \sum_{1}^{n} X_i^2 \qquad \text{and} \qquad Y_2 = \sum_{1}^{n} X_i$$

are joint complete sufficient statistics for θ_1 and θ_2. Since the relations

$$Z_1 = \frac{Y_2}{n} = \bar{X}, \qquad Z_2 = \frac{Y_1 - Y_2^2/n}{n - 1} = \frac{\sum (X_i - \bar{X})^2}{n - 1}$$

define a one-to-one transformation, Z_1 and Z_2 are also joint complete sufficient statistics for θ_1 and θ_2. Moreover,

$$E(Z_1) = \theta_1 \qquad \text{and} \qquad E(Z_2) = \theta_2.$$

From completeness, we have that Z_1 and Z_2 are the only functions of Y_1 and Y_2 that are unbiased estimators of θ_1 and θ_2, respectively.

A p.d.f.

$$f(x; \theta_1, \theta_2, \ldots, \theta_m)$$
$$= \exp\left[\sum_{j=1}^{m} p_j(\theta_1, \theta_2, \ldots, \theta_m)K_j(x) + S(x) + q(\theta_1, \theta_2, \ldots, \theta_m)\right],$$

$$x = a_1, a_2, a_3, \ldots,$$

zero elsewhere, is said to represent a regular case of the exponential class of probability density functions of the discrete type if

(a) the set $\{x; x = a_1, a_2, \ldots\}$ does not depend upon any or all of the parameters $\theta_1, \theta_2, \ldots, \theta_m$,

(b) the $p_j(\theta_1, \theta_2, \ldots, \theta_m)$, $j = 1, 2, \ldots, m$, are nontrivial, functionally independent, continuous functions of θ_j, $\gamma_j < \theta_j < \delta_j$, $j = 1, 2, \ldots, m$,

(c) the $K_j(x)$, $j = 1, 2, \ldots, m$, are nontrivial functions of x on the set $\{x; x = a_1, a_2, \ldots\}$ and no one is a linear function of the others.

Let $X_1, X_2, \ldots, X_n$ denote a random sample from a discrete-type distribution that represents a regular case of the exponential class. Then the statements made above in connection with the random variable of the continuous type are also valid here.

Not always do we sample from a distribution of one random variable X. We could, for instance, sample from a distribution of two random variables V and W with joint p.d.f. $f(v, w; \theta_1, \theta_2, \ldots, \theta_m)$. Recall that by a random sample $(V_1, W_1), (V_2, W_2), \ldots, (V_n, W_n)$ from a distribution of this sort, we mean that the joint p.d.f. of these $2n$ random variables is given by

$$f(v_1, w_1; \theta_1, \ldots, \theta_m)f(v_2, w_2; \theta_1, \ldots, \theta_m) \cdots f(v_n, w_n; \theta_1, \ldots, \theta_m).$$

In particular, suppose that the random sample is taken from a distribution that has the p.d.f. of V and W of the exponential class

$$(3) \quad f(v, w; \theta_1, \ldots, \theta_m)$$

$$= \exp\left[\sum_{j=1}^{m} p_j(\theta_1, \ldots, \theta_m)K_j(v, w) + S(v, w) + q(\theta_1, \ldots, \theta_m)\right]$$

for $a < v < b, c < w < d$, and equals zero elsewhere, where a, b, c, d do not depend on the parameters and conditions similar to (a), (b), (c), and (d), p. 366, are imposed. Then the m statistics

$$Y_1 = \sum_{i=1}^{n} K_1(V_i, W_i), \ldots, \quad Y_m = \sum_{i=1}^{n} K_m(V_i, W_i)$$

are joint complete sufficient statistics for the m parameters $\theta_1, \theta_2, \ldots, \theta_m$.

EXERCISES

10.33. Let $Y_1 < Y_2 < Y_3$ be the order statistics of a random sample of size 3 from the distribution with p.d.f.

$$f(x; \theta_1, \theta_2) = \frac{1}{\theta_2}\exp\left(-\frac{x - \theta_1}{\theta_2}\right), \qquad \theta_1 < x < \infty, -\infty < \theta_1 < \infty, 0 < \theta_2 < \infty,$$

zero elsewhere. Find the joint p.d.f. of $Z_1 = Y_1$, $Z_2 = Y_2$, and $Z_3 = Y_1 + Y_2 + Y_3$. The corresponding transformation maps the space $\{(y_1, y_2, y_3); \theta_1 < y_1 < y_2 < y_3 < \infty\}$ onto the space $\{(z_1, z_2, z_3); \theta_1 < z_1 < z_2 < (z_3 - z_1)/2 < \infty\}$. Show that Z_1 and Z_3 are joint sufficient statistics for θ_1 and θ_2.

10.34. Let $X_1, X_2, \ldots, X_n$ be a random sample from a distribution that has a p.d.f. of form (1) of this section. Show that $Y_1 = \sum_{i=1}^{n} K_1(X_i), \ldots,$ $Y_m = \sum_{i=1}^{n} K_m(X_i)$ have a joint p.d.f. of form (2) of this section.

10.35. Let $(X_1, Y_1), (X_2, Y_2), \ldots, (X_n, Y_n)$ denote a random sample of size n from a bivariate normal distribution with means μ_1 and μ_2, positive variances σ_1^2 and σ_2^2, and correlation coefficient ρ. Show that $\sum_1^n X_i$, $\sum_1^n Y_i$, $\sum_1^n X_i^2$, $\sum_1^n Y_i^2$, and $\sum_1^n X_i Y_i$ are joint sufficient statistics for the five parameters. Are $\bar{X} = \sum_1^n X_i/n$, $\bar{Y} = \sum_1^n Y_i/n$, $S_1^2 = \sum_1^n (X_i - \bar{X})^2/n$, $S_2^2 = \sum_1^n (Y_i - \bar{Y})^2/n$, and $\sum_1^n (X_i - \bar{X})(Y_i - \bar{Y})/nS_1S_2$ also joint sufficient statistics for these parameters?

10.36. Let the p.d.f. $f(x; \theta_1, \theta_2)$ be of the form

$$\exp\left[p_1(\theta_1, \theta_2)K_1(x) + p_2(\theta_1, \theta_2)K_2(x) + S(x) + q(\theta_1, \theta_2)\right], \qquad a < x < b,$$

zero elsewhere. Let $K_1'(x) = cK_2'(x)$. Show that $f(x; \theta_1, \theta_2)$ can be written in the form

$$\exp\left[p(\theta_1, \theta_2)K(x) + S(x) + q_1(\theta_1, \theta_2)\right], \qquad a < x < b,$$

zero elsewhere. This is the reason why it is required that no one $K_j'(x)$ be a linear homogeneous function of the others, that is, so that the number of sufficient statistics equals the number of parameters.

10.37. Let $Y_1 < Y_2 < \cdots < Y_n$ be the order statistics of a random sample $X_1, X_2, \ldots, X_n$ of size n from a distribution of the continuous type with p.d.f. $f(x)$. Show that the ratio of the joint p.d.f. of $X_1, X_2, \ldots, X_n$ and that of $Y_1 < Y_2 < \cdots < Y_n$ is equal to $1/n!$, which does not depend upon the underlying p.d.f. This suggests that $Y_1 < Y_2 < \cdots < Y_n$ are joint sufficient statistics for the unknown "parameter" f.

Chapter II

Further Topics in Statistical Inference

11.1 The Rao–Cramér Inequality

In this section we establish a lower bound for the variance of an unbiased estimator of a parameter.

Let $X_1, X_2, \ldots, X_n$ denote a random sample from a distribution with p.d.f. $f(x; \theta)$, $\theta \in \Omega = \{\theta; \gamma < \theta < \delta\}$, where γ and δ are known. Let $Y = u(X_1, X_2, \ldots, X_n)$ be an unbiased estimator of θ. We shall show that the variance of Y, say σ_Y^2, satisfies the inequality

$$(1) \qquad \sigma_Y^2 \geq \frac{1}{nE\{[\partial \ln f(X; \theta)/\partial \theta]^2\}}.$$

Throughout this section, unless otherwise specified, it will be assumed that we may differentiate, with respect to a parameter, under an integral or a summation symbol. This means, among other things, that the domain of positive probability density does not depend upon θ.

We shall now give a proof of inequality (1) when X is a random variable of the continuous type. The reader can easily handle the discrete case by changing integrals to sums. Let $g(y; \theta)$ denote the p.d.f. of the unbiased statistic Y. We are given that

$$1 = \int_{-\infty}^{\infty} f(x_i; \theta) \, dx_i, \qquad i = 1, 2, \ldots, n,$$

and

$$\theta = \int_{-\infty}^{\infty} yg(y; \theta) \, dy$$

$$= \int_{-\infty}^{\infty} \cdots \int_{-\infty}^{\infty} u(x_1, x_2, \ldots, x_n)f(x_1; \theta) \cdots f(x_n; \theta) \, dx_1 \cdots dx_n.$$

The final form of the right-hand member of the second equation is justified by the discussion in Section 4.7. If we differentiate both members of each of these equations with respect to θ, we have

$$0 = \int_{-\infty}^{\infty} \frac{\partial f(x_i; \theta)}{\partial \theta} \, dx_i = \int_{-\infty}^{\infty} \frac{\partial \ln f(x_i; \theta)}{\partial \theta} f(x_i; \theta) \, dx_i,$$

(2)

$$1 = \int_{-\infty}^{\infty} \cdots \int_{-\infty}^{\infty} u(x_1, x_2, \ldots, x_n) \left[\sum_{1}^{n} \frac{1}{f(x_i; \theta)} \frac{\partial f(x_i; \theta)}{\partial \theta} \right]$$

$$\times f(x_1; \theta) \cdots f(x_n; \theta) \, dx_1 \cdots dx_n$$

$$= \int_{-\infty}^{\infty} \cdots \int_{-\infty}^{\infty} u(x_1, x_2, \ldots, x_n) \left[\sum_{1}^{n} \frac{\partial \ln f(x_i; \theta)}{\partial \theta} \right]$$

$$\times f(x_1; \theta) \cdots f(x_n; \theta) \, dx_1 \cdots dx_n.$$

Define the random variable Z by $Z = \sum_{1}^{n} [\partial \ln f(X_i; \theta)/\partial\theta]$. In accordance with the first of Equations (2) we have $E(Z) = \sum_{1}^{n} E[\partial \ln f(X_i; \theta)/\partial\theta] = 0$. Moreover, Z is the sum of n mutually stochastically independent random variables each with mean zero and consequently with variance $E\{[\partial \ln f(X; \theta)/\partial\theta]^2\}$. Hence the variance of Z is the sum of the n variances,

$$\sigma_Z^2 = nE\left[\left(\frac{\partial \ln f(X; \theta)}{\partial \theta} \right)^2 \right].$$

Because $Y = u(X_1, \ldots, X_n)$ and $Z = \sum_{1}^{n} [\partial \ln f(X_i; \theta)/\partial\theta]$, the second of Equations (2) shows that $E(YZ) = 1$. Recall (Section 2.3) that

$$E(YZ) = E(Y)E(Z) + \rho\sigma_Y\sigma_Z,$$

where ρ is the correlation coefficient of Y and Z. Since $E(Y) = \theta$ and $E(Z) = 0$, we have

$$1 = \theta \cdot 0 + \rho\sigma_Y\sigma_Z \quad \text{or} \quad \rho = \frac{1}{\sigma_Y\sigma_Z}.$$

Now $\rho^2 \le 1$. Hence

$$\frac{1}{\sigma_Y^2\sigma_Z^2} \le 1 \quad \text{or} \quad \frac{1}{\sigma_Z^2} \le \sigma_Y^2.$$

If we replace σ_Z^2 by its value, we have inequality (1),

$$\sigma_Y^2 \geq \frac{1}{nE\left[\left(\dfrac{\partial \ln f(X; \theta)}{\partial \theta}\right)^2\right]}.$$

Inequality (1) is known as the *Rao–Cramér inequality*. It provides, in cases in which we can differentiate with respect to a parameter under an integral or summation symbol, a lower bound on the variance of an unbiased estimator of a parameter, usually called the Rao–Cramér lower bound.

We now make the following definitions.

Definition 1. Let Y be an unbiased estimator of a parameter θ in such a case of point estimation. The statistic Y is called an *efficient estimator* of θ if and only if the variance of Y attains the Rao–Cramér lower bound.

It is left as an exercise to show, in these cases of point estimation, that $E\{[\partial \ln f(X; \theta)/\partial \theta]^2\} = -E[\partial^2 \ln f(X; \theta)/\partial \theta^2]$. In some instances the latter is much easier to compute.

Definition 2. In cases in which we can differentiate with respect to a parameter under an integral or summation symbol, the ratio of the Rao–Cramér lower bound to the actual variance of any unbiased estimator of a parameter is called the *efficiency* of that statistic.

Example 1. Let $X_1, X_2, \ldots, X_n$ denote a random sample from a Poisson distribution that has the mean $\theta > 0$. It is known that $\bar{X}$ is a maximum likelihood estimator of θ; we shall show that it is also an efficient estimator of θ. We have

$$\frac{\partial \ln f(x; \theta)}{\partial \theta} = \frac{\partial}{\partial \theta}(x \ln \theta - \theta - \ln x!)$$

$$= \frac{x}{\theta} - 1 = \frac{x - \theta}{\theta}.$$

Accordingly,

$$E\left[\left(\frac{\partial \ln f(X; \theta)}{\partial \theta}\right)^2\right] = \frac{E(X - \theta)^2}{\theta^2} = \frac{\sigma^2}{\theta^2} = \frac{\theta}{\theta^2} = \frac{1}{\theta}.$$

The Rao–Cramér lower bound in this case is $1/[n(1/\theta)] = \theta/n$. But θ/n is the variance $\sigma_{\bar{X}}^2$ of $\bar{X}$. Hence $\bar{X}$ is an efficient estimator of θ.

Example 2. Let S^2 denote the variance of a random sample of size $n > 1$ from a distribution that is $n(\mu, \theta)$, $0 < \theta < \infty$. We know that

$E[nS^2/(n-1)] = \theta$. What is the efficiency of the estimator $nS^2/(n-1)$? We have

$$\ln f(x; \theta) = -\frac{(x-\mu)^2}{2\theta} - \frac{\ln(2\pi\theta)}{2},$$

$$\frac{\partial \ln f(x; \theta)}{\partial \theta} = \frac{(x-\mu)^2}{2\theta^2} - \frac{1}{2\theta},$$

and

$$\frac{\partial^2 \ln f(x; \theta)}{\partial \theta^2} = -\frac{(x-\mu)^2}{\theta^3} + \frac{1}{2\theta^2}.$$

Accordingly,

$$-E\left[\frac{\partial^2 \ln f(X; \theta)}{\partial \theta^2}\right] = \frac{\theta}{\theta^3} - \frac{1}{2\theta^2} = \frac{1}{2\theta^2}.$$

Thus the Rao–Cramér lower bound is $2\theta^2/n$. Now nS^2/θ is $\chi^2(n-1)$, so the variance of nS^2/θ is $2(n-1)$. Accordingly, the variance of $nS^2/(n-1)$ is $2(n-1)[\theta^2/(n-1)^2] = 2\theta^2/(n-1)$. Thus the efficiency of the estimator $nS^2/(n-1)$ is $(n-1)/n$.

Example 3. Let $X_1, X_2, \ldots, X_n$ denote a random sample of size $n > 2$ from a distribution with p.d.f.

$$f(x; \theta) = \theta x^{\theta-1} = \exp(\theta \ln x - \ln x + \ln \theta), \qquad 0 < x < 1,$$

$$= 0 \text{ elsewhere.}$$

It is easy to verify that the Rao–Cramér lower bound is θ^2/n. Let $Y_i = -\ln X_i$. We shall indicate that each Y_i has a gamma distribution. The associated transformation $y_i = -\ln x_i$, with inverse $x_i = e^{-y_i}$, is one-to-one and the transformation maps the space $\{x_i; 0 < x_i < 1\}$ onto the space $\{y_i; 0 < y_i < \infty\}$. We have $|J| = e^{-y_i}$. Thus Y_i has a gamma distribution with $\alpha = 1$ and $\beta = 1/\theta$. Let $Z = -\sum_1^n \ln X_i$. Then Z has a gamma distribution with $\alpha = n$ and $\beta = 1/\theta$. Accordingly, we have $E(Z) = \alpha\beta = n/\theta$. This suggests that we compute the expectation of $1/Z$ to see if we can find an unbiased estimator of θ. A simple integration shows that $E(1/Z) = \theta/(n-1)$. Hence $(n-1)/Z$ is an unbiased estimator of θ. With $n > 2$, the variance of $(n-1)/Z$ exists and is found to be $\theta^2/(n-2)$, so that the efficiency of $(n-1)/Z$ is $(n-2)/n$. This efficiency tends to 1 as n increases. In such an instance, the estimator is said to be *asymptotically efficient*.

The concept of joint efficient estimators of several parameters has been developed along with the associated concept of joint efficiency of several estimators. But limitations of space prevent their inclusion in this book.

EXERCISES

11.1. Prove that $\bar{X}$, the mean of a random sample of size n from a distribution that is $n(\theta, \sigma^2)$, $-\infty < \theta < \infty$, is, for every known $\sigma^2 > 0$, an efficient estimator of θ.

11.2. Show that the mean $\bar{X}$ of a random sample of size n from a distribution which is $b(1, \theta)$, $0 < \theta < 1$, is an efficient estimator of θ.

11.3. Given $f(x; \theta) = 1/\theta$, $0 < x < \theta$, zero elsewhere, with $\theta > 0$, formally compute the reciprocal of

$$nE\left\{\left[\frac{\partial \ln f(X; \theta)}{\partial \theta}\right]^2\right\}.$$

Compare this with the variance of $(n + 1)Y_n/n$, where Y_n is the largest item of a random sample of size n from this distribution. Comment.

11.4. Given the p.d.f.

$$f(x; \theta) = \frac{1}{\pi[1 + (x - \theta)^2]}, \qquad -\infty < x < \infty, -\infty < \theta < \infty.$$

Show that the Rao–Cramér lower bound is $2/n$, where n is the size of a random sample from this Cauchy distribution.

11.5. Show, with appropriate assumptions, that

$$E\left\{\left[\frac{\partial \ln f(X; \theta)}{\partial \theta}\right]^2\right\} = -E\left[\frac{\partial^2 \ln f(X; \theta)}{\partial \theta^2}\right].$$

Hint. Differentiate with respect to θ the first equation in display (2) of this section,

$$0 = \int_{-\infty}^{\infty} \frac{\partial \ln f(x; \theta)}{\partial \theta} f(x; \theta)\, dx.$$

11.2 The Sequential Probability Ratio Test

In Section 7.2 we proved a theorem that provided us with a method for determining a best critical region for testing a simple hypothesis against an alternative simple hypothesis. The theorem was as follows. Let $X_1, X_2, \ldots, X_n$ be a random sample with fixed sample size n from a distribution that has p.d.f. $f(x; \theta)$, where $\theta \in \{\theta; \theta = \theta', \theta''\}$ and θ' and θ'' are known numbers. Let the joint p.d.f. of $X_1, X_2, \ldots, X_n$ be denoted by

$$L(\theta, n) = f(x_1; \theta)f(x_2; \theta)\cdots f(x_n; \theta),$$

a notation that reveals both the parameter θ and the sample size n. If we reject H_0: $\theta = \theta'$ and accept H_1: $\theta = \theta''$ when and only when

$$\frac{L(\theta', n)}{L(\theta'', n)} \leq k,$$

where $k > 0$, then this is a best test of H_0 against H_1.

Let us now suppose that the sample size n is *not* fixed in advance. In fact, let the sample size be a random variable N with sample space $\{n; n = 1, 2, 3, \ldots\}$. An interesting procedure for testing the simple hypothesis H_0: $\theta = \theta'$ against the simple hypothesis H_1: $\theta = \theta''$ is the following. Let k_0 and k_1 be two positive constants with $k_0 < k_1$. Observe the mutually stochastically independent outcomes $X_1, X_2, X_3, \ldots$ in sequence, say $x_1, x_2, x_3, \ldots$, and compute

$$\frac{L(\theta', 1)}{L(\theta'', 1)}, \frac{L(\theta', 2)}{L(\theta'', 2)}, \frac{L(\theta', 3)}{L(\theta'', 3)}, \ldots.$$

The hypothesis H_0: $\theta = \theta'$ is rejected (and H_1: $\theta = \theta''$ is accepted) if and only if there exists a positive integer n so that $(x_1, x_2, \ldots, x_n)$ belongs to the set

$$C_n = \left\{ (x_1, \ldots, x_n); k_0 < \frac{L(\theta', j)}{L(\theta'', j)} < k_1, j = 1, \ldots, n - 1, \right.$$

$$\left. \text{and} \quad \frac{L(\theta', n)}{L(\theta'', n)} \leq k_0 \right\}.$$

On the other hand, the hypothesis H_0: $\theta = \theta'$ is accepted (and H_1: $\theta = \theta''$ is rejected) if and only if there exists a positive integer n so that $(x_1, x_2, \ldots, x_n)$ belongs to the set

$$B_n = \left\{ (x_1, \ldots, x_n); k_0 < \frac{L(\theta', j)}{L(\theta'', j)} < k_1, j = 1, \ldots, n - 1, \right.$$

$$\left. \text{and} \quad \frac{L(\theta', n)}{L(\theta'', n)} \geq k_1 \right\}.$$

That is, we continue to observe sample items as long as

(1) $$k_0 < \frac{L(\theta', n)}{L(\theta'', n)} < k_1.$$

We stop these observations in one of two ways:

(a) With rejection of H_0: $\theta = \theta'$ as soon as

$$\frac{L(\theta', n)}{L(\theta'', n)} \le k_0,$$

or

(b) with acceptance of H_0: $\theta = \theta'$ as soon as

$$\frac{L(\theta', n)}{L(\theta'', n)} \ge k_1.$$

A test of this kind is called a *sequential probability ratio test*. Now, frequently inequality (1) can be conveniently expressed in an equivalent form

$$c_0(n) < u(x_1, x_2, \ldots, x_n) < c_1(n),$$

where $u(X_1, X_2, \ldots, X_n)$ is a statistic and $c_0(n)$ and $c_1(n)$ depend on the constants k_0, k_1, θ', θ'', and on n. Then the observations are stopped and a decision is reached as soon as

$$u(x_1, x_2, \ldots, x_n) \le c_0(n) \qquad \text{or} \qquad u(x_1, x_2, \ldots, x_n) \ge c_1(n).$$

We now give an illustrative example.

Example 1. Let X have a p.d.f.

$$f(x; \theta) = \theta^x(1 - \theta)^{1-x}, \qquad x = 0, 1,$$

$$= 0 \text{ elsewhere.}$$

In the preceding discussion of a sequential probability ratio test, let H_0: $\theta = \frac{1}{3}$ and H_1: $\theta = \frac{2}{3}$; then, with $\sum x_i = \sum_1^n x_i$,

$$\frac{L(\frac{1}{3}, n)}{L(\frac{2}{3}, n)} = \frac{(\frac{1}{3})^{\sum x_i}(\frac{2}{3})^{n-\sum x_i}}{(\frac{2}{3})^{\sum x_i}(\frac{1}{3})^{n-\sum x_i}} = 2^{n - 2\sum x_i}.$$

If we take logarithms to the base 2, the inequality

$$k_0 < \frac{L(\frac{1}{3}, n)}{L(\frac{2}{3}, n)} < k_1,$$

with $0 < k_0 < k_1$, becomes

$$\log_2 k_0 < n - 2\sum_1^n x_i < \log_2 k_1,$$

or, equivalently,

$$c_0(n) = \frac{n}{2} - \frac{1}{2}\log_2 k_1 < \sum_1^n x_i < \frac{n}{2} - \frac{1}{2}\log_2 k_0 = c_1(n).$$

Note that $L(\frac{1}{3}, n)/L(\frac{2}{3}, n) \leq k_0$ if and only if $c_1(n) \leq \sum_1^n x_i$; and $L(\frac{1}{3}, n)/L(\frac{2}{3}, n) \geq k_1$ if and only if $c_0(n) \geq \sum_1^n x_i$. Thus we continue to observe outcomes as long as $c_0(n) < \sum_1^n x_i < c_1(n)$. The observation of outcomes is discontinued with the first value n of N for which either $c_1(n) \leq \sum_1^n x_i$ or $c_0(n) \geq \sum_1^n x_i$. The inequality $c_1(n) \leq \sum_1^n x_i$ leads to the rejection of H_0: $\theta = \frac{1}{3}$ (the acceptance of H_1), and the inequality $c_0(n) \geq \sum_1^n x_i$ leads to the acceptance of H_0: $\theta = \frac{1}{3}$ (the rejection of H_1).

Remarks. At this point, the reader undoubtedly sees that there are many questions which should be raised in connection with the sequential probability ratio test. Some of these questions are possibly among the following:

(a) What is the probability of the procedure continuing indefinitely?

(b) What is the value of the power function of this test at each of the points $\theta = \theta'$ and $\theta = \theta''$?

(c) If θ'' is one of several values of θ specified by an alternative composite hypothesis, say H_1: $\theta > \theta'$, what is the power function at each point $\theta \geq \theta'$?

(d) Since the sample size N is a random variable, what are some of the properties of the distribution of N? In particular, what is the expected value $E(N)$ of N?

(e) How does this test compare with tests that have a fixed sample size n?

A course in sequential analysis would investigate these and many other problems. However, in this book our objective is largely that of acquainting the reader with this kind of test procedure. Accordingly, we assert that the answer to question (a) is zero. Moreover, it can be proved that if $\theta = \theta'$ or if $\theta = \theta''$, $E(N)$ is smaller, for this sequential procedure, than the sample size of a fixed-sample-size test which has the same values of the power function at those points. We now consider question (b) in some detail.

In this section we shall denote the power of the test when H_0 is true by the symbol α and the power of the test when H_1 is true by the symbol $1 - \beta$. Thus α is the probability of committing a type I error (the rejection of H_0 when H_0 is true), and β is the probability of committing a type II error (the acceptance of H_0 when H_0 is false). With the sets C_n and B_n as previously defined, and with random variables of the continuous type, we then have

$$\alpha = \sum_{n=1}^{\infty} \int_{C_n} L(\theta', n), \qquad 1 - \beta = \sum_{n=1}^{\infty} \int_{C_n} L(\theta'', n).$$

Since the probability is 1 that the procedure will terminate, we also have

$$1 - \alpha = \sum_{n=1}^{\infty} \int_{B_n} L(\theta', n), \qquad \beta = \sum_{n=1}^{\infty} \int_{B_n} L(\theta'', n).$$

If $(x_1, x_2, \ldots, x_n) \in C_n$, we have $L(\theta', n) \leq k_0 L(\theta'', n)$; hence it is clear that

$$\alpha = \sum_{n=1}^{\infty} \int_{C_n} L(\theta', n) \leq \sum_{n=1}^{\infty} \int_{C_n} k_0 L(\theta'', n) = k_0(1 - \beta).$$

Because $L(\theta', n) \geq k_1 L(\theta'', n)$ at each point of the set B_n, we have

$$1 - \alpha = \sum_{n=1}^{\infty} \int_{B_n} L(\theta', n) \geq \sum_{n=1}^{\infty} \int_{B_n} k_1 L(\theta'', n) = k_1 \beta.$$

Accordingly, it follows that

(2)
$$\frac{\alpha}{1 - \beta} \leq k_0, \qquad k_1 \leq \frac{1 - \alpha}{\beta},$$

provided that β is not equal to zero or 1.

Now let α_a and β_a be preassigned proper fractions; some typical values in the applications are 0.01, 0.05, and 0.10. If we take

$$k_0 = \frac{\alpha_a}{1 - \beta_a}, \qquad k_1 = \frac{1 - \alpha_a}{\beta_a},$$

then inequalities (2) become

(3)
$$\frac{\alpha}{1 - \beta} \leq \frac{\alpha_a}{1 - \beta_a}, \qquad \frac{1 - \alpha_a}{\beta_a} \leq \frac{1 - \alpha}{\beta};$$

or, equivalently,

$$\alpha(1 - \beta_a) \leq (1 - \beta)\alpha_a, \qquad \beta(1 - \alpha_a) \leq (1 - \alpha)\beta_a.$$

If we add corresponding members of the immediately preceding inequalities, we find that

$$\alpha + \beta - \alpha\beta_a - \beta\alpha_a \leq \alpha_a + \beta_a - \beta\alpha_a - \alpha\beta_a$$

and hence

$$\alpha + \beta \leq \alpha_a + \beta_a.$$

That is, the sum $\alpha + \beta$ of the probabilities of the two kinds of errors is bounded above by the sum $\alpha_a + \beta_a$ of the preassigned numbers.

Moreover, since α and β are positive proper fractions, inequalities (3) imply that

$$\alpha \leq \frac{\alpha_a}{1 - \beta_a}, \qquad \beta \leq \frac{\beta_a}{1 - \alpha_a};$$

consequently, we have an upper bound on each of α and β. Various investigations of the sequential probability ratio test seem to indicate that in most practical cases, the values of α and β are quite close to α_a and β_a. This prompts us to approximate the power function at the points $\theta = \theta'$ and $\theta = \theta''$ by α_a and $1 - \beta_a$, respectively.

Example 2. Let X be $n(\theta, 100)$. To find the sequential probability ratio test for testing $H_0: \theta = 75$ against $H_1: \theta = 78$ such that each of α and β is approximately equal to 0.10, take

$$k_0 = \frac{0.10}{1 - 0.10} = \frac{1}{9}, \qquad k_1 = \frac{1 - 0.10}{0.10} = 9.$$

Since

$$\frac{L(75, n)}{L(78, n)} = \frac{\exp\left[-\sum (x_i - 75)^2/2(100)\right]}{\exp\left[-\sum (x_i - 78)^2/2(100)\right]} = \exp\left(-\frac{6 \sum x_i - 459n}{200}\right),$$

the inequality

$$k_0 = \frac{1}{9} < \frac{L(75, n)}{L(78, n)} < 9 = k_1$$

can be rewritten, by taking logarithms, as

$$-\ln 9 < -\frac{6 \sum x_i - 459n}{200} < \ln 9.$$

This inequality is equivalent to the inequality

$$c_0(n) = \tfrac{153}{2}n - \tfrac{100}{3} \ln 9 < \sum_1^n x_i < \tfrac{153}{2}n + \tfrac{100}{3} \ln 9 = c_1(n).$$

Moreover, $L(75, n)/L(78, n) \leq k_0$ and $L(75, n)/L(78, n) \geq k_1$ are equivalent to the inequalities $\sum_1^n x_i \geq c_1(n)$ and $\sum_1^n x_i \leq c_0(n)$, respectively. Thus the observation of outcomes is discontinued with the first value n of N for which either $\sum_1^n x_i \geq c_1(n)$ or $\sum_1^n x_i \leq c_0(n)$. The inequality $\sum_1^n x_i \geq c_1(n)$ leads to the rejection of $H_0: \theta = 75$, and the inequality $\sum_1^n x_i \leq c_0(n)$ leads to the acceptance of $H_0: \theta = 75$. The power of the test is approximately 0.10 when H_0 is true, and approximately 0.90 when H_1 is true.

Remark. It is interesting to note that a sequential probability ratio test can be thought of as a *random-walk procedure*. For illustrations, the final inequalities of Examples 1 and 2 can be rewritten as

$$-\log_2 k_1 < \sum_{1}^{n} 2(x_i - 0.5) < -\log_2 k_0$$

and

$$-\frac{100}{3}\ln 9 < \sum_{1}^{n} (x_i - 76.5) < \frac{100}{3}\ln 9,$$

respectively. In each instance we can think of starting at the point zero and taking random steps until one of the boundaries is reached. In the first situation the random steps are $2(X_1 - 0.5)$, $2(X_2 - 0.5)$, $2(X_3 - 0.5)$, ... and hence are of the same length, 1, but with random directions. In the second instance, both the length and the direction of the steps are random variables, $X_1 - 76.5$, $X_2 - 76.5$, $X_3 - 76.5$, ...

EXERCISES

11.6. Let X be $n(0, \theta)$ and, in the notation of this section, let $\theta' = 4$, $\theta'' = 9$, $\alpha_a = 0.05$, and $\beta_a = 0.10$. Show that the sequential probability ratio test can be based upon the statistic $\sum_{1}^{n} X_i^2$. Determine $c_0(n)$ and $c_1(n)$.

11.7. Let X have a Poisson distribution with mean θ. Find the sequential probability ratio test for testing $H_0: \theta = 0.02$ against $H_1: \theta = 0.07$. Show that this test can be based upon the statistic $\sum_{1}^{n} X_i$. If $\alpha_a = 0.20$ and $\beta_a = 0.10$, find $c_0(n)$ and $c_1(n)$.

11.8. Let the stochastically independent random variables Y and Z be $n(\mu_1, 1)$ and $n(\mu_2, 1)$, respectively. Let $\theta = \mu_1 - \mu_2$. Let us observe mutually stochastically independent items from each distribution, say Y_1, Y_2, ... and Z_1, Z_2, ... To test sequentially the hypothesis $H_0: \theta = 0$ against $H_1: \theta = \frac{1}{2}$, use the sequence $X_i = Y_i - Z_i$, $i = 1, 2, \ldots$. If $\alpha_a = \beta_a = 0.05$, show that the test can be based upon $\bar{X} = \bar{Y} - \bar{Z}$. Find $c_0(n)$ and $c_1(n)$.

11.3 Multiple Comparisons

Consider b mutually stochastically independent random variables that have normal distributions with unknown means $\mu_1, \mu_2, \ldots, \mu_b$, respectively, and with unknown but common variance σ^2. Let $k_1, k_2, \ldots, k_b$ represent b known real constants that are not all zero. We want to find a confidence interval for $\sum_{1}^{b} k_j\mu_j$, a linear function of the means $\mu_1, \mu_2, \ldots, \mu_b$. To do this, we take a random sample X_{1j}, X_{2j},

$\ldots$, X_{aj} of size a from the distribution $n(\mu_j, \sigma^2)$, $j = 1, 2, \ldots, b$. If we denote $\sum\limits_{i=1}^{a} X_{ij}/a$ by $\bar{X}_{.j}$, then we know that $\bar{X}_{.j}$ is $n(\mu_j, \sigma^2/a)$, that $\sum\limits_{i=1}^{a} (X_{ij} - \bar{X}_{.j})^2/\sigma^2$ is $\chi^2(a - 1)$, and that the two random variables are stochastically independent. Since the random samples are taken from mutually independent distributions, the $2b$ random variables $\bar{X}_{.j}$, $\sum\limits_{i=1}^{a} (X_{ij} - \bar{X}_{.j})^2/\sigma^2$, $j = 1, 2, \ldots, b$, are mutually stochastically independent. Moreover, $\bar{X}_{.1}, \bar{X}_{.2}, \ldots, \bar{X}_{.b}$ and

$$\sum_{j=1}^{b} \sum_{i=1}^{a} \frac{(X_{ij} - \bar{X}_{.j})^2}{\sigma^2}$$

are mutually stochastically independent and the latter is $\chi^2[b(a - 1)]$. Let $Z = \sum\limits_{1}^{b} k_j \bar{X}_{.j}$. Then Z is normal with mean $\sum\limits_{1}^{b} k_j \mu_j$ and variance $\left(\sum\limits_{1}^{b} k_j^2\right)\sigma^2/a$, and Z is stochastically independent of

$$V = \frac{1}{b(a - 1)} \sum_{j=1}^{b} \sum_{i=1}^{a} (X_{ij} - \bar{X}_{.j})^2.$$

Hence the random variable

$$T = \frac{\dfrac{\sum\limits_{1}^{b} k_j \bar{X}_{.j} - \sum\limits_{1}^{b} k_j \mu_j}{\sqrt{\left(\sum\limits_{1}^{b} k_i^2\right)\sigma^2/a}}}{\sqrt{V/\sigma^2}} = \frac{\sum\limits_{1}^{b} k_j \bar{X}_{.j} - \sum\limits_{1}^{b} k_j \mu_j}{\sqrt{\left(\sum\limits_{1}^{b} k_j^2\right)V/a}}$$

has a t distribution with $b(a - 1)$ degrees of freedom. A positive number c can be found in Table IV in Appendix B, for certain values of α, $0 < \alpha < 1$, such that $\Pr(-c \leq T \leq c) = 1 - \alpha$. It follows that the probability is $1 - \alpha$ that

$$\sum_{1}^{b} k_j \bar{X}_{.j} - c\sqrt{\left(\sum_{1}^{b} k_j^2\right)\frac{V}{a}} \leq \sum_{1}^{b} k_j \mu_j \leq \sum_{1}^{b} k_j \bar{X}_{.j} + c\sqrt{\left(\sum_{1}^{b} k_j^2\right)\frac{V}{a}}.$$

The experimental values of $\bar{X}_{.j}$, $j = 1, 2, \ldots, b$, and V will provide a $100(1 - \alpha)$ per cent confidence interval for $\sum\limits_{1}^{b} k_j \mu_j$.

It should be observed that the confidence interval for $\sum\limits_{1}^{b} k_j \mu_j$ depends upon the particular choice of $k_1, k_2, \ldots, k_b$. It is conceivable that we may be interested in more than one linear function of μ_1, μ_2, $\ldots$, μ_b, such as $\mu_2 - \mu_1$, $\mu_3 - (\mu_1 + \mu_2)/2$, or $\mu_1 + \cdots + \mu_b$. We can, of course, find for each $\sum_1^b k_j \mu_j$ a random interval that has a preassigned probability of including that particular $\sum_1^b k_j \mu_j$. But how can we compute the probability that *simultaneously* these random intervals include their respective linear functions of μ_1, μ_2, $\ldots$, μ_b? The following procedure of multiple comparisons, due to Scheffé, is one solution to this problem.

The random variable

$$\frac{\sum\limits_{j=1}^{b} (\overline{X}_{\cdot j} - \mu_j)^2}{\sigma^2/a}$$

is $\chi^2(b)$ and, because it is a function of $\overline{X}_{\cdot 1}$, $\ldots$, $\overline{X}_{\cdot b}$ alone, it is stochastically independent of the random variable

$$V = \frac{1}{b(a-1)} \sum_{j=1}^{b} \sum_{i=1}^{a} (X_{ij} - \overline{X}_{\cdot j})^2.$$

Hence the random variable

$$F = \frac{a \sum\limits_{j=1}^{b} (\overline{X}_{\cdot j} - \mu_j)^2/b}{V}$$

has an F distribution with b and $b(a-1)$ degrees of freedom. From Table V in Appendix B, for certain values of α, we can find a constant d such that $\Pr(F \le d) = 1 - \alpha$ or

$$\Pr\left[\sum_{j=1}^{b} (\overline{X}_{\cdot j} - \mu_j)^2 \le bd\frac{V}{a} \right] = 1 - \alpha.$$

Note that $\sum\limits_{j=1}^{b} (\overline{X}_{\cdot j} - \mu_j)^2$ is the square of the distance, in b-dimensional space, from the point $(\mu_1, \mu_2, \ldots, \mu_b)$ to the random point $(\overline{X}_{\cdot 1}, \overline{X}_{\cdot 2}, \ldots, \overline{X}_{\cdot b})$. Consider a space of dimension b and let $(t_1, t_2, \ldots, t_b)$ denote the coordinates of a point in that space. An equation of a hyperplane that passes through the point $(\mu_1, \mu_2, \ldots, \mu_b)$ is given by

(1) $k_1(t_1 - \mu_1) + k_2(t_2 - \mu_2) + \cdots + k_b(t_b - \mu_b) = 0,$

where not all the real numbers k_j, $j = 1, 2, \ldots, b$, are equal to zero. The square of the distance from this hyperplane to the point $(t_1 = \overline{X}_{\cdot 1}$, $t_2 = \overline{X}_{\cdot 2}, \ldots, t_b = \overline{X}_{\cdot b})$ is

(2) $\dfrac{[k_1(\overline{X}_{\cdot 1} - \mu_1) + k_2(\overline{X}_{\cdot 2} - \mu_2) + \cdots + k_b(\overline{X}_{\cdot b} - \mu_b)]^2}{k_1^2 + k_2^2 + \cdots + k_b^2}.$

From the geometry of the situation it follows that $\sum_1^b (\bar{X}_{\cdot j} - \mu_j)^2$ is equal to the maximum of expression (2) with respect to $k_1, k_2, \ldots, k_b$. Thus the inequality $\sum_1^b (\bar{X}_{\cdot j} - \mu_j)^2 \leq (bd)(V/a)$ holds if and only if

(3)
$$\frac{\left[\sum_{j=1}^b k_j (\bar{X}_{\cdot j} - \mu_j)\right]^2}{\sum_{j=1}^b k_j^2} \leq bd\,\frac{V}{a},$$

for every real $k_1, k_2, \ldots, k_b$, not all zero. Accordingly, these two equivalent events have the same probability, $1 - \alpha$. However, inequality (3) may be written in the form

$$\left|\sum_1^b k_j \bar{X}_{\cdot j} - \sum_1^b k_j \mu_j\right| \leq \sqrt{bd\left(\sum_1^b k_j^2\right)}\,\frac{V}{a}.$$

Thus the probability is $1 - \alpha$ that simultaneously, for *all* real $k_1, k_2, \ldots, k_b$, not all zero,

(4)
$$\sum_1^b k_j \bar{X}_{\cdot j} - \sqrt{bd\left(\sum_1^b k_j^2\right)}\,\frac{V}{a} \leq \sum_1^b k_j \mu_j \leq \sum_1^b k_j \bar{X}_{\cdot j} + \sqrt{bd\left(\sum_1^b k_j^2\right)}\,\frac{V}{a}.$$

Denote by A the event where inequality (4) is true for all real $k_1, \ldots, k_b$, and denote by B the event where that inequality is true for a finite number of b-tuples $(k_1, \ldots, k_b)$. If the event A occurs, certainly the event B occurs. Hence $P(A) \leq P(B)$. In the applications, one is often interested only in a finite number of linear functions $\sum_1^b k_j \mu_j$. Once the experimental values are available, we obtain from (4) a confidence interval for each of these linear functions. Since $P(B) \geq P(A) = 1 - \alpha$, we have a confidence coefficient of at least $100(1 - \alpha)$ per cent that the linear functions are in these respective confidence intervals.

Remarks. If the sample sizes, say $a_1, a_2, \ldots, a_b$, are unequal, inequality (4) becomes

(4′)
$$\sum_1^b k_j \bar{X}_{\cdot j} - \sqrt{bd\sum_1^b \frac{k_j^2}{a_j}}\,V \leq \sum_1^b k_j \mu_j \leq \sum_1^b k_j \bar{X}_{\cdot j} + \sqrt{bd\sum_1^b \frac{k_j^2}{a_j}}\,V,$$

where

$$\bar{X}_{\cdot j} = \frac{\sum_{i=1}^{a_j} X_{ij}}{a_j}, \qquad V = \frac{\sum_{j=1}^b \sum_{i=1}^{a_j} (X_{ij} - \bar{X}_{\cdot j})^2}{\sum_1^b (a_j - 1)},$$

and d is selected from Table V with b and $\sum_1^b (a_j - 1)$ degrees of freedom. Inequality (4') reduces to inequality (4) when $a_1 = a_2 = \cdots = a_b$.

Moreover, if we restrict our attention to linear functions of the form $\sum_1^b k_j\mu_j$ with $\sum_1^b k_j = 0$ (such linear functions are called *contrasts*), the radical in inequality (4') is replaced by

$$\sqrt{d(b - 1) \sum_1^b \frac{k_j^2}{a_j} V},$$

where d is now found in Table V with $b - 1$ and $\sum_1^b (a_j - 1)$ degrees of freedom.

In these multiple comparisons, one often finds that the length of a confidence interval is much greater than the length of a $100(1 - \alpha)$ per cent confidence interval for a particular linear function $\sum_1^b k_j\mu_j$. But this is to be expected because in one case the probability $1 - \alpha$ applies to just one event, and in the other it applies to the simultaneous occurrence of many events. One reasonable way to reduce the length of these intervals is to take a larger value of α, say 0.25, instead of 0.05. After all, it is still a very strong statement to say that the probability is 0.75 that *all* these events occur.

EXERCISES

11.9. If $A_1, A_2, \ldots, A_k$ are events, prove, by induction, Boole's inequality $P(A_1 \cup A_2 \cup \cdots \cup A_k) \le \sum_1^k P(A_i)$. Then show that

$$P(A_1^* \cap A_2^* \cap \cdots \cap A_k^*) \ge 1 - \sum_1^k P(A_i).$$

11.10. In the notation of this section, let $(k_{i1}, k_{i2}, \ldots, k_{ib})$, $i = 1, 2, \ldots, m$, represent a finite number of b-tuples. The problem is to find simultaneous confidence intervals for $\sum_{j=1}^b k_{ij}\mu_j$, $i = 1, 2, \ldots, m$, by a method different from that of Scheffé. Define the random variable T_i by

$$\left(\sum_{j=1}^b k_{ij}\bar{X}_{\cdot j} - \sum_{j=1}^b k_{ij}\mu_j \right) \Big/ \sqrt{\left(\sum_{j=1}^b k_{ij}^2 \right) V/a}, \qquad i = 1, 2, \ldots, m.$$

(a) Let the event A_i^* be given by $-c_i \le T_i \le c_i$, $i = 1, 2, \ldots, m$. Find the random variables U_i and W_i such that $U_i \le \sum_{j=1}^b k_{ij}\mu_j \le W_i$ is equivalent to A_i^*.

(b) Select c_i such that $P(A_i^*) = 1 - \alpha/m$; that is, $P(A_i) = \alpha/m$. Use the results of Exercise 11.9 to determine a lower bound on the probability that simultaneously the random intervals $(U_1, W_1), \ldots, (U_m, W_m)$ include $\sum_{j=1}^{b} k_{1j}\mu_j, \ldots, \sum_{j=1}^{b} k_{mj}\mu_j$, respectively.

(c) Let $a = 3$, $b = 6$, and $\alpha = 0.05$. Consider the linear functions $\mu_1 - \mu_2$, $\mu_2 - \mu_3$, $\mu_3 - \mu_4$, $\mu_4 - (\mu_5 + \mu_6)/2$, and $(\mu_1 + \mu_2 + \cdots + \mu_6)/6$. Here $m = 5$. Show that the lengths of the confidence intervals given by the results of part (b) are shorter than the corresponding ones given by the method of Scheffé, as described in the text. If m becomes sufficiently large, however, this is not the case.

11.4 Classification

The problem of classification can be described as follows. An investigator makes a number of measurements on an item and wants to place it into one of several categories (or classify it). For convenience in our discussion, we assume that only two measurements, say X and Y, are made on the item to be classified. Moreover, let X and Y have a joint p.d.f. $f(x, y; \theta)$, where the parameter θ represents one or more parameters. In our simplification, suppose that there are only two possible joint distributions (categories) for X and Y, which are indexed by the parameter values θ' and θ'', respectively. In this case, the problem then reduces to one of observing $X = x$ and $Y = y$ and then testing the hypothesis $\theta = \theta'$ against the hypothesis $\theta = \theta''$, with the classification of X and Y being in accord with which hypothesis is accepted. From the Neyman–Pearson theorem, we know that a best decision of this sort is of the form: If

$$\frac{f(x, y; \theta')}{f(x, y; \theta'')} \leq k,$$

choose the distribution indexed by θ''; that is, we classify (x, y) as coming from the distribution indexed by θ''. Otherwise, choose the distribution indexed by θ'; that is, we classify (x, y) as coming from the distribution indexed by θ'.

In order to investigate an appropriate value of k, let us consider a Bayesian approach to the problem (see Section 6.6). We need a p.d.f. $h(\theta)$ of the parameter, which here is of the discrete type since the parameter space Ω consists of but two points θ' and θ''. So we have that $h(\theta') + h(\theta'') = 1$. Of course, the conditional p.d.f. $g(\theta|x, y)$ of the parameter, given $X = x$, $Y = y$, is proportional to the product of $h(\theta)$ and $f(x, y; \theta)$,

$$g(\theta|x, y) \propto h(\theta)f(x, y; \theta).$$

In particular, in this case,

$$g(\theta|x, y) = \frac{h(\theta)f(x, y; \theta)}{h(\theta')f(x, y; \theta') + h(\theta'')f(x, y; \theta'')}.$$

Let us introduce a loss function $\mathscr{L}[\theta, w(x, y)]$, where the decision function $w(x, y)$ selects decision $w = \theta'$ or decision $w = \theta''$. Because the pairs $(\theta = \theta', w = \theta')$ and $(\theta = \theta'', w = \theta'')$ represent correct decisions, we always take $\mathscr{L}(\theta', \theta') = \mathscr{L}(\theta'', \theta'') = 0$. On the other hand, positive values of the loss function should be assigned for incorrect decisions; that is, $\mathscr{L}(\theta', \theta'') > 0$ and $\mathscr{L}(\theta'', \theta') > 0$.

A Bayes' solution to the problem is defined to be such that the conditional expected value of the loss $\mathscr{L}[\theta, w(x, y)]$, given $X = x$, $Y = y$, is a minimum. If $w = \theta'$, this conditional expectation is

$$\sum_{\Omega} \mathscr{L}(\theta, \theta')g(\theta|x, y) = \frac{\mathscr{L}(\theta'', \theta')h(\theta'')f(x, y; \theta'')}{h(\theta')f(x, y; \theta') + h(\theta'')f(x, y; \theta'')}$$

because $\mathscr{L}(\theta', \theta') = 0$; and if $w = \theta''$, it is

$$\sum_{\Omega} \mathscr{L}(\theta, \theta'')g(\theta|x, y) = \frac{\mathscr{L}(\theta', \theta'')h(\theta')f(x, y; \theta')}{h(\theta')f(x, y; \theta') + h(\theta'')f(x, y; \theta'')}$$

because $\mathscr{L}(\theta'', \theta'') = 0$. Accordingly, a Bayes' solution is one that decides $w = \theta''$ if the latter ratio is less than or equal to the former; or, equivalently, if

$$\mathscr{L}(\theta', \theta'')h(\theta')f(x, y; \theta') \leq \mathscr{L}(\theta'', \theta')h(\theta'')f(x, y; \theta'').$$

That is, the decision $w = \theta''$ is made if

$$\frac{f(x, y; \theta')}{f(x, y; \theta'')} \leq \frac{\mathscr{L}(\theta'', \theta')h(\theta'')}{\mathscr{L}(\theta', \theta'')h(\theta')} = k;$$

otherwise, the decision $w = \theta'$ is made. Hence, if prior probabilities $h(\theta')$ and $h(\theta'')$ and losses $\mathscr{L}(\theta = \theta', w = \theta'')$ and $\mathscr{L}(\theta = \theta'', w = \theta')$ can be assigned, the constant k of the Neyman–Pearson theorem can be found easily from this formula.

Example 1. Let (x, y) be an observation of the random pair (X, Y), which has a bivariate normal distribution with parameters $\mu_1, \mu_2, \sigma_1^2, \sigma_2^2$, and ρ. In Section 3.5 that joint p.d.f. is given by

$$f(x, y; \mu_1, \mu_2, \sigma_1^2, \sigma_2^2, \rho) = \frac{1}{2\pi\sigma_1\sigma_2\sqrt{1 - \rho^2}} e^{-q(x, y; \mu_1, \mu_2)/2},$$

$$-\infty < x < \infty, \quad -\infty < y < \infty,$$

where $\sigma_1 > 0$, $\sigma_2 > 0$, $-1 < \rho < 1$, and

$$q(x, y; \mu_1, \mu_2) = \frac{1}{1 - \rho^2}\left[\left(\frac{x - \mu_1}{\sigma_1}\right)^2 - 2\rho\left(\frac{x - \mu_1}{\sigma_1}\right)\left(\frac{y - \mu_2}{\sigma_2}\right) + \left(\frac{y - \mu_2}{\sigma_2}\right)^2\right].$$

Assume that σ_1^2, σ_2^2, and ρ are known but that we do not know whether the respective means of (X, Y) are (μ_1', μ_2') or (μ_1'', μ_2''). The inequality

$$\frac{f(x, y; \mu_1', \mu_2', \sigma_1^2, \sigma_2^2, \rho)}{f(x, y; \mu_1'', \mu_2'', \sigma_1^2, \sigma_2^2, \rho)} \leq k$$

is equivalent to

$$\tfrac{1}{2}[q(x, y; \mu_1'', \mu_2'') - q(x, y; \mu_1', \mu_2')] \leq \ln k.$$

Moreover, it is clear that the difference in the left-hand member of this inequality does not contain terms involving x^2, xy, and y^2. In particular, this inequality is the same as

$$(1)\quad \frac{1}{1 - \rho^2}\left\{\left[\frac{\mu_1' - \mu_1''}{\sigma_1^2} - \frac{\rho(\mu_2' - \mu_2'')}{\sigma_1\sigma_2}\right]x + \left[\frac{\mu_2' - \mu_2''}{\sigma_2^2} - \frac{\rho(\mu_1' - \mu_1'')}{\sigma_1\sigma_2}\right]y\right\}$$

$$\leq \ln k + \tfrac{1}{2}[q(0, 0; \mu_1', \mu_2') - q(0, 0; \mu_1'', \mu_2'')],$$

or, for brevity,

$$ax + by \leq c.$$

That is, if this linear function of x and y in the left-hand member of inequality (1) is less than or equal to a certain constant, we would classify that (x, y) as coming from the bivariate normal distribution with means μ_1'' and μ_2''. Otherwise, we would classify (x, y) as arising from the bivariate normal distribution with means μ_1' and μ_2'. Of course, if the prior probabilities and losses are given, k and thus c can be found easily; this will be illustrated in Exercise 11.11.

Once the rule for classification is established, the statistician might be interested in the two probabilities of misclassifications using that rule. The first of these two is associated with the classification of (x, y) as arising from the distribution indexed by θ'' if, in fact, it comes from that index by θ'. The second misclassification is similar, but with the interchange of θ' and θ''. In the previous example, the probabilities of these respective misclassifications are

$$\Pr(aX + bY \leq c; \mu_1', \mu_2') \qquad \text{and} \qquad \Pr(aX + bY > c; \mu_1'', \mu_2'').$$

Fortunately, the distribution of $Z = aX + bY$ is easy to determine, so each of these probabilities is easy to calculate. The moment-generating function of Z is

$$E(e^{tZ}) = E[e^{t(aX + bY)}] = E(e^{atX + btY}).$$

Hence in the joint moment-generating function of X and Y found in Section 3.5, simply replace t_1 by at and t_2 by bt, to obtain

$$E(e^{tZ}) = \exp\left[\mu_1 at + \mu_2 bt + \frac{\sigma_1^2(at)^2 + 2\rho\sigma_1\sigma_2(at)(bt) + \sigma_2^2(bt)^2}{2}\right]$$

$$= \exp\left[(a\mu_1 + b\mu_2)t + \frac{(a^2\sigma_1^2 + 2ab\rho\sigma_1\sigma_2 + b^2\sigma_2^2)t^2}{2}\right].$$

However, this is the moment-generating function of the normal distribution

$$n(a\mu_1 + b\mu_2, \; a^2\sigma_1^2 + 2ab\rho\sigma_1\sigma_2 + b^2\sigma_2^2).$$

With this information, it is easy to compute the probabilities of mis-classifications, and this will also be demonstrated in Exercise 11.11.

One final remark must be made with respect to the use of the important classification rule established in Example 1. In most instances the parameter values μ_1', μ_2' and μ_1'', μ_2'' as well as σ_1^2, σ_2^2, and ρ are unknown. In such cases the statistician has usually observed a random sample (frequently called a *training sample*) from each of the two distributions. Let us say the samples have sizes n' and n'', respectively, with sample characteristics

$$\bar{x}', \bar{y}', (s_x')^2, (s_y')^2, r' \qquad \text{and} \qquad \bar{x}'', \bar{y}'', (s_x'')^2, (s_y'')^2, r''.$$

Accordingly, if in inequality (1) the parameters μ_1', μ_2', μ_1'', μ_2'', σ_1^2, σ_2^2, and $\rho\sigma_1\sigma_2$ are replaced by the unbiased estimates

$$\bar{x}', \bar{y}', \bar{x}'', \bar{y}'', \frac{n'(s_x')^2 + n''(s_x'')^2}{n' + n'' - 2}, \frac{n'(s_y')^2 + n''(s_y'')^2}{n' + n'' - 2},$$

$$\frac{n'r's_x's_y' + n''r''s_x''s_y''}{n' + n'' - 2},$$

the resulting expression in the left-hand member is frequently called Fisher's *linear discriminant function*. Since those parameters have been estimated, the distribution theory associated with $aX + bY$ is not appropriate for Fisher's function. However, if n' and n'' are large, the distribution of $aX + bY$ does provide an approximation.

Although we have considered only bivariate distributions in this section, the results can easily be extended to multivariate normal distributions after a study of Chapter 12.

EXERCISES

11.11. In Example 1 let $\mu_1' = \mu_2' = 0$, $\mu_1'' = \mu_2'' = 1$, $\sigma_1^2 = 1$, $\sigma_2^2 = 1$, and $\rho = \frac{1}{2}$.

(a) Evaluate inequality (1) when the prior probabilities are $h(\mu_1', \mu_2') = \frac{1}{3}$ and $h(\mu_1'', \mu_2'') = \frac{2}{3}$ and the losses are $\mathscr{L}[\theta = (\mu_1', \mu_2'), w = (\mu_1'', \mu_2'')] = 4$ and $\mathscr{L}[\theta = (\mu_1'', \mu_2''), w = (\mu_1', \mu_2')] = 1$.

(b) Find the distribution of the linear function $aX + bY$ that results from part (a).

(c) Compute $\Pr(aX + bY \le c; \mu_1' = \mu_2' = 0)$ and $\Pr(aX + bY > c; \mu_1'' = \mu_2'' = 1)$.

11.12. Let X and Y have the joint p.d.f.

$$f(x, y; \theta_1, \theta_2) = \frac{1}{\theta_1 \theta_2} \exp\left(-\frac{x}{\theta_1} - \frac{y}{\theta_2}\right), \qquad 0 < x < \infty, \, 0 < y < \infty,$$

zero elsewhere, where $0 < \theta_1$, $0 < \theta_2$. An observation (x, y) arises from the joint distribution with parameters equal to either $\theta_1' = 1$, $\theta_2' = 5$ or $\theta_1'' = 3$, $\theta_2'' = 2$. Determine the form of the classification rule.

11.13. Let X and Y have a joint bivariate normal distribution. An observation (x, y) arises from the joint distribution with parameters equal to either

$$\mu_1' = \mu_2' = 0, \; (\sigma_1^2)' = (\sigma_2^2)' = 1, \; \rho' = \tfrac{1}{2}$$

or

$$\mu_1'' = \mu_2'' = 1, \; (\sigma_1^2)'' = 4, \; (\sigma_2^2)'' = 9, \; \rho'' = \tfrac{1}{2}.$$

Show that the classification rule involves a second-degree polynomial in x and y.

11.5 Sufficiency, Completeness, and Stochastic Independence

In Chapter 10 we noted that if we have a sufficient statistic Y_1 for a parameter θ, $\theta \in \Omega$, then $h(z|y_1)$, the conditional p.d.f. of another statistic Z, given $Y_1 = y_1$, does not depend upon θ. If, moreover, Y_1 and Z are stochastically independent, the p.d.f. $g_2(z)$ of Z is such that $g_2(z) = h(z|y_1)$, and hence $g_2(z)$ must not depend upon θ either. So the stochastic independence of a statistic Z and the sufficient statistic Y_1 for a parameter θ means that the distribution of Z does not depend upon $\theta \in \Omega$.

It is interesting to investigate a converse of that property. Suppose that the distribution of a statistic Z does not depend upon θ; then, are Z and the sufficient statistic Y_1 for θ stochastically independent? To begin our search for the answer, we know that the joint p.d.f. of Y_1 and Z is $g_1(y_1; \theta)h(z|y_1)$, where $g_1(y_1; \theta)$ and $h(z|y_1)$ represent the marginal p.d.f. of Y_1 and the conditional p.d.f. of Z given $Y_1 = y_1$, respectively. Thus the marginal p.d.f. of Z is

$$\int_{-\infty}^{\infty} g_1(y_1; \theta)h(z|y_1)\, dy_1 = g_2(z),$$

which, by hypothesis, does not depend upon θ. Because

$$\int_{-\infty}^{\infty} g_2(z)g_1(y_1; \theta)\, dy_1 = g_2(z),$$

it follows, by taking the difference of the last two integrals, that

$$(1) \qquad \int_{-\infty}^{\infty} [g_2(z) - h(z|y_1)]g_1(y_1; \theta)\, dy_1 = 0$$

for all $\theta \in \Omega$. Since Y_1 is a sufficient statistic for θ, $h(z|y_1)$ does not depend upon θ. By assumption, $g_2(z)$ and hence $g_2(z) - h(z|y_1)$ do not depend upon θ. Now if the family $\{g_1(y_1; \theta); \theta \in \Omega\}$ is complete, Equation (1) would require that

$$g_2(z) - h(z|y_1) = 0 \qquad \text{or} \qquad g_2(z) = h(z|y_1).$$

That is, the joint p.d.f. of Y_1 and Z must be equal to

$$g_1(y_1; \theta)h(z|y_1) = g_1(y_1; \theta)g_2(z).$$

Accordingly, Y_1 and Z are stochastically independent, and we have proved the following theorem.

Theorem 1. *Let $X_1, X_2, \ldots, X_n$ denote a random sample from a distribution having a p.d.f. $f(x; \theta)$, $\theta \in \Omega$, where Ω is an interval set. Let $Y_1 = u_1(X_1, X_2, \ldots, X_n)$ be a sufficient statistic for θ, and let the family $\{g_1(y_1; \theta); \theta \in \Omega\}$ of probability density functions of Y_1 be complete. Let $Z = u(X_1, X_2, \ldots, X_n)$ be any other statistic (not a function of Y_1 alone). If the distribution of Z does not depend upon θ, then Z is stochastically independent of the sufficient statistic Y_1.*

In the discussion above, it is interesting to observe that if Y_1 is a sufficient statistic for θ, then the stochastic independence of Y_1 and Z implies that the distribution of Z does not depend upon θ whether $\{g_1(y_1; \theta); \theta \in \Omega\}$ is or is not complete. However, in the converse, to prove the stochastic independence from the fact that $g_2(z)$ does not depend upon θ, we definitely need the completeness. Accordingly, if we are dealing with situations in which we know that the family

$\{g(y_1; \theta), \theta \in \Omega\}$ is complete (such as a regular case of the exponential class), we can say that the statistic Z is stochastically independent of the sufficient statistic Y_1 if, and only if, the distribution of Z does not depend upon θ.

It should be remarked that the theorem (including the special formulation of it for regular cases of the exponential class) extends immediately to probability density functions that involve m parameters for which there exist m joint sufficient statistics. For example, let $X_1, X_2, \ldots, X_n$ be a random sample from a distribution having the p.d.f. $f(x; \theta_1, \theta_2)$ that represents a regular case of the exponential class such that there are two joint complete sufficient statistics for θ_1 and θ_2. Then any other statistic $Z = u(X_1, X_2, \ldots, X_n)$ is stochastically independent of the joint complete sufficient statistics if and only if the distribution of Z does not depend upon θ_1 or θ_2.

We give an example of the theorem that provides an alternative proof of the stochastic independence of $\bar{X}$ and S^2, the mean and the variance of a random sample of size n from a distribution that is $n(\mu, \sigma^2)$. This proof is presented as if we did not know that nS^2/σ^2 is $\chi^2(n-1)$ because that fact and the stochastic independence were established in the same argument (see Section 4.8).

Example 1. Let $X_1, X_2, \ldots, X_n$ denote a random sample of size n from a distribution that is $n(\mu, \sigma^2)$. We know that the mean $\bar{X}$ of the sample is, for every known σ^2, a complete sufficient statistic for the parameter μ, $-\infty < \mu < \infty$. Consider the statistic

$$S^2 = \frac{1}{n} \sum_{i=1}^{n} (X_i - \bar{X})^2$$

and the one-to-one transformation defined by $W_i = X_i - \mu, i = 1, 2, \ldots, n$. Since $\bar{W} = \bar{X} - \mu$, we have that

$$S^2 = \frac{1}{n} \sum_{i=1}^{n} (W_i - \bar{W})^2;$$

moreover, each W_i is $n(0, \sigma^2)$, $i = 1, 2, \ldots, n$. That is, S^2 can be written as a function of the random variables $W_1, W_2, \ldots, W_n$, which have distributions that do not depend upon μ. Thus S^2 must have a distribution that does not depend upon μ; and hence, by the theorem, S^2 and $\bar{X}$, the complete sufficient statistic for μ, are stochastically independent.

The technique that is used in Example 1 can be generalized to situations in which there is a complete sufficient statistic for a location parameter θ. Let $X_1, X_2, \ldots, X_n$ be a random sample from a distribution that has a p.d.f. of the form $f(x - \theta)$, for every real θ; that is, θ is a location parameter. Let

$Y_1 = u_1(X_1, X_2, \ldots, X_n)$ be a complete sufficient statistic for θ. Moreover, let $Z = u(X_1, X_2, \ldots, X_n)$ be another statistic such that

$$u(x_1 + d, x_2 + d, \ldots, x_n + d) = u(x_1, x_2, \ldots, x_n),$$

for all real d. The one-to-one transformation defined by $W_i = X_i - \theta$, $i = 1, 2, \ldots, n$, requires that the joint p.d.f. of $W_1, W_2, \ldots, W_n$ be

$$f(w_1)f(w_2) \cdots f(w_n),$$

which does not depend upon θ. In addition, we have, because of the special functional nature of $u(x_1, x_2, \ldots, x_n)$, that

$$Z = u(W_1 + \theta, W_2 + \theta, \ldots, W_n + \theta) = u(W_1, W_2, \ldots, W_n)$$

is a function of $W_1, W_2, \ldots, W_n$ alone (not of θ). Hence Z must have a distribution that does not depend upon θ and thus, by the theorem, is stochastically independent of Y_1.

Example 2. Let $X_1, X_2, \ldots, X_n$ be a random sample of size n from the distribution having p.d.f.

$$f(x; \theta) = e^{-(x-\theta)}, \qquad \theta < x < \infty, -\infty < \theta < \infty.$$
$$= 0 \text{ elsewhere.}$$

Here the p.d.f. is of the form $f(x - \theta)$, where $f(x) = e^{-x}$, $0 < x < \infty$, zero elsewhere. Moreover, we know (Exercise 10.17, Section 10.3) that the first order statistic $Y_1 = \min(X_i)$ is a complete sufficient statistic for θ. Hence Y_1 must be stochastically independent of each statistic $u(X_1, X_2, \ldots, X_n)$, enjoying the property that

$$u(x_1 + d, x_2 + d, \ldots, x_n + d) = u(x_1, x_2, \ldots, x_n)$$

for all real d. Illustrations of such statistics are S^2, the sample range, and

$$\frac{1}{n} \sum_{i=1}^{n} [X_i - \min(X_i)].$$

There is a result on stochastic independence of a complete sufficient statistic for a scale parameter and another statistic that corresponds to that associated with a location parameter. Let $X_1, X_2, \ldots, X_n$ be a random sample from a distribution that has a p.d.f. of the form $(1/\theta)f(x/\theta)$, for all $\theta > 0$; that is, θ is a scale parameter. Let $Y_1 = u_1(X_1, X_2, \ldots, X_n)$ be a complete sufficient statistic for θ. Say $Z = u(X_1, X_2, \ldots, X_n)$ is another statistic such that

$$u(cx_1, cx_2, \ldots, cx_n) = u(x_1, x_2, \ldots, x_n)$$

for all $c > 0$. The one-to-one transformation defined by $W_i = X_i/\theta$, $i =$

$1, 2, \ldots, n$, requires the following: (a) that the joint p.d.f. of $W_1, W_2, \ldots, W_n$ be equal to

$$f(w_1)f(w_2) \cdots f(w_n),$$

and (b) that the statistic Z be equal to

$$Z = u(\theta W_1, \theta W_2, \ldots, \theta W_n) = u(W_1, W_2, \ldots, W_n).$$

Since neither the joint p.d.f. of $W_1, W_2, \ldots, W_n$ nor Z contain θ, the distribution of Z must not depend upon θ and thus, by the theorem, Z is stochastically independent of the complete sufficient statistic Y_1 for the parameter θ.

Example 3. Let X_1 and X_2 denote a random sample of size $n = 2$ from a distribution with p.d.f.

$$f(x; \theta) = \frac{1}{\theta} e^{-x/\theta}, \qquad 0 < x < \infty, 0 < \theta < \infty,$$

$$= 0 \text{ elsewhere.}$$

The p.d.f. is of the form $(1/\theta)f(x/\theta)$, where $f(x) = e^{-x}$, $0 < x < \infty$, zero elsewhere. We know (Section 10.4) that $Y_1 = X_1 + X_2$ is a complete sufficient statistic for θ. Hence Y_1 is stochastically independent of every statistic $u(X_1, X_2)$ with the property $u(cx_1, cx_2) \doteq u(x_1, x_2)$. Illustrations of these are X_1/X_2 and $X_1/(X_1 + X_2)$, statistics that have F and beta distributions, respectively.

Finally, the location and the scale parameters can be combined in a p.d.f. of the form $(1/\theta_2)f[(x - \theta_1)/\theta_2]$, $-\infty < \theta_1 < \infty, 0 < \theta_2 < \infty$. Through a one-to-one transformation defined by $W_i = (X_i - \theta_1)/\theta_2$, $i = 1, 2, \ldots, n$, it is easy to show that a statistic $Z = u(X_1, X_2, \ldots, X_n)$ such that

$$u(cx_1 + d, \ldots, cx_n + d) = u(x_1, \ldots, x_n)$$

for $-\infty < d < \infty$, $0 < c < \infty$, has a distribution that does not depend upon θ_1 and θ_2. Thus, by the extension of the theorem, the joint complete sufficient statistics Y_1 and Y_2 for the parameters θ_1 and θ_2 are stochastically independent of Z.

Example 4. Let $X_1, X_2, \ldots, X_n$ denote a random sample from a distribution that is $n(\theta_1, \theta_2)$, $-\infty < \theta_1 < \infty, 0 < \theta_2 < \infty$. In Example 2, Section 10.6, it was proved that the mean $\bar{X}$ and the variance S^2 of the sample are joint complete sufficient statistics for θ_1 and θ_2. Consider the statistic

$$Z = \frac{\sum_1^{n-1} (X_{i+1} - X_i)^2}{\sum_1^n (X_i - \bar{X})^2} = u(X_1, X_2, \ldots, X_n),$$

which satisfies the property that $u(cx_1 + d, \ldots, cx_n + d) = u(x_1, \ldots, x_n)$. That is, Z is stochastically independent of both $\bar{X}$ and S^2.

Let $n(\theta_1, \theta_3)$ and $n(\theta_2, \theta_4)$ denote two independent normal distributions. Recall that in Example 2, Section 7.4, a statistic, which was denoted by T, was used to test the hypothesis that $\theta_1 = \theta_2$, provided the unknown variances θ_3 and θ_4 were equal. The hypothesis that $\theta_1 = \theta_2$ is rejected if the computed $|T| \geq c$, where the constant c is selected so that $\alpha_2 = \Pr\left(|T| \geq c; \theta_1 = \theta_2, \theta_3 = \theta_4\right)$ is the assigned significance level of the test. We shall show that, if $\theta_3 = \theta_4$, F of Example 3, Section 7.4, and T are stochastically independent. Among other things, this means that if these two tests are performed sequentially, with respective significance levels α_1 and α_2, the probability of accepting both these hypotheses, when they are true, is $(1 - \alpha_1)(1 - \alpha_2)$. Thus the significance level of this joint test is $\alpha = 1 - (1 - \alpha_1)(1 - \alpha_2)$.

The stochastic independence of F and T, when $\theta_3 = \theta_4$, can be established by an appeal to sufficiency and completeness. The three statistics $\bar{X}$, $\bar{Y}$, and $\sum_1^n (X_i - \bar{X})^2 + \sum_1^m (Y_i - \bar{Y})^2$ are joint complete sufficient statistics for the three parameters θ_1, θ_2, and $\theta_3 = \theta_4$. Obviously, the distribution of F does not depend upon θ_1, θ_2, and $\theta_3 = \theta_4$, and hence F is stochastically independent of the three joint complete sufficient statistics. However, T is a function of these three joint complete sufficient statistics alone, and, accordingly, T is stochastically independent of F. It is important to note that these two statistics are stochastically independent whether $\theta_1 = \theta_2$ or $\theta_1 \neq \theta_2$, that is, whether T is or is not central. This permits us to calculate probabilities other than the significance level of the test. For example, if $\theta_3 = \theta_4$ and $\theta_1 \neq \theta_2$, then

$$\Pr\left(c_1 < F < c_2, |T| \geq c\right) = \Pr\left(c_1 < F < c_2\right) \Pr\left(|T| \geq c\right).$$

The second factor in the right-hand member is evaluated by using the probabilities for a noncentral t distribution. Of course, if $\theta_3 = \theta_4$ and the difference $\theta_1 - \theta_2$ is large, we would want the preceding probability to be close to 1 because the event $\{c_1 < F < c_2, |T| \geq c\}$ leads to a correct decision, namely accept $\theta_3 = \theta_4$ and reject $\theta_1 = \theta_2$.

EXERCISES

11.14. Let $Y_1 < Y_2 < Y_3 < Y_4$ denote the order statistics of a random sample of size $n = 4$ from a distribution having p.d.f. $f(x; \theta) = 1/\theta$, $0 < x < \theta$, zero elsewhere, where $0 < \theta < \infty$. Argue that the complete sufficient statistic Y_4 for θ is stochastically independent of each of the statistics Y_1/Y_4

and $(Y_1 + Y_2)/(Y_3 + Y_4)$. *Hint.* Show that the p.d.f. is of the form $(1/\theta)f(x/\theta)$, where $f(x) = 1$, $0 < x < 1$, zero elsewhere.

11.15. Let $Y_1 < Y_2 < \cdots < Y_n$ be the order statistics of a random sample from the normal distribution $n(\theta, \sigma^2)$, $-\infty < \theta < \infty$. Show that the distribution of $Z = Y_n - \bar{Y}$ does not depend upon θ. Thus $\bar{Y} = \sum_1^n Y_i/n$, a complete sufficient statistic for θ, is stochastically independent of Z.

11.16. Let $X_1, X_2, \ldots, X_n$ be a random sample from the normal distribution $n(\theta, \sigma^2)$, $-\infty < \theta < \infty$. Prove that a necessary and sufficient condition that the statistics $Z = \sum_1^n a_i X_i$ and $Y = \sum_1^n X_i$, a complete sufficient statistic for θ, be stochastically independent is that $\sum_1^n a_i = 0$.

11.17. Let X and Y be random variables such that $E(X^k)$ and $E(Y^k) \neq 0$ exist for $k = 1, 2, 3, \ldots$. If the ratio X/Y and its denominator Y are stochastically independent, prove that $E[(X/Y)^k] = E(X^k)/E(Y^k)$, $k = 1, 2, 3, \ldots$. *Hint.* Write $E(X^k) = E[Y^k(X/Y)^k]$.

11.18. Let $Y_1 < Y_2 < \cdots < Y_n$ be the order statistics of a random sample of size n from a distribution that has p.d.f. $f(x; \theta) = (1/\theta)e^{-x/\theta}$, $0 < x < \infty$, $0 < \theta < \infty$, zero elsewhere. Show that the ratio $R = nY_1/\sum_1^n Y_i$ and its denominator (a complete sufficient statistic for θ) are stochastically independent. Use the result of the preceding exercise to determine $E(R^k)$, $k = 1, 2, 3, \ldots$.

11.19. Let $X_1, X_2, \ldots, X_5$ be a random sample of size 5 from the distribution that has p.d.f. $f(x) = e^{-x}$, $0 < x < \infty$, zero elsewhere. Show that $(X_1 + X_2)/(X_1 + X_2 + \cdots + X_5)$ and its denominator are stochastically independent. *Hint.* The p.d.f. $f(x)$ is a member of $\{f(x; \theta); 0 < \theta < \infty\}$, where $f(x; \theta) = (1/\theta)e^{-x/\theta}$, $0 < x < \infty$, zero elsewhere.

11.20. Let $Y_1 < Y_2 < \cdots < Y_n$ be the order statistics of a random sample from the normal distribution $n(\theta_1, \theta_2)$, $-\infty < \theta_1 < \infty$, $0 < \theta_2 < \infty$. Show that the joint complete sufficient statistics $\bar{X} = \bar{Y}$ and S^2 for θ_1 and θ_2 are stochastically independent of each of $(Y_n - \bar{Y})/S$ and $(Y_n - Y_1)/S$.

11.21. Let $Y_1 < Y_2 < \cdots < Y_n$ be the order statistics of a random sample from a distribution with the p.d.f.

$$f(x; \theta_1, \theta_2) = \frac{1}{\theta_2} \exp\left(-\frac{x - \theta_1}{\theta_2}\right)$$

$\theta_1 < x < \infty$, zero elsewhere, where $-\infty < \theta_1 < \infty$, $0 < \theta_2 < \infty$. Show that the joint complete sufficient statistics Y_1 and $\bar{X} = \bar{Y}$ for θ_1 and θ_2 are stochastically independent of $(Y_2 - Y_1)/\sum_1^n (Y_i - Y_1)$.

11.6 Robust Nonparametric Methods

Frequently, an investigator is tempted to evaluate several test statistics associated with a single hypothesis and then use the one statistic that best supports his or her position, usually rejection. Obviously, this type of procedure changes the actual significance level of the test from the nominal α that is used. However, there is a way in which the investigator can first look at the data and then select a test statistic without changing this significance level. For illustration, suppose there are three possible test statistics W_1, W_2, W_3 of the hypothesis H_0 with respective critical regions C_1, C_2, C_3 such that $\Pr(W_i \in C_i; H_0) = \alpha$, $i = 1, 2, 3$. Moreover, suppose that a statistic Q, based upon the same data, selects one and only one of the statistics W_1, W_2, W_3, and that W is then used to test H_0. For example, we choose to use the test statistic W_i if $Q \in D_i$, $i = 1, 2, 3$, where the events defined by D_1, D_2, and D_3 are mutually exclusive and exhaustive. Now if Q and each W_i are stochastically independent when H_0 is true, then the probability of rejection, using the entire procedure (selecting and testing), is, under H_0,

$$\Pr(Q \in D_1, W_1 \in C_1) + \Pr(Q \in D_2, W_2 \in C_2) + \Pr(Q \in D_3, W_3 \in C_3)$$

$$= \Pr(Q \in D_1)\Pr(W_1 \in C_1) + \Pr(Q \in D_2)\Pr(W_2 \in C_2)$$

$$+ \Pr(Q \in D_3)\Pr(W_3 \in C_3)$$

$$= \alpha[\Pr(Q \in D_1) + \Pr(Q \in D_2) + \Pr(Q \in D_3)] = \alpha.$$

That is, the procedure of selecting W_i using a stochastically independent statistic Q and then constructing a test of significance level α with the statistic W_i has overall significance level α.

Of course, the important element in this procedure is the ability to be able to find a selector Q that is independent of each test statistic W. This can frequently be done by using the fact that the complete sufficient statistics for the parameters, given by H_0, are stochastically independent of every statistic whose distribution is free of those parameters (see Section 11.5). For illustration, if random samples of sizes m and n arise from two independent normal distributions with respective means μ_1 and μ_2 and common variance σ^2, then the complete sufficient statistics $\bar{X}$, $\bar{Y}$, and

$$V = \sum_1^m (X_i - \bar{X})^2 + \sum_1^n (Y_i - \bar{Y})^2$$

for μ_1, μ_2, and σ^2 are stochastically independent of every statistic whose distribution is free of μ_1, μ_2, and σ^2 such as

$$\frac{\sum_1^m (X_i - \bar{X})^2}{\sum_1^n (Y_i - \bar{Y})^2}, \frac{\sum_1^m |X_i - \text{median } (X_i)|}{\sum_1^n |Y_i - \text{median } (Y_i)|}, \frac{\text{range } (X_1, X_2, \ldots, X_m)}{\text{range } (Y_1, Y_2, \ldots, Y_n)}.$$

Thus, in general, we would hope to be able to find a selector Q that is a function of the complete sufficient statistics for the parameters, under H_0, so that it is independent of the test statistics.

It is particularly interesting to note that it is relatively easy to use this technique in *nonparametric* methods by using the independence result based upon complete sufficient statistics for *parameters*. How can we use an argument depending on parameters in nonparametric methods? Although this does sound strange, it is due to the unfortunate choice of a name in describing this broad area of nonparametric methods. Most statisticians would prefer to describe the subject as being *distribution-free*, since the test statistics have distributions that do not depend on the underlying distribution of the continuous type, described by either the distribution function F or the p.d.f. f. In addition, the latter name provides the clue for our application here because we have many test statistics whose distributions are free of the unknown (infinite vector) "parameter" F (or f). We now must find complete sufficient statistics for the distribution function F of the continuous type. In many instances, this is easy to do.

In Exercise 10.37, Section 10.6, it is shown that the order statistics $Y_1 < Y_2 < \cdots < Y_n$ of a random sample of size n from a distribution of the continuous type with p.d.f. $F'(x) = f(x)$ are sufficient statistics for the "parameter" f (or F). Moreover, if the family of distributions contains all probability density functions of the continuous type, the family of joint probability density functions of Y_1, Y_2, ..., Y_n is also complete. We accept this latter fact without proof, as it is beyond the level of this text; but doing so, we can now say that the order statistics Y_1, Y_2, ..., Y_n are complete sufficient statistics for the parameter f (or F).

Accordingly, our selector Q will be based upon those complete sufficient statistics, the order statistics under H_0. This allows us to independently choose a distribution-free test appropriate for this type of underlying distribution, and thus increase our power. Although it is well known that distribution-free tests hold the significance level α for all underlying distributions of the continuous type, they have often been criticized because their powers are sometimes low. The independent selection of the distribution-free test to be used can help correct this. So selecting—or adapting the test to the data—provides a new dimension to nonparametric tests, which usually improves the power of the overall test.

A statistical test that maintains the significance level close to a desired significance level α for a wide variety of underlying distributions with good (not necessarily the best for any one type of distribution) power for all these distributions is described as being *robust*. As an illustration, the T used to test the equality of the means of two independent normal distributions (see Section 7.4) is quite robust *provided* that the underlying distributions are rather close to normal ones with common variance. However, if the class of distributions includes those that are not too close to normal ones, such as the Cauchy distribution, the test based upon T is *not* robust; the significance level is not maintained and the power of the T test is low with Cauchy distributions. As a matter of fact, the test based on the Mann–Whitney–Wilcoxon statistic (Section 9.6) is a much more robust test than that based upon T if the class of distributions is fairly wide (in particular, if long-tailed distributions such as the Cauchy are included).

An illustration of the adaptive distribution-free procedure that is robust is provided by considering a test of the equality of two independent distributions of the continuous type. From the discussion in Section 9.8, we know that we could construct many linear rank statistics by changing the scoring function. However, we concentrate on three such statistics mentioned explicitly in that section: that based on normal scores, say L_1; that of Mann–Whitney–Wilcoxon, say L_2; and that of the median test, say L_3. Moreover, respective critical regions C_1, C_2, and C_3 are selected so that, under the equality of the two distributions, we have

$$\alpha = \Pr\left(L_1 \in C_1\right) = \Pr\left(L_2 \in C_2\right) = \Pr\left(L_3 \in C_3\right).$$

Of course, we would like to use the test given by $L_1 \in C_1$ if the tails of the distributions are like or shorter than those of the normal distributions. With distributions having somewhat longer tails, $L_2 \in C_2$ provides an excellent test. And with distributions having very long tails, the test based on $L_3 \in C_3$ is quite satisfactory.

In order to select the appropriate test in an independent manner we let $V_1 < V_2 < \cdots < V_N$, where $N = m + n$, be the order statistics of the combined sample, which is of size N. Recall that if the two independent distributions are equal and have the same distribution function

F, these order statistics are the complete sufficient statistics for the parameter F. Hence every statistic based on $V_1, V_2, \ldots, V_N$ is stochastically independent of L_1, L_2, and L_3, since the latter statistics have distributions that do not depend upon F. In particular, the kurtosis (Exercise 1.98, Section 1.10) of the combined sample,

$$K = \frac{\dfrac{1}{N} \sum_{i=1}^{N} (V_i - \bar{V})^4}{\left[\dfrac{1}{N} \sum_{i=1}^{N} (V_i - \bar{V})^2\right]^2},$$

is stochastically independent of L_1, L_2, and L_3. From Exercise 3.56, Section 3.4, we know that the kurtosis of the normal distribution is 3; hence if the two distributions were equal and normal, we would expect K to be about 3. Of course, a longer-tailed distribution has a bigger kurtosis. Thus one simple way of defining the independent selection procedure would be by letting

$$D_1 = \{k; k \le 3\}, \qquad D_2 = \{k; 3 < k \le 8\}, \qquad D_3 = \{k; 8 < k\}.$$

These choices are not necessarily the best way of selecting the appropriate test, but they are reasonable and illustrative of the adaptive procedure. From the stochastic independence of K and (L_1, L_2, L_3), we know that the overall test has significance level α. Since a more appropriate test has been selected, the power will be relatively good throughout a wide range of distributions. Accordingly, this distribution-free adaptive test is robust.

EXERCISES

11.22. Let $F(x)$ be a distribution function of a distribution of the continuous type which is symmetric about its median ξ. We wish to test $H_0: \xi = 0$ against $H_1: \xi > 0$. Use the fact that the $2n$ values, X_i and $-X_i$, $i = 1, 2, \ldots, n$, after ordering, are complete sufficient statistics for F, provided that H_0 is true. Then construct an adaptive distribution-free test based upon Wilcoxon's statistic and two of its modifications given in Exercises 9.20 and 9.21.

11.23. Suppose that the hypothesis H_0 concerns the stochastic independence of two random variables X and Y. That is, we wish to test $H_0: F(x, y) = F_1(x)F_2(y)$, where F, F_1, and F_2 are the respective joint and marginal distribution functions of the continuous type, against all alternatives. Let $(X_1, Y_1), (X_2, Y_2), \ldots, (X_n, Y_n)$ be a random sample from the joint distribution. Under H_0, the order statistics of $X_1, X_2, \ldots, X_n$ and the order statistics of $Y_1, Y_2, \ldots, Y_n$ are, respectively, complete sufficient statistics for

F_1 and F_2. Use Spearman's statistic (Example 2, Section 9.8) and at least two modifications of it to create an adaptive distribution-free test of H_0. *Hint.* Instead of ranks, use normal and median scores (Section 9.8) to obtain two additional correlation coefficients. The one associated with the median scores is frequently called the *quadrant test*.

11.7 Robust Estimation

In Examples 2 and 4, Section 6.1, the maximum likelihood estimator $\hat{\mu} = \bar{X}$ of the mean μ of the normal distribution $n(\mu, \sigma^2)$ was found by minimizing a certain sum of squares,

$$\sum_{i=1}^{n} (x_i - \mu)^2.$$

Also, in the regression problem of Section 8.6, the maximum likelihood estimators $\hat{\alpha}$ and $\hat{\beta}$ of the α and β in the mean $\alpha + \beta(c_i - \bar{c})$ were determined by minimizing the sum of squares

$$\sum_{i=1}^{n} [x_i - \alpha - \beta(c_i - \bar{c})]^2.$$

Both of these procedures come under the general heading of the *method of least squares*, because in each case a sum of squares is minimized. More generally, in the estimation of means of normal distributions, the method of least squares or some generalization of it is always used. The problems in the analyses of variance found in Chapter 8 are good illustrations of this fact. Hence, in this sense, normal assumptions and the method of least squares are mathematical companions.

It is interesting to note what procedures are obtained if we consider distributions that have longer tails than those of a normal distribution. For illustration, in Exercise 6.1(d), Section 6.1, the sample arises from a double exponential distribution with p.d.f.

$$f(x; \theta) = \tfrac{1}{2} e^{-|x-\theta|}, \qquad -\infty < x < \infty,$$

$$= 0 \text{ elsewhere,}$$

where $-\infty < \theta < \infty$. The maximum likelihood estimator, $\hat{\theta} =$ median (X_i), is found by minimizing the sum of absolute values,

$$\sum_{i=1}^{n} |x_i - \theta|,$$

and hence this is illustrative of the *method of least absolute values*.

Possibly a more extreme case is the determination of the maximum likelihood estimator of the center θ of the Cauchy distribution with p.d.f.

$$f(x; \theta) = \frac{1}{\pi[1 + (x - \theta)^2]}, \quad -\infty < x < \infty,$$

where $-\infty < \theta < \infty$. The logarithm of the likelihood function of a random sample $X_1, X_2, \ldots, X_n$ from this distribution is

$$\ln L(\theta) = -n \ln \pi - \sum_{i=1}^{n} \ln [1 + (x_i - \theta)^2].$$

To maximize, we differentiate $\ln L(\theta)$ to obtain

$$\frac{d \ln L(\theta)}{d\theta} = \sum_{i=1}^{n} \frac{2(x_i - \theta)}{1 + (x_i - \theta)^2} = 0.$$

The solution of this equation cannot be found in closed form, but the equation can be solved by some iterative process (for example, Newton's method), of course checking that the approximate solution actually provides the maximum of $L(\theta)$, approximately.

The generalization of these three special cases is described as follows. Let $X_1, X_2, \ldots, X_n$ be a random sample from a distribution with a p.d.f. of the form $f(x - \theta)$, where θ is a location parameter such that $-\infty < \theta < \infty$. Thus

$$\ln L(\theta) = \sum_{i=1}^{n} \ln f(x_i - \theta) = - \sum_{i=1}^{n} \rho(x_i - \theta),$$

where $\rho(x) = -\ln f(x)$, and

$$\frac{d \ln L(\theta)}{d\theta} = - \sum_{i=1}^{n} \frac{f'(x_i - \theta)}{f(x_i - \theta)} = \sum_{i=1}^{n} \Psi(x_i - \theta),$$

where $\rho'(x) = \Psi(x)$. For the normal, double exponential, and Cauchy distributions, we have that these respective functions are

$$\rho(x) = \frac{1}{2} \ln 2\pi + \frac{x^2}{2}; \quad \rho(x) = \ln 2 + |x|;$$

$$\rho(x) = \ln \pi + \ln (1 + x^2),$$

and

$$\Psi(x) = x; \quad \Psi(x) = -1, x < 0; \quad \Psi(x) = \frac{2x}{1 + x^2}$$

$$= 1, \quad 0 < x.$$

Clearly, these functions are very different from one distribution to another; and hence the respective maximum likelihood estimators may differ greatly. Thus we would suspect that the maximum likelihood estimator associated with one distribution would not necessarily be a good estimator in another situation. This is true; for example, $\bar{X}$ is a very poor estimator of the median of a Cauchy distribution, as the variance of $\bar{X}$ does not even exist if the sample arises from a Cauchy distribution. Intuitively, $\bar{X}$ is not a good estimator with the Cauchy distribution, because the very small or very large values (outliers) that can arise from that distribution influence the mean $\bar{X}$ of the sample too much.

An estimator that is fairly good (small variance, say) for a wide variety of distributions (not necessarily the best for any one of them) is called a *robust* estimator. Also estimators associated with the solution of the equation

$$\sum_{i=1}^{n} \Psi(x_i - \theta) = 0$$

are frequently called *M-estimators* (denoted by $\hat{\theta}$) because they can be thought of as *m*aximum likelihood estimators. So in finding a robust *M*-estimator we must select a Ψ function which will provide an estimator that is good for each distribution in the collection under consideration. For certain theoretical reasons that we cannot explain at this level, Huber suggested a Ψ function that is a combination of those associated with the normal and double exponential distributions,

$$\Psi(x) = -k, \qquad x < -k$$
$$= x, \qquad -k \leq x \leq k,$$
$$= k, \qquad k < x.$$

In Exercise 11.25 the reader is asked to find the p.d.f. $f(x)$ so that the *M*-estimator associated with this Ψ function is the maximum likelihood estimator of the location parameter θ in the p.d.f. $f(x - \theta)$.

With Huber's Ψ function, another problem arises. Note that if we double (for illustration) each $X_1, X_2, \ldots, X_n$, estimators such as $\bar{X}$ and median (X_i) also double. This is not at all true with the solution of the equation

$$\sum_{i=1}^{n} \Psi(x_i - \theta) = 0,$$

where the Ψ function is that of Huber. One way to avoid this difficulty is to solve another, but similar, equation instead,

$$(1) \qquad \sum_{i=1}^{n} \Psi\left(\frac{x_i - \theta}{d}\right) = 0,$$

where d is a robust estimate of the scale. A popular d to use is

$$d = \text{median } |x_i - \text{median } (x_i)|/0.6745.$$

The divisor 0.6745 is inserted in the definition of d because then the expected value of the corresponding statistic D is about equal to σ, if the sample arises from a normal distribution. That is, σ can be approximated by d under normal assumptions.

That scheme of selecting d also provides us with a clue for selecting k. For if the sample actually arises from a normal distribution, we would want most of the items $x_1, x_2, \ldots, x_n$ to satisfy the inequality

$$\left| \frac{x_i - \theta}{d} \right| \leq k$$

because then

$$\Psi\left(\frac{x_i - \theta}{d} \right) = \frac{x_i - \theta}{d}.$$

That is, for illustration, if *all* the items satisfy this inequality, then Equation (1) becomes

$$\sum_{i=1}^{n} \Psi\left(\frac{x_i - \theta}{d} \right) = \sum_{i=1}^{n} \frac{x_i - \theta}{d} = 0.$$

This has the solution $\bar{x}$, which of course is most desirable with normal distributions. Since d approximates σ, popular values of k to use are 1.5 and 2.0, because with those selections most normal variables would satisfy the desired inequality.

Again an iterative process must usually be used to solve Equation (1). One such scheme, Newton's method, is described. Let $\tilde{\theta}_1$ be a first estimate of θ, such as $\tilde{\theta}_1 = \text{median} (x_i)$. Approximate the left-hand member of Equation (1) by the first two terms of Taylor's expansion about $\tilde{\theta}_1$ to obtain

$$\sum_{i=1}^{n} \Psi\left(\frac{x_i - \tilde{\theta}_1}{d} \right) + (\theta - \tilde{\theta}_1) \sum_{i=1}^{n} \Psi'\left(\frac{x_i - \tilde{\theta}_1}{d} \right)\left(-\frac{1}{d} \right) = 0,$$

approximately. The solution of this provides a second estimate of θ,

$$\tilde{\theta}_2 = \tilde{\theta}_1 + \frac{d \sum\limits_{i=1}^{n} \Psi\left(\dfrac{x_i - \tilde{\theta}_1}{d} \right)}{\sum\limits_{i=1}^{n} \Psi'\left(\dfrac{x_i - \tilde{\theta}_1}{d} \right)},$$

which is called the one-step M-estimate of θ. If we use $\tilde{\theta}_2$ in place of $\tilde{\theta}_1$, we obtain $\tilde{\theta}_3$, the two-step M-estimate of θ. This process can continue to obtain any desired degree of accuracy. With Huber's Ψ function, the denominator of the second term,

$$\sum_{i=1}^{n} \Psi'\left(\frac{x_i - \tilde{\theta}_1}{d} \right),$$

is particularly easy to compute because $\Psi'(x) = 1$, $-k \le x \le k$, and zero elsewhere. Thus that denominator simply counts the number of $x_1, x_2, \ldots, x_n$ such that $|x_i - \tilde{\theta}_1|/d \le k$.

Although beyond the scope of this text, it can be shown, under very general conditions with known $\sigma = 1$, that the limiting distribution of

$$\frac{\sqrt{n}(\hat{\theta} - \theta)}{\sqrt{E[\Psi^2(X - \theta)]/\{E[\Psi'(X - \theta)]\}^2}},$$

where $\hat{\theta}$ is the M-estimator associated with Ψ, is $n(0, 1)$. In applications, the denominator of this ratio can be approximated by the square root of

$$\frac{1}{n} \sum_{i=1}^{n} \Psi^2(x_i - \hat{\theta}) \bigg/ \left[\frac{1}{n} \sum_{i=1}^{n} \Psi'(x_i - \hat{\theta})\right]^2.$$

Moreover, after this substitution has been made, it has been discovered empirically that certain t-distributions approximate the distribution of the ratio better than does $n(0, 1)$.

These M-estimators can be extended to regression situations. In general, they give excellent protection against outliers and bad data points; yet these M-estimators perform almost as well as least-squares estimators if the underlying distributions are actually normal.

EXERCISES

11.24. Compute the one-step M-estimate $\tilde{\theta}_2$ using Huber's ψ with $k = 1.5$ if $n = 7$ and the seven observations are 2.1, 5.2, 2.3, 1.4, 2.2, 2.3, and 1.6. Here take $\tilde{\theta}_1 = 2.2$, the median of the sample. Compare $\tilde{\theta}_2$ with $\bar{x}$.

11.25. Let the p.d.f. $f(x)$ be such that the M-estimator associated with Huber's ψ function is a maximum likelihood estimator of the location parameter in $f(x - \theta)$. Show that $f(x)$ is of the form $ce^{-\rho_1(x)}$, where $\rho_1(x) = x^2/2$, $|x| \le k$ and $\rho_1(x) = k|x| - k^2/2$, $k < |x|$.

11.26. Plot the ψ functions associated with the normal, double exponential, and Cauchy distributions in addition to that of Huber. Why is the M-estimator associated with the ψ function of the Cauchy distribution called a *descending M-estimator*?

11.27. Use the data in Exercise 11.24 to find the one-step descending M-estimator $\tilde{\theta}_2$ associated with $\psi(x) = \sin(x/1.5)$, $|x| \le 1.5\pi$, zero elsewhere. This was first proposed by D. F. Andrews. Compare this to $\bar{x}$ and the one-step M-estimator of Exercise 11.24.

Chapter 12

Further Normal Distribution Theory

12.1 The Multivariate Normal Distribution

We have studied in some detail normal distributions of one and of two random variables. In this section, we shall investigate a joint distribution of n random variables that will be called a *multivariate normal* distribution. This investigation assumes that the student is familiar with elementary matrix algebra, with real symmetric quadratic forms, and with orthogonal transformations. Henceforth the expression quadratic form means a quadratic form in a prescribed number of variables whose matrix is real and symmetric. All symbols which represent matrices will be set in boldface type.

Let $\mathbf{A}$ denote an $n \times n$ real symmetric matrix which is positive definite. Let $\boldsymbol{\mu}$ denote the $n \times 1$ matrix such that $\boldsymbol{\mu}'$, the transpose of $\boldsymbol{\mu}$, is $\boldsymbol{\mu}' = [\mu_1, \mu_2, \ldots, \mu_n]$, where each μ_i is a real constant. Finally, let $\mathbf{x}$ denote the $n \times 1$ matrix such that $\mathbf{x}' = [x_1, x_2, \ldots, x_n]$. We shall show that if C is an appropriately chosen positive constant, the non-negative function

$$f(x_1, x_2, \ldots, x_n) = C \exp\left[-\frac{(\mathbf{x} - \boldsymbol{\mu})'\mathbf{A}(\mathbf{x} - \boldsymbol{\mu})}{2}\right],$$

$$-\infty < x_i < \infty, \, i = 1, 2, \ldots, n,$$

is a joint p.d.f. of n random variables $X_1, X_2, \ldots, X_n$ that are of the continuous type. Thus we need to show that

(1) $$\int_{-\infty}^{\infty} \cdots \int_{-\infty}^{\infty} f(x_1, x_2, \ldots, x_n) \, dx_1 \, dx_2 \cdots dx_n = 1.$$

Let $\mathbf{t}$ denote the $n \times 1$ matrix such that $\mathbf{t}' = [t_1, t_2, \ldots, t_n]$, where $t_1, t_2, \ldots, t_n$ are arbitrary real numbers. We shall evaluate the integral

$$(2) \quad C \int_{-\infty}^{\infty} \cdots \int_{-\infty}^{\infty} \exp\left[\mathbf{t}'\mathbf{x} - \frac{(\mathbf{x} - \boldsymbol{\mu})'\mathbf{A}(\mathbf{x} - \boldsymbol{\mu})}{2}\right] dx_1 \cdots dx_n,$$

and then we shall subsequently set $t_1 = t_2 = \cdots = t_n = 0$, and thus establish Equation (1). First we change the variables of integration in integral (2) from $x_1, x_2, \ldots, x_n$ to $y_1, y_2, \ldots, y_n$ by writing $\mathbf{x} - \boldsymbol{\mu} = \mathbf{y}$, where $\mathbf{y}' = [y_1, y_2, \ldots, y_n]$. The Jacobian of the transformation is one and the n-dimensional x-space is mapped onto an n-dimensional y-space, so that integral (2) may be written as

$$(3) \quad C \exp(\mathbf{t}'\boldsymbol{\mu}) \int_{-\infty}^{\infty} \cdots \int_{-\infty}^{\infty} \exp\left(\mathbf{t}'\mathbf{y} - \frac{\mathbf{y}'\mathbf{A}\mathbf{y}}{2}\right) dy_1 \cdots dy_n.$$

Because the real symmetric matrix $\mathbf{A}$ is positive definite, the n characteristic numbers (proper values, latent roots, or eigenvalues) $a_1, a_2, \ldots, a_n$ of $\mathbf{A}$ are positive. There exists an appropriately chosen $n \times n$ real orthogonal matrix $\mathbf{L}(\mathbf{L}' = \mathbf{L}^{-1}$, where $\mathbf{L}^{-1}$ is the inverse of $\mathbf{L}$) such that

$$\mathbf{L}'\mathbf{A}\mathbf{L} = \begin{bmatrix} a_1 & 0 & \cdots & 0 \\ 0 & a_2 & \cdots & 0 \\ \vdots & \vdots & & \vdots \\ 0 & 0 & \cdots & a_n \end{bmatrix},$$

for a suitable ordering of $a_1, a_2, \ldots, a_n$. We shall sometimes write $\mathbf{L}'\mathbf{A}\mathbf{L} = \text{diag}[a_1, a_2, \ldots, a_n]$. In integral (3), we shall change the variables of integration from $y_1, y_2, \ldots, y_n$ to $z_1, z_2, \ldots, z_n$ by writing $\mathbf{y} = \mathbf{L}\mathbf{z}$, where $\mathbf{z}' = [z_1, z_2, \ldots, z_n]$. The Jacobian of the transformation is the determinant of the orthogonal matrix $\mathbf{L}$. Since $\mathbf{L}'\mathbf{L} = \mathbf{I}_n$, where $\mathbf{I}_n$ is the unit matrix of order n, we have the determinant $|\mathbf{L}'\mathbf{L}| = 1$ and $|\mathbf{L}|^2 = 1$. Thus the absolute value of the Jacobian is one. Moreover, the n-dimensional y-space is mapped onto an n-dimensional z-space. The integral (3) becomes

$$(4) \quad C \exp(\mathbf{t}'\boldsymbol{\mu}) \int_{-\infty}^{\infty} \cdots \int_{-\infty}^{\infty} \exp\left[\mathbf{t}'\mathbf{L}\mathbf{z} - \frac{\mathbf{z}'(\mathbf{L}'\mathbf{A}\mathbf{L})\mathbf{z}}{2}\right] dz_1 \cdots dz_n.$$

It is computationally convenient to write, momentarily, $\mathbf{t}'\mathbf{L} = \mathbf{w}'$, where $\mathbf{w}' = [w_1, w_2, \ldots, w_n]$. Then

$$\exp[\mathbf{t}'\mathbf{L}\mathbf{z}] = \exp[\mathbf{w}'\mathbf{z}] = \exp\left(\sum_{1}^{n} w_i z_i\right).$$

Moreover,

$$\exp\left[-\frac{\mathbf{z}'(\mathbf{L}'\mathbf{AL})\mathbf{z}}{2}\right] = \exp\left[-\frac{\sum_1^n a_i z_i^2}{2}\right].$$

Then integral (4) may be written as the product of n integrals in the following manner:

$$(5) \quad C \exp(\mathbf{w}'\mathbf{L}'\boldsymbol{\mu}) \prod_{i=1}^n \left[\int_{-\infty}^{\infty} \exp\left(w_i z_i - \frac{a_i z_i^2}{2}\right) dz_i\right]$$

$$= C \exp(\mathbf{w}'\mathbf{L}'\boldsymbol{\mu}) \prod_{i=1}^n \left[\sqrt{\frac{2\pi}{a_i}} \int_{-\infty}^{\infty} \frac{\exp\left(w_i z_i - \frac{a_i z_i^2}{2}\right)}{\sqrt{2\pi/a_i}} dz_i\right].$$

The integral that involves z_i can be treated as the moment-generating function, with the more familiar symbol t replaced by w_i, of a distribution which is $n(0, 1/a_i)$. Thus the right-hand member of Equation (5) is equal to

$$(6) \quad C \exp(\mathbf{w}'\mathbf{L}'\boldsymbol{\mu}) \prod_{i=1}^n \left[\sqrt{\frac{2\pi}{a_i}} \exp\left(\frac{w_i^2}{2a_i}\right)\right]$$

$$= C \exp(\mathbf{w}'\mathbf{L}'\boldsymbol{\mu}) \sqrt{\frac{(2\pi)^n}{a_1 a_2 \cdots a_n}} \exp\left(\sum_1^n \frac{w_i^2}{2a_i}\right).$$

Now, because $\mathbf{L}^{-1} = \mathbf{L}'$, we have

$$(\mathbf{L}'\mathbf{AL})^{-1} = \mathbf{L}'\mathbf{A}^{-1}\mathbf{L} = \text{diag}\left[\frac{1}{a_1}, \frac{1}{a_2}, \ldots, \frac{1}{a_n}\right].$$

Thus,

$$\sum_1^n \frac{w_i^2}{a_i} = \mathbf{w}'(\mathbf{L}'\mathbf{A}^{-1}\mathbf{L})\mathbf{w} = (\mathbf{Lw})'\mathbf{A}^{-1}(\mathbf{Lw}) = \mathbf{t}'\mathbf{A}^{-1}\mathbf{t}.$$

Moreover, the determinant $|\mathbf{A}^{-1}|$ of $\mathbf{A}^{-1}$ is

$$|\mathbf{A}^{-1}| = |\mathbf{L}'\mathbf{A}^{-1}\mathbf{L}| = \frac{1}{a_1 a_2 \cdots a_n}.$$

Accordingly, the right-hand member of Equation (6), which is equal to integral (2), may be written as

$$(7) \quad C e^{\mathbf{t}'\boldsymbol{\mu}} \sqrt{(2\pi)^n |\mathbf{A}^{-1}|} \exp\left(\frac{\mathbf{t}'\mathbf{A}^{-1}\mathbf{t}}{2}\right).$$

If, in this function, we set $t_1 = t_2 = \cdots = t_n = 0$, we have the value of the left-hand member of Equation (1). Thus, we have

$$C\sqrt{(2\pi)^n |\mathbf{A}^{-1}|} = 1.$$

Accordingly, the function

$$f(x_1, x_2, \ldots, x_n) = \frac{1}{(2\pi)^{n/2}\sqrt{|\mathbf{A}^{-1}|}} \exp\left[-\frac{(\mathbf{x} - \boldsymbol{\mu})'\mathbf{A}(\mathbf{x} - \boldsymbol{\mu})}{2}\right],$$

$-\infty < x_i < \infty$, $i = 1, 2, \ldots, n$, is a joint p.d.f. of n random variables $X_1, X_2, \ldots, X_n$ that are of the continuous type. Such a p.d.f. is called a *nonsingular multivariate normal p.d.f.*

We have now proved that $f(x_1, x_2, \ldots, x_n)$ is a p.d.f. However, we have proved more than that. Because $f(x_1, x_2, \ldots, x_n)$ is a p.d.f., integral (2) is the moment-generating function $M(t_1, t_2, \ldots, t_n)$ of this joint distribution of probability. Since integral (2) is equal to function (7), the moment-generating function of the multivariate normal distribution is given by

$$M(t_1, t_2, \ldots, t_n) = \exp\left(\mathbf{t}'\boldsymbol{\mu} + \frac{\mathbf{t}'\mathbf{A}^{-1}\mathbf{t}}{2}\right).$$

Let the elements of the real, symmetric, and positive definite matrix $\mathbf{A}^{-1}$ be denoted by σ_{ij}, $i, j = 1, 2, \ldots, n$. Then

$$M(0, \ldots, 0, t_i, 0, \ldots, 0) = \exp\left(t_i\mu_i + \frac{\sigma_{ii}t_i^2}{2}\right)$$

is the moment-generating function of X_i, $i = 1, 2, \ldots, n$. Thus, X_i is $n(\mu_i, \sigma_{ii})$, $i = 1, 2, \ldots, n$. Moreover, with $i \neq j$, we see that $M(0, \ldots, 0, t_i, 0, \ldots, 0, t_j, 0, \ldots, 0)$, the moment-generating function of X_i and X_j, is equal to

$$\exp\left(t_i\mu_i + t_j\mu_j + \frac{\sigma_{ii}t_i^2 + 2\sigma_{ij}t_it_j + \sigma_{jj}t_j^2}{2}\right).$$

But this is the moment-generating function of a bivariate normal distribution, so that σ_{ij} is the covariance of the random variables X_i and X_j. Thus the matrix $\boldsymbol{\mu}$, where $\boldsymbol{\mu}' = [\mu_1, \mu_2, \ldots, \mu_n]$, is the matrix of the means of the random variables $X_1, \ldots, X_n$. Moreover, the elements on the principal diagonal of $\mathbf{A}^{-1}$ are, respectively, the variances $\sigma_{ii} = \sigma_i^2$, $i = 1, 2, \ldots, n$, and the elements not on the principal diagonal of $\mathbf{A}^{-1}$ are, respectively, the covariances $\sigma_{ij} = \rho_{ij}\sigma_i\sigma_j$, $i \neq j$, of the random variables $X_1, X_2, \ldots, X_n$. We call the matrix $\mathbf{A}^{-1}$, which is given by

$$\begin{bmatrix} \sigma_{11}, & \sigma_{12}, & \cdots, & \sigma_{1n} \\ \sigma_{12}, & \sigma_{22}, & \cdots, & \sigma_{2n} \\ \vdots & \vdots & & \vdots \\ \sigma_{1n}, & \sigma_{2n}, & \cdots, & \sigma_{nn} \end{bmatrix},$$

the *covariance matrix* of the multivariate normal distribution and henceforth we shall denote this matrix by the symbol $\mathbf{V}$. In terms of the positive definite covariance matrix $\mathbf{V}$, the multivariate normal p.d.f. is written

$$\frac{1}{(2\pi)^{n/2}\sqrt{|\mathbf{V}|}} \exp\left[-\frac{(\mathbf{x} - \boldsymbol{\mu})'\mathbf{V}^{-1}(\mathbf{x} - \boldsymbol{\mu})}{2}\right], \qquad -\infty < x_i < \infty,$$

$i = 1, 2, \ldots, n$, and the moment-generating function of this distribution is given by

$$\exp\left(\mathbf{t}'\boldsymbol{\mu} + \frac{\mathbf{t}'\mathbf{V}\mathbf{t}}{2}\right)$$

for all real values of $\mathbf{t}$.

Example 1. Let $X_1, X_2, \ldots, X_n$ have a multivariate normal distribution with matrix $\boldsymbol{\mu}$ of means and positive definite covariance matrix $\mathbf{V}$. If we let $\mathbf{X}' = [X_1, X_2, \ldots, X_n]$, then the moment-generating function $M(t_1, t_2, \ldots, t_n)$ of this joint distribution of probability is

$$(8) \qquad\qquad E(e^{\mathbf{t}'\mathbf{X}}) = \exp\left(\mathbf{t}'\boldsymbol{\mu} + \frac{\mathbf{t}'\mathbf{V}\mathbf{t}}{2}\right).$$

Consider a linear function Y of $X_1, X_2, \ldots, X_n$ which is defined by $Y = \mathbf{c}'\mathbf{X} = \sum_1^n c_i X_i$, where $\mathbf{c}' = [c_1, c_2, \ldots, c_n]$ and the several c_i are real and not all zero. We wish to find the p.d.f. of Y. The moment-generating function $M(t)$ of the distribution of Y is given by

$$M(t) = E(e^{tY}) = E(e^{t\mathbf{c}'\mathbf{X}}).$$

Now the expectation (8) exists for all real values of $\mathbf{t}$. Thus we can replace $\mathbf{t}'$ in expectation (8) by $t\mathbf{c}'$ and obtain

$$M(t) = \exp\left(t\mathbf{c}'\boldsymbol{\mu} + \frac{\mathbf{c}'\mathbf{V}\mathbf{c}t^2}{2}\right).$$

Thus the random variable Y is $n(\mathbf{c}'\boldsymbol{\mu}, \mathbf{c}'\mathbf{V}\mathbf{c})$.

EXERCISES

12.1. Let $X_1, X_2, \ldots, X_n$ have a multivariate normal distribution with positive definite covariance matrix $\mathbf{V}$. Prove that these random variables are mutually stochastically independent if and only if $\mathbf{V}$ is a diagonal matrix.

12.2. Let $n = 2$ and take

$$\mathbf{V} = \begin{bmatrix} \sigma_1^2 & \rho\sigma_1\sigma_2 \\ \rho\sigma_1\sigma_2 & \sigma_2^2 \end{bmatrix}.$$

Determine $|\mathbf{V}|$, $\mathbf{V}^{-1}$, and $(\mathbf{x} - \boldsymbol{\mu})'\mathbf{V}^{-1}(\mathbf{x} - \boldsymbol{\mu})$. Compare the bivariate normal p.d.f. with the multivariate normal p.d.f. when $n = 2$.

12.3. Let $X_1, X_2, \ldots, X_n$ have a multivariate normal distribution, where $\boldsymbol{\mu}$ is the matrix of the means and $\mathbf{V}$ is the positive definite covariance matrix. Let $Y = \mathbf{c}'\mathbf{X}$ and $Z = \mathbf{d}'\mathbf{X}$, where $\mathbf{X}' = [X_1, \ldots, X_n]$, $\mathbf{c}' = [c_1, \ldots, c_n]$, and $\mathbf{d}' = [d_1, \ldots, d_n]$ are real matrices. (a) Find $M(t_1, t_2) = E(e^{t_1 Y + t_2 Z})$ to see that Y and Z have a bivariate normal distribution. (b) Prove that Y and Z are stochastically independent if and only if $\mathbf{c}'\mathbf{V}\mathbf{d} = 0$. (c) If X_1, $X_2, \ldots, X_n$ are mutually stochastically independent random variables which have the same variance σ^2, show that the necessary and sufficient condition of part (b) becomes $\mathbf{c}'\mathbf{d} = 0$.

12.4. Let $\mathbf{X}' = [X_1, X_2, \ldots, X_n]$ have the multivariate normal distribution of Exercise 12.3. Consider the p linear functions of $X_1, \ldots, X_n$ defined by $\mathbf{W} = \mathbf{B}\mathbf{X}$, where $\mathbf{W}' = [W_1, \ldots, W_p]$, $p \leq n$, and $\mathbf{B}$ is a $p \times n$ real matrix of rank p. Find $M(v_1, \ldots, v_p) = E(e^{\mathbf{v}'\mathbf{W}})$, where $\mathbf{v}'$ is the real matrix $[v_1, \ldots, v_p]$, to see that $W_1, \ldots, W_p$ have a p-variate normal distribution which has $\mathbf{B}\boldsymbol{\mu}$ for the matrix of the means and $\mathbf{B}\mathbf{V}\mathbf{B}'$ for the covariance matrix.

12.5. Let $\mathbf{X}' = [X_1, X_2, \ldots, X_n]$ have the n-variate normal distribution of Exercise 12.3. Show that $X_1, X_2, \ldots, X_p$, $p < n$, have a p-variate normal distribution. What submatrix of $\mathbf{V}$ is the covariance matrix of $X_1, X_2, \ldots, X_p$? *Hint.* In the moment-generating function $M(t_1, t_2, \ldots, t_n)$ of $X_1, X_2, \ldots, X_n$, let $t_{p+1} = \cdots = t_n = 0$.

12.2 The Distributions of Certain Quadratic Forms

Let X_i, $i = 1, 2, \ldots, n$, denote mutually stochastically independent random variables which are $n(\mu_i, \sigma_i^2)$, $i = 1, 2, \ldots, n$, respectively. Then $Q = \sum_1^n (X_i - \mu_i)^2/\sigma_i^2$ is $\chi^2(n)$. Now Q is a quadratic form in the $X_i - \mu_i$ and Q is seen to be, apart from the coefficient $-\frac{1}{2}$, the random variable which is defined by the exponent on the number e in the joint p.d.f. of $X_1, X_2, \ldots, X_n$. We shall now show that this result can be generalized.

Let $X_1, X_2, \ldots, X_n$ have a multivariate normal distribution with p.d.f.

$$\frac{1}{(2\pi)^{n/2}\sqrt{|\mathbf{V}|}} \exp\left[-\frac{(\mathbf{x} - \boldsymbol{\mu})'\mathbf{V}^{-1}(\mathbf{x} - \boldsymbol{\mu})}{2} \right],$$

where, as usual, the covariance matrix $\mathbf{V}$ is positive definite. We shall show that the random variable Q (a quadratic form in the $X_i - \mu_i$), which is defined by $(\mathbf{x} - \boldsymbol{\mu})'\mathbf{V}^{-1}(\mathbf{x} - \boldsymbol{\mu})$, is $\chi^2(n)$. We have for the moment-generating function $M(t)$ of Q the integral

$$\int_{-\infty}^{\infty} \cdots \int_{-\infty}^{\infty} \frac{1}{(2\pi)^{n/2}\sqrt{|\mathbf{V}|}}$$

$$\times \exp\left[t(\mathbf{x} - \boldsymbol{\mu})'\mathbf{V}^{-1}(\mathbf{x} - \boldsymbol{\mu}) - \frac{(\mathbf{x} - \boldsymbol{\mu})'\mathbf{V}^{-1}(\mathbf{x} - \boldsymbol{\mu})}{2}\right] dx_1 \cdots dx_n$$

$$= \int_{-\infty}^{\infty} \cdots \int_{-\infty}^{\infty} \frac{1}{(2\pi)^{n/2}\sqrt{|\mathbf{V}|}}$$

$$\times \exp\left[-\frac{(\mathbf{x} - \boldsymbol{\mu})'\mathbf{V}^{-1}(\mathbf{x} - \boldsymbol{\mu})(1 - 2t)}{2}\right] dx_1 \cdots dx_n.$$

With $\mathbf{V}^{-1}$ positive definite, the integral is seen to exist for all real values of $t < \frac{1}{2}$. Moreover, $(1 - 2t)\mathbf{V}^{-1}$, $t < \frac{1}{2}$, is a positive definite matrix and, since $|(1 - 2t)\mathbf{V}^{-1}| = (1 - 2t)^n|\mathbf{V}^{-1}|$, it follows that

$$\frac{1}{(2\pi)^{n/2}\sqrt{|\mathbf{V}|/(1 - 2t)^n}} \exp\left[-\frac{(\mathbf{x} - \boldsymbol{\mu})'\mathbf{V}^{-1}(\mathbf{x} - \boldsymbol{\mu})(1 - 2t)}{2}\right]$$

can be treated as a multivariate normal p.d.f. If we multiply our integrand by $(1 - 2t)^{n/2}$, we have this multivariate p.d.f. Thus the moment-generating function of Q is given by

$$M(t) = \frac{1}{(1 - 2t)^{n/2}}, \qquad t < \frac{1}{2},$$

and Q is $(\chi^2 n)$, as we wished to show. This fact is the basis of the chi-square tests that were discussed in Chapter 8.

The remarkable fact that the random variable which is defined by $(\mathbf{x} - \boldsymbol{\mu})'\mathbf{V}^{-1}(\mathbf{x} - \boldsymbol{\mu})$ is $\chi^2(n)$ stimulates a number of questions about quadratic forms in normally distributed variables. We would like to treat this problem in complete generality, but limitations of space forbid this, and we find it necessary to restrict ourselves to some special cases.

Let $X_1, X_2, \ldots, X_n$ denote a random sample of size n from a distribution which is $n(0, \sigma^2)$, $\sigma^2 > 0$. Let $\mathbf{X}' = [X_1, X_2, \ldots, X_n]$ and let $\mathbf{A}$ denote an arbitrary $n \times n$ real symmetric matrix. We shall investigate the distribution of the quadratic form $\mathbf{X}'\mathbf{A}\mathbf{X}$. For instance, we know that $\mathbf{X}'\mathbf{I}_n\mathbf{X}/\sigma^2 = \mathbf{X}'\mathbf{X}/\sigma^2 = \sum_{1}^{n} X_i^2/\sigma^2$ is $\chi^2(n)$. First we shall find the moment-generating function of $\mathbf{X}'\mathbf{A}\mathbf{X}/\sigma^2$. Then we shall investigate

the conditions which must be imposed upon the real symmetric matrix
$\mathbf{A}$ if $\mathbf{X'AX}/\sigma^2$ is to have a chi-square distribution. This moment-
generating function is given by

$$M(t) = \int_{-\infty}^{\infty} \cdots \int_{-\infty}^{\infty} \left(\frac{1}{\sigma\sqrt{2\pi}}\right)^n \exp\left(\frac{t\mathbf{x'Ax}}{\sigma^2} - \frac{\mathbf{x'x}}{2\sigma^2}\right) dx_1 \cdots dx_n$$

$$= \int_{-\infty}^{\infty} \cdots \int_{-\infty}^{\infty} \left(\frac{1}{\sigma\sqrt{2\pi}}\right)^n \exp\left[-\frac{\mathbf{x'(I} - 2t\mathbf{A)x}}{2\sigma^2}\right] dx_1 \cdots dx_n,$$

where $\mathbf{I} = \mathbf{I}_n$. The matrix $\mathbf{I} - 2t\mathbf{A}$ is positive definite if we take $|t|$
sufficiently small, say $|t| < h$, $h > 0$. Moreover, we can treat

$$\frac{1}{(2\pi)^{n/2}\sqrt{|(\mathbf{I} - 2t\mathbf{A})^{-1}\sigma^2|}} \exp\left[-\frac{\mathbf{x'(I} - 2t\mathbf{A)x}}{2\sigma^2}\right]$$

as a multivariate normal p.d.f. Now $|(\mathbf{I} - 2t\mathbf{A})^{-1}\sigma^2|^{1/2} = \sigma^n/|\mathbf{I} - 2t\mathbf{A}|^{1/2}$. If we multiply our integrand by $|\mathbf{I} - 2t\mathbf{A}|^{1/2}$, we have
this multivariate p.d.f. Hence the moment-generating function of
$\mathbf{X'AX}/\sigma^2$ is given by

(1) $$M(t) = |\mathbf{I} - 2t\mathbf{A}|^{-1/2}, \qquad |t| < h.$$

It proves useful to express this moment-generating function in a
different form. To do this, let $a_1, a_2, \ldots, a_n$ denote the characteristic
numbers of $\mathbf{A}$ and let $\mathbf{L}$ denote an $n \times n$ orthogonal matrix such that
$\mathbf{L'AL} = \text{diag}[a_1, a_2, \ldots, a_n]$. Thus,

$$\mathbf{L'(I} - 2t\mathbf{A)L} = \begin{bmatrix} (1 - 2ta_1) & 0 & \cdots & 0 \\ 0 & (1 - 2ta_2) & \cdots & 0 \\ \vdots & \vdots & & \vdots \\ 0 & 0 & \cdots & (1 - 2ta_n) \end{bmatrix}.$$

Then

$$\prod_{i=1}^{n} (1 - 2ta_i) = |\mathbf{L'(I} - 2t\mathbf{A)L}| = |\mathbf{I} - 2t\mathbf{A}|.$$

Accordingly we can write $M(t)$, as given in Equation (1), in the form

2) $$M(t) = \left[\prod_{i=1}^{n} (1 - 2ta_i)\right]^{-1/2}, \qquad |t| < h.$$

Let r, $0 < r \leq n$, denote the rank of the real symmetric matrix $\mathbf{A}$.
Then exactly r of the real numbers $a_1, a_2, \ldots, a_n$, say $a_1, \ldots, a_r$, are
not zero and exactly $n - r$ of these numbers, say $a_{r+1}, \ldots, a_n$, are
zero. Thus we can write the moment-generating function of $\mathbf{X'AX}/\sigma^2$ as

$$M(t) = [(1 - 2ta_1)(1 - 2ta_2)\cdots(1 - 2ta_r)]^{-1/2}.$$

Now that we have found, in suitable form, the moment-generating
function of our random variable, let us turn to the question of the con-

ditions that must be imposed if $\mathbf{X'AX}/\sigma^2$ is to have a chi-square distribution. Assume that $\mathbf{X'AX}/\sigma^2$ is $\chi^2(k)$. Then

$$M(t) = [(1 - 2ta_1)(1 - 2ta_2) \cdots (1 - 2ta_r)]^{-1/2} = (1 - 2t)^{-k/2},$$

or, equivalently,

$$(1 - 2ta_1)(1 - 2ta_2) \cdots (1 - 2ta_r) = (1 - 2t)^k, \qquad |t| < h.$$

Because the positive integers r and k are the degrees of these polynomials, and because these polynomials are equal for infinitely many values of t, we have $k = r$, the rank of $\mathbf{A}$. Moreover, the uniqueness of the factorization of a polynomial implies that $a_1 = a_2 = \cdots = a_r = 1$. If each of the nonzero characteristic numbers of a real symmetric matrix is one, the matrix is idempotent, that is, $\mathbf{A}^2 = \mathbf{A}$, and conversely (see Exercise 12.7). Accordingly, if $\mathbf{X'AX}/\sigma^2$ has a chi-square distribution, then $\mathbf{A}^2 = \mathbf{A}$ and the random variable is $\chi^2(r)$, where r is the rank of $\mathbf{A}$. Conversely, if $\mathbf{A}$ is of rank r, $0 < r \leq n$, and if $\mathbf{A}^2 = \mathbf{A}$, then $\mathbf{A}$ has exactly r characteristic numbers that are equal to one, and the remaining $n - r$ characteristic numbers are equal to zero. Thus the moment-generating function of $\mathbf{X'AX}/\sigma^2$ is given by $(1 - 2t)^{-r/2}$, $t < \frac{1}{2}$, and $\mathbf{X'AX}/\sigma^2$ is $\chi^2(r)$. This establishes the following theorem.

Theorem 1. *Let Q denote a random variable which is a quadratic form in the items of a random sample of size n from a distribution which is $n(0, \sigma^2)$. Let $\mathbf{A}$ denote the symmetric matrix of Q and let r, $0 < r \leq n$, denote the rank of $\mathbf{A}$. Then Q/σ^2 is $\chi^2(r)$ if and only if $\mathbf{A}^2 = \mathbf{A}$.*

Remark. If the normal distribution in Theorem 1 is $n(\mu, \sigma^2)$, the condition $\mathbf{A}^2 = \mathbf{A}$ remains a necessary and sufficient condition that Q/σ^2 have a chi-square distribution. In general, however, Q/σ^2 is not $\chi^2(r)$ but, instead, Q/σ^2 has a noncentral chi-square distribution if $\mathbf{A}^2 = \mathbf{A}$. The number of degrees of freedom is r, the rank of $\mathbf{A}$, and the noncentrality parameter is $\mu'\mathbf{A}\mu/\sigma^2$, where $\mu' = [\mu, \mu, \ldots, \mu]$. Since $\mu'\mathbf{A}\mu = \mu^2 \sum_{i,j} a_{ij}$, where $\mathbf{A} = [a_{ij}]$, then, if $\mu \neq 0$, the conditions $\mathbf{A}^2 = \mathbf{A}$ and $\sum_{i,j} a_{ij} = 0$ are necessary and sufficient conditions that Q/σ^2 be central $\chi^2(r)$. Moreover, the theorem may be extended to a quadratic form in random variables which have a multivariate normal distribution with positive definite covariance matrix $\mathbf{V}$; here the necessary and sufficient condition that Q have a chi-square distribution is $\mathbf{AVA} = \mathbf{A}$.

EXERCISES

12.6. Let $Q = X_1 X_2 - X_3 X_4$, where X_1, X_2, X_3, X_4 is a random sample of size 4 from a distribution which is $n(0, \sigma^2)$. Show that Q/σ^2 does not have a chi-square distribution. Find the moment-generating function of Q/σ^2.

12.7. Let $\mathbf{A}$ be a real symmetric matrix. Prove that each of the nonzero characteristic numbers of $\mathbf{A}$ is equal to one if and only if $\mathbf{A}^2 = \mathbf{A}$. *Hint.* Let $\mathbf{L}$ be an orthogonal matrix such that $\mathbf{L'AL} = \text{diag}\,[a_1, a_2, \ldots, a_n]$ and note that $\mathbf{A}$ is idempotent if and only if $\mathbf{L'AL}$ is idempotent.

12.8. The sum of the elements on the principal diagonal of a square matrix $\mathbf{A}$ is called the trace of $\mathbf{A}$ and is denoted by $\text{tr}\,\mathbf{A}$. (a) If $\mathbf{B}$ is $n \times m$ and $\mathbf{C}$ is $m \times n$, prove that $\text{tr}\,(\mathbf{BC}) = \text{tr}\,(\mathbf{CB})$. (b) If $\mathbf{A}$ is a square matrix and if $\mathbf{L}$ is an orthogonal matrix, use the result of part (a) to show that $\text{tr}\,(\mathbf{L'AL}) = \text{tr}\,\mathbf{A}$. (c) If $\mathbf{A}$ is a real symmetric idempotent matrix, use the result of part (b) to prove that the rank of $\mathbf{A}$ is equal to $\text{tr}\,\mathbf{A}$.

12.9. Let $\mathbf{A} = [a_{ij}]$ be a real symmetric matrix. Prove that $\sum_j \sum_i a_{ij}^2$ is equal to the sum of the squares of the characteristic numbers of $\mathbf{A}$. *Hint.* If $\mathbf{L}$ is an orthogonal matrix, show that $\sum_j \sum_i a_{ij}^2 = \text{tr}\,(\mathbf{A}^2) = \text{tr}\,(\mathbf{L'A^2L}) = \text{tr}\,[(\mathbf{L'AL})(\mathbf{L'AL})]$.

12.10. Let $\overline{X}$ and S^2 denote, respectively, the mean and the variance of a random sample of size n from a distribution which is $n(0, \sigma^2)$. (a) If $\mathbf{A}$ denotes the symmetric matrix of $n\overline{X}^2$, show that $\mathbf{A} = (1/n)\mathbf{P}$, where $\mathbf{P}$ is the $n \times n$ matrix, each of whose elements is equal to one. (b) Demonstrate that $\mathbf{A}$ is idempotent and that the $\text{tr}\,\mathbf{A} = 1$. Thus $n\overline{X}^2/\sigma^2$ is $\chi^2(1)$. (c) Show that the symmetric matrix $\mathbf{B}$ of nS^2 is $\mathbf{I} - (1/n)\mathbf{P}$. (d) Demonstrate that $\mathbf{B}$ is idempotent and that $\text{tr}\,\mathbf{B} = n - 1$. Thus nS^2/σ^2 is $\chi^2(n - 1)$, as previously proved otherwise. (e) Show that the product matrix $\mathbf{AB}$ is the zero matrix.

12.3 The Independence of Certain Quadratic Forms

We have previously investigated the stochastic independence of linear functions of normally distributed variables (see Exercise 12.3). In this section we shall prove some theorems about the stochastic independence of quadratic forms. As we remarked on p. 411, we shall confine our attention to normally distributed variables that constitute a random sample of size n from a distribution that is $n(0, \sigma^2)$.

Let $X_1, X_2, \ldots, X_n$ denote a random sample of size n from a distribution which is $n(0, \sigma^2)$. Let $\mathbf{A}$ and $\mathbf{B}$ denote two real symmetric matrices, each of order n. Let $\mathbf{X'} = [X_1, X_2, \ldots, X_n]$ and consider the two quadratic forms $\mathbf{X'AX}$ and $\mathbf{X'BX}$. We wish to show that these quadratic forms are stochastically independent if and only if $\mathbf{AB} = \mathbf{0}$, the zero matrix. We shall first compute the moment-generating function $M(t_1, t_2)$ of the joint distribution of $\mathbf{X'AX}/\sigma^2$ and $\mathbf{X'BX}/\sigma^2$. We have

$$M(t_1, t_2) = \left(\frac{1}{\sigma\sqrt{2\pi}}\right)^n \int_{-\infty}^{\infty} \cdots \int_{-\infty}^{\infty}$$

$$\exp\left(\frac{t_1 \mathbf{x}'\mathbf{Ax}}{\sigma^2} + \frac{t_2 \mathbf{x}'\mathbf{Bx}}{\sigma^2} - \frac{\mathbf{x}'\mathbf{x}}{2\sigma^2}\right) dx_1 \cdots dx_n$$

$$= \left(\frac{1}{\sigma\sqrt{2\pi}}\right)^n \int_{-\infty}^{\infty} \cdots \int_{-\infty}^{\infty}$$

$$\exp\left(-\frac{\mathbf{x}'(\mathbf{I} - 2t_1\mathbf{A} - 2t_2\mathbf{B})\mathbf{x}}{2\sigma^2}\right) dx_1 \cdots dx_n.$$

The matrix $\mathbf{I} - 2t_1\mathbf{A} - 2t_2\mathbf{B}$ is positive definite if we take $|t_1|$ and $|t_2|$ sufficiently small, say $|t_1| < h_1$, $|t_2| < h_2$, where $h_1, h_2 > 0$. Then, as on p. 412, we have

$$M(t_1, t_2) = |\mathbf{I} - 2t_1\mathbf{A} - 2t_2\mathbf{B}|^{-1/2}, \qquad |t_1| < h_1, \ |t_2| < h_2.$$

Let us assume that $\mathbf{X}'\mathbf{AX}/\sigma^2$ and $\mathbf{X}'\mathbf{BX}/\sigma^2$ are stochastically independent (so that likewise are $\mathbf{X}'\mathbf{AX}$ and $\mathbf{X}'\mathbf{BX}$) and prove that $\mathbf{AB} = \mathbf{0}$. Thus we assume that

(1) $$M(t_1, t_2) = M(t_1, 0)M(0, t_2)$$

for all t_1 and t_2 for which $|t_i| < h_i$, $i = 1, 2$. Identity (1) is equivalent to the identity

(2) $$|\mathbf{I} - 2t_1\mathbf{A} - 2t_2\mathbf{B}| = |\mathbf{I} - 2t_1\mathbf{A}|\,|\mathbf{I} - 2t_2\mathbf{B}|, \qquad |t_i| < h_i, \ i = 1, 2.$$

Let $r > 0$ denote the rank of $\mathbf{A}$ and let $a_1, a_2, \ldots, a_r$ denote the r nonzero characteristic numbers of $\mathbf{A}$. There exists an orthogonal matrix $\mathbf{L}$ such that

$$\mathbf{L}'\mathbf{AL} = \begin{bmatrix} a_1 & 0 & \cdots & 0 & \vdots \\ 0 & a_2 & \cdots & 0 & \vdots & 0 \\ \vdots & \vdots & & \vdots & \vdots \\ 0 & 0 & \cdots & a_r & \vdots \\ \hline & & \mathbf{0} & & \vdots & 0 \end{bmatrix} = \begin{bmatrix} \mathbf{C}_{11} & \vdots & \mathbf{0} \\ \hline \mathbf{0} & \vdots & \mathbf{0} \end{bmatrix} = \mathbf{C}$$

for a suitable ordering of $a_1, a_2, \ldots, a_r$. Then $\mathbf{L}'\mathbf{BL}$ may be written in the identically partitioned form

$$\mathbf{L}'\mathbf{BL} = \begin{bmatrix} \mathbf{D}_{11} & \vdots & \mathbf{D}_{12} \\ \hline \mathbf{D}_{21} & \vdots & \mathbf{D}_{22} \end{bmatrix} = \mathbf{D}.$$

The identity (2) may be written as

(2') $$|\mathbf{L}'|\,|\mathbf{I} - 2t_1\mathbf{A} - 2t_2\mathbf{B}|\,|\mathbf{L}| = |\mathbf{L}'|\,|\mathbf{I} - 2t_1\mathbf{A}|\,|\mathbf{L}|\,|\mathbf{L}'|\,|\mathbf{I} - 2t_2\mathbf{B}|\,|\mathbf{L}|,$$

or as

(3) $$|\mathbf{I} - 2t_1\mathbf{C} - 2t_2\mathbf{D}| = |\mathbf{I} - 2t_1\mathbf{C}|\,|\mathbf{I} - 2t_2\mathbf{D}|.$$

The coefficient of $(-2t_1)^r$ in the right-hand member of Equation (3) is seen by inspection to be $a_1a_2\cdots a_r|\mathbf{I} - 2t_2\mathbf{D}|$. It is not so easy to find the coefficient of $(-2t_1)^r$ in the left-hand member of Equation (3). Conceive of expanding this determinant in terms of minors of order r formed from the first r columns. One term in this expansion is the product of the minor of order r in the upper left-hand corner, namely, $|\mathbf{I}_r - 2t_1\mathbf{C}_{11} - 2t_2\mathbf{D}_{11}|$, and the minor of order $n - r$ in the lower right-hand corner, namely, $|\mathbf{I}_{n-r} - 2t_2\mathbf{D}_{22}|$. Moreover, this product is the only term in the expansion of the determinant that involves $(-2t_1)^r$. Thus the coefficient of $(-2t_1)^r$ in the left-hand member of Equation (3) is $a_1a_2\cdots a_r|\mathbf{I}_{n-r} - 2t_2\mathbf{D}_{22}|$. If we equate these coefficients of $(-2t_1)^r$, we have, for all t_2, $|t_2| < h_2$,

(4) $$|\mathbf{I} - 2t_2\mathbf{D}| = |\mathbf{I}_{n-r} - 2t_2\mathbf{D}_{22}|.$$

Equation (4) implies that the nonzero characteristic numbers of the matrices $\mathbf{D}$ and $\mathbf{D}_{22}$ are the same (see Exercise 12.17). Recall that the sum of the squares of the characteristic numbers of a symmetric matrix is equal to the sum of the squares of the elements of that matrix (see Exercise 12.9). Thus the sum of the squares of the elements of matrix $\mathbf{D}$ is equal to the sum of the squares of the elements of $\mathbf{D}_{22}$. Since the elements of the matrix $\mathbf{D}$ are real, it follows that each of the elements of $\mathbf{D}_{11}$, $\mathbf{D}_{12}$, and $\mathbf{D}_{21}$ is zero. Accordingly, we can write $\mathbf{D}$ in the form

$$\mathbf{D} = \mathbf{L'BL} = \begin{bmatrix} 0 & 0 \\ \hline 0 & \mathbf{D}_{22} \end{bmatrix}.$$

Thus $\mathbf{CD} = \mathbf{L'ALL'BL} = 0$ and $\mathbf{L'ABL} = 0$ and $\mathbf{AB} = 0$, as we wished to prove. To complete the proof of the theorem, we assume that $\mathbf{AB} = 0$. We are to show that $\mathbf{X'AX}/\sigma^2$ and $\mathbf{X'BX}/\sigma^2$ are stochastically independent. We have, for all real values of t_1 and t_2,

$$(\mathbf{I} - 2t_1\mathbf{A})(\mathbf{I} - 2t_2\mathbf{B}) = \mathbf{I} - 2t_1\mathbf{A} - 2t_2\mathbf{B},$$

since $\mathbf{AB} = 0$. Thus,

$$|\mathbf{I} - 2t_1\mathbf{A} - 2t_2\mathbf{B}| = |\mathbf{I} - 2t_1\mathbf{A}|\,|\mathbf{I} - 2t_2\mathbf{B}|.$$

Since the moment-generating function of the joint distribution of $\mathbf{X'AX}/\sigma^2$ and $\mathbf{X'BX}/\sigma^2$ is given by

$$M(t_1, t_2) = |\mathbf{I} - 2t_1\mathbf{A} - 2t_2\mathbf{B}|^{-1/2}, \qquad |t_i| < h_i, \ i = 1, 2,$$

we have

$$M(t_1, t_2) = M(t_1, 0)M(0, t_2),$$

and the proof of the following theorem is complete.

Theorem 2. *Let Q_1 and Q_2 denote random variables which are quadratic forms in the items of a random sample of size n from a distribution which is $n(0, \sigma^2)$. Let $\mathbf{A}$ and $\mathbf{B}$ denote respectively the real symmetric matrices of Q_1 and Q_2. The random variables Q_1 and Q_2 are stochastically independent if and only if $\mathbf{AB} = \mathbf{0}$.*

Remark. Theorem 2 remains valid if the random sample is from a distribution which is $n(\mu, \sigma^2)$, whatever be the real value of μ. Moreover, Theorem 2 may be extended to quadratic forms in random variables that have a joint multivariate normal distribution with a positive definite covariance matrix $\mathbf{V}$. The necessary and sufficient condition for the stochastic independence of two such quadratic forms with symmetric matrices $\mathbf{A}$ and $\mathbf{B}$ then becomes $\mathbf{AVB} = \mathbf{0}$. In our Theorem 2, we have $\mathbf{V} = \sigma^2\mathbf{I}$, so that $\mathbf{AVB} = \mathbf{A}\sigma^2\mathbf{IB} = \sigma^2\mathbf{AB} = \mathbf{0}$.

We shall next prove the theorem that was used in Chapter 8 (p. 279).

Theorem 3. *Let $Q = Q_1 + \cdots + Q_{k-1} + Q_k$, where Q, $Q_1, \ldots,$ Q_{k-1}, Q_k are $k + 1$ random variables that are quadratic forms in the items of a random sample of size n from a distribution which is $n(0, \sigma^2)$. Let Q/σ^2 be $\chi^2(r)$, let Q_i/σ^2 be $\chi^2(r_i)$, $i = 1, 2, \ldots, k - 1$, and let Q_k be nonnegative. Then the random variables $Q_1, Q_2, \ldots, Q_k$ are mutually stochastically independent and, hence, Q_k/σ^2 is $\chi^2(r_k = r - r_1 - \cdots - r_{k-1})$.*

Proof. Take first the case of $k = 2$ and let the real symmetric matrices of Q, Q_1, and Q_2 be denoted, respectively, by $\mathbf{A}$, $\mathbf{A}_1$, $\mathbf{A}_2$. We are given that $Q = Q_1 + Q_2$ or, equivalently, that $\mathbf{A} = \mathbf{A}_1 + \mathbf{A}_2$. We are also given that Q/σ^2 is $\chi^2(r)$ and that Q_1/σ^2 is $\chi^2(r_1)$. In accordance with Theorem 1, p. 413, we have $\mathbf{A}^2 = \mathbf{A}$ and $\mathbf{A}_1^2 = \mathbf{A}_1$. Since $Q_2 \geq 0$, each of the matrices $\mathbf{A}$, $\mathbf{A}_1$, and $\mathbf{A}_2$ is positive semidefinite. Because $\mathbf{A}^2 = \mathbf{A}$, we can find an orthogonal matrix $\mathbf{L}$ such that

$$\mathbf{L'AL} = \begin{bmatrix} \mathbf{I}_r & \vdots & \mathbf{0} \\ \cdots & \cdots & \cdots \\ \mathbf{0} & \vdots & \mathbf{0} \end{bmatrix}.$$

If then we multiply both members of $\mathbf{A} = \mathbf{A}_1 + \mathbf{A}_2$ on the left by $\mathbf{L'}$ and on the right by $\mathbf{L}$, we have

$$\begin{bmatrix} \mathbf{I}_r & \vdots & \mathbf{0} \\ \cdots & \cdots & \cdots \\ \mathbf{0} & \vdots & \mathbf{0} \end{bmatrix} = \mathbf{L'A}_1\mathbf{L} + \mathbf{L'A}_2\mathbf{L}.$$

Now each of $\mathbf{A}_1$ and $\mathbf{A}_2$, and hence each of $\mathbf{L}'\mathbf{A}_1\mathbf{L}$ and $\mathbf{L}'\mathbf{A}_2\mathbf{L}$ is positive semidefinite. Recall that, if a real symmetric matrix is positive semidefinite, each element on the principal diagonal is positive or zero. Moreover, if an element on the principal diagonal is zero, then all elements in that row and all elements in that column are zero. Thus $\mathbf{L}'\mathbf{A}\mathbf{L} = \mathbf{L}'\mathbf{A}_1\mathbf{L} + \mathbf{L}'\mathbf{A}_2\mathbf{L}$ can be written as

$$
(5) \qquad
\left[\begin{array}{c|c} \mathbf{I}_r & \mathbf{0} \\ \hline \mathbf{0} & \mathbf{0} \end{array}\right]
=
\left[\begin{array}{c|c} \mathbf{G}_r & \mathbf{0} \\ \hline \mathbf{0} & \mathbf{0} \end{array}\right]
+
\left[\begin{array}{c|c} \mathbf{H}_r & \mathbf{0} \\ \hline \mathbf{0} & \mathbf{0} \end{array}\right].
$$

Since $\mathbf{A}_1^2 = \mathbf{A}_1$, we have

$$
(\mathbf{L}'\mathbf{A}_1\mathbf{L})^2 = \mathbf{L}'\mathbf{A}_1\mathbf{L} =
\left[\begin{array}{c|c} \mathbf{G}_r & \mathbf{0} \\ \hline \mathbf{0} & \mathbf{0} \end{array}\right].
$$

If we multiply both members of Equation (5) on the left by the matrix $\mathbf{L}'\mathbf{A}_1\mathbf{L}$, we see that

$$
\left[\begin{array}{c|c} \mathbf{G}_r & \mathbf{0} \\ \hline \mathbf{0} & \mathbf{0} \end{array}\right]
=
\left[\begin{array}{c|c} \mathbf{G}_r & \mathbf{0} \\ \hline \mathbf{0} & \mathbf{0} \end{array}\right]
+
\left[\begin{array}{c|c} \mathbf{G}_r\mathbf{H}_r & \mathbf{0} \\ \hline \mathbf{0} & \mathbf{0} \end{array}\right],
$$

or, equivalently, $\mathbf{L}'\mathbf{A}_1\mathbf{L} = \mathbf{L}'\mathbf{A}_1\mathbf{L} + (\mathbf{L}'\mathbf{A}_1\mathbf{L})(\mathbf{L}'\mathbf{A}_2\mathbf{L})$. Thus, $(\mathbf{L}'\mathbf{A}_1\mathbf{L}) \times (\mathbf{L}'\mathbf{A}_2\mathbf{L}) = \mathbf{0}$ and $\mathbf{A}_1\mathbf{A}_2 = \mathbf{0}$. In accordance with Theorem 2, Q_1 and Q_2 are stochastically independent. This stochastic independence immediately implies that Q_2/σ^2 is $\chi^2(r_2 = r - r_1)$. This completes the proof when $k = 2$. For $k > 2$, the proof may be made by induction. We shall merely indicate how this can be done by using $k = 3$. Take $\mathbf{A} = \mathbf{A}_1 + \mathbf{A}_2 + \mathbf{A}_3$, where $\mathbf{A}^2 = \mathbf{A}$, $\mathbf{A}_1^2 = \mathbf{A}_1$, $\mathbf{A}_2^2 = \mathbf{A}_2$ and $\mathbf{A}_3$ is positive semidefinite. Write $\mathbf{A} = \mathbf{A}_1 + (\mathbf{A}_2 + \mathbf{A}_3) = \mathbf{A}_1 + \mathbf{B}_1$, say. Now $\mathbf{A}^2 = \mathbf{A}$, $\mathbf{A}_1^2 = \mathbf{A}_1$, and $\mathbf{B}_1$ is positive semidefinite. In accordance with the case of $k = 2$, we have $\mathbf{A}_1\mathbf{B}_1 = \mathbf{0}$, so that $\mathbf{B}_1^2 = \mathbf{B}_1$. With $\mathbf{B}_1 = \mathbf{A}_2 + \mathbf{A}_3$, where $\mathbf{B}_1^2 = \mathbf{B}_1$, $\mathbf{A}_2^2 = \mathbf{A}_2$, it follows from the case of $k = 2$ that $\mathbf{A}_2\mathbf{A}_3 = \mathbf{0}$ and $\mathbf{A}_3^2 = \mathbf{A}_3$. If we regroup by writing $\mathbf{A} = \mathbf{A}_2 + (\mathbf{A}_1 + \mathbf{A}_3)$, we obtain $\mathbf{A}_1\mathbf{A}_3 = \mathbf{0}$, and so on.

Remark. In our statement of Theorem 3 we took $X_1, X_2, \ldots, X_n$ to be items of a random sample from a distribution which is $n(0, \sigma^2)$. We did this because our proof of Theorem 2 was restricted to that case. In fact, if $Q', Q_1', \ldots, Q_k'$ are quadratic forms in any normal variables (including multivariate normal variables), if $Q' = Q_1' + \cdots + Q_k'$, if $Q', Q_1', \ldots, Q_{k-1}'$ are central or noncentral chi-square, and if Q_k' is nonnegative, then $Q_1', \ldots, Q_k'$

are mutually stochastically independent and Q'_k is either central or noncentral chi-square.

This section will conclude with a proof of a frequently quoted theorem due to Cochran.

Theorem 4. *Let* $X_1, X_2, \ldots, X_n$ *denote a random sample from a distribution which is* $n(0, \sigma^2)$. *Let the sum of the squares of these items be written in the form*

$$\sum_1^n X_i^2 = Q_1 + Q_2 + \cdots + Q_k,$$

where Q_j *is a quadratic form in* $X_1, X_2, \ldots, X_n$, *with matrix* $\mathbf{A}_j$ *which has rank* r_j, $j = 1, 2, \ldots, k$. *The random variables* $Q_1, Q_2, \ldots, Q_k$ *are mutually stochastically independent and* Q_j/σ^2 *is* $\chi^2(r_j)$, $j = 1, 2, \ldots, k$, *if and only if* $\sum_1^k r_j = n$.

Proof. First assume the two conditions $\sum_1^k r_j = n$ and $\sum_1^n X_i^2 = \sum_1^k Q_j$ to be satisfied. The latter equation implies that $\mathbf{I} = \mathbf{A}_1 + \mathbf{A}_2 + \cdots + \mathbf{A}_k$. Let $\mathbf{B}_i = \mathbf{I} - \mathbf{A}_i$. That is, $\mathbf{B}_i$ is the sum of the matrices $\mathbf{A}_1, \ldots, \mathbf{A}_k$ exclusive of $\mathbf{A}_i$. Let R_i denote the rank of $\mathbf{B}_i$. Since the rank of the sum of several matrices is less than or equal to the sum of the ranks, we have $R_i \leq \sum_1^k r_j - r_i = n - r_i$. However, $\mathbf{I} = \mathbf{A}_i + \mathbf{B}_i$, so that $n \leq r_i + R_i$ and $n - r_i \leq R_i$. Hence $R_i = n - r_i$. The characteristic numbers of $\mathbf{B}_i$ are the roots of the equation $|\mathbf{B}_i - \lambda\mathbf{I}| = 0$. Since $\mathbf{B}_i = \mathbf{I} - \mathbf{A}_i$, this equation can be written as $|\mathbf{I} - \mathbf{A}_i - \lambda\mathbf{I}| = 0$. Thus, we have $|\mathbf{A}_i - (1 - \lambda)\mathbf{I}| = 0$. But each root of the last equation is one minus a characteristic number of $\mathbf{A}_i$. Since $\mathbf{B}_i$ has exactly $n - R_i = r_i$ characteristic numbers that are zero, then $\mathbf{A}_i$ has exactly r_i characteristic numbers that are equal to one. However, r_i is the rank of $\mathbf{A}_i$. Thus, each of the r_i nonzero characteristic numbers of $\mathbf{A}_i$ is one. That is, $\mathbf{A}_i^2 = \mathbf{A}_i$ and thus Q_i/σ^2 is $\chi^2(r_i)$, $i = 1, 2, \ldots, k$. In accordance with Theorem 3, the random variables $Q_1, Q_2, \ldots, Q_k$ are mutually stochastically independent.

To complete the proof of Theorem 4, take $\sum_1^n X_i^2 = Q_1 + Q_2 + \cdots + Q_k$, let $Q_1, Q_2, \ldots, Q_k$ be mutually stochastically independent, and let Q_j/σ^2 be $\chi^2(r_j)$, $j = 1, 2, \ldots, k$. Then $\sum_1^k Q_j/\sigma^2$ is $\chi^2\left(\sum_1^k r_j\right)$. But $\sum_1^k Q_j/\sigma^2 = \sum_1^n X_i^2/\sigma^2$ is $\chi^2(n)$. Thus, $\sum_1^k r_j = n$ and the proof is complete.

EXERCISES

12.11. Let $X_1, X_2, \ldots, X_n$ denote a random sample of size n from a distribution which is $n(0, \sigma^2)$. Prove that $\sum_1^n X_i^2$ and every quadratic form, which is nonidentically zero in $X_1, X_2, \ldots, X_n$, are stochastically dependent.

12.12 Let X_1, X_2, X_3, X_4 denote a random sample of size 4 from a distribution which is $n(0, \sigma^2)$. Let $Y = \sum_1^4 a_i X_i$, where a_1, a_2, a_3, and a_4 are real constants. If Y^2 and $Q = X_1 X_2 - X_3 X_4$ are stochastically independent, determine a_1, a_2, a_3, and a_4.

12.13. Let $\mathbf{A}$ be the real symmetric matrix of a quadratic form Q in the items of a random sample of size n from a distribution which is $n(0, \sigma^2)$. Given that Q and the mean $\overline{X}$ of the sample are stochastically independent. What can be said of the elements of each row (column) of $\mathbf{A}$? *Hint.* Are Q and $\overline{X}^2$ stochastically independent?

12.14. Let $\mathbf{A}_1, \mathbf{A}_2, \ldots, \mathbf{A}_k$ be the matrices of $k > 2$ quadratic forms $Q_1, Q_2, \ldots, Q_k$ in the items of a random sample of size n from a distribution which is $n(0, \sigma^2)$. Prove that the pairwise stochastic independence of these forms implies that they are mutually stochastically independent. *Hint.* Show that $\mathbf{A}_i \mathbf{A}_j = \mathbf{0}$, $i \neq j$, permits $E[\exp(t_1 Q_1 + t_2 Q_2 + \cdots + t_k Q_k)]$ to be written as a product of the moment-generating functions of $Q_1, Q_2, \ldots, Q_k$.

12.15. Let $\mathbf{X}' = [X_1, X_2, \ldots, X_n]$, where $X_1, X_2, \ldots, X_n$ are items of a random sample from a distribution which is $n(0, \sigma^2)$. Let $\mathbf{b}' = [b_1, b_2, \ldots, b_n]$ be a real nonzero matrix, and let $\mathbf{A}$ be a real symmetric matrix of order n. Prove that the linear form $\mathbf{b}'\mathbf{X}$ and the quadratic form $\mathbf{X}'\mathbf{A}\mathbf{X}$ are stochastically independent if and only if $\mathbf{b}'\mathbf{A} = \mathbf{0}$. Use this fact to prove that $\mathbf{b}'\mathbf{X}$ and $\mathbf{X}'\mathbf{A}\mathbf{X}$ are stochastically independent if and only if the two quadratic forms, $(\mathbf{b}'\mathbf{X})^2 = \mathbf{X}'\mathbf{b}\mathbf{b}'\mathbf{X}$ and $\mathbf{X}'\mathbf{A}\mathbf{X}$, are stochastically independent.

12.16. Let Q_1 and Q_2 be two nonnegative quadratic forms in the items of a random sample from a distribution which is $n(0, \sigma^2)$. Show that another quadratic form Q is stochastically independent of $Q_1 + Q_2$ if and only if Q is stochastically independent of each of Q_1 and Q_2. *Hint.* Consider the orthogonal transformation that diagonalizes the matrix of $Q_1 + Q_2$. After this transformation, what are the forms of the matrices of Q, Q_1, and Q_2 if Q and $Q_1 + Q_2$ are stochastically independent?

12.17. Prove that Equation (4) of this section implies that the nonzero characteristic numbers of the matrices $\mathbf{D}$ and $\mathbf{D}_{22}$ are the same. *Hint.* Let $\lambda = 1/(2t_2)$, $t_2 \neq 0$, and show that Equation (4) is equivalent to $|\mathbf{D} - \lambda\mathbf{I}| = (-\lambda)^r |\mathbf{D}_{22} - \lambda\mathbf{I}_{n-r}|$.

Appendix A

References

[1] Anderson, T. W., *An Introduction to Multivariate Statistical Analysis*, John Wiley & Sons, Inc., New York, 1958.

[2] Basu, D., "On Statistics Independent of a Complete Sufficient Statistic," *Sankhyā*, **15**, 377 (1955).

[3] Box, G. E. P., and Muller, M. A., "A Note on the Generation of Random Normal Deviates," *Ann. Math. Stat.*, **29**, 610 (1958).

[4] Carpenter, O., "Note on the Extension of Craig's Theorem to Noncentral Variates," *Ann. Math. Stat.*, **21**, 455 (1950).

[5] Cochran, W. G., "The Distribution of Quadratic Forms in a Normal System, with Applications to the Analysis of Covariance," *Proc. Cambridge Phil. Soc.*, **30**, 178 (1934).

[6] Craig, A. T., "Bilinear Forms in Normally Correlated Variables," *Ann. Math. Stat.*, **18**, 565 (1947).

[7] Craig, A. T., "Note on the Independence of Certain Quadratic Forms," *Ann. Math. Stat.*, **14**, 195 (1943).

[8] Cramér, H., *Mathematical Methods of Statistics*, Princeton University Press, Princeton, N.J., 1946.

[9] Curtiss, J. H. "A Note on the Theory of Moment Generating Functions," *Ann. Math. Stat.*, **13**, 430 (1942).

[10] Fisher, R. A., "On the Mathematical Foundation of Theoretical Statistics," *Phil. Trans. Royal Soc. London*, Series A, **222**, 309 (1921).

[11] Fraser, D. A. S., *Nonparametric Methods in Statistics*, John Wiley & Sons, Inc., New York, 1957.

[12] Graybill, F. A., *An Introduction to Linear Statistical Models*, Vol. 1, McGraw-Hill Book Company, New York, 1961.

[13] Hogg, R. V., "Adaptive Robust Procedures: A Partial Review and Some

Suggestions for Future Applications and Theory," *J. Amer. Stat. Assoc.*, **69**, 909 (1974).

[14] Hogg, R. V., and Craig, A. T., "On the Decomposition of Certain Chi-Square Variables," *Ann. Math. Stat.*, **29**, 608 (1958).

[15] Hogg, R. V., and Craig, A. T., "Sufficient Statistics in Elementary Distribution Theory," *Sankhyā*, **17**, 209 (1956).

[16] Huber, P., "Robust Statistics: A Review," *Ann. Math. Stat.*, **43**, 1041 (1972).

[17] Johnson, N. L., and Kotz, S., *Continuous Univariate Distributions*, Vols. 1 and 2, Houghton Mifflin Company, Boston, 1970.

[18] Koopman, B. O., "On Distributions Admitting a Sufficient Statistic," *Trans. Amer. Math. Soc.*, **39**, 399 (1936).

[19] Lancaster, H. O., "Traces and Cumulants of Quadratic Forms in Normal Variables," *J. Royal Stat. Soc.*, Series B, **16**, 247 (1954).

[20] Lehmann, E. L., *Testing Statistical Hypotheses*, John Wiley & Sons, Inc., New York, 1959.

[21] Lehmann, E. L., and Scheffé, H., "Completeness, Similar Regions, and Unbiased Estimation," *Sankhyā*, **10**, 305 (1950).

[22] Lévy, P., *Théorie de l'addition des variables aléatoires*, Gauthier-Villars, Paris, 1937.

[23] Mann, H. B., and Whitney, D. R., "On a Test of Whether One of Two Random Variables Is Stochastically Larger Than the Other," *Ann. Math. Stat.*, **18**, 50 (1947).

[24] Neymann, J., "Su un teorema concernente le cosiddette statistiche sufficienti," *Giornale dell' Istituto degli Attuari*, **6**, 320 (1935).

[25] Neyman, J., and Pearson, E. S., "On the Problem of the Most Efficient Tests of Statistical Hypotheses," *Phil. Trans. Royal Soc. London*, Series A, **231**, 289 (1933).

[26] Pearson, K., "On the Criterion That a Given System of Deviations from the Probable in the Case of a Correlated System of Variables Is Such That It Can Be Reasonably Supposed to Have Arisen from Random Sampling," *Phil. Mag.*, Series 5, **50**, 157 (1900).

[27] Pitman, E. J. G., "Sufficient Statistics and Intrinsic Accuracy," *Proc. Cambridge Phil. Soc.*, **32**, 567 (1936).

[28] Rao, C. R., *Linear Statistical Inference and Its Applications*, John Wiley & Sons, Inc., New York, 1965.

[29] Scheffé, H., *The Analysis of Variance*, John Wiley & Sons, Inc., New York, 1959.

[30] Wald, A., *Sequential Analysis*, John Wiley & Sons, Inc., New York, 1947.

[31] Wilcoxon, F., "Individual Comparisons by Ranking Methods," *Biometrics Bull.*, **1**, 80 (1945).

[32] Wilks, S. S., *Mathematical Statistics*, John Wiley & Sons., Inc., New York, 1962.

Appendix *B*
Tables

TABLE I

The Poisson Distribution

$$\Pr\,(X \le x) = \sum_{w=0}^{x} \frac{\mu^{w} e^{-\mu}}{w!}$$

	$\mu = E(X)$											
x	0.5	1.0	1.5	2.0	3.0	4.0	5.0	6.0	7.0	8.0	9.0	10.0
0	0.607	0.368	0.223	0.135	0.050	0.018	0.007	0.002	0.001	0.000	0.000	0.000
1	0.910	0.736	0.558	0.406	0.199	0.092	0.040	0.017	0.007	0.003	0.001	0.000
2	0.986	0.920	0.809	0.677	0.423	0.238	0.125	0.062	0.030	0.014	0.006	0.003
3	0.998	0.981	0.934	0.857	0.647	0.433	0.265	0.151	0.082	0.042	0.021	0.010
4	1.000	0.996	0.981	0.947	0.815	0.629	0.440	0.285	0.173	0.100	0.055	0.029
5		0.999	0.996	0.983	0.916	0.785	0.616	0.446	0.301	0.191	0.116	0.067
6		1.000	0.999	0.995	0.966	0.889	0.762	0.606	0.450	0.313	0.207	0.130
7			1.000	0.999	0.988	0.949	0.867	0.744	0.599	0.453	0.324	0.220
8				1.000	0.996	0.979	0.932	0.847	0.729	0.593	0.456	0.333
9					0.999	0.992	0.968	0.916	0.830	0.717	0.587	0.458
10					1.000	0.997	0.986	0.957	0.901	0.816	0.706	0.583
11						0.999	0.995	0.980	0.947	0.888	0.803	0.697
12						1.000	0.998	0.991	0.973	0.936	0.876	0.792
13							0.999	0.996	0.987	0.966	0.926	0.864
14							1.000	0.999	0.994	0.983	0.959	0.917
15								0.999	0.998	0.992	0.978	0.951
16								1.000	0.999	0.996	0.989	0.973
17									1.000	0.998	0.995	0.986
18										0.999	0.998	0.993
19										1.000	0.999	0.997
20											1.000	0.998
21												0.999
22												1.000

TABLE II

*The Chi-Square Distribution**

$$\Pr\left(X \le x\right) = \int_0^x \frac{1}{\Gamma(r/2)2^{r/2}}\, w^{r/2-1}e^{-w/2}\, dw$$

r	\multicolumn{6}{c}{Pr $(X \le x)$}
	0.01	0.025	0.050	0.95	0.975	0.99
1	0.000	0.001	0.004	3.84	5.02	6.63
2	0.020	0.051	0.103	5.99	7.38	9.21
3	0.115	0.216	0.352	7.81	9.35	11.3
4	0.297	0.484	0.711	9.49	11.1	13.3
5	0.554	0.831	1.15	11.1	12.8	15.1
6	0.872	1.24	1.64	12.6	14.4	16.8
7	1.24	1.69	2.17	14.1	16.0	18.5
8	1.65	2.18	2.73	15.5	17.5	20.1
9	2.09	2.70	3.33	16.9	19.0	21.7
10	2.56	3.25	3.94	18.3	20.5	23.2
11	3.05	3.82	4.57	19.7	21.9	24.7
12	3.57	4.40	5.23	21.0	23.3	26.2
13	4.11	5.01	5.89	22.4	24.7	27.7
14	4.66	5.63	6.57	23.7	26.1	29.1
15	5.23	6.26	7.26	25.0	27.5	30.6
16	5.81	6.91	7.96	26.3	28.8	32.0
17	6.41	7.56	8.67	27.6	30.2	33.4
18	7.01	8.23	9.39	28.9	31.5	34.8
19	7.63	8.91	10.1	30.1	32.9	36.2
20	8.26	9.59	10.9	31.4	34.2	37.6
21	8.90	10.3	11.6	32.7	35.5	38.9
22	9.54	11.0	12.3	33.9	36.8	40.3
23	10.2	11.7	13.1	35.2	38.1	41.6
24	10.9	12.4	13.8	36.4	39.4	43.0
25	11.5	13.1	14.6	37.7	40.6	44.3
26	12.2	13.8	15.4	38.9	41.9	45.6
27	12.9	14.6	16.2	40.1	43.2	47.0
28	13.6	15.3	16.9	41.3	44.5	48.3
29	14.3	16.0	17.7	42.6	45.7	49.6
30	15.0	16.8	18.5	43.8	47.0	50.9

* This table is abridged and adapted from "Tables of Percentage Points of the Incomplete Beta Function and of the Chi-Square Distribution," *Biometrika*, **32** (1941). It is published here with the kind permission of Professor E. S. Pearson on behalf of the author, Catherine M. Thompson, and of the Biometrika Trustees.

TABLE III

The Normal Distribution

$$\Pr(X \le x) = N(x) = \int_{-\infty}^{x} \frac{1}{\sqrt{2\pi}}\, e^{-w^2/2}\, dw$$

$$[N(-x) = 1 - N(x)]$$

x	N(x)	x	N(x)	x	N(x)
0.00	0.500	1.10	0.864	2.05	0.980
0.05	0.520	1.15	0.875	2.10	0.982
0.10	0.540	1.20	0.885	2.15	0.984
0.15	0.560	1.25	0.894	2.20	0.986
0.20	0.579	1.282	0.900	2.25	0.988
0.25	0.599	1.30	0.903	2.30	0.989
0.30	0.618	1.35	0.911	2.326	0.990
0.35	0.637	1.40	0.919	2.35	0.991
0.40	0.655	1.45	0.926	2.40	0.992
0.45	0.674	1.50	0.933	2.45	0.993
0.50	0.691	1.55	0.939	2.50	0.994
0.55	0.709	1.60	0.945	2.55	0.995
0.60	0.726	1.645	0.950	2.576	0.995
0.65	0.742	1.65	0.951	2.60	0.995
0.70	0.758	1.70	0.955	2.65	0.996
0.75	0.773	1.75	0.960	2.70	0.997
0.80	0.788	1.80	0.964	2.75	0.997
0.85	0.802	1.85	0.968	2.80	0.997
0.90	0.816	1.90	0.971	2.85	0.998
0.95	0.829	1.95	0.974	2.90	0.998
1.00	0.841	1.960	0.975	2.95	0.998
1.05	0.853	2.00	0.977	3.00	0.999

TABLE IV

The t Distribution*

$$\Pr\,(T \le t) = \int_{-\infty}^{t} \frac{\Gamma[(r + 1)/2]}{\sqrt{\pi r}\,\Gamma(r/2)(1 + w^2/r)^{(r+1)/2}}\, dw$$

$$[\Pr\,(T \le -t) = 1 - \Pr\,(T \le t)]$$

			Pr $(T \le t)$		
r	0.90	0.95	0.975	0.99	0.995
1	3.078	6.314	12.706	31.821	63.657
2	1.886	2.920	4.303	6.965	9.925
3	1.638	2.353	3.182	4.541	5.841
4	1.533	2.132	2.776	3.747	4.604
5	1.476	2.015	2.571	3.365	4.032
6	1.440	1.943	2.447	3.143	3.707
7	1.415	1.895	2.365	2.998	3.499
8	1.397	1.860	2.306	2.896	3.355
9	1.383	1.833	2.262	2.821	3.250
10	1.372	1.812	2.228	2.764	3.169
11	1.363	1.796	2.201	2.718	3.106
12	1.356	1.782	2.179	2.681	3.055
13	1.350	1.771	2.160	2.650	3.012
14	1.345	1.761	2.145	2.624	2.977
15	1.341	1.753	2.131	2.602	2.947
16	1.337	1.746	2.120	2.583	2.921
17	1.333	1.740	2.110	2.567	2.898
18	1.330	1.734	2.101	2.552	2.878
19	1.328	1.729	2.093	2.539	2.861
20	1.325	1.725	2.086	2.528	2.845
21	1.323	1.721	2.080	2.518	2.831
22	1.321	1.717	2.074	2.508	2.819
23	1.319	1.714	2.069	2.500	2.807
24	1.318	1.711	2.064	2.492	2.797
25	1.316	1.708	2.060	2.485	2.787
26	1.315	1.706	2.056	2.479	2.779
27	1.314	1.703	2.052	2.473	2.771
28	1.313	1.701	2.048	2.467	2.763
29	1.311	1.699	2.045	2.462	2.756
30	1.310	1.697	2.042	2.457	2.750

* This table is abridged from Table III of Fisher and Yates; *Statistical Tables for Biological, Agricultural, and Medical Research*, published by Oliver and Boyd, Ltd., Edinburgh, by permission of the authors and publishers.

TABLE V

The F Distribution*

$$\Pr(F \le f) = \int_0^f \frac{\Gamma[(r_1 + r_2)/2][(r_1/r_2)^{r_1/2}w^{r_1/2-1}}{\Gamma(r_1/2)\Gamma(r_2/2)(1 + r_1w/r_2)^{(r_1+r_2)/2}}\, dw$$

Pr(F ≤ f)	r_2	1	2	3	4	5	6	7	8	9	10	12	15
0.95	1	161	200	216	225	230	234	237	239	241	242	244	246
0.975		648	800	864	900	922	937	948	957	963	969	977	985
0.99		4052	4999	5403	5625	5764	5859	5928	5982	6023	6056	6106	6157
0.95	2	18.5	19.0	19.2	19.2	19.3	19.3	19.4	19.4	19.4	19.4	19.4	19.4
0.975		38.5	39.0	39.2	39.2	39.3	39.3	39.4	39.4	39.4	39.4	39.4	39.4
0.99		98.5	99.0	99.2	99.2	99.3	99.3	99.4	99.4	99.4	99.4	99.4	99.4
0.95	3	10.1	9.55	9.28	9.12	9.01	8.94	8.89	8.85	8.81	8.79	8.74	8.70
0.975		17.4	16.0	15.4	15.1	14.9	14.7	14.6	14.5	14.5	14.4	14.3	14.3
0.99		34.1	30.8	29.5	28.7	28.2	27.9	27.7	27.5	27.3	27.2	27.1	26.9
0.95	4	7.71	6.94	6.59	6.39	6.26	6.16	6.09	6.04	6.00	5.96	5.91	5.86
0.975		12.2	10.6	9.98	9.60	9.36	9.20	9.07	8.98	8.90	8.84	8.75	8.66
0.99		21.2	18.0	16.7	16.0	15.5	15.2	15.0	14.8	14.7	14.5	14.4	14.2
0.95	5	6.61	5.79	5.41	5.19	5.05	4.95	4.88	4.82	4.77	4.74	4.68	4.62
0.975		10.0	8.43	7.76	7.39	7.15	6.98	6.85	6.76	6.68	6.62	6.52	6.43
0.99		16.3	13.3	12.1	11.4	11.0	10.7	10.5	10.3	10.2	10.1	9.89	9.72

r_1

Pr(F≤f)	r_2	\(r_1\) 1	2	3	4	5	6	7	8	9	10	12	15
0.95	6	5.99	5.14	4.76	4.53	4.39	4.28	4.21	4.15	4.10	4.06	4.00	3.94
0.975		8.81	7.26	6.60	6.23	5.99	5.82	5.70	5.60	5.52	5.46	5.37	5.27
0.99		13.7	10.9	9.78	9.15	8.75	8.47	8.26	8.10	7.98	7.87	7.72	7.56
0.95	7	5.59	4.74	4.35	4.12	3.97	3.87	3.79	3.73	3.68	3.64	3.57	3.51
0.975		8.07	6.54	5.89	5.52	5.29	5.12	4.99	4.90	4.82	4.76	4.67	4.57
0.99		12.2	9.55	8.45	7.85	7.46	7.19	6.99	6.84	6.72	6.62	6.47	6.31
0.95	8	5.32	4.46	4.07	3.84	3.69	3.58	3.50	3.44	3.39	3.35	3.28	3.22
0.975		7.57	6.06	5.42	5.05	4.82	4.65	4.53	4.43	4.36	4.30	4.20	4.10
0.99		11.3	8.65	7.59	7.01	6.63	6.37	6.18	6.03	5.91	5.81	5.67	5.52
0.95	9	5.12	4.26	3.86	3.63	3.48	3.37	3.29	3.23	3.18	3.14	3.07	3.01
0.975		7.21	5.71	5.08	4.72	4.48	4.32	4.20	4.10	4.03	3.96	3.87	3.77
0.99		10.6	8.02	6.99	6.42	6.06	5.80	5.61	5.47	5.35	5.26	5.11	4.96
0.95	10	4.96	4.10	3.71	3.48	3.33	3.22	3.14	3.07	3.02	2.98	2.91	2.85
0.975		6.94	5.46	4.83	4.47	4.24	4.07	3.95	3.85	3.78	3.72	3.62	3.52
0.99		10.0	7.56	6.55	5.99	5.64	5.39	5.20	5.06	4.94	4.85	4.71	4.56
0.95	12	4.75	3.89	3.49	3.26	3.11	3.00	2.91	2.85	2.80	2.75	2.69	2.62
0.975		6.55	5.10	4.47	4.12	3.89	3.73	3.61	3.51	3.44	3.37	3.28	3.18
0.99		9.33	6.93	5.95	5.41	5.06	4.82	4.64	4.50	4.39	4.30	4.16	4.01
0.95	15	4.54	3.68	3.29	3.06	2.90	2.79	2.71	2.64	2.59	2.54	2.48	2.40
0.975		6.20	4.77	4.15	3.80	3.58	3.41	3.29	3.20	3.12	3.06	2.96	2.86
0.99		8.68	6.36	5.42	4.89	4.56	4.32	4.14	4.00	3.89	3.80	3.67	3.52

* This table is abridged and adapted from "Tables of Percentage Points of the Inverted Beta Distribution," *Biometrika,* **33** (1943). It is published here with the kind permission of Professor E. S. Pearson on behalf of the authors, Maxine Merrington and Catherine M. Thompson, and of the *Biometrika* Trustees.

Appendix C

Answers to Selected Exercises

CHAPTER 1

1.1 (e) $\frac{23}{32}$.

1.2 $\frac{1}{4}$.

1.3 $\frac{1}{4}$; $\frac{1}{2}$.

1.4 (a) $\{x; x = 0, 1, 2, 3, 4\}$; $\{x; x = 2\}$.
(b) $\{x; 0 < x < 3\}$; $\{x; 1 \le x < 2\}$.

1.5 (a) $\{x; 0 < x < \frac{5}{8}\}$.

1.10 (a) $\{x; 0 < x < 3\}$.
(b) $\{(x, y); 0 < x^2 + y^2 < 4\}$.

1.11 (a) $\{x; x = 2\}$.
(b) Null set.
(c) $\{(x, y); x^2 + y^2 = 0\}$.

1.12 $\frac{80}{81}$; 1.

1.13 $\frac{11}{16}$; 0; 1.

1.14 16; 7; 21; 2.

1.16 $\frac{8}{3}$; 0; $\pi/2$.

1.17 $\frac{1}{2}$; 0; $\frac{2}{9}$.

1.18 $\frac{1}{6}$; 0.

1.20 10.

1.23 $\frac{1}{4}$; $\frac{1}{13}$; $\frac{1}{52}$; $\frac{4}{13}$.

1.24 $\frac{31}{32}$; $\frac{3}{64}$; $\frac{1}{32}$; $\frac{63}{64}$.

1.25 0.3.

1.26 e^{-4}; $1 - e^{-4}$; 1.

1.27 $\frac{1}{2}$.

1.31 $\frac{9}{13}$, $\frac{1}{13}$, $\frac{1}{13}$, $\frac{1}{13}$, $\frac{1}{13}$.

1.33 (a) $\frac{3}{52}$. (b) $\frac{1}{13}$.

1.35 $\frac{3}{4}$.

1.37 $\frac{5}{8}$; $\frac{7}{8}$; $\frac{3}{8}$.

1.38 $\frac{1}{4}$.

1.40 $e^{-2} - e^{-3}$.

1.41 (a) $\frac{1}{2}$. (b) 1.

1.43 $\frac{1}{27}$, 1, $\frac{2}{9}$, $\frac{25}{36}$.

1.45 $\frac{15}{64}$; 0; $\frac{1}{2}$; $\frac{1}{2}$.

1.47 (a) 1. (b) $\frac{2}{3}$. (c) 2.

1.50 $n!$; 1; $1/n!$.

1.52 (a) $0, x < 0; 1, 0 \le x$.
(d) $0, x < 0; 1 - (1 - x)^3, 0 \le x < 1; 1, 1 \le x$.

1.53 (b) 0. (e) 2. (f) 2.5.

1.54 (a) $\frac{1}{4}$. (b) 0. (c) $\frac{1}{4}$. (d) 0.

1.56 $0, y < 0; y^2, 0 \le y < 1; 1, 1 \le y, 2y, 0 < y < 1$; 0 elsewhere.

1.58 $\frac{1}{2}$; $\frac{1}{4}$.

1.63 $1/3\sqrt{y}, 0 < y < 1; 1/6\sqrt{y},$
     $1 < y < 4; 0$ elsewhere.

1.64   (a) $\binom{6}{4}\Big/\binom{16}{4}.$

     (b) $\binom{10}{4}\Big/\binom{16}{4}.$

1.65 $1 - \binom{990}{5}\Big/\binom{1000}{5}.$

1.67   (b) $1 - \binom{10}{3}\Big/\binom{20}{3}.$

1.70 $-\ln z, 0 < z < 1;$
     0 elsewhere.

1.72 $(x - 1)(10 - x)(9 - x)/$
     $420, x = 2, 3, \ldots, 8.$

1.74 $2; 86.4; -160.8.$

1.75 $3; 11; 27.$

1.76 $\frac{1}{9}.$

1.78 $(\frac{1}{3})(\frac{2}{3}) \neq \frac{1}{4}.$

1.79   (a) $\frac{1}{2}.$ (b) $f(1) = \frac{1}{3}.$

1.80 $7.80.

1.83 $\frac{7}{3}.$

1.84   (a) 1.5, 0.75. (b) 0.5, 0.05.
     (c) 2; does not exist.

1.85 $e^t/(2 - e^t), t < \ln 2; 2; 2.$

1.92 $\frac{7}{6}; \frac{5}{36}$

1.95 $10; 0; 2; -30.$

1.97 $-\frac{2\sqrt{2}}{5}, \frac{2\sqrt{2}}{5}$

1.99 $1/2p; \frac{3}{2}; \frac{5}{2}; 5; 50.$

1.101 $\frac{31}{12}, \frac{167}{144}.$

1.106   0.84.

CHAPTER 2

2.3   (a) $\frac{1}{7}.$ (b) $\frac{5}{56}.$

     (c) $\left[\binom{3}{x}\Big/\binom{8}{x}\right][5/(8 - x)].$

2.6 $\frac{1}{3}.$

2.8 $\frac{9}{20}; \frac{2}{3}.$

2.9 $\frac{5}{14}.$

2.10 $(3x_1 + 2)/(6x_1 + 3);$
     $(6x_1^2 + 6x_1 + 1)/$
             $(2)(6x_1 + 3)^2.$

2.12 $3x_2/4; 3x_2^2/80.$

2.17   (b) $1/e.$

2.18   (a) 1. (b) $-1.$ (c) 0.

2.19 $7/\sqrt{804}.$

2.26 $b_2 = \sigma_1(\rho_{12} - \rho_{13}\rho_{23})/$
             $[\sigma_2(1 - \rho_{23}^2)];$
     $b_3 = \sigma_1(\rho_{13} - \rho_{12}\rho_{23})/$
             $[\sigma_3(1 - \rho_{23}^2)].$

2.31 $\frac{7}{8}.$

2.33 $1 - (1 - y)^{12}, 0 \le y < 1;$
     $12(1 - y)^{11}, 0 < y < 1.$

2.34 $g(y) = [y^3 - (y - 1)^3]/6^3,$
     $y = 1, 2, 3, 4, 5, 6.$

2.35 $\frac{1}{2}.$

CHAPTER 3

3.1 $\frac{40}{81}.$

3.4 $\frac{147}{512}.$

3.6   5.

3.9 $\frac{65}{81}.$

3.11 $\frac{5}{72}.$

3.14 $\frac{1}{6}.$

3.15 $\frac{24}{625}.$

3.17 $\frac{11}{6}; x_1/2; \frac{11}{6}.$

3.18 $\frac{25}{4}.$

3.19   0.09.

3.22 $4^x e^{-4}/x!, x = 0, 1, 2, \ldots.$

3.23   0.84.

3.27   (a) $\exp[-2 + e^{t_2}(1 + e^{t_1})].$
     (b) $\mu_1 = 1, \mu_2 = 2,$
        $\sigma_1^2 = 1, \sigma_2^2 = 2,$
        $\rho = \sqrt{2}/2.$
     (c) $y/2.$

3.28   0.05.

3.29   0.831, 12.8.

3.30   0.90.

3.31 $\chi^2(4)$.

3.33 $3e^{-3y}$, $0 < y < \infty$.

3.34   2, 0.95.

3.39 $\frac{11}{16}$.

3.40 $\chi^2(2)$.

3.42   0.067; 0.685.

3.44   71.3, 189.7.

3.45 $\sqrt{\ln 2/\pi}$.

3.49   0.774.

3.50 $\sqrt{2/\pi}$; $(\pi - 2)/\pi$.

3.51   0.90.

3.52   0.477.

3.53   0.461.

3.54 $n(0, 1)$.

3.55   0.433.

3.56   0; 3.

3.61 $n(0, 2)$.

3.62   (a) 0.264. (b) 0.440.
(c) 0.433. (d) 0.642.

3.64 $\rho = \frac{4}{5}$.

3.65   (38.2, 43.4).

CHAPTER 4

4.3   0.405.

4.4 $\frac{1}{8}$.

4.6 $(n + 1)/2$; $(n^2 - 1)/12$.

4.7 $a + b\bar{x}$; $b^2 s_x^2$.

4.8 $\chi^2(2)$.

4.11 $\frac{1}{2}$, $0 < y < 1$;
$1/2y^2$, $1 < y < \infty$.

4.12 y^{15}, $0 \le y < 1$; $15y^{14}$,
$0 < y < 1$.

4.13 $\frac{4}{7}$.

4.14 $\frac{1}{3}$, $y = 3, 5, 7$.

4.16 $(\frac{1}{2})^{\sqrt[3]{y}}$, $y = 1, 8, 27, \ldots$.

4.17

y_1	$g_1(y_1)$
1	$\frac{1}{36}$
2	$\frac{4}{36}$
3	$\frac{6}{36}$
4	$\frac{4}{36}$
6	$\frac{12}{36}$
9	$\frac{9}{36}$

4.20 $\frac{1}{27}$, $0 < y < 27$.

4.23 $\alpha/(\alpha + \beta)$;
$\alpha\beta/[(\alpha + \beta + 1)(\alpha + \beta)^2]$.

4.24   (a) 20. (b) 1260. (c) 495.

4.25 $\frac{10}{243}$.

4.34   0.05.

4.37   1/4.74, 3.33.

4.42 $(1/\sqrt{2\pi})^3 y_1^2 e^{-y_1^2/2} \sin y_3$,
$0 \le y_1 < \infty$, $0 \le y_2 < 2\pi$,
$0 \le y_3 \le \pi$.

4.43 $y_2 y_3^2 e^{-y_3}$, $0 < y_1 < 1$,
$0 < y_2 < 1$, $0 < y_3 < \infty$.

4.47 $1/(2\sqrt{y})$, $0 < y < 1$.

4.48 $e^{-y_1/2}/(2\pi\sqrt{y_1 - y_2^2})$,
$-\sqrt{y_1} < y_2 < \sqrt{y_1}$,
$0 < y_1 < \infty$.

4.50 $1 - (1 - e^{-3})^4$.

4.51 $\frac{1}{8}$.

4.56 $\frac{5}{16}$.

4.57 $48z_1 z_2^3 z_3^5$, $0 < z_1 < 1$,
$0 < z_2 < 1$, $0 < z_3 < 1$.

4.58 $\frac{7}{12}$.

4.63 $6uv(u + v)$,
$0 < u < v < 1$.

4.68

y	$g(y)$
2	$\frac{1}{36}$
3	$\frac{2}{36}$
4	$\frac{3}{36}$
5	$\frac{4}{36}$
6	$\frac{5}{36}$
7	$\frac{6}{36}$
8	$\frac{5}{36}$

y	$g(y)$
9	$\frac{4}{36}$
10	$\frac{3}{36}$
11	$\frac{2}{36}$
12	$\frac{1}{36}$

4.69 0.24.

4.77 0.818.

4.80 (b) -1 or 1.

(c) $Z_i = \sigma_i Y_i + \mu_i$.

4.81 $\sum_1^n a_i b_i = 0$.

4.83 6.41.

4.84 $n = 16$.

4.86 $(n-1)\sigma^2/n$; $2(n-1)\sigma^4/n^2$.

4.87 0.90.

4.89 0.78.

4.90 $\frac{8}{3}; \frac{2}{9}$.

4.91 7.

4.93 2.5; 0.25.

4.95 -5; $60 - 12\sqrt{6}$.

4.96 $\sigma_1/\sqrt{\sigma_1^2 + \sigma_2^2}$.

4.99 22.5, $\frac{261}{4}$.

4.100 $r_2 > 4$.

4.102 $\mu_2\sigma_1/\sqrt{\sigma_1^2\sigma_2^2 + \mu_1^2\sigma_2^2 + \mu_2^2\sigma_1^2}$.

4.105 $5/\sqrt{39}$.

4.109 $e^{\mu + \sigma^2/2}$; $e^{2\mu + \sigma^2}(e^{\sigma^2} - 1)$.

CHAPTER 5

5.1 Degenerate at μ.

5.2 Gamma ($\alpha = 1, \beta = 1$).

5.3 Gamma ($\alpha = 1, \beta = 1$).

5.4 Gamma ($\alpha = 2; \beta = 1$).

5.13 0.682.

5.14 (b) 0.815.

5.17 Degenerate at μ_2 $+ (\sigma_2/\sigma_1)(x - \mu_1)$.

5.18 (b) $n(0, 1)$.

5.19 (b) $n(0, 1)$.

5.21 0.954.

5.23 0.840.

5.26 0.08.

5.28 0.267.

5.29 0.682.

5.34 $n(0, 1)$.

CHAPTER 6

6.1 (a) $\bar{X}$.

(b) $-n/\ln (X_1 X_2 \cdots X_n)$.

(c) $\bar{X}$. (d) The median.

(e) The first order statistic.

6.2 The first order statistic Y_1, $\sum_1^n (X_i - Y_1)/n$.

6.4 $\frac{4}{25}, \frac{11}{25}, \frac{7}{25}$.

6.5 $Y_1 = \min (X_i)$; $n/\ln [(X_1 X_2 \cdots X_n)/Y_1^n]$.

6.7 (b) $\bar{X}/(1 - \bar{X})$. (d) $\bar{X}$.

(e) $\bar{X} - 1$.

6.11 $\frac{1}{3}, \frac{2}{3}$.

6.12 $w_1(y)$.

6.13 $b = 0$; does not exist.

6.14 Does not exist.

6.16 (77.28, 85.12).

6.17 24 or 25.

6.18 (3.7, 5.7).

6.19 160.

6.25 $(5\bar{x}/6, 5\bar{x}/4)$.

6.26 $(-3.6, 2.0)$.

6.30 135 or 136.

6.32 (0.43, 2.21).

6.34 (2.68, 9.68).

6.36 (0.71, 5.50).

6.41 $[y\tau^2 + \mu\sigma^2/n]/(\tau^2 + \sigma^2/n)$.

6.42 $\beta(y + \alpha)/(n\beta + 1)$.

CHAPTER 7

7.1 $\frac{1}{4} + \frac{3}{4}\ln \frac{3}{2}$; $\frac{7}{16} + \frac{9}{8}\ln \frac{3}{4}$.

7.2 $\frac{11}{64}$, $(31)3^8/4^9$.

7.5 $n = 19$ or 20.

7.6 $K(\frac{1}{2}) = 0.062$;
      $K(\frac{1}{12}) = 0.920$.

7.10 $\sum_1^{10} x_i^2 \geq 18.3$; yes; yes.

7.12 $3\sum_1^{10} x_i^2 + 2\sum_1^{10} x_i \geq c$.

7.13   95 or 96; 76.7.

7.14   38 or 39; 15.

7.15   0.08; 0.875.

7.16 $(1 - \theta)^9(1 + 9\theta)$.

7.17 $1, 0 < \theta \leq \frac{1}{2}$; $1/(16\theta^4)$,
      $\frac{1}{2} < \theta < 1$; $1 - 15/(16\theta^4)$,
      $1 \leq \theta$.

7.19   53 or 54, 5.6.

7.22   Reject H_0 if $\bar{x} \geq 77.564$.

7.23   26 or 27;
      reject H_0 if $\bar{x} \leq 24$.

7.24   220 or 221;
      reject H_0 if $y \geq 17$.

7.27 $t = 3 > 2.262$, reject H_0.

7.28 $|t| = 2.27 > 2.145$,
      reject H_0.

CHAPTER 8

8.1 $q_3 = \frac{176}{21} > 7.81$,
      reject H_0.

8.3 $b \leq 8$ or $32 \leq b$.

8.4 $q_3 = \frac{22}{9} < 11.3$,
      accept H_0.

8.5 $6.4 < 9.49$, accept H_0.

8.7 $\hat{p} = (X_1 + X_2/2)/$
           $(X_1 + X_2 + X_3)$.

8.16 $r + \theta, 2r + 4\theta$.

8.17 $r_2(\theta + r_1)/[r_1(r_2 - 2)]$,
      $r_2 > 2$.

8.25 $\hat{\beta} = \sum (X_i/nc_i)$,
      $\sum [(X_i - \hat{\beta}c_i)^2/nc_i^2]$.

8.29   Reject H_0.

CHAPTER 9

9.2   (a) $\frac{15}{16}$. (b) 675/1024;
      (c) $(0.8)^4$.

9.4   8.

9.6   0.954; 0.92; 0.788.

9.8   8.

9.11   (a) Beta $(n - j + 1, j)$.
      (b) Beta $(n - j + i - 1,$
                $j - i + 2)$.

9.13   0.067.

9.15   Reject H_0.

9.22   0; $4(4^n - 1)/3$; no.

9.33 $\frac{2}{99}$.

9.39   98; $\frac{686}{3}$.

CHAPTER 10

10.9 $60y_3^2(y_5 - y_3)/\theta^5$; $6y_5/5$;
      $\theta^2/7$; $\theta^2/35$.

10.10 $(1/\theta^2)e^{-y_1/\theta}$,
      $0 < y_2 < y_1 < \infty$;
      $y_1/2$; $\theta^2/2$.

10.12 $\sum X_i^2/n$; $\sum X_i/n$; $\sum X_i/n$.

10.14 X; X.

10.15 Y_1/n.

10.17 $Y_1 - 1/n$.

10.19 $Y_1 = \sum_1^n X_i$; $Y_1/4n$; yes.

10.27 $\bar{x}$.

10.28 $\bar{X}^2 - 1/n$.

10.31 $\left(\dfrac{n-1}{n}\right)^Y\left(1 + \dfrac{Y}{n-1}\right)$.

CHAPTER 11

11.3 θ^2/n; $\theta^2/n(n + 2)$.

11.6 $c_0(n) = (14.4)$
           $\times (n \ln 1.5 - \ln 9.5)$;
      $c_1(n) = (14.4)$
           $\times (n \ln 1.5 + \ln 18)$.

11.7 $c_0(n) = (0.05n - \ln 8)/$
$\ln 3.5;$
$c_1(n) = (0.05n + \ln 4.5)/$
$\ln 3.5.$

11.12 $(9y - 20x)/30 \le c.$

11.24 $2.17; 2.44.$

11.27 $2.20.$

CHAPTER 12

12.3   (a) $\exp\{(t_1\mathbf{c} + t_2\mathbf{d})'\boldsymbol{\mu} +$
$[(t_1\mathbf{c} + t_2\mathbf{d})'\mathbf{V}$
$\times (t_1\mathbf{c} + t_2\mathbf{d})]/2\}.$

12.12 $a_i = 0, i = 1, 2, 3, 4.$

12.13 $\sum\limits_{j=1}^{n} a_{ij} = 0, i = 1, 2, \ldots, n.$

Index